FOUNDATION GCSE MATHEMATICS

REVISION AND PRACTICE

6th EDITION

DAVID RAYNER

OXFORD
UNIVERSITY PRESS

OXFORD
UNIVERSITY PRESS

Great Clarendon Street, Oxford, OX2 6DP, United Kingdom

Oxford University Press is a department of the University of Oxford.
It furthers the University's objective of excellence in research, scholarship, and
education by publishing worldwide. Oxford is a registered trade mark of Oxford
University Press in the UK and in certain other countries

British Library Cataloguing in Publication Data
Data available

978-0-19-835570-0

1 3 5 7 9 10 8 6 4 2

Paper used in the production of this book is a natural, recyclable product made from
wood grown in sustainable forests. The manufacturing process conforms to the
environmental regulations of the country of origin.

Printed in Great Britain by Bell and Bain Ltd, Glasgow

Acknowledgements

The publishers would like to thank the following for permission to use their photographs:

p01: Vladimir Nenezic/Shutterstock; **p17:** JuliusKielaitis/Shutterstock; **p23:** Daniel Wilson/Shutterstock;
p41t: anyaivanova/Shutterstock; **p41b:** HamsterMan/Shutterstock; **p49t:** nito/Shutterstock; **p49b:** graja/
Shutterstock; **p50:** pixelparticle/Shutterstock; **p52:** Hal_P/Shutterstock; **p57:** WitR/Shutterstock;
p56: Coprid/Shutterstock; **p64:** Sebastian Duda/Shutterstock; **p66:** Handmade Pictures/Shutterstock;
p70: Marek Velechovsky/Shutterstock; **p79:** Neftali/Shutterstock; **p81:** SpeedKingz/Shutterstock;
p82: Bohbeh/Shutterstock; **p101:** charobnica/Shutterstock; **p140:** Michel de Wit/Shutterstock;
p147l: pbombaert/Shutterstock; **p147r:** izobrazie/Shutterstock; **p147c:** eAlisa/Shutterstock; **p154:** Stefan
Schurr/Shutterstock; **p163:** Boris Stroujko/Shutterstock; **p166:** Anton Gvozdikov/Shutterstock;
p177: Cattallina/Shutterstock; **p178:** SergeBertasiusPhotography/Shutterstock ; **p179:** Adisa/Shutterstock;
p182: Brenda Carson/Shutterstock; **p186:** Dennis van de Water/Shutterstock; **p187:** Dragon Images/
Shutterstock; **p197:** Aspen Photo/Shutterstock; **p202:** Chad McDermott/Shutterstock; **p214:** Hal_P/
Shutterstock; **p216:** BlueSkyImage/Shutterstock; **p227:** Marcio Jose Bastos Silva/Shutterstock; **p229t:** Anton
Gvozdikov/Shutterstock; **p229b:** Khakimullin Aleksandr/Shutterstock; **p230:** Vitaly Titov & Maria Sidelnikova/
Shutterstock; **p236:** Steve Collender/Shutterstock; **p242:** Twin Design/Shutterstock; **p246:** Savo Ilic/
Shutterstock; **p263:** andrea lehmkuhl/Shutterstock; **p265:** Lynn Y/Shutterstock; **p266:** Georgios Kollidas/
Shutterstock; **p274:** OlenaB/Shutterstock; **p280:** Boris Stroujko/Shutterstock; **p292:** Tatiana Popova/
Shutterstock; **p317:** Anita Ponne/Shutterstock; **p325:** Andrew Holt/Alamy; **p326t:** KevinTate/Shutterstock;
p326b: Andrei Nekrassov/Shutterstock; **p386:** kongsky/Shutterstock; **p398:** F. JIMENEZ MECA/Shutterstock;
p401: Sean Pavone/Shutterstock; **p406:** Andrea Danti/Shutterstock; **p407:** pzAxe/Shutterstock; **p410:** Africa
Studio/Shutterstock; **p411:** Stolbov Pavel/Shutterstock; **p427:** irabel8/Shutterstock; **p466t:** Michaelpuche/
Shutterstock; **p466b:** Mitch Gunn/Shutterstock.

Although we have made every effort to trace and contact all copyright holders before publication this has
not been possible in all cases. If notified, the publisher will rectify any errors or omissions at the earliest
opportunity.

About this book

This Foundation book is for candidates working through Key Stage 4 towards a GCSE in Mathematics, and has been adapted for the new specifications for first examination in 2017. It can be used both in the classroom and by students working on their own. There are explanations, worked examples and numerous exercises which, it is hoped, will help students to build up confidence. The questions are graded in difficulty throughout the exercises.

The book can be used either as a course book over the years before the GCSE examinations or as a revision text in the final year.

There is a renewed focus on problem solving throughout this edition, with extra material on proof and ratio and proportion.

Exercises have been graded according to their level of demand, with the labels 1, 2 and 3 corresponding to material found in standard, underlined and bold type in the GCSE scope of study.

Questions marked * are intended as a challenge or extension.

At the end of the book there is a specific section on problem solving and proof, and several revision exercises which provide mixed questions across the curriculum.

The author is indebted to the many students and colleagues who have assisted him in this work. He is particularly grateful to Christine Godfrey for her help and many suggestions, and to M. R. Heylings for his work on this new edition.

Thanks are also due to the following examination boards for kindly allowing the use of questions from past mathematics papers.

Edexcel (Edexcel)

AQA Education (AQA)

Oxford Cambridge and RSA Examinations (OCR)

Northern Ireland Council for the Curriculum Examinations and Assessment (CCEA)

Welsh Joint Education Committee (WJEC)

D. Rayner 2015

Contents

1 Number 1

In this unit you will:
- revise place value of digits, decimals and negative numbers
- revise factors and multiples
- learn about highest common factor and lowest common multiple
- learn about prime numbers and prime factors
- learn about square numbers, cube numbers, square roots and cube roots
- learn index notation
- learn how to use your calculator effectively
- learn about writing numbers in standard form.

Supercomputers do many billions of numerical calculations every second.

1.1 Properties of numbers

1.1.1 Place value of digits

- The number 576 means $\boxed{500}$ + $\boxed{70}$ + $\boxed{6}$

- The number 2408 means $\boxed{2000}$ + $\boxed{400}$ + $\boxed{8}$

- The number 13 416 is written as 'thirteen thousand, four hundred and sixteen'.

Exercise 1 ①

1 You can write the number 427 as $\boxed{400}$ + $\boxed{20}$ + $\boxed{7}$.
Copy and complete these numbers.

   a   613 = $\boxed{600}$ + $\boxed{10}$ + $\boxed{}$

   b   954 = $\boxed{}$ + $\boxed{50}$ + $\boxed{}$

   c $\boxed{}$ = $\boxed{500}$ + $\boxed{30}$ + $\boxed{9}$

   d   2416 = $\boxed{}$ + $\boxed{}$ + $\boxed{}$ + $\boxed{}$

2 In the number 13 485, write the value of
   a   the figure 8
   b   the figure 3.

3 When you write a cheque you write the amount in words and in figures.

Write these numbers in words.

a 523 **b** 6410 **c** 25 000

4 Write these numbers in figures

a two hundred and seventeen

b four thousand, two hundred and fifty

c five million

d six thousand and twenty.

5 The populations of three towns are

Penton 15 614   Quarkby 21 604   Roydon 11 999

Write the towns in order of their populations with the smallest first.

6 Write these numbers in order of size, smallest first.

a 314, 29, 290, 85

b 5010, 2000, 564, 645, 1666

c 60 000, 7510, ten thousand, 8888

7 a Use these cards to make four different 3-digit numbers less than 600.

b Write the largest 3-digit number you can make with these cards.

8 a Use these cards to make the smallest possible 4-digit number.

b Use these cards to make the largest possible 4-digit number.

c Use three of the cards to make the largest possible 3-digit number less than 600.

9 What four coins can you use to make 44p?

10 What number is half of fifty thousand?

11 What number is ten times as big as 411?

12 Write in figures the number five hundred and ten thousand, two hundred and twelve.

13 Add together in your head 9, 21 and 40 and write your answer.

14 a Use these cards to make the largest possible 4-digit number.
   b Use all four cards to make the smallest possible 4-digit number (you cannot start with a zero).

$\boxed{3}$ $\boxed{7}$ $\boxed{0}$ $\boxed{2}$

15 Here are five number cards.
   a Use all the cards to make the largest possible **odd** number.
   b Use all the cards to make the smallest possible **even** number.

$\boxed{3}$ $\boxed{5}$ $\boxed{2}$ $\boxed{4}$ $\boxed{7}$

16 Write the number that is 10 more than
   a 247 **b** 3211 **c** 694

17 Write the number that is 1000 more than
   a 392 **b** 25 611 **c** 256 900

18 a Prini puts a 2-digit whole number into her calculator.
   She multiplies the number by 10.

   $\boxed{\boxed{5}}$

   Write **one** other digit which you know must now be on the calculator.
   b Prini starts again with the same 2-digit number and this time she multiplies it by 1000.

   $\boxed{\boxed{2}}$

   Write all five digits on the calculator this time.

19 Write the numbers in order, from the smallest to the largest.
   a 2142 2290 2058 2136
   b 5329 5029 5299 5330
   c 25 117 25 200 25 171 25 000 25 500

*20 Here are six number cards.
   Copy and complete these sums.
   Use all six cards each time
   to make the largest possible answer.

$\boxed{3}$ $\boxed{4}$ $\boxed{6}$ $\boxed{7}$ $\boxed{8}$ $\boxed{9}$

   a $\boxed{}\boxed{} + \boxed{}\boxed{} + \boxed{}\boxed{}$
   b $\boxed{}\boxed{}\boxed{} + \boxed{}\boxed{}\boxed{}$

*21 Find a number n so that $5n + 7 = 507$.

*22 Find a number x so that $6x + 8 = 68$.

> Remember:
> $5n$ means $5 \times n$
> $6x$ means $6 \times x$

1.1.2 Factors and multiples

- The **factors** of 8 are the numbers that divide exactly into 8.
  So the factors of 8 are 1, 2, 4 and 8.
  The factors of 15 are 1, 3, 5 and 15.

Factor pairs

1×8	2×4
1×15	3×5

- The **multiples** of 4 are the numbers in the 4 times table.
  The multiples of 4 are 4, 8, 12, 16,…and so on.
  The multiples of 7 are 7, 14, 21, 28,…and so on.

Exercise 2 ①

The factors of 6 are the numbers that divide into 6 exactly.
The factors of 6 are 1, 2, 3 and 6.

1×6	2×3

1 Write the factors of
  a 9   **b** 10   **c** 12

2 Write the factors of
  a 14   **b** 20   **c** 30

EXAMPLE

The factors of 6 are 1, 2, 3, 6.
The factors of 8 are 1, 2, 4, 8.
The numbers 1 and 2 are in both lists. The numbers 1 and 2
are **common factors** of 6 and 8.

3 The table shows the factors of 10 and 15.

Number	Factors
10	1, 2, 5, 10
15	1, 3, 5, 15

Write the common factors of 10 and 15 (the numbers that are
factors of both 10 and 15).

4 a Copy and complete this table.

Number	Factors
12	
20	

  b Write the common factors of 12 and 20.

5 a The first four multiples of 5 are   5   10   15   20.

  b Write the first four multiples of

  i 3   **ii** 4   **iii** 10

In questions **6** to **9** find the 'odd one out' (the number which is not a multiple of the number given).

6 Multiples of 6: 6, 12, 16, 24.

7 Multiples of 11: 22, 44, 76, 88.

8 Multiples of 4: 8, 12, 22, 28.

9 Multiples of 7: 7, 14, 21, 34.

*10 Copy and complete this sentence.
  'An even number is a multiple of _____.'

*11 n is a odd number.

  a Is $2n$ an odd number or an even number?

  b Is $2n + 1$ an odd number, an even number or could it be either?

1.1.3 L.C.M. and H.C.F.

● The first few **multiples** of 4 are 4, 8, 12, 16, ⃝20, 24, 28…
  The first few multiples of 5 are 5, 10, 15, ⃝20, 25, 30, 35…

  ▶ The **Lowest Common Multiple** (L.C.M.) of 4 and 5 is 20.
    It is the lowest number which is in both lists.

● The **factors** of 12 are 1, 2, 3, ⃝4, 6, 12
  The factors of 20 are 1, 2, ⃝4, 5, 10, 20

  ▶ The **Highest Common Factor** (H.C.F.) of 12 and 20 is 4.
    It is the highest number which is in both lists.

Exercise 3 ①

1 a Write the first six multiples of 2.
  b Write the first six multiples of 5.
  c Write the L.C.M. of 2 and 5.

2 a Write the first four multiples of 4.
  b Write the first four multiples of 12.
  c Write the L.C.M. of 4 and 12.

> The L.C.M. is the lowest number that is in both lists of multiples.

3 Find the L.C.M. of

    a 6 and 9   **b** 8 and 12   **c** 14 and 35

    d 4 and 6   **e** 5 and 10   **f** 7 and 9

4 The table shows the factors and common factors of 24 and 36.

Number	Factors	Common Factors
24	1, 2, 3, 4, 6, 8, 12, 24	} 1, 2, 3, 4, 6, 12
36	1, 2, 3, 4, 6, 9, 12, 18, 36	

> The H.C.F. is the highest number in the list of common factors.

Write the H.C.F. of 24 and 36.

5 The table shows the factors and common factors of 18 and 24.

Number	Factors	Common Factors
18	1, 2, 3, 6, 9, 18	} 1, 2, 3, 6
24	1, 2, 3, 4, 6, 8, 12, 24	

Write the H.C.F. of 18 and 24.

6 Find the H.C.F. of

    a 12 and 18   **b** 22 and 55   **c** 45 and 72

    d 18 and 30   **e** 60 and 72   **f** 40 and 50

7 **a** Find the H.C.F. of 12 and 30.

    b Find the L.C.M. of 8 and 20.

    c Write two numbers whose H.C.F. is 11.

    d Write two numbers whose L.C.M. is 10.

> Don't confuse your L.C.M.s with your H.C.F.s!

*8 Given that $30 = 2 \times 3 \times 5$ and $165 = 3 \times 5 \times 11$, find the highest common factor of 30 and 165 (that is, the highest number that divides exactly into 30 and 165).

*9 If $315 = 3 \times 3 \times 5 \times 7$ and $273 = 3 \times 7 \times 13$, find the highest common factor of 315 and 273.

1.1.4 Square numbers and cube numbers

● **Square numbers**

    $4^2 = 4 \times 4 = 16$

    You say '4 squared equals 16'.

    $100^2 = 100 \times 100 = 10\,000$

● **Square root**

The square root of 25 is 5.

You ask 'What do I square to get 25?'

You write $\sqrt{25} = 5$.

Here are some other square roots: $\sqrt{100} = 10$, $\sqrt{1} = 1$, $\sqrt{2} = 1\cdot414$ approximately.

Notice that only **square** numbers have square roots that are whole numbers.

● **Cubes**

The number 64 can be written as

$64 = 4 \times 4 \times 4 = 4^3$

You say '4 cubed is 64' or sometimes

'4 to the power 3 is 64'.

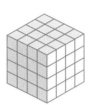

Exercise 4 ①

1 The first three square numbers are 1, 4, 9.

$1 \times 1 = ①$ $2 \times 2 = ④$ $3 \times 3 = ⑨$

Draw diagrams with labels to show the next two square numbers.

2 The square number 9 is 3×3 and can be written as 3^2 ('3 squared'). Work out

  a 5^2 **b** 10^2 **c** 6^2

  d $3^2 + 2^2$ **e** $1^2 + 7^2$ **f** $10^2 - 2^2$

3 What number do you multiply by itself to give these numbers?

  a 16   **b** 81   **c** 1

4 The symbol for square root is $\sqrt{}$ (for example $\sqrt{36} = 6$). Work out

  a $\sqrt{9}$ **b** $\sqrt{25}$ **c** $\sqrt{100}$ **d** $\sqrt{49}$

5 Copy and complete

  a $6 + \sqrt{4} = \square$ **b** $3 + \sqrt{16} = \square$ **c** $\sqrt{\square} = 6$

6 Answer 'true' or 'false'.

  a $2 \times 2 \times 2 = 8$, so 8 is a cube number.

  b 27 is a cube number.

  c $1 \times 1 = 2$

  d $2^3 + 1^3 = 9$

7 Work out

  a $10^2 + 2^2$ **b** $1^3 + 3^3$ **c** $4^2 + \sqrt{4}$

  d $10^3 - 10^2$ **e** $\sqrt{6^2}$ **f** $10^4 + 1$

8 In each part, copy the four numbers and circle the one that is **not** a square number.

  a 4, 9, 48, 64   **b** 16, 25, 36, 55   **c** 1, 8, 16, 100

  d 4, 10, 49, 81   **e** 25, 64, 120, 144 **f** 16, 36, 108, 121

*9 For each pair of numbers here, there is just one **square number** that lies between them. In each case, write that square number.

  a   6   12   **b** 18   33   **c** 27   47

  d 10   22   **e**   2   8   **f** 88   108

  g 40   50   **h** 75   95

*10 Shirin has five number cards.

  a Write two of the cards to show how Shirin can make a square number.

  b Write two of the cards to show how Shirin can make a cube number.

  c Write three of the cards to show how Shirin can make a larger cube number.

| 1 | 2 | 4 | 5 | 7 |

*11 Find the smallest value of n for which
$$1^2 + 2^2 + 3^2 + 4^2 + 5^2 + \ldots + n^2 > 800$$

> $>$ means 'is greater than'

1.1.5 Prime numbers

▶ A **prime number** is divisible only by itself and by 1.

● The first six prime numbers are 2, 3, 5, 7, 11 and 13.

● Notice that the number 1 is **not** a prime number.

Prime factor decomposition

▶ The factors of a number that are also prime numbers are called **prime factors**. You find the prime factors of any number by 'prime factor decomposition'.

> It is not as hard as it sounds!

EXAMPLE

Find the factors of 90
a with a factor tree   **b** using repeated divisions.

a

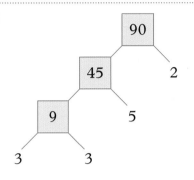

$90 = 3 \times 3 \times 5 \times 2$
or $90 = 2 \times 3 \times 3 \times 5$ (in order)

b

2	90	divide by 2
5	45	divide by 5
3	9	divide by 3
	3	

where 2, 3 and 5
are prime factors

$90 = 2 \times 5 \times 3 \times 3$

Exercise 5 ①

1 Which of these numbers are prime numbers?

| 2 | 7 | 9 | 15 | 17 |

2 Write the first ten prime numbers.

3 Write all the even prime numbers up to 100.

4 Find all the prime numbers between 30 and 40.

5 For each of these pairs of numbers, there is just one **prime number**
that lies between them. In each case, write that prime number.
  a   8   12   **b**   14   18   **c**   25   30
  d   20   28   **e**   32   40   **f**   44   52
  g   54   60   **h**   38   42

6 In each part, copy the four numbers and circle the one that is **not** a prime number.
  a   7, 9, 13, 17   **b**   2, 13, 21, 23   **c**   11, 13, 19, 27
  d   15, 19, 29, 31   **e**   31, 37,39, 41   **f**   23, 43, 47, 49

Here is a factor tree for 140.   Here is a factor tree for 40.

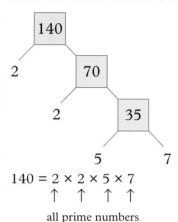   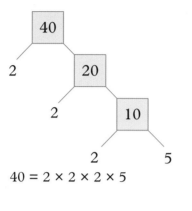

$140 = 2 \times 2 \times 5 \times 7$ $40 = 2 \times 2 \times 2 \times 5$
    ↑ ↑ ↑ ↑

all prime numbers

7 You can find prime factors using a factor tree.
Draw a factor tree for each of these numbers.

   a 36   **b** 60   **c** 216   **d** 200   **e** 1500

8 Here is the number 600 written as the product of its prime factors.
$600 = 2 \times 2 \times 2 \times 3 \times 5 \times 5$

Use this information to write 1200 as a product of
its prime factors.

9 Write each number as a product of its prime factors.
(For example: $30 = 2 \times 3 \times 5$)

   a 28   **b** 32   **c** 34   **d** 81   **e** 294

10 Copy the grid and use a pencil for your answers
(so that you can rub out mistakes).

Write every number from 1 to 9, one in each box, so that all the
numbers match the conditions for both the rows and the columns.

	Even number	**Square number**	**Prime number**
Factor of 14	2		
Multiple of 3			
Between 3 and 9			

Remember: Write
just **one** number in
each box.

11 Copy the grid and write the numbers from 1 to 9, one in each box.

	Factor of 10	More than 3	Factor of 18
Square number			
Even number			
Prime number			

*12 This one is more difficult. Copy the grid and write the numbers from 1 to 16, one in each box. There are several correct solutions. Ask another student to check yours.

	Prime number	Odd number	Factor of 16	Even number
Numbers less than 7				
Factor of 36				
Numbers less than 12				
Numbers between 11–17				

*13 The pair of numbers 11, 13 is a 'prime pair', since 11 and 13 are both prime numbers with only one other number in between. Similarly, the numbers 17, 19 form another 'prime pair'.

There are four prime pairs between 20 and 100. See if you can find them. Remember, both numbers of the pair **must be prime**.

*14 a Copy these statements and fill in the missing numbers.

$$1 \qquad\quad = 1 \quad = 1^2$$
$$1 + 3 \qquad = 4 \quad = 2^2$$
$$1 + 3 + 5 \qquad = \square \quad = \square^2$$
$$1 + 3 + 5 + 7 = \square \quad = \square^2$$

b Write the next five lines of the sequence.

*15 The numbers 1, 8, 27 are the first three **cube** numbers.

$$1 \times 1 \times 1 = 1^3 \quad = 1 \qquad \text{(say '1 cubed')}$$
$$2 \times 2 \times 2 = 2^3 \quad = 8 \qquad \text{(say '2 cubed')}$$
$$3 \times 3 \times 3 = 3^3 \quad = 27 \quad \text{(say '3 cubed')}$$

You can add the odd numbers in groups to give an interesting sequence.

$$1 \qquad\quad = 1 \quad = 1^3$$
$$3 + 5 \qquad = 8 \quad = 2^3$$
$$7 + 9 + 11 = 27 \quad = 3^3$$

Write the next three rows of the sequence to see if the sum of each row always gives a cube number.

? ***16** The letters p and q represent prime numbers.
Is $p + q$ always an even number? Explain your answer.

? ***17** Apart from 1, 3 and 5 all odd numbers less than 100 can be
written in the form $p + 2^n$ where p is a prime number and n is
greater than or equal to 2.

For example $43 = 11 + 2^5$
$$27 = 23 + 2^2$$

Write the odd numbers from 7 to 21 in the form $p + 2^n$.
Can you write all the odd numbers to 99 in this form?

> Any number $n \geq 2$
> can be written
> as a product of
> prime factors.
> This is the **unique
> factorisation
> theorem.**

Exercise 6 ①

? **Find the number!**

Each number is described in three different ways.

1 a 2-digit number; a prime number; the sum of its digits is 13

2 a 2-digit number; a multiple of both 3 and 4; the sum of its digits is 15

3 a prime number; a factor of 78; a 2-digit number

4 a 3-digit number; a square number; the product of its digits is 2

5 a multiple of 9; a 3-digit number below 200; the product of its digits is 16

6 a 2-digit number; a prime number; the product of its digits is 24

7 a square number; a 3-digit number; the product of its digits is 20

8 a 2-digit number; a factor of 184; a prime number

9 a multiple of 11; a multiple of 7; the product of its digits is 6

10 a 3-digit number; the sum of its digits is 9; a multiple of 7

1.2   Arithmetic without a calculator

1.2.1   Adding and subtracting

You can use different methods to add and subtract without a calculator.

> **EXAMPLE**
>
> **a**
> ```
>   3 6 4
>     5 3 +
>   ─────
>   4 1 7
>     1
> ```
> **b**
> ```
>   2 1 4
>   3 0 7
>     5 2 +
>   ─────
>   5 7 3
>       1
> ```
> **c**
> ```
>     4 2 7
>   5 1 8 6 +
>   ───────
>   5 6 1 3
>       1 1
> ```
> **d**
> ```
>   8 4
>   6 1 −
>   ───
>   2 3
> ```
> **e**
> ```
>   ⁶7̶¹ 4
>     2 6 −
>   ─────
>     4 8
> ```
> **f**
> ```
>   2 7 ⁷8̶¹ 4
>       6 3 5 −
>   ─────────
>   2 1 4 9
> ```

> Remember to keep
> the numbers in
> columns when you
> are adding and
> subtracting.

Exercise 7 ①

Do these additions.

1 27 + 31

2 45 + 22

3 234 + 17

4 316 + 204

5 50 + 911

6 291 + 46

7 299 + 197

8 306 + 205

9 45 + 275

10 903 + 89

11 415 + 207 + 25

12 41 + 607 + 423

13 206 + 114 + 8

14 9 + 19 + 912

15 157 + 16 + 24

16 16 + 2341 + 27

17 3047 + 265

18 274 + 5061

19 2941 + 4067

20 8046 + 147

21 401 + 609 + 21

22 506 + 2615

23 2947 + 4 + 590

24 209 + 607 + 11

25 6672 + 11 + 207

26 994 + 27

27 604 + 12 407

28 9150 + 12 694

29 53 246 + 62 141

30 19 274 + 27 + 584

Exercise 8 ①

Do these subtractions.

1 97 − 63

2 69 − 41

3 83 − 60

4 87 − 5

5 192 − 81

6 214 − 10

7 86 − 29

8 52 − 37

9 74 − 18

10 91 − 68

11 265 − 128

12 642 − 181

13 562 − 181

14 816 − 274

15 509 − 208

16 604 − 491

17 808 − 275

18 250 − 127

19 640 − 118

20 265 − 184

21 484 − 219

22 6064 − 418

23 5126 − 307

24 6417 − 29

25 8050 − 218

26 406 − 22

27 649 − 250

28 6009 − 205

29 1717 − 356

30 843 − 295

Find the answer.

31 100 − 99 + 98 − 97 + 96 − ... + 4 − 3 + 2 − 1

Exercise 9 ①

Magic squares

In a magic square you get the same number when you add across each row (↔), add down each column (↕) and add diagonally (↗, ↘). Copy and complete these magic squares.

1

3		
2		
7	0	

2

9	7	5
		10

3

8	1	6
4		

4

7		
	6	4
		5

5

12		14
	11	
8		

6

17		
12		16
13		

7

		7	14
	13	2	
	3	16	5
15			4

8

	6		4
10		16	
		2	
1	12	7	14

9

	10		
14	15		8
	4	18	
16	13		6

10

18		12	7
	6		
11	16	5	14
4			

*11

11		7	20	3
	12	25		16
	5	13		9
	18		14	
		19	2	15

*12

16		10	17	4
3	15		9	21
20		14		8
	19			
24			5	12

1.2.2 Multiplying and dividing

Questions on multiplying and dividing are much easier
if you know your multiplication facts, for example:

● '3 nines are 27',
● '4 fours are 16'.

Copy and complete this multiplication table to help you.

×	4	7	3	5	9	11	8	6	2	10
4										
7										
3					27					
5										
9										
11									22	
8		24								
6										
2										
10										

EXAMPLE

a
```
  3 2
    4 ×
-------
1 2 8
```

b
```
  6 4
    4 ×
-------
2 5 6
    1
```

c
```
  3 7 4
      6 ×
---------
2 2 4 4
    4 2
```

d
$$5\overline{)\,7\,^2 5}$$ quotient 1 5

e
$$7\overline{)\,3\,7^2\,9^1\,4}$$ quotient 5 4 2

f
$$5\overline{)\,6^1\,9^4\,4}$$ quotient 1 3 8   r 4   or $138\frac{4}{5}$

Exercise 10 ❶

Do these multiplications.

1 21 × 3	**2** 32 × 3	**3** 42 × 6	**4** 35 × 4
5 213 × 3	**6** 46 × 5	**7** 205 × 6	**8** 28 × 6
9 211 × 7	**10** 302 × 7	**11** 213 × 5	**12** 641 × 3
13 21 × 8	**14** 314 × 6	**15** 131 × 9	**16** 214 × 8
17 820 × 6	**18** 921 × 4	**19** 2141 × 6	**20** 3025 × 5
21 324 × 8	**22** 643 × 7	**23** 295 × 9	**24** 641 × 10
25 846 × 10	**26** 275 × 8	**27** 631 × 7	**28** 885 × 9
29 497 × 8	**30** 2153 × 6		

Exercise 11 ❶

Do these divisions.

1 69 ÷ 3	**2** 286 ÷ 2	**3** 844 ÷ 4	**4** 345 ÷ 3
5 712 ÷ 4	**6** 1160 ÷ 5	**7** 1581 ÷ 3	**8** 2112 ÷ 4
9 415 ÷ 5	**10** 994 ÷ 2	**11** 1092 ÷ 4	**12** 18072 ÷ 3
13 3020 ÷ 5	**14** 1626 ÷ 6	**15** 1660 ÷ 4	**16** 1915 ÷ 5
17 4944 ÷ 6	**18** 5616 ÷ 6	**19** 2247 ÷ 7	**20** 10710 ÷ 5
21 18972 ÷ 2	**22** 9256 ÷ 4	**23** 1928 ÷ 8	**24** 14010 ÷ 2
25 5859 ÷ 7	**26** 55305 ÷ 9	**27** 21104 ÷ 8	**28** 3735 ÷ 9

There are 'remainders' in these questions.

29 76 ÷ 5	**30** 87 ÷ 4	**31** 57 ÷ 2	**32** 373 ÷ 6
33 247 ÷ 6	**34** 124 ÷ 5	**35** 281 ÷ 5	**36** 1173 ÷ 9
37 2143 ÷ 4	**38** 6418 ÷ 5	**39** 6027 ÷ 4	**40** 4135 ÷ 6

Speed tests

You can do these questions either:

1 with your book open or

2 with your book closed and your teacher reading out the questions.

In either case write **the answer** only. Be as quick as possible.

Test 1	**Test 2**	**Test 3**	**Test 4**
1 30 − 8	**1** 6 + 16	**1** 8 × 5	**1** 5 × 7
2 9 × 5	**2** 32 − 5	**2** 17 + 23	**2** 36 − 18
3 40 ÷ 5	**3** 9 × 6	**3** 60 ÷ 6	**3** 103 − 20
4 24 + 34	**4** 90 ÷ 2	**4** 101 − 20	**4** 56 ÷ 7
5 11 × 7	**5** 98 + 45	**5** 49 × 2	**5** 8 × 4
6 60 − 12	**6** 16 − 7	**6** 52 + 38	**6** 53 + 36
7 9 × 4	**7** 45 ÷ 9	**7** 66 ÷ 11	**7** 51 − 22
8 27 ÷ 3	**8** 13 × 100	**8** 105 − 70	**8** 36 ÷ 3
9 55 + 55	**9** 99 + 99	**9** 13 × 4	**9** 20 × 5
10 60 − 18	**10** 67 − 17	**10** 220 − 30	**10** 99 + 55
11 8 × 6	**11** 570 ÷ 10	**11** 100 ÷ 20	**11** 200 − 145
12 49 ÷ 7	**12** 7 × 3	**12** 2 × 2 × 2	**12** 88 ÷ 8
13 99 + 17	**13** 55 − 6	**13** 91 + 19	**13** 50 × 100
14 80 − 59	**14** 19 + 18	**14** 200 − 5	**14** 199 + 26
15 9 × 100	**15** 60 ÷ 5	**15** 16 × 2	**15** 80 − 17

EXAMPLE

A van driver bought 15 litres of diesel at £1·01 per litre.
How much did he spend?

15 litres at £1·01 per litre
Total cost = 15 × 1·01 = £15·15

Exercise 12 Mixed problems ①

1 Nicki made a tower using 27 identical discs, each of thickness
  5 cm. How high was the tower?

2 What four coins have a total of 37p?

3 Thomas shares £189 equally between seven people.
  How much does each person receive?

4 At a banquet for 456 people, eight guests sat at each
  table. How many tables were there?

5 A motorist bought 9 litres of petrol at 93p per litre.
  a How much did it cost?
  b What change did she receive from £10?

6 The population of a town decreased from 8716 to 7823.
  How many people left the town?

7 A man bought five felt tip pens at 28p each and six
  pads at 84p each. How much did he spend altogether?

8 A tin has a mass of 240 g when empty.
When it is half-full of rice the total mass is 570 g.
What is its total mass when it is full?

240 g   570 g

9 A well-organised flock of sheep queue up for
their daily ration of 8 kg of grass.
The farmer at the head of the queue has only 2064 kg
of grass to give out.
How many of his flock of 300 sheep will be disappointed?

*10 In a school with 280 students there are 10 more girls than boys.
  a How many girls are there?
  b How many boys are there?
  c Check that your answers add up to 280.

*11 Find a pair of positive integers a and b for which

$8a + 65b + = 1865$

> An **integer** is a
> whole number.

Exercise 13 ①

Cross squares

Each empty square contains either a number or a mathematical operator
(+, −, ×, ÷). Copy each square and fill in the missing details.

1

11		4	→	15
×		÷		
		2	→	3
↓		↓		
66			→	132

2

9		17	→	26
×		−		
5	×		→	
↓		↓		
	÷	9	→	5

3

14	+		→	31
×				
4		23	→	92
↓		↓		
	−	40	→	

4

15			→	5
+		×		
	5	→	110	
↓		↓		
	−	15	→	22

5

	×	10	→	90
+		÷		
		→	$5\frac{1}{2}$	
↓		↓		
20	×		→	100

6

	×		→	52
−		×		
	×	4	→	
↓		↓		
8		8	→	1

7

5			→	60
×		÷		
		24	→	44
↓		↓		
	×	$\frac{1}{2}$	→	50

8

	×	6	→	42
÷		÷		
14	−		→	
↓		↓		
		2	→	1

9

	×	2	→	38
−		÷		
			→	48
↓		↓		
7	−		→	$6\frac{1}{2}$

1.2.3 Inverse operations

The word inverse means 'opposite'.

▶ The inverse of **adding** is **subtracting** $5 + 19 = 24, 5 = 24 - 19$
▶ The inverse of **subtracting** is **adding** $31 - 6 = 25, 31 = 25 + 6$
▶ The inverse of **multiplying** is **dividing** $7 \times 6 = 42, 7 = 42 \div 6$
▶ The inverse of **dividing** is **multiplying** $30 \div 3 = 10, 30 = 10 \times 3$

▶ The reciprocal of 7 is $\frac{1}{7}$

EXAMPLE

Find the missing digits.

a $\square\,4 \div 6 = 14$

b $2\,\square\,8 \times 5 = 1340$

c
$$\begin{array}{r} 3\ \square\ 7 \\ +\ 2\ 5\ \square \\ \hline \square\ 3\ 9 \end{array}$$

a Work out 14×6 because multiplying is the inverse of dividing. Since $14 \times 6 = 84$, the missing digit is 8.

b Work out $1340 \div 5$ because dividing is the inverse of multiplying. Since $1340 \div 5 = 268$, the missing digit is 6.

c Start from the right: $7 + 2 = 9$
Middle column: $8 + 5 = 13$
Check
$$\begin{array}{r} 3\ 8\ 7 \\ +\ 2\ 5\ 2 \\ \hline 6\ 3\ 9 \\ {\scriptstyle 1} \end{array}$$

Exercise 14 ①

Copy these calculations and find the missing digits.

1 a
$$\begin{array}{r} 2\ 8\ 5 \\ +\ \square\ 1\ 4 \\ \hline 7\ \square\ \square \end{array}$$

b
$$\begin{array}{r} 6\ 3\ \square \\ +\ \square\ 5\ 2 \\ \hline 8\ \square\ 9 \end{array}$$

c
$$\begin{array}{r} \square\ 3\ 5 \\ +\ 3\ 4\ \square \\ \hline 9\ \square\ 9 \end{array}$$

2 a
$$\begin{array}{r} 3\ 5\ 6 \\ +\ 5\ \square\ 6 \\ \hline \square\ 8\ \square \end{array}$$

b
$$\begin{array}{r} 2\ \square\ 4 \\ +\ 5\ 3\ 7 \\ \hline \square\ 6\ 1 \end{array}$$

c
$$\begin{array}{r} 3\ 8\ 8 \\ +\ \square\ 2\ \square \\ \hline 8\ \square\ 3 \end{array}$$

3 a
$$4\,\square \times 3 = 1\,4\,4$$

b
$$3\,\square \times 7 = 2\,3\,1$$

c
$$\square\,\square\,1 \times 5 = 1\,6\,0\,5$$

4 a $\square\,\square\,\square \div 3 = 50$ **b** $\square\,\square \times 4 = 60$

c $9 \times \square = 81$ **d** $\square\,\square\,\square\,\square \div 6 = 192$

5 a
$$\begin{array}{r} 4\ \square\ 5 \\ +\ 2\ \ 8\ \square \\ \hline \square\ \ 3\ \ 0 \end{array}$$

b
$$\begin{array}{r} 4\ \square\ 7 \\ +\ \square\ 7\ \square \\ \hline 6\ \ 0\ \ 4 \end{array}$$

c
$$\begin{array}{r} \square\ 3\ \square \\ +2\ \square\ 4 \\ \hline 7\ \ 9\ \ 9 \end{array}$$

6 a $\square\,\square \times 7 = 245$ **b** $\square\,\square \times 10 = 580$

c $32 \div \square = 8$ **d** $\square\,\square\,\square \div 5 = 190$

7 a $\square\,\square + 29 = 101$ **b** $\square\,\square\,\square - 17 = 91$

c
$$\begin{array}{r} \square\ 8\ 9 \\ -\ 3\ \square\ 6 \\ \hline 5\ \ 4\ \square \end{array}$$

d
$$\begin{array}{r} 3\ \ 3\ 5 \\ -2\ \ 1\ \square \\ \hline \square\ \square\ 7 \end{array}$$

8 There is more than one correct answer for each of these questions. Copy and complete them and ask another student to check your solution.

a $\boxed{2}\,\boxed{3} + \square\,\square - \square\,\square = 23$ **b** $\boxed{8}\,\boxed{5} - \square\,\square + \square\,\square = 86$

c $\boxed{2}\,\boxed{5} \times \square \ \div \square = 25$ **d** $\boxed{4}\,\boxed{0} \times \square\,\square \div \square = 80$

9 Each of these calculations has the same number missing from all three boxes. Copy them and find the missing number in each calculation.

a $\square \times \square - \square = 12$ **b** $\square \div \square + \square = 9$ **c** $\square \times \square + \square = 72$

10 Copy the diagrams and work out the missing numbers in these calculations.

a $5 \to \boxed{\times 6} \to \boxed{+9} \to ?$ **b** $? \to \boxed{+2} \to \boxed{\times 5} \to 40$

c $2 \to \boxed{+?} \to \boxed{\times 4} \to 36$ **d** $7 \to \boxed{\times ?} \to \boxed{-11} \to 10$

11 Copy these and write +, −, × or ÷ in the circle to make each calculation correct.

a $7 \times 4 \bigcirc 3 = 25$

b $8 \times 5 \bigcirc 2 = 20$

c $7 \bigcirc 3 - 9 = 12$

d $12 \bigcirc 2 + 4 = 10$

e $75 \div 5 \bigcirc 5 = 20$

12 Copy these and write the correct signs in the circles.

a $5 \times 4 \times 3 \bigcirc 3 = 63$

b $5 + 4 \bigcirc 3 \bigcirc 2 = 4$

c $5 \times 2 \times 3 \bigcirc 1 = 31$

1.3 Decimals

1.3.1 Decimals and fractions

▶ Decimal numbers are used as a way of writing fractions.

The decimal number 0·3 is $\dfrac{3}{10}$.

The decimal number 0·09 is $\dfrac{9}{100}$.

The decimal number 0·31 is $\dfrac{31}{100}$ or $\dfrac{3}{10} + \dfrac{1}{100}$.

The decimal number 4·27 is 4 units + $\dfrac{2}{10} + \dfrac{7}{100}$.

You can show decimal numbers in a place value table.

	T	U	.	$\dfrac{1}{10}$	$\dfrac{1}{100}$
40 =	4	0	.	0	
7 =		7	.	0	
$\dfrac{7}{10} =$		0	.	7	
$\dfrac{5}{100} =$		0	.	0	5

Exercise 15

1 What part of each shape is shaded?

Write your answer as a fraction and as decimal fraction.

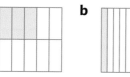

2 Write the value of the red digit in each number.

a 3·5	**b** 26·4	**c** 17·41	**d** 18·9
e 1·41	**f** 0·74	**g** 34·11	**h** 3·38

> In 6·24, the digit 2 has a value of $\frac{2}{10}$.

3 Write the next two terms in each sequence.

a 0·2   0·3   0·4   0·5   ...
b 0·2   0·4   0·6   ...
c 0·5   1·0   1·5   2·0   ...
d 0·1   0·3   0·5   0·7   ...

4 Write the numbers shown by the arrows as decimals.

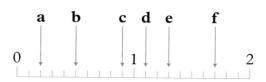

5 Write each fraction as a decimal.

a $\frac{2}{10}$ **b** $\frac{13}{100}$ **c** $\frac{2}{100}$ **d** $\frac{15}{100}$ **e** $\frac{155}{1000}$ **f** $\frac{227}{1000}$

1.3.2   Writing decimals in order of size

Method

1 Write the decimals in a column lining up the decimal points.
2 Fill any empty spaces with zeros.
3 Write the numbers in order of size.

EXAMPLE

Write 1·6, 0·51, 3 and 1·15 in order.

1 Write in a column.	**2** Put in zeros.	**3** Write in order.
1·6	1·60	0·51
0·51	0·51	1·15
3	3·00	1·60
1·15	1·15	3·00

Exercise 16

Write the numbers in order, starting with the smallest.

1 3·7, 4, 1·5, 12   **2** 31, 3·1, 1·3, 13

3 11, 0·2, 5·2, 6   **4** 0·4, 0·11, 1, 1·7

5 22, 2·2, 2, 20   **6** 0·21, 0·31, 0·12

7 0·04, 0·4, 0·35   **8** 0·67, 0·672, 0·7

9 0·05, 0·045, 0·07   **10** 0·1, 0·09, 0·089

11 0·75, 0·57, 0·705   **12** 0·41, 0·041, 0·14

13 0·809, 0·81, 0·8   **14** 0·006, 0·6, 0·059

15 0·15, 0·143, 0·2   **16** 0·04, 0·14, 0·2, 0·53

17 1·2, 0·12, 0·21, 1·12   **18** 2·3, 2·03, 0·75, 0·08

19 0·62, 0·26, 0·602, 0·3   **20** 0·5, 1·3, 1·03, 1·003

21 0·79, 0·792, 0·709, 0·97   **22** 1·23, 0·321, 0·312, 1·04

23 0·008, 0·09, 0·091, 0·007   **24** 2·05, 2·5, 2, 2·046

25 Here are numbers with letters. Put the numbers in order and write the letters to make a word.

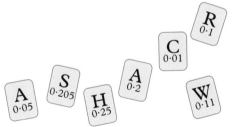

26 Add $\frac{1}{10}$ to each of these numbers.

   a   32·41   **b**   0·753   **c**   1·06

27 Write these amounts in pounds.

   a   350 pence   **b**   15 pence   **c**   3 pence

   d   10 pence   **e**   1260 pence   **f**   8 pence

28 Write each statement and say whether it is true or false.

   a   £5·4 = £5 + 40p   **b**   £0·6 = 6p

   c   5p = £0·05   **d**   50p is more than £0·42

Exercise 17

In questions **1** to **6**, write the one line that is correct.

   1 a   0·06 is equal to 0·6   **2 a**   0·04 is equal to 0·040

     b   0·06 is greater than 0·6   **b**   0·04 is greater than 0·040

     c   0·06 is less than 0·6   **c**   0·04 is less than 0·040

3 a 0·14 is equal to 0·41

  b 0·14 is greater than 0·41

  c 0·14 is less than 0·41

5 a 0·61 is equal to 0·6

  b 0·61 is greater than 0·6

  c 0·61 is less than 0·6

4 a 0·12 is equal to 0·1

  b 0·12 is greater than 0·1

  c 0·12 is less than 0·1

6 a 0·6 is equal to 0·60

  b 0·6 is greater than 0·60

  c 0·6 is less than 0·60

In questions **7–34**
> means 'is greater than' (for example 9 > 5)
< means 'is less than' (for example 7 < 10)
For each question write 'true' or 'false'.

7 0·8 = 0·08 **8** 0·7 < 0·71 **9** 0·61 > 0·16 **10** 0·08 > 0·008

11 0·5 = 0·500 **12** 0·4 < 0·35 **13** 0·613 < 0·631 **14** 0·06 > 0·055

15 8 = 8·00 **16** 7 = 0·7 **17** 0·63 > 0·36 **18** 8·2 < 8·022

19 6·04 < 6·40 **20** 0·75 = 0·075 **21** 5 = 0·5 **22** 0·001 > 0·0001

23 0·078 < 0·08 **24** 9 = 9·0 **25** 0·9 > 0·085 **26** 6·2 < 6·02

27 0·05 < 0·005 **28** 0·718 < 0·871 **29** 0·09 > 0·1 **30** 11 = 0·11

31 0·88 > 0·088 **32** 0·65 > 0·605 **33** 2·42 = 2·420 **34** 0·31 = 0·3100

In questions **35** to **38** copy the line and then put each number
from the box on the number line.

35

| 0·3 | 0·9 | 0·5 | 0·7 |

0   1

36

| 0·7 | 1·4 | 0·4 | 1·7 |

0   2

37

| 1·1 | 1·15 | 1·05 | 1·19 | 1·02 |

1·0   1·2

38

| 0·97 | 1·08 | 0·91 | 1·0 | 1·05 | 0·94 |

0·9   1·1

1.3.3 Scale readings

Exercise 18 ①

Write the value shown by the arrow.

1 4 ⸺ 5

2 0 ⸺ 1

3 9 ⸺ 10

4 0 ⸺ 2

5 0 ⸺ 20

6 10 ⸺ 11

7 17 ⸺ 19

8 120 ⸺ 140

9 0 ⸺ 1000

10 0·2 m ↓ 0·3 m   **11** 1·9 m ↓ 2 m   **12** 3 mg ↓ 6 mg

13 0 cm ↓ 0·1 cm   **14** 1·7 mm ↓ 1·9 mm   **15** 3·1 kg ↓ 3·2 kg

16 80 km ↓ 120 km   **17** 0 g ↓ 400 g   **18** 0 kg ↓ 200 kg

1.3.4 Adding and subtracting decimals

Remember to line up the decimal points when adding or subtracting.

> **EXAMPLE**
>
> **a** 2·4 + 3·3   **b** 2·7 + 16·5   **c** 3·64 + 4
>
> ....
>
> **a** $\begin{array}{r} 2{\cdot}4 \\ +\ 3{\cdot}3 \\ \hline 5{\cdot}7 \end{array}$ **b** $\begin{array}{r} 2{\cdot}7 \\ +\ 16{\cdot}5 \\ \hline 19{\cdot}2 \\ \small{1} \end{array}$ **c** $\begin{array}{r} 3{\cdot}64 \\ +\ 4{\cdot}00 \\ \hline 7{\cdot}64 \end{array}$

> **EXAMPLE**
>
> **a** 3·34 − 1·84   **b** 0·4 − 0·17   **c** 5·03 − 3·47
>
> ....
>
> **a** $\begin{array}{r} {}^2\cancel{3}{\cdot}\,{}^1\cancel{3}4 \\ -\ 1{\cdot}\ 84 \\ \hline 1{\cdot}\ 50 \end{array}$ **b** $\begin{array}{r} 0{\cdot}\,{}^3\cancel{4}\ 0 \\ -\ 0{\cdot}\ 1\ 7 \\ \hline 0{\cdot}\ 2\ 3 \end{array}$ **c** $\begin{array}{r} {}^4\cancel{5}{\cdot}\,{}^9\cancel{0}\,{}^1 3 \\ -\ 3{\cdot}\ 4\ 7 \\ \hline 1{\cdot}\ 5\ 6 \end{array}$

Exercise 19 ①

Do these additions.

1 $\begin{array}{r} 4{\cdot}5 \\ +\ 2{\cdot}3 \\ \hline \end{array}$ **2** $\begin{array}{r} 7{\cdot}2 \\ +\ 1{\cdot}6 \\ \hline \end{array}$ **3** $\begin{array}{r} 8{\cdot}7 \\ +\ 3{\cdot}0 \\ \hline \end{array}$ **4** $\begin{array}{r} 6{\cdot}7 \\ +\ 8{\cdot}2 \\ \hline \end{array}$ **5** $\begin{array}{r} 3{\cdot}7 \\ +\ 2{\cdot}9 \\ \hline \end{array}$

6 $\begin{array}{r} 8{\cdot}5 \\ +\ 4{\cdot}31 \\ \hline \end{array}$ **7** $\begin{array}{r} 3{\cdot}9 \\ +\ 4{\cdot}87 \\ \hline \end{array}$ **8** $\begin{array}{r} 8{\cdot}35 \\ +\ 1{\cdot}84 \\ \hline \end{array}$ **9** $\begin{array}{r} 11{\cdot}7 \\ +\ 2{\cdot}84 \\ \hline \end{array}$ **10** $\begin{array}{r} 4{\cdot}62 \\ 1{\cdot}14 \\ +\ 3{\cdot}31 \\ \hline \end{array}$

Now do these.

11 2·84 + 7·3

12 18·6 + 2·34

13 25·96 + 0·75

14 212·7 + 4·25

15 3·6 + 6

16 7 + 16·1

17 8 + 3·4 + 0·85

18 12 + 5·32 + 0·08

19 0·004 + 0·058

20 7·77 + 77·7

21 1·9 + 19·1 + 7

22 15 + 6·02 + 6

23 0·24 + 0·2 + 2

24 245 + 27·9 + 3

25 67·1 + 29 + 0·7

26 0·07 + 0·008 + 12

27 4·76 + 1 + 0·07

28 17 + 0·61 + 5

29 513 + 47·2 + 0·157

30 2·6 + 26·6 + 26

31 47·4 + 11

32 0·055 + 5 + 15

33 3·24 + 32

34 9·09 + 999

35 2·63 + 19 + 0·4

36 251 + 0·1 + 6·3

37 19·7 + 0·8 + 15

38 27 + 2·07 + 0·59

39 16·4 + 27 + 0·15

40 374 + 200·6 + 9

Exercise 20

Do these subtractions.

1   8·6
    − 1·4
  _____

2   7·7
    − 2·3
  _____

3   8·8
    − 1·9
  _____

4   7·4
    − 1·0
  _____

5   8·4
    − 1·6
  _____

6   7·2
    − 4·5
  _____

7   6·3
    − 1·8
  _____

8   5·4
    − 2·7
  _____

9   17·25
    − 8·14
  _____

10   8·27
    − 1·19
  _____

Now do these.

11 4·81 − 3·7

12 6·92 − 2·56

13 8·27 − 5·86

14 19·7 − 8·9

15 3·6 − 2·24

16 8·4 − 2·17

17 8·24 − 5·78

18 19·6 − 7·36

19 15·4 − 7

20 23·96 − 8

21 8 − 5·2

22 9 − 6·8

23 13 − 2·7

24 25 − 3·2

25 0·325 − 0·188

26 0·484 − 0·43

27 7 − 0·35

28 6 − 1·28

29 2·38 − 1·81

30 11 − 7·4

By adding and subtracting, find the value of

*31 4·24 + 7·0 − 4·01

*32 8·36 + 1·28 − 3·11

*33 0·35 + 0·63 − 0·55

*34 5·27 + 0·761 − 1·23

*35 4·76 + 0·09 − 3·55

*36 2·6 + 9 − 3·64

*37 3·5 + 8 − 2·16

*38 0·31 + 8 − 5·88

*39 14·9 + 19 − 17·25

*40 11 + 0·73 − 4·2

41 Sue buys three games: *Grand Prix 9*, *Soccer Stars* and *Hold Up*.

How much change does she get from £50?

Tiger Golf
£14.95

Moon Wars
£8.99

Grand Prix 9
£23.99

Hold Up
£5.60

Get Rich Quick
£9.50

Soccer Stars
£11.45

1.3.5 Multiplying and dividing by 10, 100, 1000

▶ To **multiply**, move the decimal point to the **right**.

▶ To **divide**, move the decimal point to the **left**.

$$3·24 \times 10 = 32·4$$
$$10·61 \times 10 = 106·1$$
$$4·134 \times 100 = 413·4$$
$$8·2 \times 100 = 820$$

$$15·2 \div 10 = 1·52$$
$$624·9 \div 100 = 6·249$$
$$509 \div 1000 = 0·509$$

Exercise 21 ①

Do these multiplications.

1 $0·634 \times 10$ **2** $0·838 \times 10$ **3** $0·815 \times 100$ **4** $0·074 \times 100$

5 $7·245 \times 1000$ **6** $0·032 \times 1000$ **7** $0·63 \times 10$ **8** $1·42 \times 100$

9 $0·041 \times 100$ **10** $0·3 \times 100$ **11** $0·71 \times 1000$ **12** $3·95 \times 10$

Do these divisions.

13 $6·24 \div 10$ **14** $8·97 \div 10$ **15** $17·5 \div 100$ **16** $23·6 \div 100$

17 $127 \div 1000$ **18** $705 \div 1000$ **19** $13 \div 10$ **20** $0·8 \div 10$

21 $0·7 \div 100$ **22** $218 \div 10$ **23** $35 \div 1000$ **24** $8·6 \div 1000$

Now do these.

25 $0·95 \times 100$ **26** $11·11 \times 10$ **27** $3·2 \times 10$ **28** $0·07 \times 1000$

29 $57·6 \div 10$ **30** $999 \div 100$ **31** 66×10 **32** $100 \div 100$

33 $42 \div 1000$ **34** $0·62 \times 10\,000$ **35** $0·9 \div 100$ **36** $555 \div 10\,000$

37 Here are some number cards.

| 0 | 1 | 2 | 3 | 4 | 5 | . | 0 |

a Jason picks the cards 1 3 and 4 and arranges them
to make the number 314. What extra card could he take
to make a number ten times as big as 314?

*b Mel chose three cards to make the number 5·2.
i What cards could she take to make a number ten times as big as 5·2?
ii What cards could she take to make a number 100 times as big as 5·2?
iii What cards could she take to make a number which is $\frac{1}{100}$ of 5·2?

1.3.6 Multiplying decimals by whole numbers

EXAMPLE

a 3·2 × 4 **b** 1·8 × 2 **c** 5·4 × 6 **d** 0·71 × 5

a	3·2	**b**	1·8	**c**	5·4	**d**	0·75
	× 4		× 2		× 6		× 5
	12·8		3·6		32·4		3·55
			1		2		

EXAMPLE

Pete buys six CDs, each costing £4·50.
What change does he get from £30?

Total cost = 6 × 4·5 = £27
30 − 27 = 3. He has £3 change.

Exercise 22 ①

Do these multiplications.

1 1·3 × 3	**2** 2·4 × 2	**3** 5·1 × 4	**4** 6·1 × 5
5 7·2 × 4	**6** 1·3 × 6	**7** 0·7 × 7	**8** 3·3 × 3
9 0·84 × 4	**10** 1·63 × 5	**11** 3·1 × 7	**12** 0·14 × 6
13 11·7 × 7	**14** 2·14 × 8	**15** 1·73 × 5	**16** 2·72 × 3
17 50·1 × 4	**18** 70·7 × 7	**19** 11·9 × 2	**20** 2·09 × 4

21 A woman buys five books, each costing £3·25. What is the total cost?

22 Mo buys 100 stamps, each costing £0·31. What is the total cost?

23 At a restaurant, seven people each pay £7·15 for food and
£2·30 for drinks. What is the total bill for the seven people?

24 Calculate the area of this picture. You will need a ruler.

> Remember:
> Area = length × width,
> for a rectangle

1.3.7 Multiplying decimals by decimals

The answer has the same number of decimal places as the
total number of decimal places in the question.

EXAMPLE

a 0·2 × 0·8	**b** 0·4 × 0·07	**c** 5 × 0·06
a 0·2 × 0·8	**b** 0·4 × 0·07	**c** 5 × 0·06
(2 × 8 = 16)	(4 × 7 = 28)	(5 × 6 = 30)
So 0·2 × 0·8 = 0·16	So 0·4 × 0·07 = 0·028	So 5 × 0·06 = 0·3

Exercise 23 ①

Do these multiplications without a calculator.

1 0·2 × 0·3	**2** 0·5 × 0·3	**3** 0·4 × 0·3	**4** 0·2 × 0·03
5 0·6 × 3	**6** 0·7 × 5	**7** 0·9 × 2	**8** 8 × 0·1
9 0·4 × 0·9	**10** 0·02 × 0·7	**11** 2·1 × 0·6	**12** 4·7 × 0·5
13 21·3 × 0·4	**14** 5·2 × 0·6	**15** 4·2 × 0·03	**16** 212 × 0·6
17 0·85 × 0·2	**18** 3·27 × 0·1	**19** 12·6 × 0·01	**20** 0·02 × 17
21 0·05 × 1·1	**22** 52 × 0·01	**23** 65 × 0·02	**24** 0·5 × 0·002

1.3.8 Dividing decimals by whole numbers

EXAMPLE

a 5·6 ÷ 4   **b** 0·7 ÷ 5   **c** 52·5 ÷ 6

a
$$\begin{array}{r} 1\cdot4 \\ 4\overline{)5\cdot{}^16} \end{array}$$

b
$$\begin{array}{r} 0\cdot14 \\ 5\overline{)0\cdot7\,{}^20} \end{array}$$

c
$$\begin{array}{r} 8\cdot75 \\ 6\overline{)52\cdot{}^45\,{}^30} \end{array}$$

Exercise 24

Do these divisions without a calculator.

1 8·4 ÷ 4	**2** 9·2 ÷ 4	**3** 7·5 ÷ 3	**4** 7·5 ÷ 5
5 91·4 ÷ 2	**6** 20·7 ÷ 6	**7** 7·6 ÷ 2	**8** 13·5 ÷ 5
9 17·2 ÷ 8	**10** 10·8 ÷ 9	**11** 9·2 ÷ 5	**12** 7·8 ÷ 6
13 16·8 ÷ 7	**14** 29·4 ÷ 7	**15** 23·4 ÷ 9	**16** 18·6 ÷ 3
17 34·0 ÷ 5	**18** 51·2 ÷ 8	**19** 27·6 ÷ 6	**20** 25·2 ÷ 7
21 0·9 ÷ 5	**22** 0·7 ÷ 5	**23** 0·7 ÷ 2	**24** 0·6 ÷ 4

1.3.9 Dividing decimals by decimals

EXAMPLE

a 9·36 ÷ 0·4

b 0·0378 ÷ 0·07

a Multiply both numbers by 10 so that you can divide by a **whole number.** (Move the decimal points to the right.)
So work out 93·6 ÷ 4

$$\begin{array}{r} 2\ 3\cdot 4 \\ 4\overline{)\ 9\ ^13\cdot\ ^16} \end{array}$$

b Multiply both numbers by 100 so that you can divide by a whole number. (Move the decimal points to the right.)
So work out 3·78 ÷ 7

$$\begin{array}{r} 0.\ 5\ 4 \\ 7\overline{)\ 3\cdot\ ^37\ ^28} \end{array}$$

EXAMPLE

A stamp costs £0·70. How many stamps can you buy for £11·20?

11·20 ÷ 0·70 ⟶ Work out 112 ÷ 7
You can buy 16 stamps.

$$\begin{array}{r} 1\ 6 \\ 6\overline{)\ 11\ ^42} \end{array}$$

Exercise 25

Do these divisions without a calculator.

1 0·84 ÷ 0·4	**2** 0·93 ÷ 0·3	**3** 0·872 ÷ 0·2	**4** 0·8 ÷ 0·2
5 2·8 ÷ 0·7	**6** 1·25 ÷ 0·5	**7** 8 ÷ 0·5	**8** 40 ÷ 0·2
9 7 ÷ 0·1	**10** 0·368 ÷ 0·04	**11** 0·915 ÷ 0·03	**12** 0·248 ÷ 0·04
13 0·625 ÷ 0·05	**14** 8·54 ÷ 0·07	**15** 1·272 ÷ 0·006	**16** 4·48 ÷ 0·08
17 0·12 ÷ 0·002	**18** 7·5 ÷ 0·005	**19** 0·09 ÷ 0·3	**20** 0·77 ÷ 1·1

21 $0.055 \div 0.11$ **22** $21.28 \div 7$ **23** $22.48 \div 4$ **24** $3.12 \div 4$

25 $0.7 \div 5$ **26** $3 \div 0.8$ **27** $0.3 \div 4$ **28** $1.2 \div 8$

29 $0.732 \div 0.6$ **30** $0.1638 \div 0.001$ **31** $1.05 \div 0.6$ **32** $7.52 \div 0.4$

33 A cake weighing 4.8 kg is cut into several pieces each weighing 0.6 kg. How many pieces are there?

34 A phone call costs £0.04. How many calls can I make if I have £3.52?

*35 A sheet of paper is 0.01 cm thick. How many sheets are there in a pile of paper 5.8 cm thick?

Exercise 26 ①

1 $3.7 + 0.62$ **2** $8.45 - 2.7$ **3** $11.3 - 2.14$ **4** 2.52×0.4

5 $3.74 \div 5$ **6** $17 + 3.24$ **7** $12 - 1.8$ **8** $23.6 \div 8$

9 82.1×0.06 **10** 0.034×1000 **11** $62.1 \div 100$ **12** $11.4 - 3.16$

13 0.153×0.8 **14** $2.16 + 9.99$ **15** $18.606 \div 7$ **16** 6.042×11

17 34.1×1000 **18** $0.41 \div 100$ **19** 52.6×0.04 **20** $0.365 - 0.08$

21 $2.32956 \div 9$ **22** 654×0.005 **23** $0.7 + 0.77 + 0.777$ **24** $54 \div 100$

25 27×0.001 **26** 6.007×1.1 **27** $8.2 - 1.64$ **28** $47.04 \div 6$

Exercise 27 ①

Cross numbers

Make three copies of the cross number grid. Label them **A**, **B** and **C** and then fill in the answers using the clues.

A

Across	Down
1 13×7	**1** $101 - 7$
2 $0.214 \times 10\,000$	**2** $2.7 \div 0.1$
4 $265 - 248$	**3** $44.1 + 0.9$
5 $2 \times 2 \times 2 \times 2 \times 2 \times 2$	**4** $(2 \times 9) - (8 \div 2)$
7 $90 - (9 \times 9)$	**6** 9^2
8 14×5	**8** $6523 + 917$
9 $2226 \div 7$	**9** $418 \div 11$
11 $216 \div (18 \div 3)$	**10** $216 + (81 \times 100)$
12 $800 - 363$	**13** $2 \times 2 \times 2 \times 3 \times 3$
14 $93 - (6 \times 2)$	
15 0.23×100	
16 $8 \times 8 - 1$	

B

Across	Down
1 $2·4 \times 40$	**1** $558 \div 6$
2 $1600 - 27$	**2** $6·4 \div 0·4$
4 $913 - 857$	**3** $0·071 \times 1000$
5 $2 + (9 \times 9)$	**4** $11·61 + 4·2 + 37·19$
7 $0·4 \div 0·05$	**6** $(7 - 3·1) \times 10$
8 $27 \times 5 - 69$	**8** $8 \times 8 \times 100 - 82$
9 $4158 \div 7$	**9** $0·08 \times 700$
11 $2^6 + 6$	**10** $40 \times 30 \times 4 - 1$
12 $5·22 \div 0·03$	**13** $\frac{1}{5}$ of 235
14 $201 - 112$	
15 7 million $\div 100\,000$	
16 $\frac{1}{4}$ of 372	

C

Across	Down
1 $2·6 \times 10$	**1** $0·2 \times 100$
2 $6·314 \times 1000$	**2** $6·7 \div 0·1$
4 $600 - 563$	**3** $1800 \div 100$
5 $0·25 \times 100$	**4** $21 \div 0·6$
7 $3 \div 0·5$	**6** $420 \times 0·05$
8 $0·08 \times 1000$	**8** $0·8463 \times 10\,000$
9 $3·15 \div 0·01$	**9** $0·032 \times 1000$
11 $1·1 \times 70$	**10** $5·706 \div 0·001$
12 $499 + 103$	**13** 5^2
14 $1 \div 0·1$	
15 $0·01 \times 5700$	
16 $1000 - 936$	

1.3.10 Using one calculation to find another

Once you know the answer to one calculation, you can use it to solve others.

> **EXAMPLE**
>
> You are told that $42 \times 15 = 630$. Use this to work out $4·2 \times 15$.
>
> Since $4·2$ is $42 \div 10$, you can see that $4·2 \times 15 = 63$.

Similarly you can use the fact that $6·4 \times 27 = 172·8$ to find these answers.

a $6·4 \times 2·7 = 17·28$ $(2·7 = 27 \div 10)$ **b** $0·64 \times 27 = 17·28$ $(0·64 = 6·4 \div 10)$

c $172·8 \div 27 = 6·4$ ($\div$ is the inverse of $\times$)

Exercise 28 ①

1 You are told that $32 \times 1·9 = 60·8$. Find

   a 32×19 **b** 320×19 **c** $3·2 \times 1·9$

2 You are told that $37·6 \times 54 = 2030·4$. Find

   a $3·76 \times 54$ **b** $37·6 \times 5·4$ **c** 376×54

3 You are told that $82·3 \times 2·3 = 189·29$. Find

   a $823 \times 2·3$ **b** $8·23 \times 23$ **c** $82·3 \times 0·23$

4 You are told that $59·3 \times 61 = 3617·3$. Find

   a $5·93 \times 6·1$ **b** $0·593 \times 6·1$ **c** $59·3 \times 0·0061$

5 You are told that $36·2 \times 134 = 4850·8$. Find

   a 3620×134 **b** $36·2 \times 1·34$ **c** $\dfrac{4850·8}{134}$

6 You are told that $81·6 \times 215 = 17544$. Find

   a $8·16 \times 2150$ **b** $81·6 \times 2·15$ **c** $\dfrac{17544}{81·6}$

1.4 Negative numbers

1.4.1 Using negative numbers

● If the weather is very cold and the temperature is
3 degrees below zero, it is written −3°.

● If a golfer is 5 under par for his round, the scoreboard will show −5.

An easy way to begin calculations with negative numbers is to
think about changes in temperature.

> **EXAMPLE**
>
> **a** The temperature is −2° and it rises by 7°.
> The new temperature is 5°.
> You can write −2 + 7 = 5.
> **b** The temperature is −3° and it falls by 6°.
> The new temperature is −9°.
> You can write −3 − 6 = −9.

Exercise 29 ①

In questions **1** to **12** move up or down the thermometer to find
the new temperature.

1 The temperature is +8° and it falls by 3°.

2 The temperature is +4° and it falls by 5°.

3 The temperature is +2° and it falls by 6°.

4 The temperature is −1° and it falls by 6°.

5 The temperature is −5° and it rises by 1°.

6 The temperature is −8° and it rises by 4°.

7 The temperature is −3° and it rises by 7°.

8 The temperature is +4° and it rises by 8°.

9 The temperature is +9° and it falls by 14°.

10 The temperature is −13° and it rises by 13°.

11 The temperature is −6° and it falls by 5°.

12 The temperature is −25° and it rises by 10°.

13 Write these temperatures from the coldest to the hottest.

   a −2°C, 5°C, −7°C   **b** −1°C, 2°C, 0°C

   c −8°C, 3°C, −11°C   **d** −4°C, −1°C, −2°C, −7°C

   e 4°C, −5°C, −2°C, −4°C

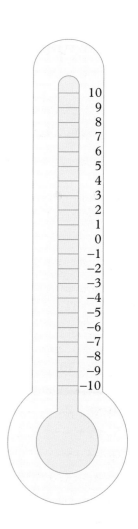

14 Write these numbers in order of size, starting with the lowest.
  a 6, −3, −5, 2, 0   **b** −8, 11, −6, 3, −1
  c −10, −15, 5, −20, −2 **d** 23, −10, −5, −15, 18

15 Copy these sequences and write the next three numbers in each.
  a 10, 8, 6, 4, —, —, —   **b** 12, 9, 6, 3, —, —, —
  c 3, 2, 1, 0, −1, —, —, — **d** 4, 2, 0, −2, —, —, —
  e 12, 6, 0, —, —, —   **f** −3, −2, −1, —, —, —
  g −8, −6, −4, —, —, —   **h** 10, 6, 2, —, —, —

> These can all be solved by adding or subtracting numbers.

1.4.2 Adding and subtracting with negative numbers

For adding and subtracting you can use a number line.

EXAMPLE

  a −1 + 4   **b** −2 − 3   **c** 4 − 6

a

−1 + 4 = 3
start here / go right / 4 places

b
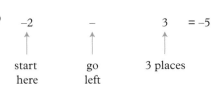
−2 − 3 = −5
start here / go left / 3 places

c
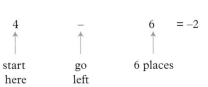
4 − 6 = −2
start here / go left / 6 places

Exercise 30 ①

Do these calculations.
  1 −6 + 2   **2** −7 − 5   **3** −3 − 8   **4** −5 + 2
  5 −6 + 1   **6** 8 − 4   **7** 4 − 9   **8** 11 − 19
  9 4 + 15   **10** −7 − 10   **11** 16 − 20   **12** −7 + 2
  13 −6 − 5   **14** 10 − 4   **15** −4 + 0   **16** −6 + 12

17 −7 + 7   **18** 2 − 20   **19** 8 − 11   **20** −6 − 5

21 −8 − 4   **22** −3 + 7   **23** −6 + 10   **24** −5 + 5

25 −11 + 3   **26** 7 − 10   **27** −5 + 8   **28** −12 + 0

29 −1 + 19   **30** 20 − 25   **31** −6 − 60   **32** −2 + 100

● When you have two + or two − signs together, use the rule in this table.

++ = +	+− = −
−− = +	−+ = −

EXAMPLE

a 3 − (−6) = 3 + 6 = 9   **d** 8 + (−8)   = 8 − 8   = 0
b −4 + (−5) = −4 − 5 = −9   **e** 0 − (−11)   = 0 + 11   = 11
c −5 − (+7) = −5 − 7 = −12   **f** −10 − (−3) = −10 + 3 = −7

Exercise 31 1

Do these calculations.

1 −3 + (−5)   **2** −5 − (+2)   **3** 4 − (+3)

4 −3 − (−4)   **5** 6 − (−3)   **6** 16 + (−5)

7 −4 + (−4)   **8** 20 − (−22)   **9** −6 − (−10)

10 95 + (−80)   **11** −3 − (+4)   **12** −5 − (+4)

13 6 + (−7)   **14** −4 + (−3)   **15** −7 − (−7)

16 3 − (−8)   **17** −8 + (−6)   **18** 7 − (+7)

19 12 − (−5)   **20** 9 − (+6)   **21** −3 − (−2)

22 8 + (−11)   **23** 10 − (−2)   **24** −7 + (−2)

25 9 − (+6)   **26** 7 + (−7)   **27** 0 − (−8)

28 −6 − (−8)

29 Copy and complete these addition squares.

a

+	−5	1	6	−2
3	−2	4		
−2				
6				
−10				

b

+			−4	
5			1	
		0		5
	7		6	17
			−4	

1.4.3 Multiplying and dividing with negative numbers

▶ When two numbers with the same sign are multiplied together, the answer is **positive**.

$+7 \times (+3) = +21$
$-6 \times (-4) = +24$

▶ When two numbers with different signs are multiplied together, the answer is **negative**.

$-8 \times (+4) = -32$
$+7 \times (-5) = -35$

▶ When dividing numbers, the rules are the same as in multiplication.

$-70 \div (-2) = +35$
$+12 \div (-3) = -4$
$-20 \div (+4) = -5$

Exercise 32 ①

1 $-3 \times (+2)$ 2 $-4 \times (+1)$ 3 $+5 \times (-3)$ 4 $-3 \times (-3)$

5 $-4 \times (2)$ 6 $-5 \times (3)$ 7 $6 \times (-4)$ 8 $3 \times (2)$

9 $-3 \times (-4)$ 10 $6 \times (-3)$ 11 $-7 \times (3)$ 12 $-5 \times (-5)$

13 $6 \times (-10)$ 14 $-3 \times (-7)$ 15 $8 \times (6)$ 16 $-8 \times (2)$

17 $-7 \times (6)$ 18 $-5 \times (-4)$ 19 $-6 \times (7)$ 20 $11 \times (-6)$

21 $8 \div (-2)$ 22 $-9 \div (3)$ 23 $-6 \div (-2)$ 24 $10 \div (-2)$

25 $-12 \div (-3)$ 26 $-16 \div (4)$ 27 $4 \div (-1)$ 28 $8 \div (-8)$

29 $16 \div (-8)$ 30 $-20 \div (-5)$ 31 $-16 \div (1)$ 32 $18 \div (-9)$

33 $36 \div (-9)$ 34 $-45 \div (-9)$ 35 $-70 \div (7)$ 36 $-11 \div (-1)$

37 $-16 \div (-1)$ 38 $100 \div (-10)$ 39 $-2 \div (-2)$ 40 $50 \div (-10)$

41 $-8 \times (-8)$ 42 $-9 \times (3)$ 43 $10 \times (-60)$ 44 $-8 \times (-5)$

45 $-12 \div (-6)$ 46 $-18 \times (-2)$ 47 $-8 \div (4)$ 48 $-80 \div (10)$

49 Copy and complete these multiplication squares.

a

×	4	−3	0	−2
−5				
2				
10				
−1				

b

×			−1	
3			−3	
		−15		18
		−14		−7	−42
			10		

Questions on negative numbers are more difficult when different operations are mixed together. The questions in these three short tests are mixed.

Test 1

1 $-8 - 8$	**2** $-8 \times (-8)$	**3** -5×3	**4** $-5 + 3$
5 $8 - (-7)$	**6** $20 - 2$	**7** $-18 \div (-6)$	**8** $4 + (-10)$
9 $-2 + 13$	**10** $+8 \times (-6)$	**11** $-9 + (+2)$	**12** $-2 - (-11)$
13 $-6 \times (-1)$	**14** $2 - 20$	**15** $-14 - (-4)$	**16** $-40 \div (-5)$
17 $5 - 11$	**18** -3×10	**19** $9 + (-5)$	**20** $7 \div (-7)$

Test 2

1 $-2 \times (+8)$	**2** $-2 + 8$	**3** $-7 - 6$	**4** $-7 \times (-6)$
5 $+36 \div (-9)$	**6** $-8 - (-4)$	**7** $-14 + 2$	**8** $5 \times (-4)$
9 $11 + (-5)$	**10** $11 - 11$	**11** $-9 \times (-4)$	**12** $-6 + (-4)$
13 $3 - 10$	**14** $-20 \div (-2)$	**15** $16 + (-10)$	**16** $-4 - (+14)$
17 $-45 \div 5$	**18** $18 - 3$	**19** $-1 \times (-1)$	**20** $-3 - (-3)$

Test 3

1 $-10 \times (-10)$	**2** $-10 - 10$	**3** $-8 \times (+1)$	**4** $-8 + 1$
5 $5 + (-9)$	**6** $15 - 5$	**7** $-72 \div (-8)$	**8** $-12 - (-2)$
9 $-1 + 8$	**10** $-5 \times (-7)$	**11** $-10 + (-10)$	**12** $-6 \times (+4)$
13 $6 - 16$	**14** $-42 \div (+6)$	**15** $-13 + (-6)$	**16** $-8 - (-7)$
17 $5 \times (-1)$	**18** $2 - 15$	**19** $21 + (-21)$	**20** $-16 \div (-2)$

1.5 Order of operations

1.5.1 The BIDMAS rule

Mathematicians all over the world regularly exchange their ideas and the results of their theories, even though much of the time they are unable to speak the same language! They can communicate mathematically because it has been agreed that everyone follows certain rules.

● Think about this question:
  'What is five add seven multiplied by three?'
  If you add first, you get: $5 + 7 \times 3$
                                  $= 12 \times 3$
                                  $= 36$
  If you multiply first, you get: $5 + 7 \times 3$
                                  $= 5 + 21$
                                  $= 26$

If everyone came up with different answers to the same mathematical question, life would be rather stressful as people would argue constantly over who was correct.

This table shows the order in which to do mathematical operations to ensure you all get the same answers.

B rackets	( )	do first	**'B'**
I ndices	3^2	do next	**'I'**
D ivision	÷	do this pair next	**'D'**
M ultiplication	×		**'M'**
A ddition	+	do this pair last	**'A'**
S ubtraction	−		**'S'**

Indices mean powers and roots, see Section 1.6

Remember the word **'B I D M A S'**.

EXAMPLE

a $40 ÷ 5 × 2$
$= 8 × 2$
$= 16$

For ÷ and × do in the order they appear

b $9 + 8 − 7$
$= 17 − 7$
$= 10$

For + and − do in the order they appear

c $5 + 2 × 3$
$= 5 + 6$
$= 11$

× before +

d $10 − 8 ÷ 2$
$= 10 − 4$
$= 6$

÷ before −

Exercise 33 ①

Do these. Show every step in your working.

1 $5 + 3 × 2$ **2** $4 − 1 × 3$ **3** $7 − 4 × 3$

4 $2 + 2 × 5$ **5** $9 + 2 × 6$ **6** $13 − 11 × 1$

7 $7 × 2 + 3$ **8** $9 × 4 − 12$ **9** $2 × 8 − 7$

10 $4 × 7 + 2$ **11** $13 × 2 + 4$ **12** $8 × 5 − 15$

13 $6 + 10 ÷ 5$ **14** $7 − 16 ÷ 8$ **15** $8 − 14 ÷ 7$

16 $5 + 18 ÷ 6$ **17** $2 × 18 ÷ 6$ **18** $6 − 12 ÷ 4$

19 $20 ÷ 4 + 2$ **20** $15 ÷ 3 − 7$ **21** $24 ÷ 6 − 8$

22 $30 ÷ 6 + 9$ **23** $8 ÷ 2 + 9$ **24** $28 ÷ 7 − 4$

25 $13 + 3 × 13$ **26** $9 + 26 ÷ 13$ **27** $10 × 8 − 70$

28 $96 ÷ 4 − 4$ **29** $36 ÷ 9 + 1$ **30** $1 × 2 + 3$

EXAMPLE

a $8 + 3 \times 4 - 6$

$= 8 + (3 \times 4) - 6$
$= 8 + 12 - 6$
$= 14$

x and ÷ before
+ and –

b $3 \times 2 - 8 \div 4$

$= (3 \times 2) - (8 \div 4)$
$= 6 - 2$
$= 4$

c $\dfrac{8 + 6}{2} = \dfrac{14}{2}$

$= 7$

A horizontal line
acts as a bracket.

Notice that the brackets make the working easier.

Exercise 34 ❶

Evaluate these. Show every step in your working.

1 $2 + 3 \times 4 + 1$

2 $4 + 8 \times 2 - 10$

3 $7 + 2 \times 2 - 6$

4 $25 - 7 \times 3 + 5$

5 $17 - 3 \times 5 + 9$

6 $11 - 9 \times 1 - 1$

7 $1 + 6 \div 2 + 3$

8 $6 + 28 \div 7 - 2$

9 $8 + 15 \div 3 - 5$

10 $5 - 36 \div 9 + 3$

11 $6 - 24 \div 4 + 0$

12 $8 - 30 \div 6 - 2$

13 $3 \times 4 + 1 \times 6$

14 $4 \times 4 + 14 \div 7$

15 $2 \times 5 + 8 \div 4$

16 $21 \div 3 + 5 \times 4$

17 $10 \div 2 + 1 \times 3$

18 $15 \div 5 + 18 \div 6$

19 $5 \times 5 - 6 \times 4$

20 $2 \times 12 - 4 \div 2$

21 $7 \times 2 - 10 \div 2$

22 $35 \div 7 - 5 \times 1$

23 $36 \div 3 - 1 \times 7$

24 $42 \div 6 - 56 \div 8$

25 $72 \div 9 + 132 \div 11$

26 $19 + 35 \div 5 - 16$

27 $50 - 6 \times 7 + 8$

28 $30 - 9 \times 2 + 40$

*29 $4 \times 11 - 28 \div 7$

*30 $13 \times 11 - 4 \times 8$

In questions **31** to **50**, remember to do the operations in the
brackets first.

31 $3 + (6 \times 8)$

32 $(3 \times 8) + 6$

33 $(8 \div 4) + 9$

34 $3 \times (9 \div 3)$

35 $(5 \times 9) - 17$

36 $10 + (12 \times 8)$

37 $(16 - 7) \times 6$

38 $48 \div (14 - 2)$

39 $64 \div (4 \times 4)$

40 $81 + (9 \times 8)$

41 $67 - (24 \div 3)$

42 $(12 \times 8) + 69$

43 $(6 \times 6) + (7 \times 7)$

44 $(12 \div 3) \times (18 \div 6)$

45 $(5 \times 12) - (3 \times 9)$

46 $(20 - 12) \times (17 - 9)$

47 $100 - (99 \div 3)$

48 $1001 + (57 \times 3)$

49 $(3 \times 4 \times 5) - (72 \div 9)$

50 $(2 \times 5 \times 3) \div (11 - 6)$

51 $\dfrac{15 - 7}{2}$

52 $\dfrac{160}{7 + 3}$

53 $\dfrac{19 + 13}{6 - 2}$

54 $\dfrac{5 \times 7 - 9}{13}$

1.6 Powers and roots

1.6.1 Indices

▶ Indices are a short way of writing products.
$2 \times 2 \times 2 \times 2 = 2^4$ (2 to the power 4)
$5 \times 5 \times 5 = 5^3$ (5 to the power 3)
$3 \times 3 \times 3 \times 3 \times 3 \times 10 \times 10 = 3^5 \times 10^2$

● Numbers like 3^2, 5^2, 11^2 are **square numbers**.
Numbers like 2^3, 6^3, 11^3 are **cube numbers**.

● To work out $3 \cdot 2^2$ on a calculator, press $\boxed{3\cdot2}$ $\boxed{x^2}$ $\boxed{=}$

To work out 3^4 on a calculator, press
$\boxed{3}$ $\boxed{x^y}$ $\boxed{4}$ $\boxed{=}$ or $\boxed{3}$ $\boxed{\wedge}$ $\boxed{4}$ $\boxed{=}$

> You should learn all the square numbers 1, 4, 9, 16, ...up to 225.

Exercise 35

Write these using indices.
1 $3 \times 3 \times 3 \times 3$ **2** 5×5 **3** $6 \times 6 \times 6$

4 $10 \times 10 \times 10 \times 10 \times 10$ **5** $1 \times 1 \times 1 \times 1 \times 1 \times 1 \times 1$ **6** $8 \times 8 \times 8 \times 8$

7 $7 \times 7 \times 7 \times 7 \times 7 \times 7$ **8** $2 \times 2 \times 2 \times 5 \times 5$ **9** $3 \times 3 \times 7 \times 7 \times 7 \times 7$

10 $3 \times 3 \times 10 \times 10 \times 10$ **11** $5 \times 5 \times 5 \times 5 \times 11 \times 11$ **12** $2 \times 3 \times 2 \times 3 \times 3$

13 $5 \times 3 \times 3 \times 5 \times 5$ **14** $2 \times 2 \times 3 \times 3 \times 3 \times 11 \times 11$

15 Work these out without using a calculator.
    a 4^2 **b** 6^2 **c** 10^2 **d** 3^3 **e** 10^3

16 Use the $\boxed{x^2}$ button on a calculator to work out these.
    a 9^2 **b** 21^2 **c** $1 \cdot 2^2$ **d** $0 \cdot 2^2$ **e** $3 \cdot 1^2$
    f 100^2 **g** 25^2 **h** $8 \cdot 7^2$ **i** $0 \cdot 9^2$ **j** $81 \cdot 4^2$

17 Find the areas of these squares.

   a   **b**   **c**

13 cm

2·5 cm

11·4 cm

*18 Write these in index form.
    a $a \times a \times a$ **b** $n \times n \times n \times n$ **c** $s \times s \times s \times s \times s \times s$
    d $p \times p \times q \times q \times q$ **e** $b \times b \times b \times b \times b \times b \times b$

*19 Use a calculator to work out these.

    a 6^3 **b** 2^8 **c** 3^5 **d** 10^5 **e** 4^3

    f $0{\cdot}1^3$ **g** $1{\cdot}7^4$ **h** $3^4 \times 7$ **i** $5^3 \times 10$

*20 A scientist has a dish containing 10^9 germs.
One day later there are 10 times as many germs.
How many germs are in the dish now?

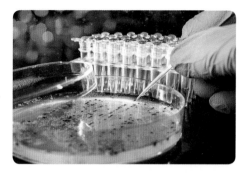

*21 A large garden has 2^8 daisies growing on the grass.
A weedkiller removes half of the daisies.
How many daisies are left?

*22 A maths teacher won the National Lottery and,
as a leaving present, he decided to set a final test
to a class of 25 students.
The person coming 25th won 2p, the 24th
won 4p, the 23rd 8p, the 22nd 16p and
so on, doubling the amount each time.

    a Write 2, 4, 8, 16 as powers of 2.

    b How much, in pounds, would be given
       to the person who came first in the test?

*23 Sean says 'If you work out the product of any four consecutive
numbers and then add one, the answer will be a square number.'
For example: $1 \times 2 \times 3 \times 4 = 24$
                       $24 + 1 = 25$,
which is a square number.
Is Sean right? Test his theory on four (or more) sets of four
consecutive numbers.

1.6.2   Square roots and cube roots

EXAMPLE

A square has an area of 529 cm².
How long is a side of the square?

In other words, what number **multiplied by itself** makes 529?
The answer is the **square root** of 529.

On a calculator press $\boxed{\sqrt{}}$ $\boxed{529}$ $\boxed{=}$

(On older calculators you may need to press $\boxed{529}$ $\boxed{\sqrt{}}$.)
The side of the square is 23 cm.

529 cm² ?

?

EXAMPLE

A cube has a volume of $512\,\text{cm}^3$.
How long is a side of the cube?

....

The answer is the **cube root** of 512.

On a calculator press $\boxed{\sqrt[3]{}}$ $\boxed{512}$ $\boxed{=}$

The side of the cube is 8 cm. (Check $8 \times 8 \times 8 = 512$)

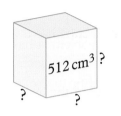

$512\,\text{cm}^3$?

Exercise 36 ②

1 Work out, without a calculator.

  a $\sqrt{16}$ **b** $\sqrt{36}$ **c** $\sqrt{1}$ **d** $\sqrt{100}$

2 Find the sides of the squares.

  a

Area $= 81\,\text{cm}^2$ x

x

  b

Area $= 49\,\text{cm}^2$ y

y

  c

Area $= 144\,\text{cm}^2$ z

z

3 Use a calculator to find these square roots, correct to 1 dp.

  a $\sqrt{10}$ **b** $\sqrt{29}$ **c** $\sqrt{107}$ **d** $\sqrt{19\cdot7}$

  e $\sqrt{2406}$ **f** $\sqrt{58\cdot6}$ **g** $\sqrt{0\cdot15}$ **h** $\sqrt{0\cdot727}$

4 A square photo has an area of $150\,\text{cm}^2$. Find the length of each side of the photo, correct to the nearest mm.

5 A square field has an area of 20 hectares. How long is each side of the field, correct to the nearest m? (1 hectare $= 10\,000\,\text{m}^2$)

6 The area of square A is equal to the sum of the areas of squares B and C. Find the length x, correct to 1 dp.

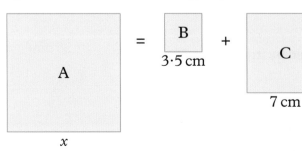

A $=$ B $+$ C

x $3\cdot5$ cm   7 cm

7 Find these cube roots.

   a $\sqrt[3]{64}$ **b** $\sqrt[3]{125}$ **c** $\sqrt[3]{1000}$

8 A cube has a volume of $200\,\text{cm}^3$.

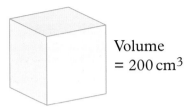

Volume
$= 200\,\text{cm}^3$

Find the length of the sides of the cube, correct to 1 dp.

*9 A challenge!

The reciprocal of 2 is $\frac{1}{2}$. The reciprocal of 7 is $\frac{1}{7}$. The reciprocal of x is $\frac{1}{x}$.

Find the square root of the reciprocal of the square root of the reciprocal of ten thousand.

1.6.3 Multiplying and dividing

▶ To multiply powers of the same number, **add** the indices.

EXAMPLE

$3^2 \times 3^4 = (3 \times 3) \times (3 \times 3 \times 3 \times 3) = 3^6$
$2^3 \times 2^2 = (2 \times 2 \times 2) \times (2 \times 2) = 2^5$
$7^3 \times 7^5 = 7^8$ Add the indices

▶ To divide powers of the same number, **subtract** the indices.

EXAMPLE

$2^4 \div 2^2 = \dfrac{2 \times 2 \times 2 \times 2}{2 \times 2} = 2^2$

$5^6 \div 5^2 = 5^4$ Subtract the indices
$7^8 \div 7^3 = 7^5$

Exercise 37 ②

Write these in simpler form.

1 $5^2 \times 5^4$	**2** $6^3 \times 6^2$	**3** $10^4 \times 10^5$	**4** $7^5 \times 7^3$
5 $3^6 \times 3^4$	**6** $8^3 \times 8^3$	**7** $2^3 \times 2^{10}$	**8** $3^6 \times 3^2$
9 $5^4 \times 5^1$	**10** $7^7 \times 7^3$	**11** $5^3 \times 5^5$	**12** $3^2 \times 3^2$
13 $6^3 \times 6^8$	**14** $5^2 \times 5^8$	**15** $7^3 \times 7^9$	**16** $7^4 \div 7^2$

17 $6^7 \div 6^2$ **18** $8^5 \div 8^4$ **19** $5^{10} \div 5^2$ **20** $10^7 \div 10^5$

21 $9^{10} \div 9^8$ **22** $3^8 \div 3^6$ **23** $2^6 \div 2^2$ **24** $3^3 \div 3$

25 $7^8 \div 7^5$ **26** $3^5 \div 3^2$ **27** $5^9 \div 5^2$ **28** $8^{10} \div 8^4$

29 $5^4 \div 5^1$ **30** $6^5 \div 6^2$ **31** $3^4 \div 3^4$ **32** $5^2 \div 5^2$

33 $\dfrac{3^4 \times 3^5}{3^2}$ **34** $\dfrac{2^8 \times 2^4}{2^5}$ **35** $\dfrac{7^3 \times 7^3}{7^4}$ **36** $\dfrac{5^9 \times 5^{10}}{5^7}$

Exercise 38 ②

Solve the equations to find the value of x.

1 $x^2 = 9$ **2** $x^5 = 1$ **3** $x^3 = 27$ **4** $x^5 = 0$

5 $2^x = 8$ **6** $3^x = 3$ **7** $5^x = 25$ **8** $10^x = 1000$

9 $2^x = 32$ **10** $4^x = 64$ **11** $2x^2 = 18$ **12** $3x^3 = 24$

> There is more on solving equations in Section 2.2

1.7   Using a calculator

1.7.1   Money and time on a calculator

● To work out £27·30 ÷ 7, key in

$$\boxed{2}\ \boxed{7}\ \boxed{\cdot}\ \boxed{3}\ \boxed{0}\ \boxed{\div}\ \boxed{7}\ \boxed{=}$$

The answer is 3·9. Remember this means £3·90.

> **EXAMPLE**
>
> A machine takes 15 minutes to make one toy. How long will it take to make 1627 toys?
> ....
> 15 minutes is one quarter of an hour and $\dfrac{1}{4} = 0·25$ as a decimal.
>
> Key in $\boxed{0·25}\ \boxed{\times}\ \boxed{1627}\ \boxed{=}$
>
> The answer is 406·75.
> It will take 406·75 hours or 406 hours 45 minutes.

● 6 minutes is $\dfrac{6}{60}$ of an hour. $\dfrac{6}{60} = \dfrac{1}{10} = 0·1$ hours.

In the same way, 27 minutes $= \dfrac{27}{60}$ of an hour.

$$\dfrac{27}{60} = 0·45 \text{ hours.}$$

Exercise 39 ❶

1 Do these calculations and give your answer in **pounds**.

    a £1·22 × 5   **b** £153·60 ÷ 24   **c** £12·35 − £7·65

    d 20p × 580   **e** 6p × 2155   **f** £10 ÷ 250

2 Write these time intervals in hours, as decimals.

    a 2 h 30 min   **b** 4 h 15 min   **c** 3 h 45 min

    d 6 min   **e** 12 min   **f** 15 min

> $6 \text{ min} = \dfrac{6}{60}$ hours
> $= 0·1$ hours
> $15 \text{ min} = \dfrac{15}{60}$ hours
> $= 0·25$ hours

3 Do these calculations and give your answer in **hours**.

    a 2 h 45 min × 9   **b** 3 h 15 min × 7   **c** 14 h 30 min ÷ 5

    d 15 min × 11   **e** 30 min × 5   **f** 5 h 15 min ÷ 3

1.7.2   Order of operations

Where there is a mixture of operations to do, remember **BIDMAS**.

▶ Work out brackets first   **B**

▶ Work out indices next   **I**

▶ Work out ÷, ×   **DM**

▶ Then +, −   **AS**

It is a good idea to check your answers by estimation or by performing the inverse operation.

Exercise 40 ❶

Use a calculator and give the answers correct to one decimal place.

  1 2·5 × 1·67   **2** 19·6 − 3·7311   **3** $0·792^2$

  4 0·13 + 8·9 − 3·714   **5** $2·4^2 − 1·712$ **6** 5·3 × 1·7 + 3·7

  7 0·71 × 0·92 − 0·15   **8** 9·6 ÷ 1·72   **9** 8·17 − 1·56 + 7·4

10 $\sqrt{4·52}$ **11** $\sqrt{198}$ **12** $\sqrt{\dfrac{2·63}{1·9}}$

In questions **13** to **30**, remember BIDMAS.

13 2·5 + 3·1 × 2·4   **14** 7·81 + 0·7 × 1·82   **15** 8·73 + 9 ÷ 11

16 11·7 ÷ 9 − 0·74   **17** 7 ÷ 0·32 + 1·15   **18** 2·6 + 5·2 × 1·7

19 $2·9 + \dfrac{8·3}{1·83}$ **20** $1·7^2 + 2·62$ **21** $5·2 + \dfrac{11·7}{1·85}$

22 9·64 + 26 ÷ 12·7   **23** $1·27 + 3·1^2$ **24** $4·2^2 ÷ 9·4$

25 $0{\cdot}151 + 1{\cdot}4 \times 9{\cdot}2$ **26** $1{\cdot}7^3$ **27** $8{\cdot}2 + 3{\cdot}2 \times 3{\cdot}3$

28 $3{\cdot}2 + \dfrac{1{\cdot}41}{6{\cdot}72}$ **29** $\dfrac{1{\cdot}9 + 3{\cdot}71}{2{\cdot}3}$ **30** $\dfrac{8{\cdot}7 - 5{\cdot}371}{1{\cdot}14}$

1.7.3 Using brackets

Most calculators have brackets buttons like these $\boxed{(}$ $\boxed{)}$.

When you use the bracket buttons, the calculator does the

calculation inside the brackets $\boxed{(}$ $\boxed{)}$ first.

Try it.

Don't forget to press the $\boxed{=}$ button at the end to give the

final answer.

EXAMPLE

a $8{\cdot}72 - (\,1{\cdot}4 \times 1{\cdot}7\,)$

b $\dfrac{8{\cdot}51}{(1{\cdot}94 - 0{\cdot}711)}$

....

a $\boxed{8{\cdot}72}$ $\boxed{-}$ $\boxed{(}$ $\boxed{1{\cdot}4}$ $\boxed{\times}$ $\boxed{1{\cdot}7}$ $\boxed{)}$ $\boxed{=}$

$8{\cdot}72 - (1{\cdot}4 \times 1{\cdot}7) = 6{\cdot}3$ to 1 dp

b $\boxed{8{\cdot}51}$ $\boxed{\div}$ $\boxed{(}$ $\boxed{1{\cdot}94}$ $\boxed{-}$ $\boxed{0{\cdot}711}$ $\boxed{)}$ $\boxed{=}$

$\dfrac{8{\cdot}51}{(1{\cdot}94 - 0{\cdot}711)} = 6{\cdot}9$ to 1 dp

---

1.7.4 Using the $\boxed{\text{ANS}}$ button

You can use the $\boxed{\text{ANS}}$ button as a short-term memory. It holds
the answer from the previous calculation.

EXAMPLE

Work out $\dfrac{5}{1{\cdot}2 - 0{\cdot}761}$, using the $\boxed{\text{ANS}}$ button.

....

Find the bottom line first

$\boxed{1{\cdot}2}$ $\boxed{-}$ $\boxed{0{\cdot}761}$ $\boxed{=}$ $\boxed{5}$ $\boxed{\div}$ $\boxed{\text{ANS}}$ $\boxed{=}$

The calculator reads $11{\cdot}389\,521\,64$.

Notice that the
$\boxed{\text{EXE}}$ button
works the same as
the $\boxed{=}$ button on
some calculators.
($\boxed{\text{EXE}}$ means
'execute'.)

Exercise 41 ①

Do these and give the answers correct to 1 decimal place.

Use the brackets buttons or the $\boxed{\text{ANS}}$ button.

1 $18\cdot41 - (7\cdot2 \times 1\cdot3)$

2 $11\cdot01 + (2\cdot6 \div 7)$

3 $(1\cdot27 + 5\cdot6) \div 1\cdot4$

4 $9\cdot6 + (11\cdot2 \div 4)$

5 $(8\cdot6 \div 3) - 1\cdot4$

6 $11\cdot7 - (2\cdot6 \times 2\cdot7)$

7 $7\cdot41 - \left(\dfrac{7\cdot3}{1\cdot4}\right)$

8 $\left(\dfrac{8\cdot91}{1\cdot7}\right) - 2\cdot63$

9 $\dfrac{1\cdot41}{(1\cdot7 + 0\cdot21)}$

10 $(1\cdot56 + 1\cdot9) \div 2\cdot45$

11 $3\cdot2 \times (1\cdot9 - 0\cdot74)$

12 $8\cdot9 \div (1\cdot3 - 0\cdot711)$

13 $(8\cdot72 \div 1\cdot4) \times 1\cdot49$

14 $(2\cdot67 + 1\cdot2 + 5) \times 1\cdot13$

15 $23 - (9\cdot2 \times 1\cdot85)$

16 $\dfrac{(8\cdot41 + 1\cdot73)}{1\cdot47}$

17 $\dfrac{7\cdot23}{(8\cdot2 \times 0\cdot91)}$

18 $\dfrac{(11\cdot4 - 7\cdot87)}{17}$

In questions **19** to **40**, use the $\boxed{x^2}$ button when you need it.

19 $2\cdot6^2 - 1\cdot4$

20 $8\cdot3^2 \times 1\cdot17$

21 $7\cdot2^2 \div 6\cdot67$

22 $(1\cdot4 + 2\cdot67)^2$

23 $(8\cdot41 - 5\cdot7)^2$

24 $(2\cdot7 \times 1\cdot31)^2$

25 $8\cdot2^2 - (1\cdot4 + 1\cdot73)$

26 $\dfrac{2\cdot6^2}{(1\cdot3 + 2\cdot99)}$

27 $4\cdot1^2 - \left(\dfrac{8\cdot7}{3\cdot2}\right)$

28 $\dfrac{(2\cdot7 + 6\cdot04)}{(1\cdot4 + 2\cdot11)}$

29 $\dfrac{(8\cdot71 - 1\cdot6)}{(2\cdot4 + 9\cdot73)}$

30 $\left(\dfrac{2\cdot3}{1\cdot4}\right)^2$

31 $9\cdot72^2 - (2\cdot9 \times 2\cdot7)$

32 $(3\cdot3 + 1\cdot3^2) \times 9$

33 $(2\cdot7^2 - 2\cdot1) \div 5$

34 $\left(\dfrac{2\cdot84}{7}\right) + \left(\dfrac{7}{11\cdot2}\right)$

35 $\dfrac{(2\cdot7 \times 8\cdot1)}{(12 - 8\cdot51)}$

36 $\left(\dfrac{2\cdot3}{1\cdot5}\right) - \left(\dfrac{6\cdot3}{8\cdot9}\right)$

*37 $(1\cdot31 + 2\cdot705) - 1\cdot3^2$

*38 $(2\cdot71 - 0\cdot951) \times 5\cdot62$

*39 $\dfrac{(8\cdot5 \times 1\cdot952)}{(7\cdot2 - 5\cdot96)}$

*40 $\left(\dfrac{80\cdot7}{30\cdot3}\right) - \left(\dfrac{11\cdot7}{10\cdot2}\right)$

1.7.5   Standard form

a   Using a calculator, work out 4 000 000 multiplied by 30 000 000.
The answer is 120 000 000 000 000. On a calculator the answer
may be displayed as ⌊ 1 . 2 ⌋ ⌊ 1 3 ⌋

The calculator does not show the answer in full because there
are too many zeros. The display ⌊ 1 . 2 ⌋ ⌊ 1 3 ⌋ is short for $1 \cdot 2 \times 10^{13}$
which is '1·2 times 10 to the power 13'.

b   Similarly for the division $0 \cdot 008 \div 400 000$ the calculator will give
the answer as ⌊ 2 − 0 8 ⌋. This is how the calculator shows 2×10^{-8}.

▶   The numbers $1 \cdot 2 \times 10^{13}$ and 2×10^{-8} are written in **standard form**.
Standard form is used to represent very large or very small numbers.

▶   The number $a \times 10^n$ is in **standard form** when $1 \le a < 10$ and n is an
integer.

EXAMPLE

Here are examples of changing numbers into standard form.

a $30000 = 3 \times 10000 = 3 \times 10 \times 10 \times 10 \times 10 = \mathbf{3 \times 10^4}$
$855000 = 8 \cdot 55 \times 100000 = \mathbf{8 \cdot 55 \times 10^5}$

b $0 \cdot 007 = \dfrac{7}{1000} = 7 \times \dfrac{1}{10^3} = \mathbf{7 \times 10^{-3}}$

$0 \cdot 00028 = \dfrac{2 \cdot 8}{10000} = 2 \cdot 8 \times \dfrac{1}{10^4} = \mathbf{2 \cdot 8 \times 10^{-4}}$

Notice: In large
numbers the power
of 10 is positive.
In small numbers
the power of 10 is
negative.

●   The numbers in bold type above are all in standard form.

Exercise 42 ①

Write these numbers in standard form.

1 6000	**2** 8200	**3** 50 000	**4** 200 000
5 2 million	**6** 550	**7** 618 000	**8** 3 billion
9 222 000	**10** 6180	**11** 100 million	**12** 700 000
13 0·007	**14** 0·055	**15** 0·00081	**16** 0·000 07
17 0·000004	**18** 0·7	**19** 0·0111	**20** 0·02

Write these as ordinary numbers.

21 7×10^3	**22** $8 \cdot 2 \times 10^2$	**23** 6×10^{-2}	**24** 8×10^{-1}
25 $4 \cdot 8 \times 10^5$	**26** $1 \cdot 3 \times 10^{-2}$	**27** $8 \cdot 26 \times 10^5$	**28** $3 \cdot 82 \times 10^{-3}$

29 The height of the Eiffel Tower in Paris is $3·01 \times 10^2$ metres. Write this height as an ordinary number in metres.

30 The average distance of the Moon from the Earth is 384 400 km. Write this distance in standard form: **a** in km   **b** in metres

31 A large ants nest contains $2·3 \times 10^5$ ants. Write this as an ordinary number.

32 Write these numbers in order of size, smallest first:
22 000 3×10^4 6×10^{-2} 10^5 1×10^{-3}

Exercise 43 ① ②

1 A pile of 1000 sheets of card is one metre high. What is the thickness of one sheet of card in metres? Write your answer in standard form.

2 The four wooden bricks each have dimensions 3 cm × 3 cm × 10 cm. Work out the total volume of 20 000 bricks, giving your answer in cm^3 in standard form.

3 Write in order of size, smallest first:
$c = 3 \times 10^4$ $d = 2·5 \times 10^3$ $e = 7 \times 10^{-3}$ $f = 3600$

4 Write these numbers in standard form:
a 56200   **b** 0·0032   **c** 120 million   **d** 0·000037

5 A cuboid has dimensions 200 cm by 3500 cm by 400 cm. Calculate the volume of the cuboid in cm^3.

6 The annual budget for a large supermarket is £5 623 000. Write this number in standard form.

7 Which of the following has the largest value if
$c = 3 \times 10^4$ and $d = 4 \times 10^{-2}$?
i dc **ii** c^2 **iii** $\dfrac{c}{d^2}$ **iv** $\dfrac{1}{c^2}$

8 An insect crawls a distance of 20 m in 8 minutes 20 seconds. At what average speed is it crawling?

$$\text{Speed} = \frac{\text{Distance}}{\text{Time}}$$

Standard form calculations

A large number like 470 000 000 000 has too many digits to type into a calculator.
In standard form the number is $4·7 \times 10^{11}$.
We can use the EXP button on a calculator:
$4·7 \times 10^{11}$ is entered as 4 . 7 EXP 1 1

Notice that you do **not** press the x button after the EXP button.

> **EXAMPLE**
>
> **Without a calculator**
>
> **a Multiplication**
>
> Multiply the numbers and add the power of 10.
>
> $(1{\cdot}5 \times 10^4) \times (2 \times 10^7) = 3 \times 10^{11}$
>
> ↑
>
> $\boxed{1{\cdot}5 \times 2}$
>
> **b Division**
>
> Divide the numbers and subtract the powers.
>
> $(8 \times 10^7) \div (2 \times 10^2)$
> $= 4 \times 10^5$
> $= 400\,000$

Exercise 44 ① ②

1 Use a calculator to work out these and write your answers in standard form.

a $(4 \times 10^5) \times (7 \times 10^3)$ **b** $(3{\cdot}2 \times 10^6) \times (2 \times 10^5)$

c $(1{\cdot}5 \times 10^4) \times (3 \times 10^8)$ **d** $(1{\cdot}1 \times 10^8) \times (3 \times 10^4)$

e $(8 \times 10^6) \times (2 \times 10^{-2})$ **f** $(3{\cdot}1 \times 10^{-3}) \times (2 \times 10^{-2})$

g $(9 \times 10^5) \div (3 \times 10^2)$ **h** $(8{\cdot}4 \times 10^4) \div (2{\cdot}1 \times 10)$

i $(7 \times 10^8) \div (3{\cdot}5 \times 10^5)$ **j** $(4{\cdot}2 \times 10^{11}) \div (2 \times 10^5)$

k $(8 \times 10^7) \times (8 \times 10^3)$ **l** $(4{\cdot}5 \times 10^3) \div (1{\cdot}5 \times 10^{-5})$

2 Given that $x = 3 \times 10^2$ and $y = 5 \times 10^{-1}$, work out

a xy **b** $\dfrac{x}{y}$ **c** $x + y$

3 Write 274 billion pounds in pence. Give your answer in standard form.

4 Write these numbers in order of size, smallest first,
2×10^3, 5×10^{-3}, $(2 \times 10^2)^2$, $(8 \times 10^4) \div (2 \times 10^2)$

5 A distant star is $5{\cdot}2 \times 10^{22}$ km from the Earth.
A spaceship travels to the star at a speed of
$1{\cdot}3 \times 10^{13}$ km per hour.
How long will the journey take? Give your
answer in hours.

6 If $p = 615 \times 10^2$ $q = 0{\cdot}421 \times 10^4$ $r = 0{\cdot}00013 \times 10^6$
Write p, q and r in order of size, smallest first.

7 Write out these numbers in full.

    a $7{\cdot}3 \times 10^5$ **b** $10{\cdot}3 \times 10^{-1}$ **c** $71{\cdot}3 \times 10^{11}$

    d 4×10^8 **e** $6{\cdot}5 \times 10^{-2}$ **f** $(2 \times 10^5)^3$

8 Work out $(300)^2 \times 200\,000$ and write your answer in standard form.

9 Water is leaking from a swimming pool at a rate of 0.2 litres per minute. Initially the pool contained $500\,\text{m}^3$ of water. How long will it take for all the water to leak from the pool?

> $1\ \text{m}^3 = 1000\ \text{litres}$

10 Work out the value of these and write the letters a, b and c in order of size, smallest first.

    $a = (4 \times 10^{-3})^2$ $b = 1{\cdot}2 \times 10^{-2}$ $c = (5 \times 10^3) \div (2 \times 10^{-4})$

1.8 Solving numerical problems 1

1.8.1 Mixed questions

Exercise 45

1 Write the reading on each scale.

0   1   2·4   2·5   2g   4g

2 There are 5 people in a team. How many teams can you make from 118 people?

3 When James Wilkinson was born, his dad planted a tree. James died in 1920, aged 75. How old was the tree in 1975?

4 Washing-up liquid is sold in 200 ml containers. Each container costs 57p. How much will it cost to buy 10 litres of the liquid?

5 A train is supposed to leave London at 11:24 and arrive in Brighton at 12:40. The train was delayed and arrived $2\frac{1}{4}$ hours late. At what time did the train arrive?

6 Big Ben stopped for repairs at 17:15 on Tuesday and restarted at 08:20 on Wednesday.
For how long had it been stopped?

7 How much would I pay for nine litres of paint if two litres cost £2·30?

8 Here is a 'magic square'. The numbers in each row, column and diagonal add up to the same number.

8	1	6
3	5	7
4	9	2

Copy and complete these magic squares.

a

6	13	
	9	
		12

b

11			10
2	13	16	
		4	
7	12		6

9 A piece of wire 48 cm long is bent to form a rectangle in which the length is twice the width. Find the area of the rectangle.

10 A rectangular floor 3 m by 4 m is being covered with square
tiles, each of side length 50 cm.
A pack of 10 tiles costs £6·95.
How much will it cost to buy all the tiles for the floor?

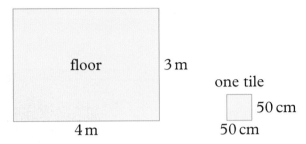

floor   3 m

one tile

50 cm

4 m   50 cm

Exercise 46

1 Copy each calculation and find the missing digits.

a
```
    5 ☐ 3
  + 3 4 ☐
  ───────
  ☐ 5 9
```

b
```
    3 3 4
  + ☐ 4 ☐
  ───────
    6 ☐ 0
```

c ☐ ☐ ☐ ÷ 5 = 32

2 The fourteenth number in the sequence 1, 2, 4, 8, 16…is 8192. What is
a the fifteenth number
b the thirteenth number?

3 The manager of a toy shop bought 10 000 model cars for £4950.
Half the models were sold at 95p each.
But half the models were sold for 85p each.
What was the total profit made?

4 Work out these, without using a calculator.
a 0·6 + 2·72   **b** 3·21 − 1·6
c 2·8 − 1·34   **d** 8 − 3·6
e 100 × 0·062   **f** 27·4 ÷ 10

5 Six people can travel in one car and there are altogether 106 people to transport. How many car journeys are needed?

6 Copy and complete these 'magic squares'.

a

−1	−2	
	0	
		1

b

		−1
	2	
5	0	

c

0		−4
	−1	
2		

7 Write these numbers in order of size, smallest first.

0·2   0·5   0·05   0·201   0·21

8 The scale shows temperatures in °C.
Write the temperatures for the arrows marked A, B, C and D.

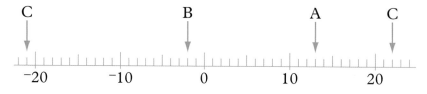

9 A baby falls asleep at 20:55 on Monday night and wakes up at 06:10 on Tuesday morning. For how long was the baby asleep?

10 Here are five number cards.

6   3   9   2   5

What is the largest **odd** number you can make with these five cards?

Exercise 47 ① ②

1 Copy and complete this bill.

$6\frac{1}{2}$ kg of potatoes at 40p per kg = £

2 kg of beef at per kg = £7·20

 jars of coffee at 95p per jar = £6·65

Total = £ □

2 Write in index form (for example $2 \times 2 \times 2 = 2^3$).

a $3 \times 3 \times 3 \times 3$ **b** $1 \times 1 \times 1 \times 1 \times 1 \times 1 \times 1$

c $7 \times 7 \times 7 \times 7 \times 7$ **d** half of 2^5

e 10 times 10^5 **f** 1% of 1 million

3 Do these without using a calculator.

   a $(-3) \times (-3)$ **b** $-7+2$ **c** $12 \div (-2)$

   d $8 - 18$ **e** $5 - (-2)$ **f** $5 \times (-2)$

4 How many 50 ml bottles can be filled from a jar containing one litre of liquid?

5 a Which four coins make a total of 77p?

   b Which five coins make a total of 86p?

   c Which five coins make a total of £1·57?

*6 Two numbers m and z are such that z is greater than 10 and m is less than 8. Arrange the numbers 9, z and m in order of size, starting with the smallest.

*7 One day, a third of the class is absent and 16 children are present. How many children are in the class when no one is absent?

*8 Copy the table and then write the numbers 1 to 9, one in each box, so that all the numbers satisfy the conditions for both the row and the column.

	Prime number	Multiple of 3	Factor of 16
Number greater than 5			
Odd number			
Even number			

*9 A man is 35 cm taller than his daughter, who is 5 cm shorter than her mother. The man was born in 1949 and is 1·80 m tall. How tall is the wife?

> Start by writing the alphabet with all the code numbers.

10 In a simple code A = 1, B = 2, C = 3,…Z = 26. Decode these messages.

   a 23, 8, 1, 20

      20, 9, 13, 5

       4, 15

     23, 5

     6, 9, 14, 9, 19, 8.

   b 19, 4^2, (3×7), 18, $(90 - 71)$

      1^3, (9×2), $(2^2 + 1^2)$

      18, $(\frac{1}{5}$ of 105$)$, 2, $(1 \div \frac{1}{2})$, 3^2, 19, 2^3.

   c 23, $(100 \div 20)$

      1, $(2 \times 3 \times 3)$, $(2^2 + 1^2)$

      21, $(100 - 86)$, $(100 \div 25)$, 5, $(2^4 + 2)$

      1, (5×4), $(10 \div \frac{1}{2})$, 1, $(27 \div 9)$, $(99 \div 9)$.

Exercise 48 ①

1 A special new cheese is on offer at £3·48 per kilogram. Mrs Mann buys half a kilogram. How much change does she receive if she pays with a £5 note?

2 A cup and a saucer together cost £2·80. The cup costs 60p more than the saucer. How much does the cup cost?

3 A garden 9 m by 12 m is to be treated with fertiliser. One cup of fertiliser covers an area of 2 m^2 and one bag of fertiliser is sufficient for 18 cups. Find the number of bags of fertiliser needed.

4 Copy and complete this pattern.

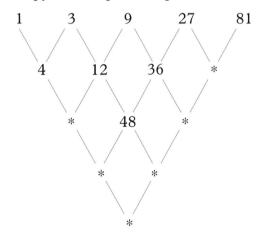

5 Six lamp posts lie at equal distances from each other along a straight road. If the distance between each pair of lamp posts is 20 m, how far is it from the first lamp post to the sixth?

6 Here is a list of numbers.

2   5   8   22   25   27   44   49

Write the numbers that are

a factors of 15

b multiples of 9

c prime numbers

d square numbers

e cube numbers.

7 A journey by boat started at 20:30 on Tuesday and finished at 07:00 on the next day.
How long was the journey in hours and minutes?

8 Work out, without using a calculator.

a $0.6 - 0.06$ b 0.04×1000

c $0.4 \div 100$ d $7.2 - 5$

e   10% of £90   f   25% of £160.

9 Here are three number cards.

You can use the three cards to make the number 728.

a   Use the three cards to make a number which is **more** than 728.
b   Use the three cards to make a number which is **less** than 728.
c   Use the three cards to make an **odd** number.

10 Find two numbers which

a   multiply to give 12 and add up to 7
b   multiply to give 42 and add up to 13
c   multiply to give 32 and add up to 12
d   multiply to give 48 and add up to 26.

Exercise 49 ①

1 In a simple code, A = 1, B = 2, C = 3 and so on. When the word 'BAT' is written in code, its total score is
(2 + 1 + 20) = 23.

a   Find the score for the word 'ZOOM'.
b   Find the score for the word 'ALPHABET'.
c   Find a word with a score of 40.

2 A cube of edge 1 cm weighs 65 grams.
This box has internal edges 5 cm by 8 cm by 3 cm and weighs 355 grams. Find the total weight of the box when it is filled with cubes.

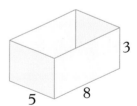

3 A swimming pool 20 m by 12 m contains water to a uniform depth of $1\frac{1}{2}$ m. 1 m³ of water weighs 1000 kg. What is the weight of the water in the pool?

4 The houses in a street are numbered from 1 to 60.
How many times does the number '2' appear?

5 Draw a large copy of this square.

1	2	3	4

Your task is to fill up all 16 squares using four 1s, four 2s, four 3s and four 4s. Each number may appear only once in any row (↔) or column (↕). The first row has been drawn already. Ask another student to check your solution.

6 Between the times 11:57 and 12:27, the milometer of a car changes from 23 793 miles to 23 825 miles.
At what average speed is the car travelling?

7 Copy these shapes and find which ones you can draw without going over any line twice and without taking the pencil from the paper? Write 'yes' or 'no' for each shape.

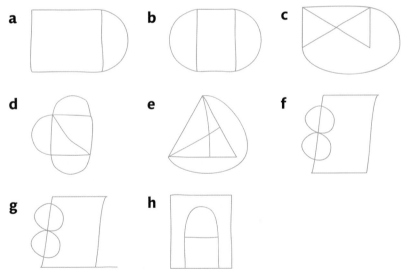

8 A generous, but not very bright, teacher decides to award 1p to the person coming 10th in a test, 2p to the person coming 9th, 4p to the person coming 8th and so on, doubling the amount each time. How much does the teacher award to the person who came first?

1.8.2 Timetables

Exercise 50 ①

1 For how many minutes do each of these programmes last
  a 'The Money Programme'
  b 'Fawlty Towers'
  c 'Face the Press'
  d '100 Great Sporting Moments'?

2 How much of a video tape would you use if you recorded 'Time Team Special' and 'Big Brother: Live'?

3 At what time on the 24-hour clock does 'Comrades' start?

4 There were four films on the two channels. What was the title of the shortest film?

5 Sam records 3 hours of TV. How much of her recording is left if she records both films in the afternoon on Channel 4?

6 How much time is devoted to sport on BBC 2? (Include 'Under Sail'.)

7 What is the starting time on the 24-hour clock of the programme in which 'Basil' appears?

8 How many programmes were repeats?

9 For how long is BBC News 24 broadcast on BBC 2?

10 What is the starting time on the 24-hour clock of the programme in which the 'Dolphins' appear?

11 How much of a two-hour video tape is left after taping 'Windmill' and 'The Natural World'?

BBC 2

1.0	**BBC News 24**
10.20	**OPEN UNIVERSITY.**
11.25	**BBC News 24**
11.50	**COLLECTAHOLICS**
12.15	**WINDMILL:** Archive film on animals.
1.10	**STATES OF MIND:** Jonathan Miller talks to Professor Richard Gregory (rpt.).
2.0	**SUPER LEAGUE SHOW:** Leeds vs Huddersfield.
2.30	**TENNIS:** Cup Final.
4.15	**UNDER SAIL:** New series.
4.35	**RACHMANINOV MASTERCLASS.**
5.20	**THINKING ALOUD:** James Bond joins a discussion on espionage.
6.0	**NEWS REVIEW:** with Nick Robinson
6.30	**THE MONEY PROGRAMME:** Guns for Sale. A look at Britain's defence industry.
7.15	**THE NATURAL WORLD:** City of Coral. A voyage beneath the Caribbean.
8.5	**COMRADES:** Educating Rita. The first of 12 films about life in Russia profiles a young trainee teacher.
8.50	**100 GREAT SPORTING MOMENTS:** Kelly Holmes: Golds in the Athens Olympics.
9.10	**FAWLTY TOWERS:** Basil and Sybil fall out over alterations to the hotel (rpt.).
9.40	**FILM:** 12 Years A Slave (see Film Guide).
11.50	**TENNIS:** Cup Final.
11.55	**MUSIC AT NIGHT.**

CHANNEL 4

1.15	**IRISH ANGLE – HANDS:** Basket Maker.
1.30	**FACE THE PRESS:** Tiger Woods, golfer, questioned by Ian Wooldridge of the Daily Mail and Brian Glanville of the Sunday Times.
2.0	**HOLLYOAKS**
2.30	**FILM*:** Journey Together (see Film Guide).
4.15	**FILM*:** Mr Turner, with Timothy Spall (see Film Guide).
5.15	**NEWS; WEATHER**, followed by **THE BUSINESS PROGRAMME.**
6.0	**AMERICAN FOOTBALL:** Dallas Cowboys and Miami Dolphins.
7.15	**THE HEART OF THE DRAGON:** Understanding (rpt.).
8.15	**TIME TEAM SPECIAL** (rpt.).
9.15	**BIG BROTHER: LIVE**
10.25	**FILM*:** Seven Days to Noon (see Film Guide).

Test yourself

1 Here is a diagram to find the factors of 12.

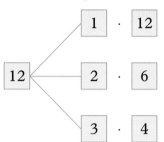

The factors of 12 are 1, 2, 3, 4, 6, 12

Copy and complete the diagram below to find the factors of 20.

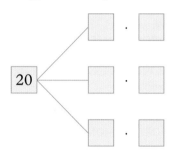

(AQA, 2007)

2 Look at this list of numbers.

3,   4,   6,   8,   10,   16,   27,   35

  a   A number and its square root both appear in the list.
       What is the number and what is its square root?

  b   A number and its cube root both appear in the list.
       What is the number and what is its cube root?

(CCEA)

3 Write a number in each box to make the calculation correct.

  a $\square$ + 28 = 45   b   100 − $\square$ = 45

  c   5 × $\square$ = 45   d $\square$ ÷ 4 = 45

(AQA, 2012)

4 a   Write these numbers in order of size.
       Start with the smallest number.
       17   6   168   24

  b   Write these numbers in order of size.
       Start with the smallest number.
       1·8   3·71   0·5   12·4

(Edexcel, 2008)

5 At 7 am the temperature was −3°C.
By noon the temperature was 11°C higher.
   a Write down the temperature at noon.

At 9 pm the temperature was −5°C.
By midnight the temperature had gone down by 7°C.
   b Write down the temperature at midnight.

(Edexcel, 2013)

6 a Write the number **two thousand and eighty five** in figures.
   b Write the number 5108 in words.
   c Write down the value of **9** in the number 2·96
   d Write down 157 correct to the nearest 10.

(Edexcel, 2013)

7 Express 252 as a product of its prime factors.

(Edexcel, 2008)

8 a Write the number 3187 to the nearest thousand.
   b Write the number **four thousand six hundred and eighty one** in figures.
   c Write the number 5060 in words.

(Edexcel, 2008)

9 At a school disco, cans of drink cost 50p and bags of crisps cost 40p.
Kiron buys 3 cans of drink and some bags of crisps.
He pays with a £5 note and gets £1·90 change.
How many bags of crisps did he buy?

(OCR, 2008)

10 Here is a list of numbers.

   6   7   9   11   13   20   26   47   51

   a From this list, write down
      i an even number,
      ii a square number,
      iii two numbers that add to give 37,
      iv two numbers that subtract to give 25.
   b i From the same list, write down a multiple of 5.
      ii Explain how you know that this is a multiple of 5.
   c i Which number in the list is a factor of 33?
      ii Explain how you know that this is a factor of 33.

(OCR, 2008)

11 A ticket for a seat at a school play costs £2·95

There are 21 rows of seats.
There are 39 seats in each row.

The school will sell all the tickets.

Work out an estimate for the total money the school will get.

(Edexcel, 2013)

12 a Copy the square. Fill in the empty boxes so that
each row, each column and each diagonal adds up to 0.
  b Multiply together the three numbers in the top row
and write down your answer.

(CIE)

−1	4	−3
	0	

13 A train timetable is shown.

Southampton	10:15	11:45	13:15
Plymouth	14:54	16:24	17:57
Devonport	14:58	16:28	18:01

  a William catches the 10:15 from Southampton.
He arrives in Devonport 4 minutes late.

  What time does he arrive in Devonport?

  b How long is William's total journey?

  c Kate catches the 11:45 from Southampton.
She arrives in Plymouth on time.

  She goes shopping.
She gets back to Plymouth station 90 minutes later.

  Is she back in time to catch the 17:57 train?
You **must** show your working.

(AQA, 2012)

14 There are 40 people at a meeting.
Each person travelled to the meeting either by car or by train.

13 of the people are male.
10 females travelled by train.
8 males travelled by car.

Work out the total number of people who travelled by car.

(Edexcel, 2013)

15 The diagram shows a car fuel gauge at the start of a journey and at the end of the journey.

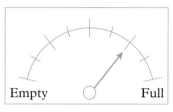

Start   **End**

There are 80 litres of fuel in the fuel tank when it is full.

a   Work out how many litres of fuel the car used on this journey.

On a different journey, the car went 180 kilometres.
The car went 15 kilometres for each litre of fuel used.

b   How many litres of fuel did the car use?

(Edexcel, 2013)

16 Jam is sold in two sizes.
A large pot of jam costs 88p and weighs 822 g.
A small pot of jam costs 47p and weighs 454 g.
Which pot of jam is better value for money?
You **must** show all your working.

(AQA)

17 a   Work out $2 \times (8 - 3)$
  b   Work out $3^2 + 4 \times 5$
  c   Find the value of 5^3
  d   Find the square root of 16

(Edexcel, 2013)

18 Use your calculator to work out
$$\frac{22 \cdot 4 \times 14 \cdot 5}{8 \cdot 5 \times 3 \cdot 2}$$
Write down all the figures on your calculator display.

(Edexcel, 2008)

19 Jess ran a race in 54·34 seconds.
The winning time was 2 hundredths of a second less than her time.

What was the winning time?

(AQA, 2013)

20 a   Work out
  i $3 + 4 \times 6$ **ii** $30 - 5 \times (3 + 1)$
  b   Put brackets into these sums so that the answer is correct.
  i $15 - 6 - 4 = 13$ **ii** $2 + 2 \times 3 + 8 = 24$

(OCR, 2013)

2 Algebra 1

In this unit you will:
- use letters for numbers in expressions
- simplify expressions
- use brackets in algebra calculations
- learn the main operations using algebra
- learn how to solve equations
- substitute a value into a formula or expression
- revise sequences and identify and use sequence rules
- find the general term of a sequence.

You can use equations to model real-life situations such as the growth of a plant.

2.1 Basic algebra

2.1.1 Using letters for numbers in expressions

- Suppose there is an unknown number of people on a bus. Call this number x. If two more people get on the bus there will be $x + 2$ people on the bus.

▶ $x + 2$ is an **expression**. An expression has no equals sign whereas an equation does have an equals sign.

- Suppose a piece of cheese weighs w kilograms. If I cut off 2 kilograms, I have $w - 2$ kilograms left.
- If I start with a number n and then double it, I will have $2n$ ($2n$ means $2 \times n$). If I then add 8, I will have $2n + 8$.

▶ When you multiply, write the number before the letter. So write $2n$, not $n2$, for $2 \times n$.

Exercise 1 ①

Find the expression that I am left with.

1 I start with n and then double it.

2 I start with y and then add 4.

> You can draw a function machine to help you.

3 I start with x and then take away 7.

4 I start with x and then add 100.

5 I start with y and then multiply by 5.

6 I start with s and multiply by 100.

7 I start with t and then treble it.

8 I start with z and then add 11.

9 I start with p and then take away 9.

10 I start with n and then add x.

11 I start with n and then multiply by 4.

12 I start with x, double it and then add 3.

13 I start with n, double it and then take away 12.

14 I start with m, treble it and then add 2.

15 I start with y and multiply by 20.

16 I start with x, treble it and then add 3.

17 I start with y, double it and then take away 7.

18 I start with k, treble it and then add 10.

EXAMPLE

Find the expression that I am left with if
a I start with n, add 2 and then multiply the result by 4.
b I start with n, subtract 3 and then divide the result by 7.
c I start with n, add 6 and then square the result.

a $n \rightarrow n + 2 \rightarrow 4(n + 2)$

b $n \rightarrow n - 3 \rightarrow \dfrac{n - 3}{7}$

c $n \rightarrow n + 6 \rightarrow (n + 6)^2$

Exercise 2 ①

Find the expression that I am left with.

1 I start with x, add 4 and then multiply the result by 3.

2 I start with x, add 3 and then multiply the result by 5.

3 I start with y, add 11 and then multiply the result by 6.

4 I start with x, add 3 and then divide the result by 4.

Use brackets.

Write this as $\dfrac{x + 3}{4}$.

5 I start with x, subtract 7 and then divide the result by 3.

6 I start with y, subtract 8 and then divide the result by 5.

7 I start with $4a$, add 3, multiply the result by 2 and then divide the final result by 4.

8 I start with m, subtract 6, multiply the result by 3 and the divide the final result by 4.

9 I start with x, square it and then subtract 6.

10 I start with x, square it, add 3 and then divide the result by 4.

11 I start with n, add 2 and then square the result.

12 I start with w, subtract x and then square the result.

> Use brackets.

13 I start with x, square it, subtract 7 and then divide the result by 3.

14 I start with x, subtract 9, square the result and then add 10.

15 I start with y, add 7, square the result and then divide by x.

16 I start with a, subtract x, cube the result and then divide by y.

17 A piece of wood is l cm long. If I cut off a piece 3 cm long, how much wood remains?

18 A piece of string is 15 cm long. How much remains after I cut off a piece of length x cm?

19 A delivery van weighs l kg. At a depot it picks up goods weighing 200 kg and later delivers goods weighing m kg. How much does it weigh after making the delivery?

20 A box usually contains n cookies. The shopkeeper puts an extra 2 cookies into each box. A girl buys 4 boxes. How many cookies does she have?

21 A brick weighs w kg. How much do six bricks weigh?

22 A sack weighs l kg. How much do x sacks weigh?

23 A man shares a sum of n pence equally between six children. How much does each child receive?

24 A sum of £p is shared equally between you and four others. How much does each person receive?

25 A cake weighing 12 kg is cut into n equal pieces. How much does each piece weigh?

2.1.2 Simplifying expressions

You can **simplify** the expression $3n + 2n$ to $5n$. This is because $3n + 2n$ means '$n + n + n$ plus $n + n$', which is equivalent to $5n$.

You can write the expression $8x + x$ as $8x + 1x$ and then simplify it to $9x$.

▶ You can only simplify **like terms**. These are parts of an expression that have the same letter.

The expression $3x + 8x$ has two **like terms**. You can simplify it to $11x$.
The expression $5x + 2y$ has two **unlike terms** and you cannot simplify it.

Here are some more examples.

EXAMPLE

> **a** $8n + 2n - n = 9n$ **b** $2x + 3y + 4x = 6x + 3y$
> **c** $4x + 3 + 2x - 1 = 6x + 2$ **d** $n + m - n + 3m = 4m$

Exercise 3 ①

Collect like terms together.

1 $2a + 3a$

2 $6a + 5a$

3 $2a + 3a + 4a$

4 $5n + n$

5 $7n - 2n$

6 $10n - n$

7 $3x + 2x - x$

8 $4x + 10x + 2x$

9 $3x - x + x$

10 $4a + b + 2a + 3b$

11 $6a + 4b + 3a + 2b$

12 $3x + y + 2x + 4y$

13 $2x + 5y + 7x + 2y$

14 $4m + 2n + m + n$

15 $6m + 3n + 10m - n$

16 $3x + 4 + 2x + 7$

17 $11x + 12 + 2x - 4$

18 $7x + 8 + x + 2$

19 $12x + 3x + 4y + x$

20 $2x + 3 + 3x + 5$

21 $4x + 8 + 5x - 3$

22 $5x - 3 + 2x + 7$

23 $6x + 1 + x + 3$

24 $4x - 3 + 2x + 10 + x$

25 $5x + 8 + x + 4 + 2x$

26 $7x - 9 + 2x + 3 + 3x$

27 $5x + 7 - 3x - 2$

28 $4x - 6 - 2x + 1$

29 $10x + 5 - 9x - 10 + x$

30 $4a + 6b + 3 + 9a - 3b - 4$

31 $8m - 3n + 1 + 6n + 2m + 7$

32 $6p - 4 + 5q - 3p - 4 - 7q$

33 $12s - 3t + 2 - 10s - 4t + 12$

34 $a - 2b - 7 + a + 2b + 8$

35 $3x + 2y + 5z - 2x - y + 2z$

36 $6x - 5y + 3z - x + y + z$

37 $2k - 3m + n + 3k - m - n$

38 $12a - 3 + 2b - 6 - 8a + 3b$

39 $3a + x + e - 2a - 5x - 6e$

40 $m + 7n - 5 - 4n + 8 - m$

In questions **41** and **46**, find the perimeter of each shape.
Give the answers in the simplest form.

Perimeter = sum of all the sides.

41

42

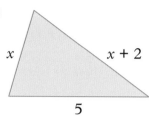

43

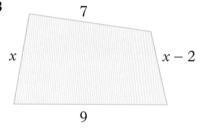

44

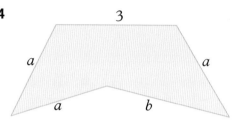

*45

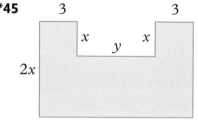

*46

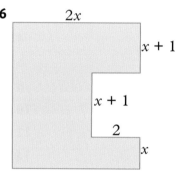

2.1.3 Multiplying with letters and numbers

$3n$ means 3 lots of n which is $3 \times n$.
$6a$ means 6 lots of a which is $6 \times a$.

When you multiply a letter by a number, you write them next to each other.

You write $5 \times n$ as $5n$ You write $a \times b$ as ab
You write $4 \times m \times n$ as $4mn$ You write $3 \times a \times n$ as $3an$

▶ You can simplify expressions by multiplying them by each other.

$2a \times 5b = 10ab$ $(2 \times 5 = 10, a \times b = ab)$
$4m \times 3n = 12mn$ $(4 \times 3 = 12, m \times n = mn)$
$2n \times 3n = 6n^2$ $(2 \times 3 = 6, \; n \times n = n^2)$

Exercise 4 ①

1 Write in a simpler form. Part **a** is done for you.

 a $a \times b = ab$ **b** $n \times m$ **c** $x \times y$

 d $h \times t$ **e** $a \times d \times n$ **f** $3 \times a \times b$

 g $4 \times n \times m$ **h** $3 \times a \times b \times c$

2 Write in a simpler form. Part **a** is done for you.

 a $2a \times 3b = 6ab$ **b** $3a \times 5b$ **c** $2c \times 6d$

 d $5n \times 2m$ **e** $2p \times 7q$ **f** $3a \times 10n$

 g $10s \times 2t$ **h** $12a \times 3b$ **i** $4u \times 8v$

3 Write in a simpler form. Part **a** is done for you.

 a $3y \times y = 3 \times y \times y = 3y^2$ **b** $y \times y$ **c** $2x \times 3x$

 d $4t \times 6t$ **e** $6a \times a$ **f** $5y \times 2y$

 g $x \times 2x$ **h** $y \times 3y$ **i** $10x \times 10x$

2.1.4 Using brackets

The expression $3(a + b)$ means $3 \times a + 3 \times b$.

   So $3(a + b) = 3a + 3b$.

▶ This is called **expanding the brackets** or removing the brackets.

Here are some more examples.

> **EXAMPLE**
>
> $2(x + 2y) = 2x + 4y$ $5(3a + 2b) = 15a + 10b$
> $5(x + 2) = 5x + 10$ $4(2x - 1) = 8x - 4$
> $n(n + 2) = n^2 + 2n$ $2n(3n + 1) = 6n^2 + 2n$

Exercise 5 ①

Expand the brackets.

 1 $2(x + 3)$ **2** $3(x + 5)$ **3** $4(x + 6)$

 4 $2(2x + 1)$ **5** $5(2x + 3)$ **6** $4(3x - 1)$

 7 $6(2x - 2)$ **8** $3(5x - 2)$ **9** $5(3x - 4)$

10 $7(2x - 3)$ **11** $2(2x + 3)$ **12** $3(2x + 1)$

13 $5(x + 4)$ **14** $6(2x + 2)$ **15** $2(4x - 1)$

16 $2(a + 3b)$ **17** $3(2a + 5b)$ **18** $5(2m + 3n)$

19 $7(2a - 3b)$ **20** $11(a + 2b)$ **21** $8(3a + 2b)$

22 $x(x + 5)$ **23** $x(x - 2)$ **24** $x(x - 3)$

25 $x(2x + 1)$ **26** $x(3x - 2)$ **27** $x(3x + 5)$

28 $2x(x - 1)$ **29** $2x(x + 2)$ **30** $3x(2x + 3)$

Remove the brackets and simplify.

31 $3(x + 2) + 4(x + 1)$ **32** $5(x − 2) + 3(x + 4)$ **33** $2(a − 3) + 3(a + 1)$

34 $5(a + 1) + 6(a + 2)$ **35** $7(a − 2) + (a + 4)$ **36** $3(t − 2) + 5(2 + t)$

37 $3(x + 2) + 2(x + 1)$ **38** $4(x + 3) + 3(x + 2)$ **39** $5(x − 2) + 3(x − 2)$

40 $4(a − 2) + 2(2a + 1)$ **41** $x(2x + 1) + 3(x + 2)$ **42** $x(2x − 3) + 5(x + 1)$

43 $a(3a + 2) + 2(2a − 2)$ **44** $y(5y + 1) + 3(y − 1)$ **45** $x(2x + 1) + x(3x + 1)$

46 $a(2a + 3) + a(a + 1)$

2.1.5 Subtracting terms in brackets

$−(a + b)$ means $−1 × (a + b) = −a − b$

$−(2a − b)$ means $−1 × (2a − b) = −2a + b$

$$3a + 2b − (2a + b) = 3a + 2b − 2a − b = a + b$$

Exercise 6 ①

1 Remove the brackets. Part **a** is done for you.

   a $−(m + n) = −m − n$ **b** $−(a + b)$ **c** $−(2a + b)$

   d $−(m − n)$ **e** $−(a − b)$ **f** $−(3a − b)$

   g $−(a + b − c)$ **h** $−(2a + b − 2c)$ **i** $−(3x − y − 2)$

Remove the brackets and simplify.

2 $2a + 5b − (a + b)$ **3** $5a + 2b − (a + b)$ **4** $6a + 8b − (2a + b)$

5 $3a + 7b − (2a + 3b)$ **6** $2(a + 2b) − (a + b)$ **7** $3(2a + b) − (a + 2b)$

8 $3(m + n) − (2m + n)$ **9** $5(m + 2n) − (m − n)$ **10** $7(2m + n) − 3(m + n)$

11 $6(m + 3n) − 10n$ **12** $3x − 2(x + y)$ **13** $10x − 3(2x + y)$

14 $5(a + 3b) + 4(a + 5b)$ **15** $2(3x + 4y) − 3(x − y)$

16 Find an expression for the perimeter of each rectangle.

a
3
2a + 1

b
4
5x − 2

c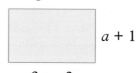
a + 1
3a − 2

17 Find an expression for
   a the perimeter of this picture
   b the area of this picture.

3x + y

2

18 The three rods, A, B and C, in the diagram have lengths of x, $x + 1$ and $x - 2$ cm.

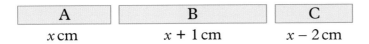

A x cm   B $x + 1$ cm   C $x - 2$ cm

In these diagrams, write the length l in terms of x.
Give your answers in their simplest form.

a A   C l **b** B   B l

c B / A l **d** B / C / A l

e A / B / B l **f** A   C / B l

g A / C / B l **h** B / C / A l

2.1.6 Overview of Algebra

● You can multiply letters and numbers.

$$4 \times n = 4n$$
$$a \times b = ab$$
$$n \times (a + b) = n (a + b)$$
$$= na + nb$$

● You can add like terms.

$$2a + 3b + 6a + b = 8a + 4b$$
$$a^2 + 2a^2 = 3a^2$$
$$3x + 8 + 2x - 2 = 5x + 6$$

> Like terms have the same letter or letters, e.g. $7n$ and $10n$
> a^2 and $3a^2$

● You can cancel fractions.

$$\frac{\cancel{5} \times 2}{\cancel{5}} = 2 \qquad \frac{6 \times \cancel{n}}{\cancel{n}} = 6$$

$$\frac{\cancel{a} \times a \times a}{\cancel{a}} = a^2 \qquad \frac{8n}{2} = 4n$$

$$\frac{n + n + n}{n} = \frac{3\cancel{n}}{\cancel{n}} = 3$$

Exercise 7 ①

In questions **1** to **18**, write those statements that are always true.

1 $6 \times a = 6a$ **2** $3 \times n = 3 + n$ **3** $n \times n = n^2$

4 $a + b = b + a$ **5** $n \times n \times n = 3n$ **6** $a \times 7 = 7a$

7 $n - m = m - n$ **8** $a + a = 2a$ **9** $n + n + n = n^3$

10 $m + m^2 = m^3$ **11** $3c - c = 3$ **12** $a + 2b = 2b + a$

13 $a(m + n) = am + an$ **14** $n \div 2 = 2 \div n$ **15** $n \times n \times m = n^2 m$

16 $2n^2 = 4n^2$ **17** $(2n)^2 = 4n^2$ **18** $2n + 2n = 4n^2$

19 Here are some algebra cards.

| $n + n$ | $n \times n^2$ | $3n \div 3$ | $n \div 4$ |

| $n \times n \times n$ | $n^2 \div n$ | $4 \div n$ | $5n - n$ | $4n - 2n$ |

a   Which cards will always be the same as $2n$?

b   Which cards will always be the same as n^3 ?

c   Which cards will always be the same as n ?

d   Which card will always be the same as $\dfrac{4}{n}$?

e   Write your own card that will always be the same as $n^2 + n^2$

> Expressions that are always equal can be written as an identity: $n + n \equiv 2n$
> $\equiv$ is the identity symbol.

20 In the expression $3n + 7$, two operations are done in the following order:

$n \longrightarrow \boxed{\times 3} \!-\! \boxed{+ 7} \longrightarrow 3n + 7$

Draw similar diagrams to show the correct order of operations for these expressions.

a $6n - 1$ **b** $8n + 10$ **c** $\dfrac{n}{2} + 3$

d $3(2n + 5)$ **e** $5(2n - 4)$ **f** $\dfrac{(n + 4)}{7}$

In questions **21** to **35**, simplify the expressions.

21 $\dfrac{3n}{n}$ **22** $\dfrac{a}{a}$ **23** $\dfrac{n^2}{n}$

24 $6n - 5n$ **25** $a + b + c + a$ **26** $3n^2 - n^2$

27 $mn + mn$ **28** $\dfrac{n \times n \times n}{n}$ **29** $\dfrac{n + n + n}{n}$

30 $a \times a^2$ **31** $6n \div 6$ **32** $3t + 4 - 3p - 1$

33 $\dfrac{2a}{2a}$ **34** $2n + 2(n + 1)$ **35** $n + 4 + 4 + n$

2.2 Solving equations

2.2.1 Using letters for numbers in equations

Many problems in mathematics are easier to solve if you use letters instead of numbers. This is called using **algebra**.

- Here is an equation.   ***n* + 10 = 75**
  In this equation, the letter *n* represents a definite number so that '*n* plus 10 equals 75'.
  So, in this equation, the value of *n* is 65.

- Here is another equation.   **2*x* + 1 = 7**
  In this equation, the letter *x* represents a definite number so that 'two times *x* plus one equals 7'.
  So, in this equation, the value of *x* must be 3.

Equations are like weighing scales that are balanced. The scales remain balanced if the same weight is added or taken away from both sides.

EXAMPLE

On the left is an unknown weight, *x*, plus a 2 kg weight.
On the right there is a 2 kg weight and a 3 kg weight.

If the two 2 kg weights are taken away, the scales are still balanced. So the weight *x* is 3 kg.
You can write this as the equation $x + 2 = 2 + 3$
$$x = 3$$

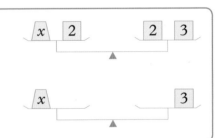

Exercise 8 ❶

The scales are balanced. Work out the weight of the object, *x*, in each case. Each small weight ☐ is 1 kg.

1

2

3

4

5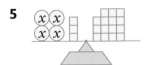

6

7

8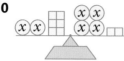

9

10

2.2.2 Rules for solving equations

You solve an equation in the same way as you solve the weighing scale problems.

> The main rule when solving equations is
> ▶ 'Do the same thing to both sides'.

You can **add** the same quantity to both sides.
You can **subtract** the same quantity from both sides.
You can **multiply** both sides by the same quantity.
You can **divide** both sides by the same quantity.

EXAMPLE

Solve the equations.

a $x + 7 = 12$
   (-7) (-7) Subtract 7 from both sides.
   $x = 5$

b $x - 3 = 11$
   $(+3)$ $(+3)$ Add 3 to both sides.
   $x = 14$

c $2x = 12$
   $(\div 2)$ $(\div 2)$ Divide both sides by 2.
   $x = 6$

d $3x + 1 = 19$
   (-1) (-1) Subtract 1 from both sides.
   $3x = 18$
   $(\div 3)$ $(\div 3)$ Divide both sides by 3.
   $x = 6$

Exercise 9 ①

Solve these equations.

1 $x + 7 = 10$

2 $x + 3 = 20$

3 $x - 7 = 7$

4 $x - 5 = 11$

5 $6 + x = 13$

6 $8 + x = 15$

7 $7 = x + 4$

8 $7 = x - 6$

9 $1 = x - 3$

10 $7 + x = 7$

11 $x - 11 = 20$

12 $14 = 6 + x$

13 $2x = 20$

14 $3x = 24$

15 $5x = 40$

16 $3x + 1 = 10$

17 $4x + 2 = 22$

18 $5x + 3 = 18$

19 $2n - 3 = 1$

20 $3n - 4 = 8$

21 $6n - 1 = 5$

22 $4a + 3 = 3$

23 $6a + 7 = 19$

24 $4 + 2a = 6$

> In question **7**,
> $7 = x + 4$ is the same
> as $x + 4 = 7$.

Exercise 10 ①

Solve these equations.

Some of the answers are fractions.

1 $5x - 3 = 1$	**2** $3x - 7 = 0$	**3** $2x + 5 = 20$
4 $6x - 9 = 2$	**5** $7x + 6 = 6$	**6** $9x - 4 = 1$
7 $11x - 10 = 1$	**8** $15y + 2 = 5$	**9** $7y + 8 = 10$
10 $4y - 11 = -8$	**11** $3z - 8 = -6$	**12** $4p + 25 = 30$
13 $5t - 6 = 0$	**14** $9m - 13 = 1$	**15** $4 + 3x = 5$
16 $7 + 2x = 8$	**17** $5 + 20x = 7$	**18** $3 + 8x = 0$
19 $50y - 7 = 2$	**20** $200y - 51 = 49$	**21** $5u - 13 = -10$
22 $9x - 7 = -11$	**23** $11t + 1 = 1$	**24** $3 + 8y = 40$
25 $12 + 7x = 2$	**26** $6 = 3x - 1$	**27** $8 = 4x + 5$
28 $9 = 2x + 7$	**29** $11 = 5x - 7$	**30** $0 = 3x - 1$
31 $40 = 11 + 14x$	**32** $-4 = 5x + 1$	**33** $-8 = 6x - 3$
34 $13 = 4x - 20$	**35** $-103 = 2x + 7$	

2.2.3 Equations with x on both sides

EXAMPLE

Solve the equations

a $8x - 3 = 3x + 1$ 　　　　　　　**b** $3x + 9 = 18 - 7x$

a $8x - 3 \; = 3x + 1$ Add 3 and subtract $3x$

　　$8x - 3x = 1 + 3$

　　　$5x = 4$

　　　　$x = \dfrac{4}{5}$ 　　　Divide by 5

b $3x + 9 \; = 18 - 7x$

　　$3x + 7x = 18 - 9$ Add $7x$ and subtract 9

　　　$10x = 9$

　　　　$x = \dfrac{9}{10}$ 　　　Divide by 10

Exercise 11 ②

Solve these equations.

1 $7x - 3 = 3x + 8$	**2** $5x + 4 = 2x + 9$	**3** $6x - 2 = x + 8$
4 $8x + 1 = 3x + 2$	**5** $7x - 10 = 3x - 8$	**6** $5x - 12 = 2x - 6$
7 $4x - 23 = x - 7$	**8** $8x - 8 = 3x - 2$	**9** $11x + 7 = 6x + 7$
10 $9x + 8 = 10$	**11** $5 + 3x = x + 8$	**12** $4 + 7x = x + 5$
13 $6x - 8 = 4 - 3x$	**14** $5x + 1 = 7 - 2x$	**15** $6x - 3 = 1 - x$
16 $3x - 10 = 2x - 3$	**17** $5x + 1 = 6 - 3x$	**18** $11x - 20 = 10x - 15$
19 $6 + 2x = 8 - 3x$	**20** $7 + x = 9 - 5x$	**21** $3y - 7 = y + 1$
22 $8y + 9 = 7y + 8$	**23** $7y - 5 = 2y$	**24** $3z - 1 = 5 - 4z$
25 $8 = 13 - 4x$	**26** $10 = 12 - 2x$	**27** $13 = 20 - 9x$
28 $8 = 5 - 2x$	**29** $5 + x = 7 - 8x$	**30** $3x + 11 = 2 - 3x$

EXAMPLE

EXAMPLE

Solve the equations

a $3(x - 1) = 2(x + 7)$ **b** $5(2x + 1) = 3(x - 2) + 20$

a $3(x - 1) = 2(x + 7)$ **b** $5(2x + 1) = 3(x - 2) + 20$

   $3x - 3 = 2x + 14$ $10x + 5 = 3x - 6 + 20$

   $3x - 2x = 14 + 3$ $10x - 3x = -6 + 20 - 5$

   $x = 17$ $7x = 9$

   $x = 1\frac{2}{7}$

Exercise 12 ②

Solve these equations.

1 $2(x + 1) = x + 5$ **2** $4(x - 2) = 2(x + 1)$

3 $5(x - 3) = 3(x + 2)$ **4** $3(x + 2) = 2(x - 1)$

5 $5(x - 3) = 2(x - 7)$ **6** $6(x + 2) = 2(x - 3)$

7 $10(x - 3) = x$ **8** $3(2x - 1) = 4(x + 1)$

9 $4(2x + 1) = 5(x + 3)$ **10** $3(x - 1) + 7 = 2(x + 1)$

11 $5(x + 1) + 3 = 3(x - 1)$ **12** $7(x - 2) - 3 = 2(x + 2)$

13 $5(2x + 1) - 5 = 3(x + 1)$ **14** $3(4x - 1) - 3 = x + 1$

15 $2(x - 10) = 4 - 3x$ **16** $3x + 2(x + 1) = 3x + 12$

17 $4x - 2(x + 4) = x + 1$ **18** $2x - 3(x + 2) = 2x + 1$

19 $5x - 2(x - 2) = 6 - 2x$ **20** $3(x + 1) + 2(x + 2) = 10$

21 $4(x + 3) + 2(x - 1) = 4$ **22** $3(x - 2) - 2(x + 1) = 5$

23 $5(x - 3) + 3(x + 2) = 7x$ **24** $3(2x + 1) - 2(2x + 1) = 10$

25 $4(3x - 1) - 3(3x + 2) = 0$

2.2.4   Equations with fractions

EXAMPLE

Solve the equations

a $\dfrac{7}{x} = 8$ **b** $\dfrac{3x}{4} = 2$

a $\dfrac{7}{x} = 8$ **b** $\dfrac{3x}{4} = 2$

   $7 = 8x$ Multiply by x $3x = 8$ Multiply by 4

   $\dfrac{7}{8} = x$ Divide by 8 $x = \dfrac{8}{3}$ Divide by 3

   $x = 2\frac{2}{3}$

Exercise 13 ②

Solve these equations.

1 $\dfrac{3}{x} = 5$ **2** $\dfrac{4}{x} = 7$ **3** $\dfrac{11}{x} = 12$ **4** $\dfrac{6}{x} = 11$

5 $\dfrac{2}{x} = 3$ **6** $\dfrac{5}{y} = 9$ **7** $\dfrac{7}{y} = 9$ **8** $\dfrac{4}{t} = 3$

9 $\dfrac{3}{a} = 6$ **10** $\dfrac{8}{x} = 12$ **11** $\dfrac{3}{p} = 1$ **12** $\dfrac{15}{q} = 10$

13 $\dfrac{x}{4} = 6$ **14** $\dfrac{x}{5} = 3$ **15** $\dfrac{y}{5} = -2$ **16** $\dfrac{a}{7} = 3$

17 $\dfrac{t}{3} = 7$ **18** $\dfrac{m}{4} = \dfrac{2}{3}$ **19** $\dfrac{x}{7} = \dfrac{5}{8}$ **20** $\dfrac{2x}{3} = 1$

21 $\dfrac{4x}{5} = 3$ **22** $\dfrac{x+1}{4} = 3$ **23** $\dfrac{x-3}{2} = 5$ **24** $\dfrac{4+x}{7} = 2$

25 $\dfrac{4}{x+1} = 3$ **26** $3 = \dfrac{9}{x+2}$ **27** $\dfrac{5}{x-2} = 3$ **28** $\dfrac{7}{x} = \dfrac{1}{4}$

29 $\dfrac{3}{x} = \dfrac{2}{3}$ **30** $\dfrac{2x+1}{3} = 1$ **31** $\dfrac{x}{2} = 110$ **32** $\dfrac{500}{y} = -1$

33 $-99 = \dfrac{98}{f}$ **34** $\dfrac{x}{3} + 5 = 7$ **35** $\dfrac{x}{5} - 2 = 4$ **36** $\dfrac{2x}{3} + 4 = 5$

37 $\dfrac{x}{6} - 10 = 4$ **38** $\dfrac{6}{x} + 1 = 2$ **39** $\dfrac{5}{x} - 7 = 0$ **40** $5 + \dfrac{3}{x} = 10$

2.2.5 Setting up equations

EXAMPLE

If you multiply a 'mystery' number by 2 and then add 3, the answer is 14. Find the 'mystery' number.

Let the mystery number be x.

Then $2x + 3 = 14$

$\qquad\quad 2x = 11$

$\qquad\qquad x = 5\dfrac{1}{2}$

The 'mystery' number is $5\dfrac{1}{2}$.

Exercise 14 ①

Find the 'mystery' number in each question by forming an equation and then solving it.

1 If you multiply the number by 3 and then add 4, the answer is 13.

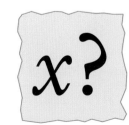

2 If you multiply the number by 4 and then add 5, the answer is 8.

3 If you multiply the number by 2 and then subtract 5, the answer is 4.

4 If you multiply the number by 10 and then add 19, the answer is 16.

5 If you add 3 to the number and then multiply the result by 4, the answer is 10.

6 If you subtract 11 from the number and then treble the result, the answer is 20.

7 If you double the number, add 4 and then multiply the result by 3, the answer is 13.

8 If you treble the number, take away 6 and then multiply the result by 2, the answer is 18.

9 If you double the number and subtract 7 you get the same answer as when you add 5 to the number.

10 If you multiply the number by 5 and subtract 4, you get the same answer as when you add 3 to the number and then double the result.

11 If you multiply the number by 6 and add 1, you get the same answer as when you add 5 to the number and then treble the result.

12 If you add 5 to the number and then multiply the result by 4, you get the same answer as when you add 1 to the number and then multiply the result by 2.

EXAMPLE

The length of a rectangle is twice the width.
If the perimeter is 36 cm, find the width.

a Let the width of the rectangle be x cm.
Then the length of the rectangle is $2x$ cm.

b Set up an equation for the perimeter.
$x + 2x + x + 2x = 36$

c Solve $\qquad 6x = 36$
$\qquad\qquad x = 6$

The width of the rectangle is 6 cm.

Exercise 15 ① ②

Answer these questions by setting up an equation
and then solving it.

1 Find x if the perimeter is 7 cm.

x cm

$x + 2$ cm

2 Find x if the perimeter is 10 cm.

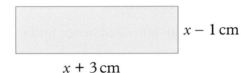

$x - 1$ cm

$x + 3$ cm

3 The length of a rectangle is 3 times its width. If the perimeter
of the rectangle is 11 cm, find its width.

> Let the width be x cm.

4 The length of a rectangle is 4 cm more than its width. If its
perimeter is 13 cm, find its width.

5 The width of a rectangle is 5 cm less than its length. If the
perimeter of the rectangle is 18 cm, find its length.

6 Find x in these shapes.

a
Area = 18 cm² | x cm

5 cm

b
Area =
15 cm² | $x + 3$ cm

5 cm

c
Area = 35 cm²

x

10 cm

7 Find x in these triangles.

a
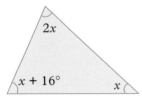
$2x$
$x + 16°$
x

b
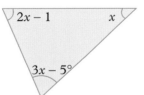
$2x - 1$ x
$3x - 5°$

c
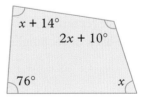
$x + 14°$
$2x + 10°$
$76°$ x

8 The angles of a triangle are $32°$, x and $4x + 3°$. Find the
value of x.

9 The sum of three consecutive whole numbers is 168. Let the
first number be x. Form an equation and hence find the
three numbers.

> Consecutive means
> 'following each
> other'. For example:
> 3, 4, 5.

10 The sum of four consecutive whole numbers is 170.
Find the numbers.

11 In this triangle, AB = x cm.
BC is 3 cm shorter than AB.
AC is twice as long as BC.
    a   Write, in terms of x, the lengths of
       i   BC
      ii AC.
The perimeter of the triangle is 41 cm.
    b   Write an equation in x and solve it to find x.

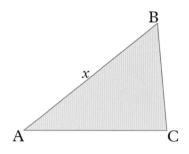

12 This is a rectangle. Work out x and hence find the
perimeter of the rectangle.

$3x - 1$

7

$2x + 7$

13 Here are four expressions involving an unknown
number n.

     A   B   C   D

   $2n + 1$ $n - 5$ $2n + 3$ $3n + 1$

    a   Find the value of n if the expressions A and B are equal.
    b   Find the value of n if the expressions C and D are equal.
    c   Which two expressions could never be equal for **any** value of n?

14 Find the length of the sides of this equilateral triangle.

15 Petra has £12 and Suki has nothing. They both receive
the same money for doing a delivery job.
Now Petra has three times as much as Suki.
How much did they get for the job?

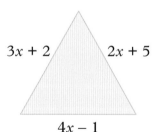

$3x + 2$ $2x + 5$

$4x - 1$

16 The area of rectangle A is
twice the area of rectangle B.
Find x.

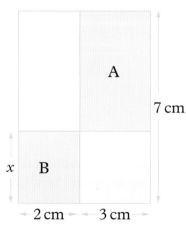

A

7 cm

x B

2 cm   3 cm

2.3 Using a formula

2.3.1 What is a formula?

Here is a **formula** for calculating a person's wage.

> Wage earned = Hours worked × Rate per hour

This formula will work for **any** value of 'hours worked' and for **any** value of 'rate per hour'.

For example, if Anna works for 40 hours at a rate per hour of £8, then

Anna's wage earned = 40 × £8
$$= £320$$

Exercise 16 ❷

1 The formula for calculating a person's wage is
   Wage earned = Hours worked × Rate per hour.
   a Calculate Amir's wage if he worked 30 hours and his rate per hour is £8.
   b Calculate Sam's wage if she worked 40 hours and her rate per hour is £11.

2 The formula for the circumference of a circle is
   Circumference ≈ diameter × 3
   Calculate the approximate circumference of a circle with diameter 20 cm.

> ≈ means 'is approximately equal to'.

3 The formula for the perimeter of an equilateral triangle is
   Perimeter = 3 × length of side
   a Find the perimeter of an equilateral triangle whose side length is 5 cm.
   b Find the perimeter of an equilateral triangle whose side length is 11 cm.

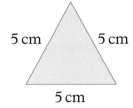

4 The formula for the area A of a triangle is
   $$A = \frac{1}{2} \times \text{base} \times \text{height}$$
   Find the area of each of these triangles.
   a base = 6 cm, height = 10 cm
   b base = 4 cm, height = 8 cm
   c base = 10 cm, height = 7 cm

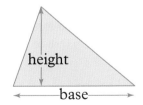

5 You can work out a baby's weight with the formula

Current weight = weight at birth + 500 grams
per month after the birth.

Calculate the weights of these babies
a weight at birth = 2500 grams, age 3 months
b weight at birth = 3100 grams, age 6 months
c weight at birth = 2750 grams, age 10 months.

6 The cost per month of Dave's phone is

Cost = (ten pounds) + (number of texts × 5 pence).

Find the cost for a month when he sent
a 200 texts
b 1000 texts.

2.3.2 Using letters in a formula

You can use letters for the quantities in a formula.
Here is a square of side s.
The formula for the perimeter is $P = s + s + s + s = 4s$.
The formula for the area is $A = s \times s = s^2$

So, for a square with side 9 cm, $P = 4 \times 9 = 36$ cm
and $A = 9 \times 9 = 81$ cm^2.

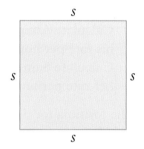

Exercise 17 ②

1 The formula for the perimeter of this rectangle is
$P = 2a + 2b$
Find the perimeter of a rectangle when
a $a = 4$ cm   and $b = 3$ cm
b $a = 7$ cm   and $b = 1$ cm.

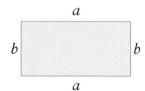

2 The formula for the area, A, of the rectangle in question **1** is

$$A = ab$$

Find the area of a rectangle where
a $a = 5$ cm   and $b = 2$ cm
b $a = 7$ cm   and $b = 3$ cm.

3 The formula for the height of a tree is $h = 5t + 3$.
Find the value of h when
a $t = 2$

b $t = 10$

c $t = 6$.

4 The formula for the volume, V, of a rectangular box with dimensions l, w and h (for length, width and height) is

$$V = lwh$$

Copy and complete the table.

	l	*w*	*h*	*V*
a	4	5	3	
b	7	2	10	
c	8	11	1	

5 Use the formulae and values given to find the value of c.

a $c = mx$ $m = 8$, $x = 7$

b $c = ny + a$ $n = 3$, $y = 4$, $a = 5$

c $c = 2ht$ $h = 6$, $t = 10$

d $c = a^2 + b^2$ $a = 5$, $b = 7$

> Formulae is the plural of formula.

6 A formula for estimating the volume of a cylinder of radius r and height h is

$$V = 3r^2h$$

a Find the value of V when $r = 10$ and $h = 2$.

b Find the value of V when $r = 5$ and $h = 4$.

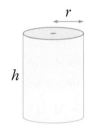

? *7 A builder uses a rule to work out how long it will take him to tile a wall.

> I need 50 minutes to prepare and then 8 minutes for each tile.

a How long does he need to lay 5 tiles?

b Write this rule in algebra using T for the total time in minutes that he needs to lay n tiles.

EXAMPLE

A formula connecting velocity with acceleration and time is $v = u + at$.

Find the value of v when $u = 3$

$a = 4$

$t = 6$.

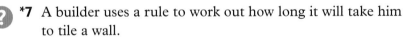

$v = u + at$

$v = 3 + (4 \times 6)$

$v = 27$

> **Velocity** is speed in a certain direction.

Exercise 18 ②

1 A formula involving force, mass and acceleration is $F = ma$.
Find the value of F when $m = 12$ and $a = 3$.

2 The height of a growing tree is given by the formula
$h = 2t + 15$. Find the value of h when $t = 7$.

3 The time, T, required to cook a joint of meat is given by the
formula $T = (\text{mass of joint}) \times 3 + \dfrac{1}{2}$
Find the value of T when the mass of joint $= 2\dfrac{1}{2}$.

4 An important formula in physics states that $I = mv - mu$.
Find the value of I when $m = 6$, $v = 8$, $u = 5$.

5 The distance travelled by an accelerating car is given by the
formula $s = \left(\dfrac{u + v}{2}\right)t$. Find the value of s when $u = 17$,
$v = 25$ and $t = 4$.

6 The formula for the total surface area, A, of a solid cuboid is
$A = 2bc + 2ab + 2ac$
Find the value of A when $a = 2$, $b = 3$, $c = 4$.

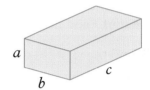

7 The formula for the height of a stone thrown upwards is
$h = ut - 5t^2$
Find the value of h when $u = 70$ and $t = 3$.

*8 The formula for the speed of an accelerating particle is $v^2 = u^2 + 2as$
Find the value of v when $u = 11$, $a = 5$ and $s = 6$.

*9 The time period, T, of a simple pendulum is given by the

| Use a calculator. |

formula $T = 2\pi\sqrt{\left(\dfrac{l}{g}\right)}$, where l is the length of the pendulum
and g is the gravitational acceleration.

Find T when $l = 0.65$, $g = 9.81$ and $\pi = 3.142$.

*10 The sum, S, of the squares of the integers from 1 to n is given
by $S = \dfrac{1}{6}n(n + 1)(2n + 1)$. Find S when $n = 12$.

2.3.3 Substituting into expressions

An algebraic **expression** has letters and numbers.
For example, $2n$, $5n + 1$ and $a^2 + b^2$ are all expressions.

> ▶ You can find the value of an expression by replacing each letter with a
> number.

Note that there is
no equals sign in an
expression.

Exercise 19 1

In questions **1** to **10**, find the value of each expression. The first one is done for you.

1 $2x + 1$ if $x = 5$

   $2 \times 5 + 1 = 10 + 1 = 11$

2 $3x - 1$ if $x = 10$

3 $5x - 2$ if $x = 7$

4 $4x + 3$ if $x = 0$

5 $10 + a$ if $a = 35$

6 $7 - a$ if $a = 1$

7 $12 - b$ if $b = 3$

8 $16 + b$ if $b = 30$

9 $4 + 3c$ if $c = 100$

10 $20 - 2c$ if $c = 5$

11 Find the value of these expressions when $n = 3$.

   a $n^2 + 2$ **b** $2n^2$ **c** n^3

> Remember:
> $2n^2 = 2(n^2)$

12 Find the value of these expressions when $a = 0.5$.

   a $4a + 1$ **b** $4 - a$ **c** $3(a + 0.5)$

*13 Find the value of these expressions when $x = 2$.

   a $\dfrac{x + 2}{x}$ **b** $\dfrac{x + 4}{x - 1}$ **c** $3x^2$

*14 Find the value of these expressions when $n = 2$.

   a $n^2 + 5$ **b** $3n - 6$ **c** $(n - 1)^2$

   d $\dfrac{5n}{2}$ **e** $\dfrac{n + 10}{2}$ **f** $n^2 + 3n$

EXAMPLE

If $n = 5$ and $x = -2$, work out the value of these expressions.

a $2n + x$
   $= 10 + -2$
   $= 10 - 2$
   $= 8$

b nx
   $= 5 \times (-2)$
   $= -10$

c $3 - x$
   $= 3 - (-2)$
   $= 3 + 2$
   $= 5$

Exercise 20 1

In questions **1** to **10**, find the value of the expressions when $m = 4$ and $t = -2$.

1 $t + m$ **2** $2(m + 3)$ **3** $5(m - 4)$ **4** t^2 **5** $2m + t$

6 $3t$ **7** $m^2 + 1$ **8** $2(t + 3)$ **9** $\dfrac{m + 12}{m}$ **10** $(t + 3)^2$

In questions **11** to **34**, find the value of the expressions when $a = 5$

$b = 4$

$c = 1$

$d = -2$.

11 $5a - c$ **12** $2b + a$ **13** $a + d$ **14** $3c - b$

15 $4b + c$ **16** $2d - a$ **17** $5b + 10$ **18** $a + b + c$

19 $b - c$ **20** $7 - 2a$ **21** $25 + 5b$ **22** $3a - 4d$

23 $a^2 + b^2$ **24** $ac + b$ **25** $6 - 2c$ **26** $d^2 + 4$

27 $ab + c$ **28** $5d - 2c$ **29** $b^2 + cd$ **30** $5a + b + d$

31 $bd + c^2$ **32** $2(a - c)$ **33** $3(a + d)$ **34** $a(c + b)$

▶   Reminders

$a^2 = a \times a$ $a^3 = a \times a \times a$

$2a^2 = 2(a^2) = 2 \times a^2$ $(2a)^2 = 2a \times 2a$

$a(b - c)$ Work out the term in brackets first.

$\dfrac{a + b}{c}$ The division line works like a bracket, so work out $a + b$ first.

Exercise 21 ①

Work out the values of these expressions when $m = 2$

$t = -2$

$x = -3$

$y = 4$.

1 m^2 **2** t^2 **3** x^2 **4** y^2

5 m^3 **6** t^3 **7** x^3 **8** y^3

9 $2m^2$ **10** $(2m)^2$ **11** $2t^2$ **12** $(2t)^2$

13 $2x^2$ **14** $(2x)^2$ **15** $3y^2$ **16** $4m^2$

17 $5t^2$ **18** $6x^2$ **19** $(3y)^2$ **20** $3m^3$

21 $x^2 + 4$ **22** $y^2 - 6$ **23** $t^2 - 3$ **24** $m^3 + 10$

25 $x^2 + t^2$ **26** $2x^2 + 1$ **27** $m^2 + xt$ **28** my^2

29 $(mt)^2$ **30** $(xy)^2$ **31** $(xt)^2$ **32** yx^2

33 $m - t$ **34** $t - x$ **35** $y - m$ **36** $m - y^2$

37 $t + x$ **38** $2m + 3x$ **39** $3t - y$ **40** $xt + y$

41 $3(m + t)$ **42** $4(x + y)$ **43** $5(m + 2y)$ **44** $2(y - m)$

45 $m(t + x)$ **46** $y(m + x)$ **47** $x(y - m)$ **48** $t(2m + y)$

49 $m^2(y - x)$ **50** $t^2(x^2 + m)$

Exercise 22 ②

If $w = -2$, $x = 3$, $y = 0$, $z = 2$, work out

1 $\dfrac{w}{z} + x$ **2** $\dfrac{w + x}{z}$ **3** $y\left(\dfrac{x + z}{w}\right)$ **4** $x^2 (z + wy)$

5 $x(x + wz)$ **6** $w^2 \sqrt{(z^2 + y^2)}$ **7** $2(w^2 + x^2 + y^2)$ **8** $2x(w - z)$

9 $\dfrac{z}{w} + x$ **10** $\dfrac{z + w}{x}$ **11** $\dfrac{x + w}{z^2}$ **12** $\dfrac{y^2 - w^2}{xz}$

13 $z^2 + 4z + 5$ **14** $\dfrac{1}{w} + \dfrac{1}{z} + \dfrac{1}{x}$ **15** $\dfrac{4}{z} + \dfrac{10}{w}$ **16** $\dfrac{yz - xw}{xz - w}$

2.4 Sequences

2.4.1 What is a sequence?

A sequence is a list of numbers that has a pattern to it. Each number in a sequence is called a **term**. Here are the first four terms of three sequences.

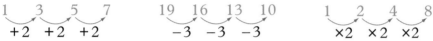

You find the next term by adding 2.

You find the next term by subtracting 3.

You find the next term by multiplying by 2.

Exercise 23 ①

1 The numbers in boxes form a sequence. Copy them and find the next term.

a | 8 | 10 | 12 | 14 | |

b | 2 | 5 | 8 | 11 | |

c | 11 | 9 | 7 | 5 | |

d | 1 | 5 | 9 | 13 | |

In questions **2** to **17**, write the sequence and find the next term.

2 1, 4, 7, 10, __   **3** 4, 8, 12, 16, __

4 2, 5, 8, 11, __   **5** 21, 17, 13, 9, __

6 5, 10, 15, 20, __   **7** 15, 13, 11, 9, __

8 2, 8, 14, 20, __   **9** 9, 18, 27, 36, __

10 31, 26, 21, 16, __   **11** 1, 2, 4, 8, 16, __

12 1, 3, 9, 27, __   **13** 2, 20, 200, 2000, __

14 80, 40, 20, 10, __   **15** 200, 100, 50, 25, __

16 88, 99, 110, __   **17** 39, 36, 33, 30, __

Write each sequence and find the missing number.

18 3   8   13   ☐   23   **19** 3   ☐   12   24   48

20 ☐   12   8   4   0   **21** 3   6   ☐   12   15

22 100   50   ☐ $12\frac{1}{2}$ **23** 1   ☐   9   27   ☐

24 ☐   8   16   24   ☐   **25** ☐   1   10   100   ☐

2.4.2 Sequence rules

- For the sequence 9, 13, 17, 21, 25..., the first term is 9 and the term-to-term rule is 'add 4'.
- For the sequence 3, 6, 12, 24, 48, ..., the first term is 3 and the term-to-term rule is 'double' or 'multiply by 2'.

Exercise 24 ❶

1 Write the first five terms for each sequence.
   a Start with 2 and add 3 each time.
   b Start with 30 and subtract 5 each time.
   c Start with 1 and multiply by 2 each time.
   d Start with 1 and multiply by 10 each time.
   e Start with 35 and subtract 7 each time.
   f Start with 64 and divide by 2 each time.
   g Start with −10 and add 2 each time.

2 The first term of a sequence is 11 and the term-to-term rule is 'add 5'. Write the first five terms of the sequence.

3 The table shows the first term and the rule of several sequences. Write the first five terms of each sequence.

	First term	Rule
a	96	add 2
b	100	subtract 11
c	10	multiply by 2
d	−6	add 3

4 Write the rule for each sequence.
   a 2, 5, 8, 11, 14,...
   b 30, 25, 20, 15, 10,...
   c 3, 6, 12, 24,...
   d 1, 8, 15, 22, 29,...
   e 21, 17, 13, 9, 5,...

5 Write each sequence and find the missing number.

a 1, 6, 11, ☐, 21

b 80, 40, 20, ☐

c 3, 30, 300, ☐, 30 000

d ☐, 5, 8, 11, 14

e 2, 4, 8, ☐, 32

f 12, 8, 4, ☐, −4

g 2, 8, ☐, 20, 26

h −1, 2, 5, ☐, 11

6 Write the rule for each sequence.

a 1, 5, 9, 13, 17,...

b 65, 55, 45, 35,...

c 1, 4, 16, 64,...

d 84, 42, 21, $10\frac{1}{2}$,...

e 37, 32, 27, 22,...

f 2, 3, 4, 5,...

g 200, 100, 50, 25,...

h 5500, 550, 55, 5·5,...

2.4.3 Harder sequences

Here are three sequences where the rule is more complicated.

● 5   8   12   17   ... so the next term is $17 + 6 = 23$.
   +3   +4   +5

● 15   14   16   13   17 ... so the next term is $17 − 5 = 12$.
   −1   +2   −3   +4

● 1   1   2   3   5   8   ... the rule is **add the previous two numbers each time**.
The next terms are 13 and 21.

> This is called the
> **Fibonacci sequence**.

Exercise 25 ① ②

1 Write each sequence and find the next term.

a 1, 3, 6, 10, __

b 0, 4, 9, 15, 22, __

c 5, 7, 10, 14, __

d 1, 2, 4, 7, 11, __

e 50, 49, 47, 44, __

f 10, 11, 9, 12, 8, __

g 8, 4, 2, 1, __

h 8, 10, 6, 12, 4, __

2 This is a Fibonacci sequence: 1, 1, 2, 3, 5, 8,...
You make each term by adding the previous two terms each time.
Write the sequence and find the next four terms.

3 Here is the start of a sequence: 1, 3, 4,...
You get each new term by adding the previous two terms.
For example, $4 = 1 + 3$. The next term will be 7.

a Write the next six terms.

b Use the same rule to write the next four terms of a sequence which starts 2, 5, 7,...

4 **a** Write the next two lines of this sequence.

$3 \times 4 = 3 + 3^2$

$4 \times 5 = 4 + 4^2$

$5 \times 6 = 5 + 5^2$

$=$

$=$

b Complete these lines.

$10 \times 11 =$

$30 \times 31 =$

5 Copy the pattern and write the next three lines.

$1 + 9 \times 0 \quad = \quad 1$

$2 + 9 \times 1 \quad = \quad 11$

$3 + 9 \times 12 \quad = \quad 111$

$4 + 9 \times 123 \quad = 1111$

$5 + 9 \times 1234 =$

6 For the sequence 2, 3, 8,... you get each new term by squaring the last term and then subtracting 1.

Write the next two terms.

7 To get the sequence 3, 3, 5, 4, 4 you count the letters in 'one, two, three, four, five,...'.

Write the next three terms.

8 Here is the sequence of the first six odd and even numbers.

	1st	2nd	3rd	4th	5th	6th
odd	1	3	5	7	9	11
even	2	4	6	8	10	12

Find **a** the 8th even number

b the 8th odd number

c the 13th even number

d the 13th odd number.

9 **a** Write the next three lines of this pattern.

$1^3 = 1^2 \qquad\qquad = 1$

$1^3 + 2^3 = (1 + 2)^2 \quad = 9$

$1^3 + 2^3 + 3^3 = (1 + 2 + 3)^2 = 36$

b Work out as simply as possible

$1^3 + 2^3 + 3^3 + 4^3 + 5^3 + 6^3 + 7^3 + 8^3 + 9^3 + 10^3$.

10 You can add the odd numbers 1, 3, 5, 7, 9,…to give an interesting sequence.

$$\begin{aligned}
1 &= 1 = 1 \times 1 \times 1 \\
3 + 5 &= 8 = 2 \times 2 \times 2 \\
7 + 9 + 11 &= 27 = 3 \times 3 \times 3 \\
13 + 15 + 17 + 19 &= 64 = 4 \times 4 \times 4
\end{aligned}$$

1, 8, 27, 64 are **cube** numbers.
You write $2^3 = 8$ ('two cubed equals eight')
$\qquad\qquad 4^3 = 64$
Or the other way round:
$\sqrt[3]{8} = 2$ ('the cube root of eight equals two')
$\sqrt[3]{27} = 3$

a Continue adding the odd numbers in the same way as before.
Do you **always** get a cube number?

b Write the value of

   i $\sqrt[3]{125}$ **ii** $\sqrt[3]{1000}$ **iii** 11^3

2.4.4 Using algebra to find the *n*th term of a sequence

● For the sequence 3, 6, 9, 12, 15,… the rule is 'add 3'.

Position   1   2   3   4   ⑤ ⟵ This is the 5th position.

Term   3   6   9   12   ⑮ ⟵ This is the 5th term.
               +3   +3   +3   +3

Here is the **mapping diagram** for the sequence.

Position		Term
1	⟶	3
2	⟶	6
3	⟶	9
4	⟶	12
⋮		⋮
10	⟶	30

You can also find each term by multiplying the position number by 3.
So the 10th term is 30 and the 33rd term is 99.
You can draw this rule as a function machine.

Position ⟶ ×3 ⟶ Term

▶ A **general** term in the sequence is the *n*th term, where *n* stands for any number.
▶ The *n*th term of this sequence is 3*n*.

You can find the nth term by putting n into the function machine.

- Here is a more difficult sequence: 5, 9, 13, 17,...

  The term-to-term rule is 'add 4' so, in the mapping diagram, there is a column for 4 times the position.

Position, n	4n	Term
1	⟶ 4 ⟶	5
2	⟶ 8 ⟶	9
3	⟶ 12 ⟶	13
4	⟶ 16 ⟶	17

You can see that each term is 1 more than $4n$. The position-to-term rule is

Position ⟶ ×4 ⟶ +1 ⟶ Term

So, the 10th term is $(4 \times 10) + 1 = 41$
the 15th term is $(4 \times 15) + 1 = 61$
the nth term is $(4 \times n) + 1 = 4n + 1$

Exercise 26 ① ②

1 Write each sequence and match it with the correct formula for the nth term.

  a 2, 4, 6, 8,...
  b 10, 20, 30, 40,...
  c 3, 6, 9, 12,...
  d 11, 22, 33, 44,...
  e 100, 200, 300, 400,...
  f 6, 12, 18, 24,...
  g 22, 44, 66, 88,...
  h 30, 60, 90, 120,...

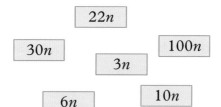

$22n$ $30n$ $100n$ $3n$ $6n$ $10n$ $2n$ $11n$

> You can draw a function machine to help you.

2 Copy and complete these mapping diagrams.

a

Position, n		Term
1	⟶	5
2	⟶	10
3	⟶	15
4	⟶	20
⋮		⋮
11	⟶	☐
⋮		⋮
n	⟶	☐

b

Position, n		Term
1	⟶	9
2	⟶	18
3	⟶	27
4	⟶	36
⋮		⋮
20	⟶	☐
⋮		⋮
n	⟶	☐

c

Position, n		Term
1	⟶	100
2	⟶	200
3	⟶	300
⋮		⋮
12	⟶	☐
⋮		⋮
n	⟶	☐

3 Copy and complete these mapping diagrams. Note that an extra column has been written.

a

Position, n		$2n$		Term
1	→	2	→	5
2	→	4	→	7
3	→	6	→	9
4	→	8	→	11
⋮		⋮		⋮
10	→	☐	→	☐
⋮		⋮		⋮
n	→	☐	→	☐

b

Position, n		$3n$		Term
1	→	3	→	4
2	→	6	→	7
3	→	9	→	10
4	→	12	→	13
⋮		⋮		⋮
20	→	☐	→	☐
⋮		⋮		⋮
n	→	☐	→	☐

4 Here you are given the nth term. Copy and complete the diagrams.

a

Position, n		$6n$		Term
1	→	6	→	8
2	→	12	→	14
3	→	☐	→	☐
4	→	☐	→	☐
⋮		⋮		⋮
n	→	$6n$	→	$6n + 2$

b

Position, n		$5n$		Term
1	→	5	→	3
2	→	10	→	8
3	→	☐	→	☐
4	→	☐	→	☐
⋮		⋮		⋮
n	→	$5n$	→	$5n - 2$

5 Here are the first five terms of a sequence:
6, 11, 16, 21, 26,...
  a Draw a mapping diagram like those in question **4**.
  b Write
     i the 10th term   **ii** the nth term.

> The term-to-term rule for the sequence is 'add 5', so write a column for '$5n$' in the diagram.

6 Here are three sequences.
A : 1, 4, 7, 10, 13,...
B : 6, 10, 14, 18, 22,...
C : 5, 12, 19, 26, 33,...
For each sequence
  a draw a mapping diagram as in question **4**
  b state the position-to-term rule in words
  c find the nth term.

7

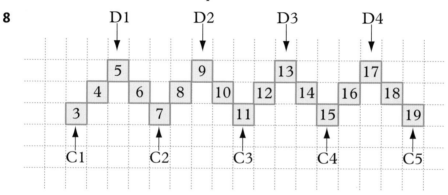

Look at the sequence made from A1, A2, A3, A4,…and
the sequence made from B1, B2, B3, B4,…

Write

a A10

b B10

c the nth term in the 'A' sequence

d the nth term in the 'B' sequence.

8

Look at the sequence made from C1, C2, C3, C4,… and the sequence
made from D1, D2, D3, D4,…

Write

a C20

b D20

c the nth term in the 'C' sequence

d the nth term in the 'D' sequence.

2.4.5 Finding a formula or rule for a sequence

Here is a sequence of 'houses' made from matches.

The table records the number of houses, h, and the number of matches, m.

The number in the h column goes up one at a time.
Look at the number in the m column.
In this case, the numbers in the m column go up by 4 each time.
This suggests that a column for $4h$ might help.

Now you can see that m is one more that $4h$.

So the formula linking m and h is **$m = 4h + 1$**

h	m	
1	5	4 more matches each time
2	9	
3	13	
4	17	

$\times 4 \qquad + 1$

h	4h	m
1	4	5
2	8	9
3	12	13
4	16	17

EXAMPLE

The table shows how r changes with n.
What is the formula that links r with n?

n	r
2	3
3	8
4	13
5	18

n 2   3   4   5

r 3   8   13   18

$+5 \quad +5 \quad +5$

Because r goes up by 5 each time,
write another column for $5n$.
The table shows that r is always
7 less than $5n$, so the formula linking r with n is $r = 5n - 7$.

n	5n	r
2	10	3
3	15	8
4	20	13
5	25	18

If the numbers in the first column do not go up by one each time, this method does not work. In that case you have to think of something more clever!

Exercise 27 ①

1 This sequence shows patterns of red tiles, r, and white tiles, w.
The table shows the numbers of red and white tiles.

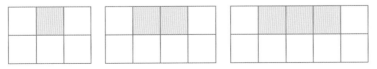

r	w
1	5
2	6
3	7
4	8

What is the formula for w in terms of r? Write it as '$w = ...$'

2 This is a different sequence with red tiles, *r*, and white tiles, *w*, and the related table.

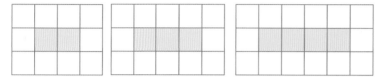

What is the formula? Write it as '*w* =...'

r	w
2	10
3	12
4	14
5	16

3 Here is a sequence of letter Is.

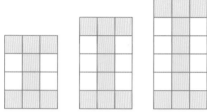

Make your own table for red tiles, *r*, and white tiles, *w*.
What is the formula for *w* in terms of *r*?

4 In this sequence, there are matches, *m*, and triangles, *t*.

Make a table for *t* and *m*. It starts like this.

Continue the table up to *t* = 10 and find a formula for *m* in terms of *t*. Write '*m* = ...'

t	m
1	3
2	5
⋮	⋮

5 Here is a different sequence of matches and triangles.

Make a table and find a formula connecting *m* and *t*.

6 In this sequence, there are triangles, *t*, and squares, *s*, around the outside of the triangles.

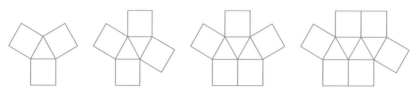

What is the formula connecting *t* and *s*?

7 Look at these tables. In each case, find a formula connecting the two letters.

a

n	p
1	3
2	8
3	13
4	18

Write 'p = ...'

b

n	k
2	17
3	24
4	31
5	38

Write 'k = ...'

c

n	w
3	17
4	19
5	21
6	23

Write 'w = ...'

*8 This is one part of a sequence of cubes, c, made from matches, m.

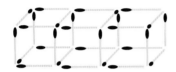

Find a formula connecting m and c.

Test yourself

1 **a** Expand $3(x - 6)$

  b Factorise $5y - 10$

  c Expand and simplify $3(4w + 1) - 5(3w - 2)$

(AQA, 2012)

2 Match each of the algebra expressions on the left with the one that is the same on the right. One has been done for you.

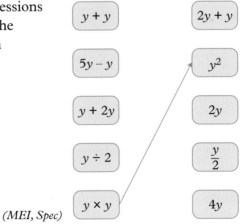

(MEI, Spec)

3 **a** Solve.

  i $x + 6 = 13$ **ii** $4x = 12$ **iii** $2x + 5 = 10$

  b **i** Use the formula $y = 3t + 2$ to work out the value of y when $t = 13$.

  ii Use the formula $M = A + 4B$ to work out the value of M when $A = 12$ and $B = -2$.

(OCR, 2008)

4 **a** Simplify $5p + p - 4p$.

   b Solve each of the following equations.

     i $\dfrac{x}{5} = 15$ **ii** $3y + 11 = 17$

   c What is the output from the following machine when the input is -3?

   d Use the formula $W = 2L + 3M$ to find M when $W = 35$ and $L = 4$.

(WJEC, 2011)

5 **a** Solve $x + 9 = 19$ **b** Solve $2y = 17$

   c Solve $\dfrac{w}{4} = 8$ **d** Expand $3(2 + t)$

(Edexcel, 2013)

6 **a** Simplify $d + d + d + d + d$ **b** Simplify $y^2 + y^2$

   c Expand $4(3a - 7)$ **d** Simplify $t \times t^2$

   e Simplify $m^5 \div m^3$

(Edexcel, 2008)

7 **a** **i** Write down the next number in this sequence.

        1   7   13   19

    ii Describe the rule for continuing the sequence.

   b **i** Write down the next number in this sequence.

        3   6   12   24

    ii Describe the rule for continuing the sequence.

(OCR, 2009)

8 **a** Write the perimeter of the pentagon, in terms of a and b, in its simplest form.

   b **i** Write the volume of the cuboid, in terms of x and y, in its simplest form.

    ii Find the total area of the faces of the cuboid, in terms of x and y, in its simplest form.

(CCEA)

9 **a** Write, in symbols, the rule

    'To find y, multiply k by 3 and then subtract 1.'

   b Work out the value of k when $y = 14$.

(Edexcel)

10 a Write as simply as possible an expression for the perimeter of this shape.

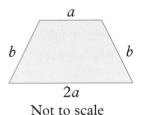

a

b   *b*

2a

Not to scale

   b Solve the equation $2x + 3 = 16$.

   c When $y = 4x + 1$

      i find the value of y when $x = -2$

      ii find the value of x when $y = 19$.

(MEI, Spec)

11 a Find the value of $3x + 5y$ when $x = -2$ and $y = 4$.

   b Find the value of $3a^2 + 5$ when $a = 4$.

   c k is an even number.

     Jo says that $\frac{1}{2}k + 1$ is always even.

     Give an example to show that Jo is wrong.

(AQA, 2004)

12 Here is a number pattern.

$$1 + 3 = 4$$
$$1 + 3 + 5 = 9$$
$$1 + 3 + 5 + 7 = 16$$
$$\dots\dots\dots\dots\dots = \dots\dots\dots$$
$$\dots\dots\dots\dots\dots = \dots\dots\dots$$

   a Write down the next two lines in the pattern.

   b What special name is given to the numbers 4, 9 and 16?

(AQA, 2007)

13 Tayub said, 'When $x = 3$, then the value of $4x^2$ is 144.'

Bryani said, 'When $x = 3$, then the value of $4x^2$ is 36.'

   a Who was right? Explain why.

   b Work out the value of $4(x + 1)^2$ when $x = 3$.

(Edexcel, 2003)

14 Angela and Michelle both work as waitresses at the same restaurant.
This formula is used to work out the total amount of money each waitress gets.

Total amount = £6·50 × number of hours worked + tips

The table shows the number of hours Angela and Michelle each worked last Saturday. It also shows the tips they got.

	Number of hours worked	**Tips**
Angela	8	£12
Michelle	7	£15

Who got the higher total amount of money last Saturday?
You must show clearly how you got your answer.

(Edexcel, 2013)

15 Work out the length of the longest side of the triangle when $x = 5$

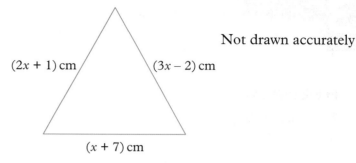

Not drawn accurately

$(2x + 1)$ cm

$(3x - 2)$ cm

$(x + 7)$ cm

(AQA, 2013)

16 $P = 2a + 3b$

Work out the value of P when $a = 11$ and $b = 5$

(AQA, 2013)

17 a The rule for continuing a sequence is

Double the previous term and add 5

A sequence starts   5   15   35   ......
Work out the next term in this sequence.

b A different sequence follows the same rule.

Double the previous term and add 5

The **third** term of this sequence is 27.
Work out the **first** term.

(AQA, 2013)

18 a This formula gives the total cost of some items bought from a cake shop.

Total cost in pence = 36 × number of cupcakes + 31 × number of scones

i Sarah buys 1 cupcake and 2 scones.
What is the total cost of Sarah's shopping?

ii Colin buys 8 cupcakes and 4 scones.
What is the total cost, in pounds, of Colin's shopping?

b Here is a formula

$$R = 3x - 7y$$

Work out the value of R when $x = 9$ and $y = 3$.

(OCR, 2013)

3 Geometry 1

In this unit you will:

- revise plotting coordinates
- revise drawing, estimating and constructing angles and shapes
- learn about the properties of shapes, including congruency and symmetry
- revise area and perimeter
- learn to draw shapes as isometric drawings.

Graphic design uses angles, shapes and projections to create detailed plans.

3.1 Coordinates

3.1.1 x- and y-coordinates

- To get to the point P on this grid you go **across** 1 and **up** 3 from the bottom left-hand corner.

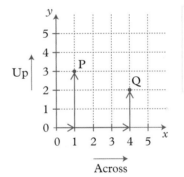

Notice also that the **lines** are numbered, **not** the squares.

The position of P is (1, 3).
The numbers 1 and 3 are called the **coordinates** of P.
The coordinates of Q are (4, 2).
The **origin** is at (0, 0).

▷ The first coordinate is the **x-coordinate** and the second coordinate is the **y-coordinate**.

▷ The **across** coordinate is always **first** and the **up** coordinate is always **second**.

Remember: 'Along the corridor and up the stairs'.

Exercise 1 ①

1 Write the coordinates of all the points
A to K like this: A(5, 1) B(1, 4)
Don't forget the brackets.

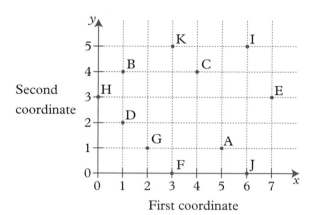

2 Write the coordinates of the points
which make up the '2' and the 'S'
in the diagram. You must give the
points in the correct order.

In questions **3** to **6**, plot the points and
join them up in order. You will get a
picture in each case.

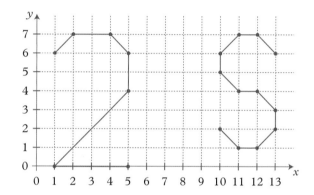

3 Draw x- and y-axes from 0 to 10.

A: (3, 2), (4, 2), (5, 3), (3, 5), (3, 6), (2, 7), (1, 6), (1, 8), (2, 9),
    (3, 9), (5, 7), (4, 6), (4, 5), (6, 4), (8, 4), (8, 5), (6, 7), (5, 7).
B: (7, 4), (9, 2), (8, 1), (7, 3), (5, 3).
C: (1, 6), (2, 8), (2, 9), (2, 7).
D: Draw a dot at (3, 8).

4 Draw x- and y-axes from 0 to 10.

A: (6, 5), (7, 6), (9, 5), (10, 3), (9, 1), (1, 1), (3, 3), (3, 4), (4, 5),
    (5, 4), (4, 3), (6, 4), (8, 4), (9, 3).
B: (8, 3), (8, 2), (7, 1).
C: (6, 3), (6, 2), (5, 1).
D: (5, 2), (4, 1).
E: Draw a dot at (3, 2).

5 Draw x- and y-axes from 0 to 8.
A: (6, 6), (1, 6), (2, 7), (7, 7), (6, 6), (6, 1), (7, 2), (7, 7).
B: (1, 6), (1, 1), (6, 1).
C: (3, 5), (3, 3), (2, 2), (2, 5), (5, 5), (5, 2), (2, 2), (3, 3), (5, 3).

6 Draw the *x*-axis from 0 to 8 and the *y*-axis from 0 to 4.
   A: (7, 1), (8, 1), (7, 2), (6, 2), (5, 3), (3, 3), (2, 2), (6, 2), (1, 2), (1, 1), (2, 1).
   B: (3, 1), (6, 1).
   C: (3, 3), (3, 2).
   D: (4, 3), (4, 2).
   E: (5, 3), (5, 2).
   F: Draw a circle of radius $\frac{1}{2}$ unit with centre at $\left(2\frac{1}{2}, 1\right)$.

   G: Draw a circle of radius $\frac{1}{2}$ unit with centre at $\left(6\frac{1}{2}, 1\right)$.

7 Use the grid to work out the joke written in coordinates.
   Work down each column. The gaps are spaces between words.

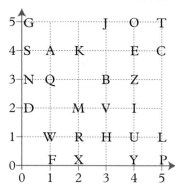

(1, 1)	(4, 0)	(1, 4)	(1, 1)	(0, 4)	(0, 2)
(3, 1)	(4, 5)		(4, 2)	(5, 0)	(4, 5)
(1, 4)	(4, 1)	(2, 2)	(5, 5)	(1, 4)	(4, 1)
(5, 5)		(1, 4)	(3, 1)	(0, 2)	(0, 5)
	(5, 4)	(0, 3)		(4, 4)	
(0, 2)	(1, 4)		(1, 4)		
(4, 5)	(5, 1)				
(5, 1)					

Exercise 2 ①

In this exercise, some coordinates will have
negative numbers.

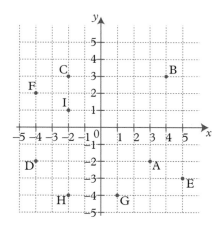

1 Write the coordinates of all the points A to I
   like this:
   A(3, −2)   B(4, 3).

2 Plot the points and join them up in order.
   You should produce a picture.
   Draw axes with *x* from −4 to +4 and *y* from 0 to 10.
   A: (3, 5), (2, 7), (0, 8), (−1, 8), (−2, 7), (−3, 7), (−4, 8),
      (−2, 9), (0, 9), (2, 8), (3, 7), (3, 2), (1, 1), (0, 3), (−2, 2),
      (−2, 4), (−3, 4), (−2, 6), (−1, 6), (−1, 5), (−2, 6), (−2, 7).
   B: (−1, 3), (−2, 3).
   C: (1, 3), (0, 3).

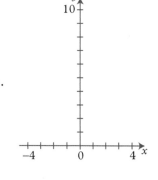

3 The graph shows several incomplete
quadrilaterals. Copy the diagram and
complete the shapes.
  a Write the coordinates of the fourth
    vertex of each shape.
  b Write the coordinates of the
    centre of each shape.

> A vertex is a corner
> of a shape.

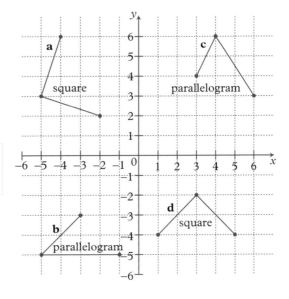

4 Copy this graph.
  a A, B and C are three corners of a square.
    Write the coordinates of the other corner.
  b C, A and D are three corners of another square.
    Write the coordinates of the other corner.
  c B, D and E are three corners of a rectangle.
    Write the coordinates of the other corner.
  d C, F and G are three vertices of a parallelogram.
    Write the coordinates of the other vertex.

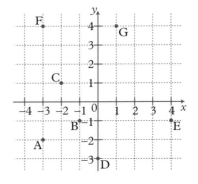

> ▶ A 'straight line' has infinite length. It goes on for ever!
> ▶ A 'line segment' has finite length. It has two end points.

For example, the line segment PQ has end points P and Q.

EXAMPLE

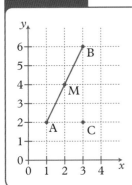

The diagram shows the line segment AB
joining A(1, 2) and B(3, 6).

You find the coordinates of the midpoint,
M, by adding the x- and y-coordinates of
A and B and then dividing by 2.

So M is the point $\left(\dfrac{1+3}{2}, \dfrac{2+6}{2}\right)$.

M has coordinates (2, 4).

*5 Find the coordinates of the midpoint of the line joining C(2, 5) and D(6, 7).

*6 Use the method in question **5** to find the midpoints of these line segments.
   a A(2, 1) and B(8, 3)
   b C(1, 3) and D(5, 0)
   c E(2, 3) and F(10, 7)
   d G(0, 8) and H(4, 2)
   e I(2, −2) and J(6, 4)
   f K(4, −3) and L(0, 7)

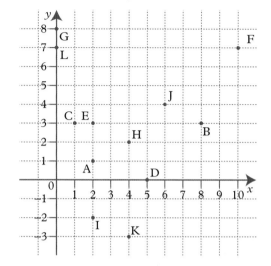

3.2 Using angles

3.2.1 Estimating angles

- Angles are measured in **degrees**.

▶ Any angle between 0° and 90° is called an **acute** angle.   acute

▶ Any angle between 90° and 180° is called an **obtuse** angle.   obtuse

▶ Any angle bigger than 180° is called a **reflex** angle.   reflex

▶ An angle of 90° is called a **right angle**. There is a special symbol for a right angle.   right angle symbol

▶ When the angle between two lines is 90°, you say the lines are **perpendicular**.

A full turn = 360°
A half turn = 180°
A quarter turn = 90°

Exercise 3 ❶

Write whether these angles are correctly or incorrectly labelled.
Do **not** measure the angles – estimate them! Where the angles are
clearly incorrect, write an estimate for the correct angle.

1 45°
    2 90°
    3 130°
    4 98°

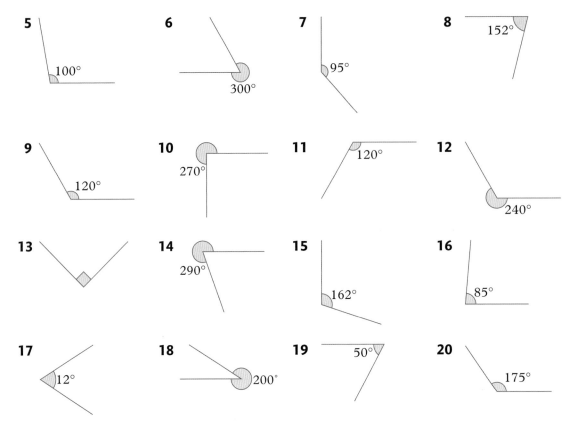

5 100°

6 300°

7 95°

8 152°

9 120°

10 270°

11 120°

12 240°

13

14 290°

15 162°

16 85°

17 12°

18 200°

19 50°

20 175°

Exercise 4 ❶

For each angle in Exercise **3**, write whether the angle **marked** is acute, obtuse, reflex or a right angle.

3.2.2 Angle facts

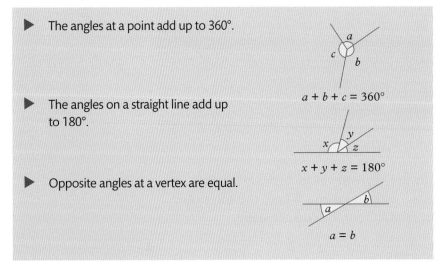

▶ The angles at a point add up to 360°.

$a + b + c = 360°$

▶ The angles on a straight line add up to 180°.

$x + y + z = 180°$

▶ Opposite angles at a vertex are equal.

$a = b$

EXAMPLE

Find the angles marked with letters.

a

b

a $x + x + 150° + 100° = 360°$
$2x + 250° = 360°$
$2x = 360° - 250°$
$x = 55°$

b $3a + 90° = 180°$
$3a = 90°$
$a = 30°$

Exercise 5 ①

Find the angles marked with letters. The line segments AB and CD in the diagrams from question **9** onwards are straight.

1

2

3

4

5

6

7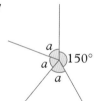

8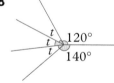

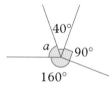

9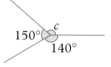

10 A B

11 A 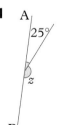 B

12 A 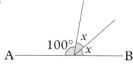 B

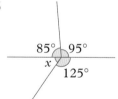
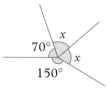

13 A 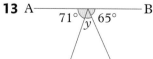 B

14 A  B

15 A 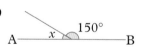 B

16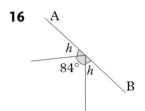

17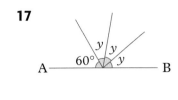

18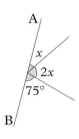

19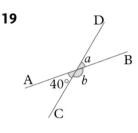

20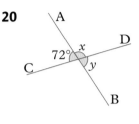

3.2.3 Angles in a triangle

Draw a triangle of any shape on a piece of card and cut it out accurately. Now tear off the three corners as shown.

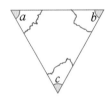

When the angles *a*, *b* and *c* are placed together they form a straight line.

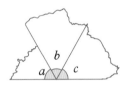

> This is a **demonstration** that the angles add up to 180°. It is not a mathematical **proof**. A proof of this result is on page 113.

▶ The angles in a triangle add up to 180°.

Isosceles and equilateral triangles

An **isosceles** triangle has two equal sides and two equal angles. It has a line of symmetry down the middle.

The sides AB and AC are equal (marked with a dash) and angles B and C are also equal.

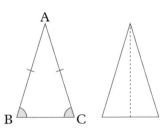

An **equilateral** triangle has three equal sides and three equal angles (all 60°). It has three lines of symmetry.

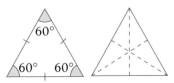

EXAMPLE

Find the angles marked with letters.

a

b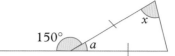

....

a $y + 80° + 40° = 180°$
      $y + 120° = 180°$
      $y = 60°$

b $a = 180° - 150° = 30°$
      The triangle is isosceles, so $x + x + 30° = 180°$
                              $2x = 150°$
                              $x = 75°$

Exercise 6 ①

Find the angles marked with letters. For the more difficult
questions it is helpful to draw a diagram.

1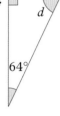

2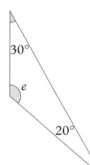

3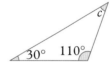

> **1** is a **scalene**
> triangle as all the
> angles are different.

4

5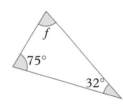

6

> **4** is a **right-angled**
> triangle because
> it contains a right
> angle.

7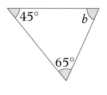

8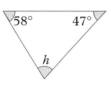

9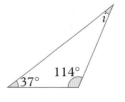

10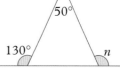

11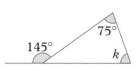

12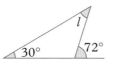

13

14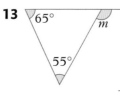

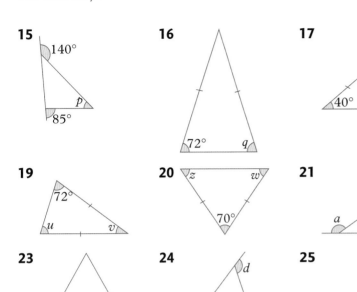

15 140° p 85°

16 72° q

17 40° r

18 55° t s

19 72° u v

20 z w 70°

21 40° a

22 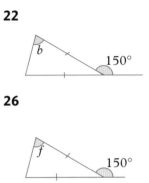 b 150°

23 c

24 d y 122°

25 90° e

26 f 150°

27 3x 2x x

28 a 38°

3.2.4 Parallel lines

> ▶ Parallel lines are always the same distance apart. They never meet.
> ▶ When a line cuts a pair of parallel lines, all the acute angles
>   are equal and all the obtuse angles are equal.

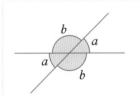

These are two pairs of
vertically opposite
angles.

EXAMPLE

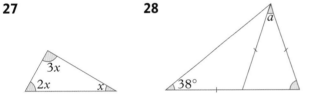

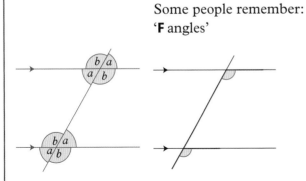

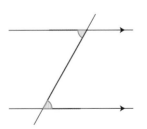

Some people remember:
'**F** angles'   and '**Z** angles'.

These are called
corresponding angles.

These are called
alternate angles.

Exercise 7 ❶

Find the angles marked with letters.

1

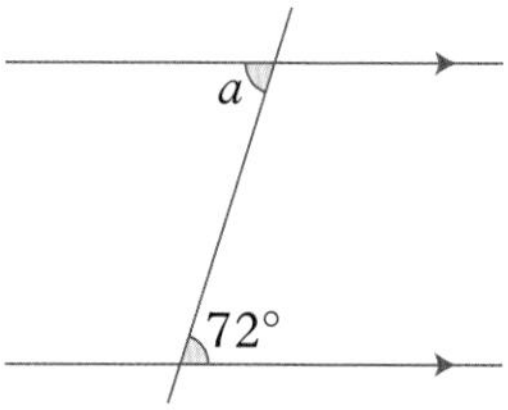

a

72°

2

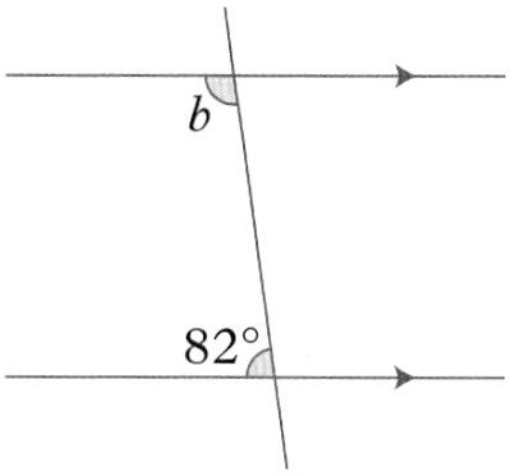

b

82°

3

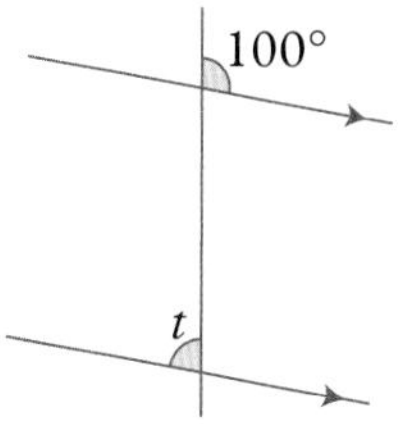

100°

t

4

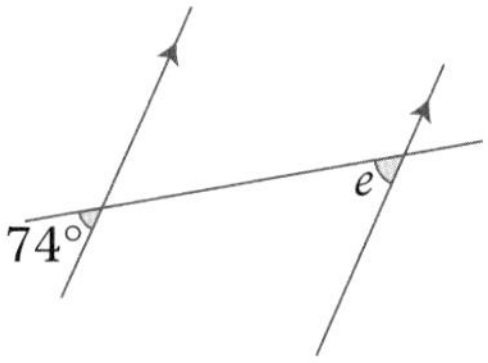

74°

e

5

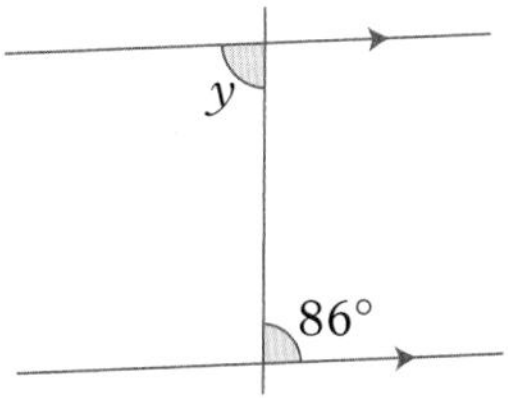

y

86°

6

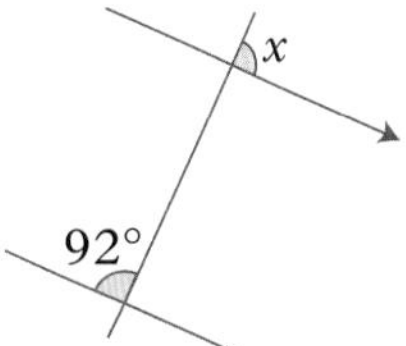

x

92°

7

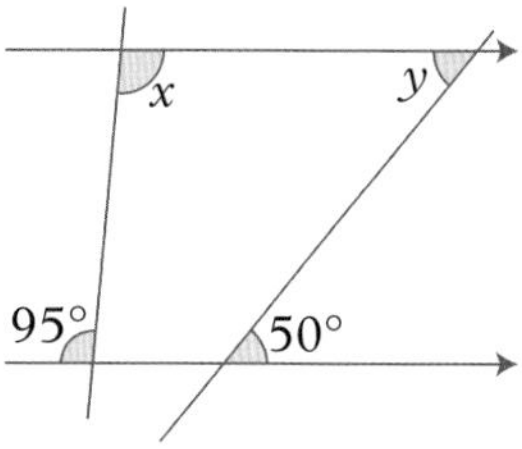

x   y

95°   50°

8

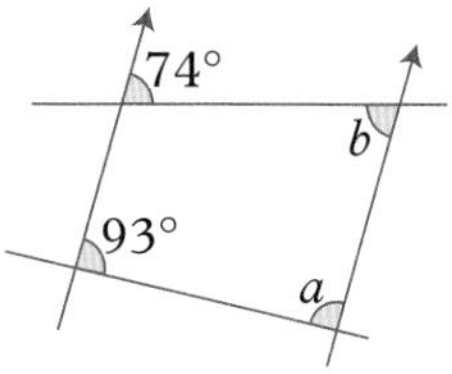

74°

b

93°

a

9

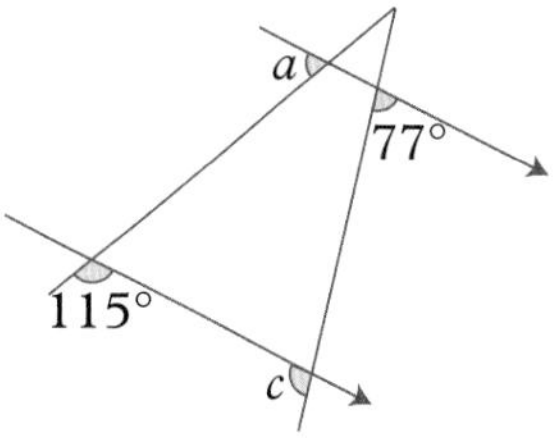

a

77°

115°

c

10

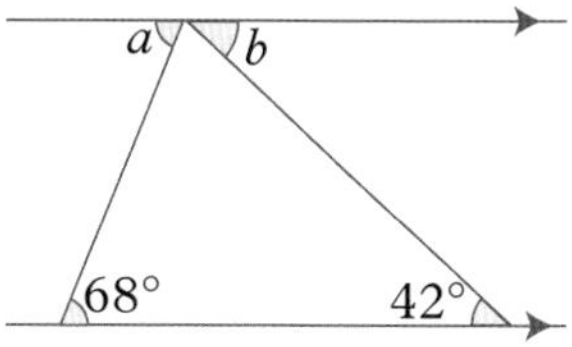

a b

68°   42°

11

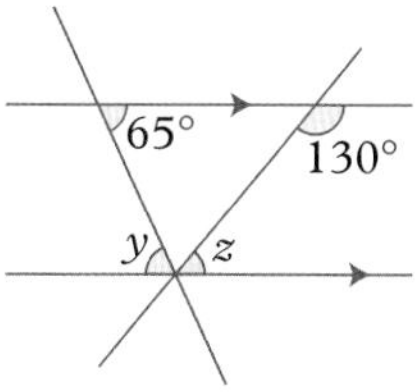

65°

130°

y z

12

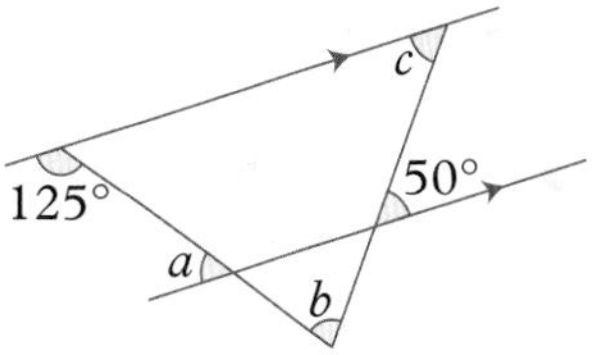

c

125°   50°

a   b

Exercise 8 ❶

The next exercise contains questions that use all the angle facts
you have learnt.
Find the angles marked with letters.

1

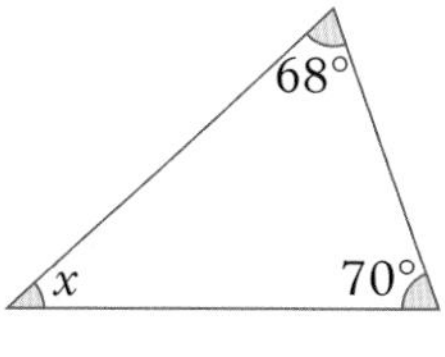

68°

x   70°

2

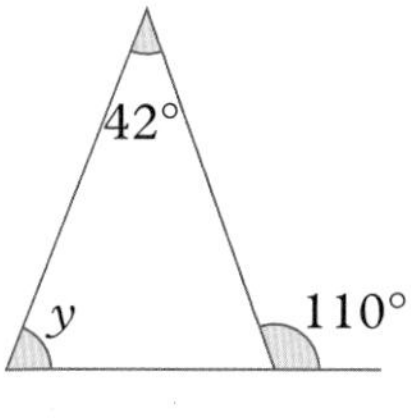

42°

y   110°

3

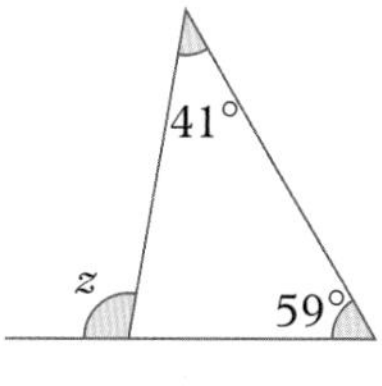

41°

z   59°

4

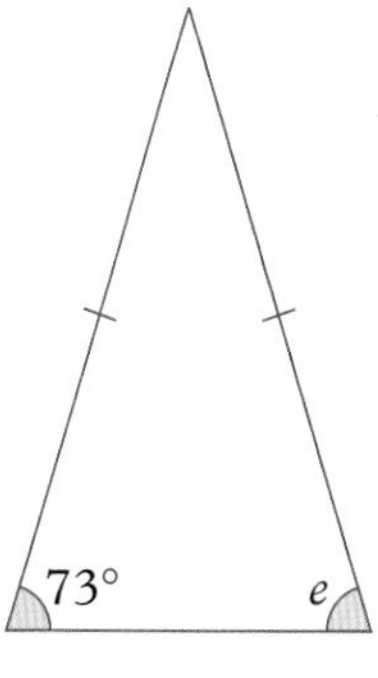

73°   e

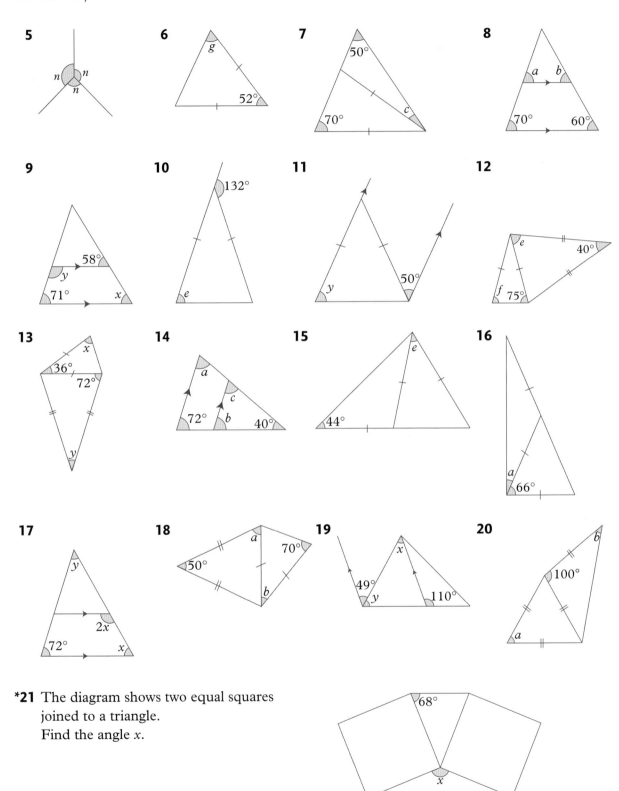

5

6

7

8

9

10

11

12

13

14

15

16

17

18

19

20

*21 The diagram shows two equal squares
joined to a triangle.
Find the angle x.

*22 Find the angle a between the diagonals
of the parallelogram.

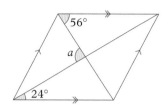

3.2.5 Proving results

On page 108, you **demonstrated** that the sum of the angles in a
triangle is 180°, by cutting out the angles and rearranging them.
When you **prove** a result it means that the result is true for every
possible shape. You can often prove one simple result and then
use that result to prove further results.

● Here is a **proof** that the sum of the angles in a triangle is 180°.

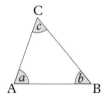

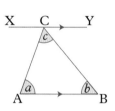

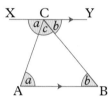

| Δ is a symbol that means triangle. |

Here is ΔABC.

Draw line XCY
parallel to AB.

∠ABC = ∠YCB (alternate angles)
∠BAC = ∠ACX (alternate angles)
$a + b + c = 180°$ (angles on a straight line)
So, angles in a triangle: $a + b + c = 180°$

This proves that the sum of the angles in a triangle is 180°.

Exercise 9 ❶ ❷

Teacher's note: Proof is not an easy topic for most students.
Some teachers may choose to go through these proofs on the
board so that students can copy them down. It is desirable that
students should have an **understanding** of the proof rather than
just the ability to reproduce it.

1 Copy and complete this proof that the sum of the angles
in a quadrilateral is 360°.
Draw any quadrilateral ABCD with diagonal BD.
Now $a + b + c = \square$ (angles in a Δ)
and $d + e + f = \square$ (angles in a Δ)
So, $a + b + c + d + e + f = \square$
This proves the result.

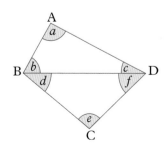

2 The exterior angle of a triangle is equal to the sum of the two interior opposite angles.

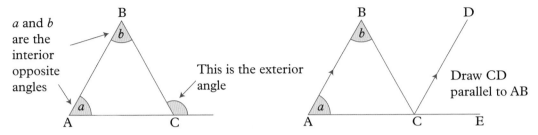

a and *b* are the interior opposite angles

This is the exterior angle

Draw CD parallel to AB

Copy and complete this proof:

∠BAC = ∠DCE (corresponding angles) ('F' angles)
∠ABC = ∠☐ (alternate angles)   ('Z' angles)
So, ☐ = ☐ + ☐

3 Prove that opposite angles of a parallelogram are equal.
(Use alternate and corresponding angles.)

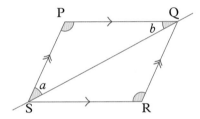

3.3   Accurate drawing

3.3.1   Using a protractor and compasses

Some questions involving bearings or irregular shapes are easy to solve if you draw an accurate diagram.

To improve the accuracy of your work, follow these guidelines.

- Use a **sharp** HB pencil.
- Don't press too hard.
- If you are drawing an **acute** angle make sure your angle is less than 90°.
- If you use a pair of compasses make sure they are fairly stiff so the radius does not change accidentally.

Using a protractor

You use a protractor to measure angles accurately.

> Remember: When you measure an **acute** angle the answer must be less than 90°. When you measure an **obtuse** angle the answer must be more than 90°.

Exercise 10 ①

Measure these angles.

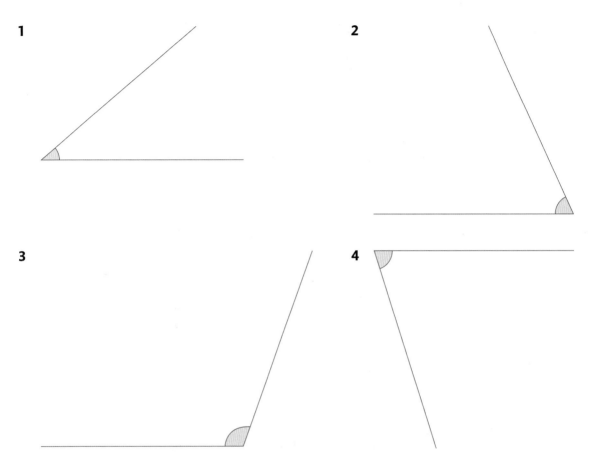

1

2

3

4

In question **5, 6, 7** and **8** measure all the angles and all the sides in each triangle.

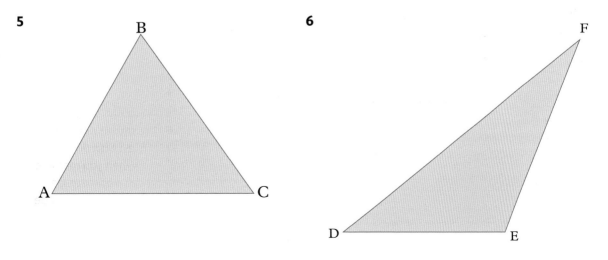

5

6

7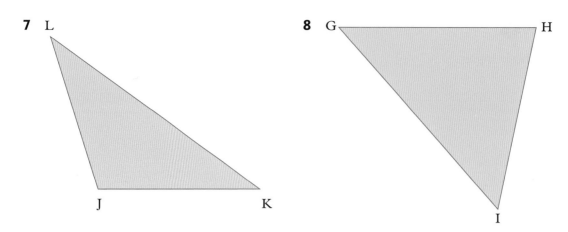

8

9 Draw these angles **accurately**.

    a 40°   **b** 62°   **c** 110°   **d** 35°   **e** 99°   **f** 122°

3.3.2 Constructing triangles

EXAMPLE

Draw the triangle ABC full size and measure the length x.

a Draw a base line **longer than 8·5 cm.** Measure and mark 8·5 cm. Label points A and B.

b Put the centre of the protractor on A and measure an angle of 64°. Draw line AP.

c Similarly draw line BQ at an angle of 40° to AB.

d You have drawn the triangle! Measure $x = 5·6$ cm

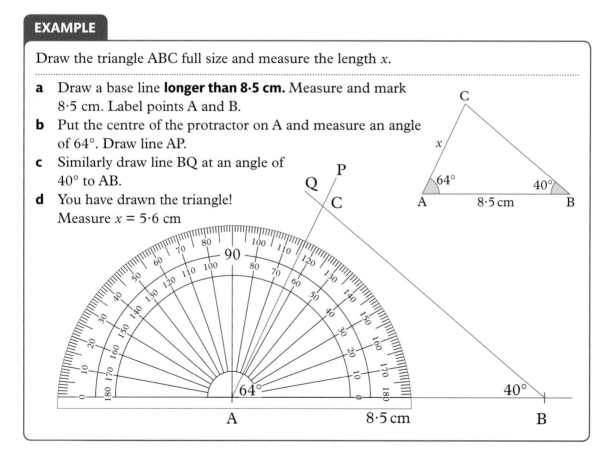

Exercise 11 ❶

Use a protractor and ruler to draw full size diagrams and measure the sides marked with letters.

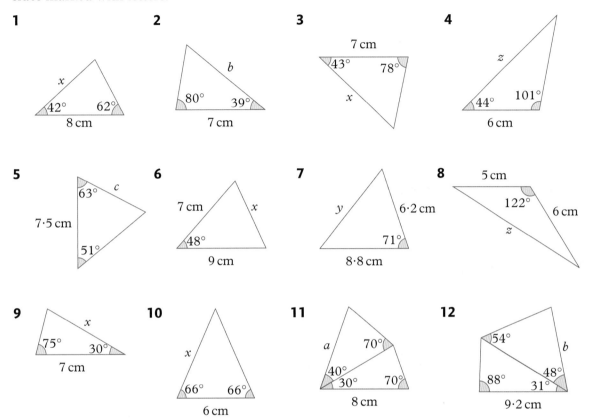

1

2

3

4

5

6

7

8

9

10

11

12

Exercise 12 ❷

This exercise will show you how to draw a triangle, if you are given three sides.

1 Follow these steps to draw a triangle with sides 7 cm, 5 cm, 6 cm.

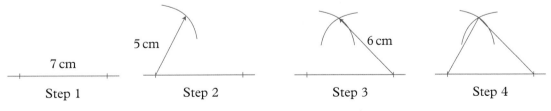

Step 1
Draw line, 7 cm.

Step 2
Draw arc, radius 5 cm.

Step 3
Draw arc, radius 6 cm.

Step 4
Complete the triangle by drawing the third side.

2 **a** Construct an equilateral triangle with sides of 6 cm.
   b Construct an equilateral triangle with sides of 7·5 cm.

In questions **3** to **6**, construct the triangles using compasses.
Measure the angles marked with letters.

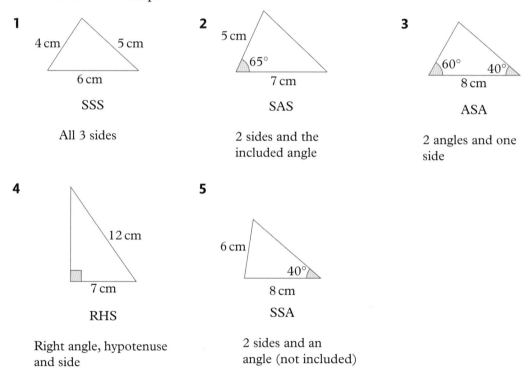

3
6 cm x 9 cm
8 cm

4
6 cm y 5 cm
7·5 cm

5
7·2 cm
9 cm z
8·1 cm

6
9 cm
4·3 cm
a
6 cm

Constructing unique triangles

You can describe a triangle according to the information you have.

You use: S   when you know a side,
A   when you know an angle,
R   when you know there is a right angle,
H   when you know the hypotenuse of a right-angled triangle.

● Here are five examples.

1
4 cm   5 cm
6 cm

SSS

All 3 sides

2
5 cm
65°
7 cm

SAS

2 sides and the
included angle

3
60°   40°
8 cm

ASA

2 angles and one
side

4
12 cm
7 cm

RHS

Right angle, hypotenuse
and side

5
6 cm
40°
8 cm

SSA

2 sides and an
angle (not included)

Exercise 13 ②

1 Using a ruler, protractor and compasses, construct each
of the triangles **1**, **2**, **3** and **4**.
Label the triangles SSS, SAS, ASA, RHS.

2 Construct triangle **5** and, using compasses, show that it is
possible to construct two different triangles with the sides and angle given.

3 Construct the triangle in the diagram. You are given SSA.
Show that you can construct two different triangles with
the sides and angle given.

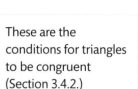

4 Copy and complete these two sentences.
'When you know SSS, SAS, ☐ or ☐ the constructed triangle
is unique.
When you know ☐ the triangle is not unique and it is
sometimes possible to construct two different triangles.'

> These are the
> conditions for triangles
> to be congruent
> (Section 3.4.2.)

3.3.3 Nets

If the cube here was made of cardboard, and you
cut along some of the edges and laid it out flat,
you would have the **net** of the cube.

A cube has: 8 vertices (corners)
                      6 faces
                      12 edges.

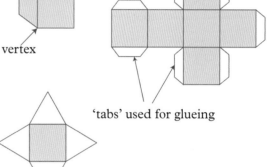

vertex

Here is the net for a square-based pyramid.
This pyramid has: 5 vertices
                               5 faces
                               8 edges.

'tabs' used for glueing

Exercise 14 ❶

You will need pencil, ruler, scissors and either glue or sticky tape.

1 Which of these nets could you use to make a cube?

a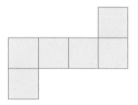

b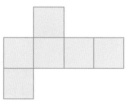

> If you want to create
> a 3-D shape you need
> to add tabs to your
> net to glue your shape
> together.

c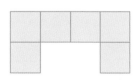

d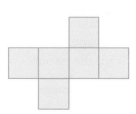

> Can you find a rule
> that tells you where to
> put the tabs?

2 The numbers on opposite faces of a dice add up to 7. Take one
of the possible nets for a cube from question **1** and show the
number of dots on each face.

3 Here is the start of the net of a cuboid (a closed rectangular box) measuring 4 cm × 3 cm × 1 cm. Copy and then complete the net.

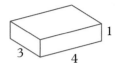

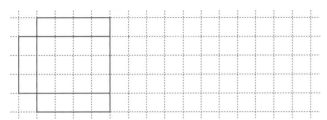

4 This diagram needs one more square to complete the net of a cube. Find the four possible positions of the square and draw the **four** possible nets which would make a cube.

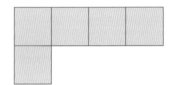

*5 You can make a solid from each of these nets.

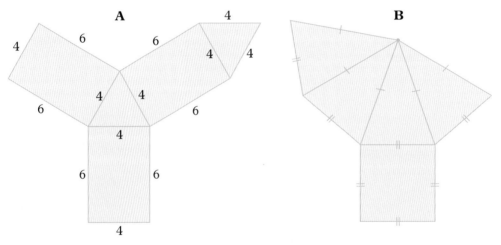

a State the number of vertices and faces for each solid.
b One object is a square-based pyramid and the other is a triangular-based prism. Which one is which?

6 Some interesting objects can be made using isometric dotty paper. The basic shape for the nets is an equilateral triangle. With the paper as shown, the triangles are easy to draw.

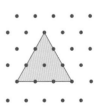

Make the sides of the triangles 3 cm long so that the solids are easy to make. Here is the net of a regular tetrahedron. Draw it and then cut it out. Now fold it to make a pyramid.

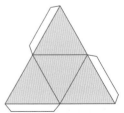

3.3.4 Constructions with a ruler and compasses

1 Perpendicular bisector of a line segment AB

With centres A and B, draw two arcs with your compass. Keep the same radius. Join the points where the arcs intersect (the dotted line).

This dotted line is the perpendicular bisector of AB.

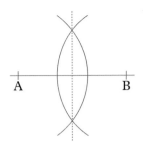

An arc is part of the circumference of a circle.

2 Perpendicular from point P to a line

P.

line

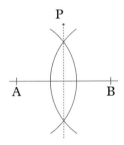

With centre P, draw an arc to cut the line at A and B.

Construct the perpendicular bisector of AB. Use a smaller radius on your compasses.

3 Bisector of an angle

With centre A, draw arc PQ.
With centres at P and Q, draw two more arcs.
Keep the same radius.
Join the point of intersection of the two arcs to A.
This line AB is the bisector of angle A.

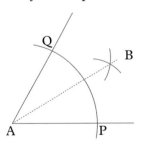

Exercise 15 ②

Use only plain unlined paper, a pencil, a ruler and compasses.

1 Draw a line AB of length 6 cm. Construct the perpendicular bisector of AB.

2 Draw a line CD of length 8 cm. Construct the perpendicular bisector of CD.

3 Draw a line and mark a point P about 4 cm from the line. Construct the line which passes through P and is perpendicular to the line.

See Construction 2.

4 a Using a set square or protractor, draw a right-angled triangle ABC as shown. For greater accuracy draw lines slightly longer than 8 cm and 6 cm and **then** mark the points A, B and C.

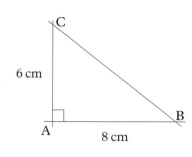

b **Construct** the perpendicular bisector of AB.

c **Construct** the perpendicular bisector of AC.

d If you do it accurately, your two lines from **b** and **c** should cross exactly on the line BC.

5 This is the construction of a perpendicular from a point P on a line, using ruler and compasses.

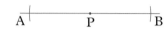

a With centre P, draw arcs to cut the line at A and B.

b Now construct the perpendicular bisector of AB. (Use a larger radius, and join the points where the arcs meet.)

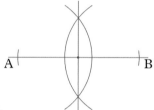

6 Draw an angle of about 60°. Construct the bisector of the angle.

7 Draw an angle of about 80°. Construct the bisector of the angle.

*8 Investigation

Draw any triangle ABC and then construct the bisectors of angles A, B and C. If you do it accurately the three bisectors should all pass through one point.

If they do **not** pass through one point (or very nearly), do this question again with a new triangle ABC.

Does this work with every type of triangle, for example, scalene, isosceles?

See Construction 3 on page 121.

3.4 Congruent and similar shapes

3.4.1 What are congruent and similar shapes?

▶ **Congruent** shapes are exactly the same in shape and size. Shapes are congruent if one shape fits exactly over the other.

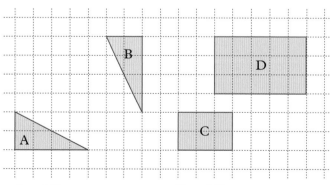

A and B are congruent   C and D are not congruent

▶ Shapes which are mathematically **similar** have the same shape, but different sizes. All **corresponding angles** in similar shapes are equal and **corresponding lengths** are in the same ratio.

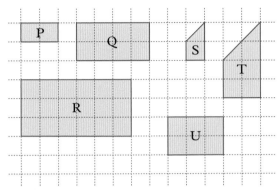

U is **not** similar to any of the shapes P, Q and R.

P, Q and R are similar.   S and T are similar.

If two shapes are similar, one shape is an **enlargement** of the other. Note that all **circles** are similar to each other and all **squares** are similar to each other.

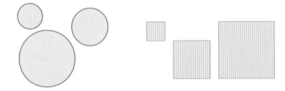

Exercise 16 ① ②

1 Write the pairs of shapes that are congruent.

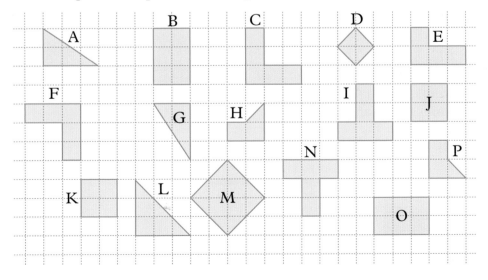

2 Copy the diagram onto squared paper and colour in the congruent shapes with the same colour.

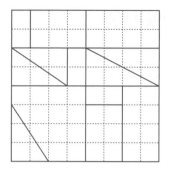

3 Copy shape X onto squared paper. Draw another shape which is congruent to shape X but turned into a different position.

4 Make a list of pairs of shapes which are similar.

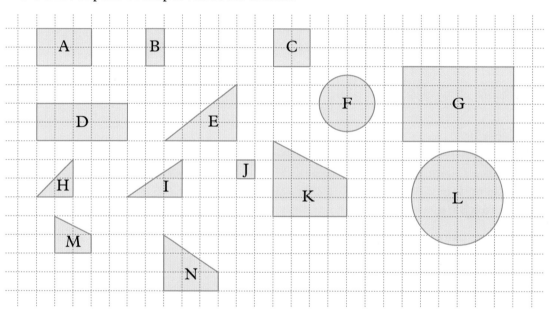

5 Draw shape I from question **4** on squared paper. Now draw a shape which is similar to shape I.

6 Here are two rectangles. Explain why they are not similar.

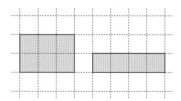

3.4.2 Congruent triangles

There are four types of congruence for triangles.

▶ Two sides and the included angle (SAS)

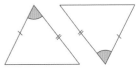

▶ Two angles and a corresponding side (ASA)

▶ Three sides (SSS)

▶ Right angle, hypotenuse and one other side (RHS)

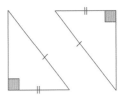

Proof using congruent triangles

EXAMPLE

PQRS is a square. M is the midpoint of PQ.
N is the midpoint of RQ.
Prove that SN = RM.

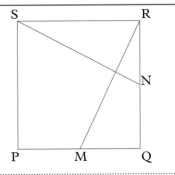

∠SRQ = ∠PQR   (Both right angles)
  SR = RQ   (Sides of a square)
  RN = MQ   (Given)
So triangle SRN is congruent to triangle RQM. (SAS)
You can write △SRN ≡ △RQM
The sign ≡ means 'is identical to'.
Since the triangles are congruent, it follows that SN = RM.

> Note that the order of letters shows corresponding vertices in the triangles.

Exercise 17 ❷

For questions **1** to **6**, decide whether the triangles in each pair of triangles are congruent. If they are congruent, state which conditions for congruency are satisfied.

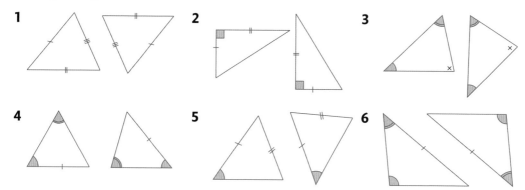

1   **2**   **3**

4   **5**   **6**

7 ABCDE is a regular pentagon. List all the triangles which are congruent to triangle BCE.

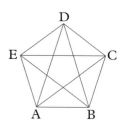

8 By construction show that it is possible to draw two **different** triangles with the angle and sides shown.

This shows that 'SSA' is not a condition for triangles to be congruent.

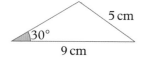

5 cm

30°

9 cm

9 Draw triangle ABC with AB = BC and point D at the midpoint of AC.

Prove that triangles ABD and CBD are congruent and state which case of congruency applies.
Hence prove that angles A and C are equal.

10 Tangents TA and TB touch the circle with centre O.
Use congruent triangles to prove that the two tangents are the same length.

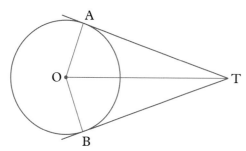

A tangent is perpendicular to the radius at the point where it touches the circle.

11 Triangle LMN is isosceles with LM = LN; X and Y are points on LM, LN respectively such that LX = LY. Prove that triangles LMY and LNX are congruent.

Draw diagrams for questions **11**, **12** and **13**.

12 ABCD is a quadrilateral and a line through A parallel to BC meets DC at X. If ∠D = ∠C, prove that △ADX is isosceles.

13 XYZ is a triangle with XY = XZ. The bisectors of angles Y and Z meet the opposite sides in M and N respectively. Prove that YM = ZN.

14 In the diagram, DX = XC, DV = ZC and the lines AB and DC are parallel. Prove that
  a AX = BX
  b AC = BD
  c triangles DBZ and CAV are congruent.

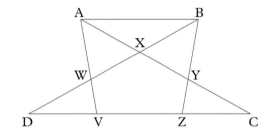

15 Draw a parallelogram ABCD and prove that its diagonals bisect each other.

16 Draw a square and label it PQRS. Draw equilateral triangles PQA and RQB so that A is inside and B is outside the square.

By considering triangles PQR and AQB, prove that AB is equal to PR.

3.5 Tessellations

- A tessellation is formed when a shape (or shapes) fit together without gaps to cover a surface. Here are some examples.

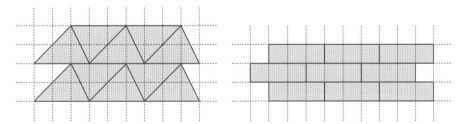

Exercise 18 ❶

1 Use squared paper to show that each of the shapes below tessellates.

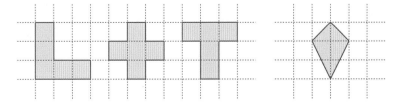

2 a Draw any irregular triangle or quadrilateral and cut twenty or so copies from cardboard or paper. Fit them together, like a jigsaw puzzle, to cover a plane.

b Say whether the statements below are true or false:
i 'all triangles tessellate'
ii 'all quadrilaterals tessellate'.

3 Explain why regular hexagons tessellate and why regular pentagons do not.

4 Show, by drawing or otherwise, whether these combinations of polygons tessellate.
a Regular octagons and squares.
b Regular hexagons and triangles.

3.6 Symmetry

3.6.1 Line and rotational symmetry

Line symmetry

The letter M has one line of symmetry, shown dotted. Line symmetry is sometimes called **reflection symmetry**.

Rotational symmetry

The shape may be turned about its centre into three identical positions. It has rotational symmetry of order **three**.

| A line of symmetry is like a mirror line. |

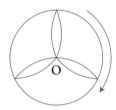

A shape with no rotational symmetry needs to be rotated a full 360° about its centre to look the same. It has rotational symmetry of order 1.

Exercise 19 ❶

For each shape write
a the number of lines of symmetry, if any
b the order of rotational symmetry, if there is any.

1 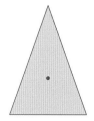 **2**  **3**

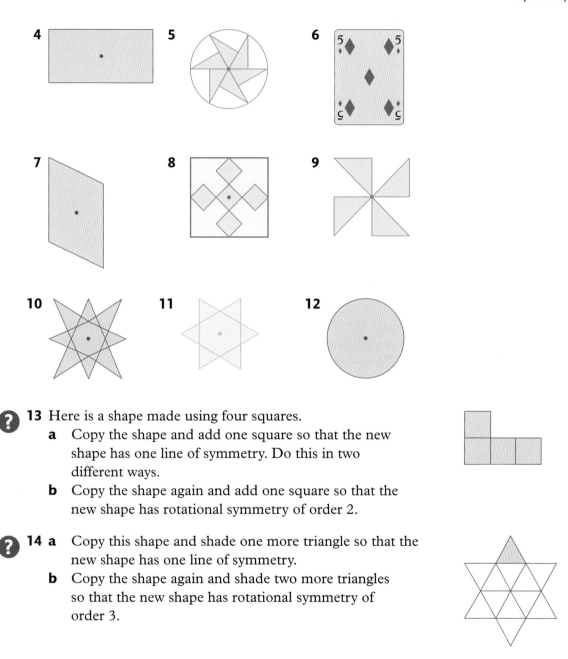

4

5

6

7

8

9

10

11

12

13 Here is a shape made using four squares.
  a Copy the shape and add one square so that the new shape has one line of symmetry. Do this in two different ways.
  b Copy the shape again and add one square so that the new shape has rotational symmetry of order 2.

14 a Copy this shape and shade one more triangle so that the new shape has one line of symmetry.
  b Copy the shape again and shade two more triangles so that the new shape has rotational symmetry of order 3.

15 Look at this shape.
  a Copy the shape and add one square so that the shape has rotational symmetry of order 2.
  b Copy the shape again and add one square so that the shape has one line of symmetry.

Exercise 20 ❶

In questions **1** to **8**, the broken lines are lines of symmetry. In each diagram, only **part of the shape** is given. Copy the part shapes onto square grid paper and then carefully complete them.

1 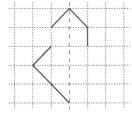   **2** 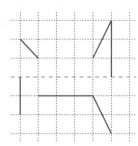   **3**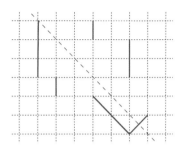

4 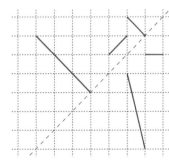   **5** 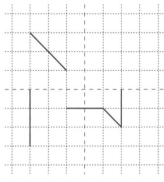   **6**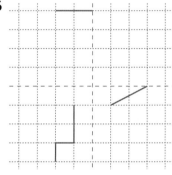

7 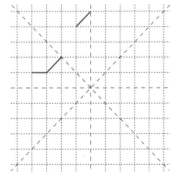   **8**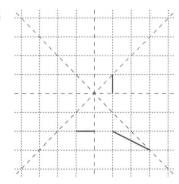

9 Fold a piece of paper twice and cut out any shape from the corner. Open the shape and stick it in your book stating the number of lines of symmetry and the order of rotational symmetry.

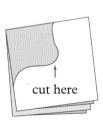

cut here

3.7 Quadrilaterals

3.7.1 Properties of quadrilaterals

▶ The angles in a quadrilateral always add up to 360°.

Title	Properties	
Square	Four equal sides Four angles 90° Four lines of symmetry Rotational symmetry of order 4	
Rectangle (not square)	Two pairs of equal and parallel sides Four angles 90° Two lines of symmetry Rotational symmetry of order 2	
Rhombus	Four equal sides Opposite sides parallel Diagonals bisect at right angles Diagonals bisect angles of rhombus Two lines of symmetry Rotational symmetry of order 4	
Parallelogram	Two pairs of equal and parallel sides Opposite angles equal No lines of symmetry Rotational symmetry of order 2	
Trapezium	One pair of parallel sides Rotational symmetry of order 1 No lines of symmetry apart from an isosceles trapezium, which has one line of symmetry.	
Kite	AB = AD, CB = CD Diagonals meet at 90° One line of symmetry Rotational symmetry of order 1	

Exercise 21 ①

1 Write the correct names for these five quadrilaterals.

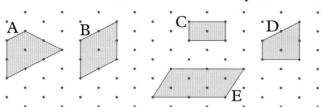

> These are drawn on isometric paper.

2 Copy the table and fill all the boxes with either ticks or crosses.

	Diagonals always equal	Diagonals always perpendicular	Diagonals always bisect the angles	Diagonals always bisect each other
Square	✓			
Rectangle				
Parallelogram				
Rhombus				
Trapezium				

3 Find the angle x.

a
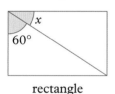
rectangle

b
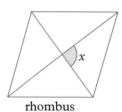
rhombus

c
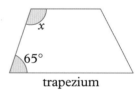
trapezium

4 Copy each diagram onto square grid paper and mark the fourth vertex with a cross.

a
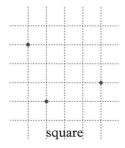
square

b
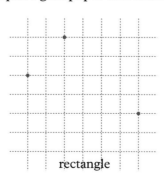
rectangle

c
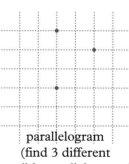
parallelogram
(find 3 different possible parallelograms)

5 What is the name for all shapes with four sides?

6 What four-sided shape has all sides the same length and all angles equal?

7 What quadrilateral has two pairs of sides the same length and all angles equal?

8 Which quadrilateral has only one pair of parallel sides?

9 Name the triangle with three equal sides and angles.

10 What is the name of the triangle with only two equal angles?

11 Which quadrilateral has all four sides the same length but only opposite angles equal?

12 True or false: 'All squares are rectangles.'

13 True or false: 'Any quadrilateral can be cut into two equal triangles.'

14 Name the shapes made by joining these points.
   a A B F G
   b C E F I
   c A B E H
   d A B D I

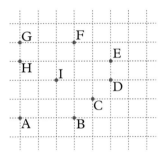

15 Name the shapes made by joining these points.
   a B I G E
   b A B E H
   c B C D F
   d C J G D
   e C J E

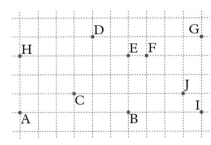

16 On square grid paper, draw a quadrilateral with just two right angles and only one pair of parallel sides.

*17 ABCD is a rhombus whose diagonals intersect at M.
Find the coordinates of C and D, the other two vertices of the rhombus.

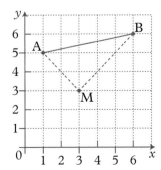

Exercise 22 ❶

In these questions, begin by drawing a diagram and remember to put the letters around the shape in alphabetical order.

1 In a rectangle KLMN, ∠LNM = 34°.
Calculate
a ∠KLN
b ∠KML.

2 In a trapezium ABCD, ∠ABD = 35°,
∠BAD = 110° and AB is parallel to DC.
Calculate
a ∠ADB **b** ∠BDC.

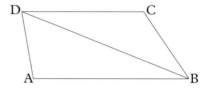

3 In a parallelogram WXYZ, ∠WXY = 72°, ∠ZWY = 80°.
Calculate
a ∠WZY **b** ∠XWZ **c** ∠WYX.

4 In a kite ABCD, AB = AD, BC = CD, ∠CAD = 40° and ∠CBD = 60°.
Calculate
a ∠BAC **b** ∠BCA **c** ∠ADC.

5 In a rhombus ABCD, ∠ABC = 64°.
Calculate
a ∠BCD
b ∠ADB
c ∠BAC.

6 In a rectangle WXYZ, M is the midpoint
of WX and ∠ZMY = 70°. Calculate
a ∠MZY **b** ∠YMX.

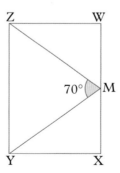

7 In a trapezium ABCD, AB is parallel to DC,
AB = AD, BD = DC and ∠BAD = 128°.
Find
a ∠ABD
b ∠BDC
c ∠BCD.

3.8 Isometric drawing

3.8.1 Drawing cuboids and solids made from cuboids

Here are two pictures of the same cuboid, measuring 4 × 3 × 2 units.

> When you draw a solid on paper, you make a 2-D representation of a 3-D object.

1 On ordinary square paper

2 On isometric paper (a grid of equilateral triangles)

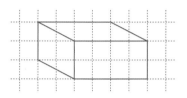

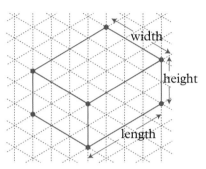

The dimensions of the cuboid cannot be taken from the first picture but they can be taken from the picture drawn on isometric paper. Instead of isometric paper you can also use 'triangular dotty' paper like this. Be careful to use it the right way round (as shown here).

Exercise 23 ❶ ❷

In questions **1** to **3**, the solids are made from 1 cm cubes joined together. Draw each solid on isometric paper (or 'triangular dotty' paper). Questions **1** and **2** are already drawn on isometric paper.

1

2

3

4 The diagram shows one edge of a cuboid. Complete the drawings of **two** possible cuboids, each with a volume of 12 cm³. Start with this edge both times.

5 Here are two shapes made using four multilink cubes.

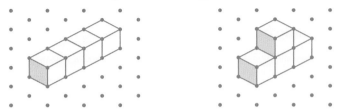

Using four cubes, make and then draw four more shapes that are different from the two above.

6 The shape shown falls over on to the blue shaded face.
Draw the shape after it has fallen over.

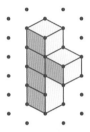

7 You need 16 small cubes.
Make the two solids in the diagram
and then arrange them into a
4 × 4 × 1 cuboid by adding
a third solid, which you have to find.
Draw the third solid on isometric paper.
There are **two** possible solids.

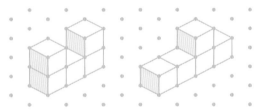

8 Make the letters of your initials from cubes and then draw them on isometric paper.

9 The side view and plan view (from above) of object A are shown.

> The side and front views are sometimes called the **side elevation** and the **front elevation**.

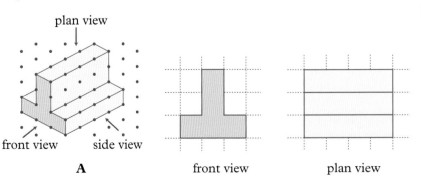

Draw the front view, side view and plan view of objects B and C.

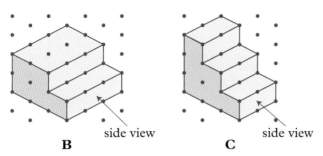

10 to **13** you are given three views of a shape. Use the information to make the shape using centimetre cubes and then draw a sketch of the solid.

10

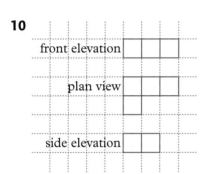

11

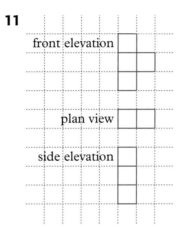

12

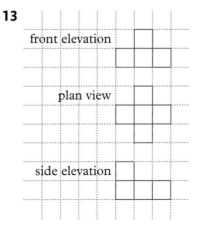

13

3.9 Area

3.9.1 Counting squares

▶ Area describes how much **surface** a shape encloses.

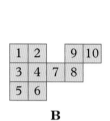

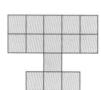

B   **C**   **D**

B contains 10 squares.   C has an area of $12\frac{1}{2}$ squares.

B has an area of 10 squares.   D has an area of 12 squares.

▶ A square that is one centimetre by one centimetre has
an area of one square centimetre. You write this
as $1\,\text{cm}^2$.

1 cm

1 cm

Exercise 24 ①

In the diagrams, each square represents $1\,\text{cm}^2$.
Copy each shape and find its area by counting squares.

1   **2**   **3**   **4**

5   **6**   **7**   **8**

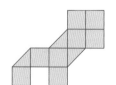

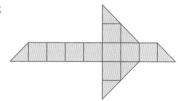

9 You can **estimate** the area
of a curved shape like this.

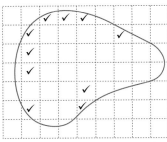

Each square
represents 1 cm².

a Count the whole squares
inside the shape and
mark them ✓.

b Count the parts where
there is half a square
or more and mark them ✓.

c Ignore the parts where there is less
than half a square inside the shape.

Some of the squares on the shape shown have been ticked
according to the instructions **a, b** and **c**.

Use a photocopy of this page to find an estimate for the
area of the shape.

10 Use the method in question **9** to estimate the area of each of these
shapes.

(Ask your teacher for a photocopy of this page.)

a 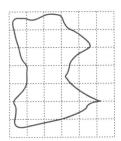   b 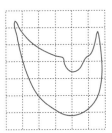   c 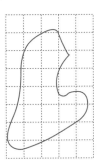   d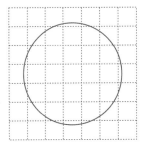

3.9.2 Areas of rectangles

You can find the area of a rectangle by counting
squares. This rectangle has an area of 30 squares.
If each square is 1 cm², this rectangle has an
area of 30 cm².

1	2	3	4	5	6
7	8	9	10	11	12
13	14	15	16	17	18
19	20	21	22	23	24
25	26	27	28	29	30

5 cm (height), 6 cm (width)

It is easier to multiply the length by the width of
the rectangle than to count squares.

Area of rectangle = length × width
$$= 6 \times 5 \, \text{cm}^2$$
$$= 30 \, \text{cm}^2$$

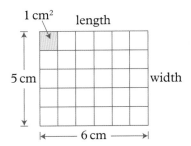

1 cm² length

5 cm width

6 cm

▶ Area of a rectangle = length × width

A square is a special rectangle in which the length and width are equal.

▶ Area of a square = length × length

Exercise 25 ①

Calculate the area of these rectangles. All length are in centimeters

1

4

|← 6 →|

2

8

|← 3 →|

3

4

|← 5 →|

4

2

|← 4 →|

5

5

← 2 →

6

5

← 5 →

7

3·5

← 10 →

8 Find the area of a rectangular table measuring 60 cm by 100 cm.

9 A farmer's field is a square of side 200 m. Find the area of the field in m².

10 Here is Kieran's passport photo.
Measure the height and width with a ruler and then work out its area in cm².

11 Work out the area of the shaded rectangle.

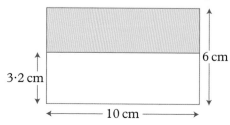

3·2 cm

6 cm

← 10 cm →

Exercise 26 ❶

All lengths are in cm.

1 This shape is made from two
rectangles joined together.
Find the total area of the shape.

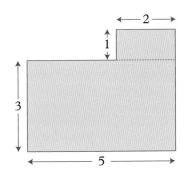

Start by finding
the area of each
rectangle and then
add them together.

The shapes in questions **2** to **12** are made of rectangles joined
together. Find the area of each shape.

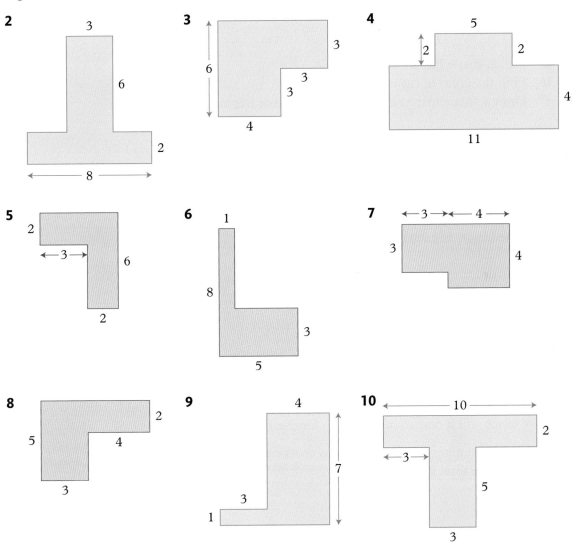

11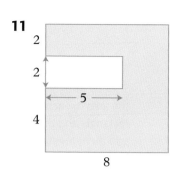

12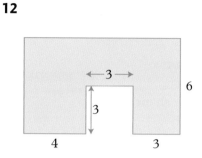

13 Here are four shapes made with centimetre squares.

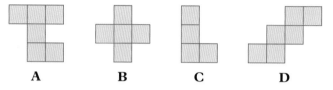

    A   **B**   **C**   **D**

  a Which shape has an area of 5 cm²?
  b Find the areas of the other three shapes.
  c Draw a shape using centimetre squares that has an area of 7cm².

14 A rectangular pond, measuring 10 m by 6 m, is completely
surrounded by a path which is 1m wide.
Find the area of the path.

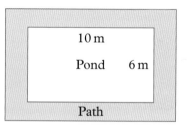

15 Find the height of each rectangle.

a   **b**   **c**

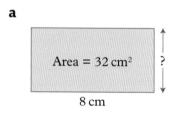   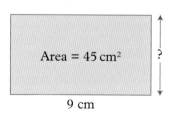

16 A wall measuring 3 m by 5 m is to be covered with tiles that are 20 cm squares.
A box of 10 tiles costs £8·95.
How much will it cost to buy the tiles for the wall?

17 The diagram shows a garden with two crossing paths.
Calculate the total area of both paths.

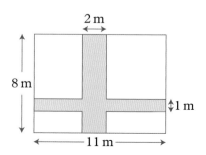

*18 A wooden cuboid has the dimensions shown.
Calculate the total surface area.

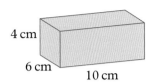

4 cm

6 cm

10 cm

> You find the surface area of a 3-D shape by adding the areas of all the faces.

3.9.3 Areas of triangles

This triangle has base 6 cm, height 4 cm and a right angle at B.

Area of rectangle ABCD = 6×4 cm^2
$= 24$ cm^2
Area of triangle ABC = area of triangle ADC
Area of triangle ABC = $24 \div 2$
$= 12$ cm^2

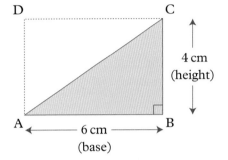

D   C

4 cm (height)

A   6 cm   B

(base)

You can show that for any triangle

▶ Area = $\frac{1}{2}$(base × height) or Area = $\frac{(\text{base} \times \text{height})}{2}$

Learn this formula.

Exercise 27 ❶

Find the area of each triangle. Lengths are in cm.

1

2
5

2
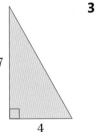
7
4

3
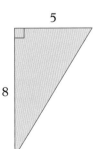
5
8

4
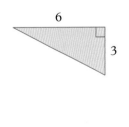
6
3

5
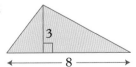
3
8

6
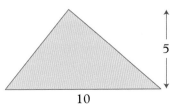
10

7
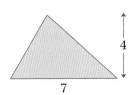
5
4
7

In questions **8** to **10**, find the total area of each shape.

8

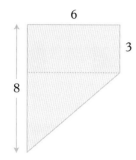

9

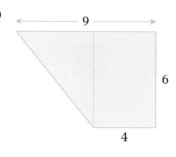

10

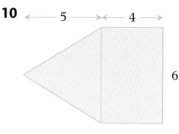

In questions **11** to **13** the outline of each shape is a rectangle.

11 Find the total shaded area.

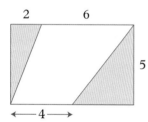

12 Find the total shaded area.

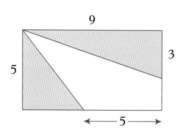

13 Find the shaded area.

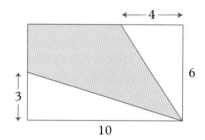

14 Find the height of each triangle.

a

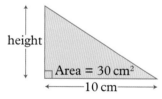

b

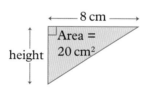

c

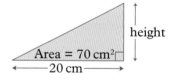

15 The triangle on the right has an area of $2\,cm^2$.

  a On square dotty paper, draw a triangle with an area of $3\,cm^2$.

  b Draw a triangle, different from the one shown, with an area of $2\,cm^2$.

16 Joe said 'There are $100\,cm^2$ in $1m^2$.' Mark said 'That's not right because one square metre is $100\,cm$ by $100\,cm$, so there are $10\,000\,cm^2$ in $1\,m^2$.' Who is right?

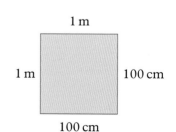

Exercise 28 ①

1 **a** Copy the diagram.

b Work out the areas of triangles A, B and C.
Give the answer in square units.

c Work out the area of the square enclosed
by the broken lines.

d Hence work out the area of the shaded triangle.

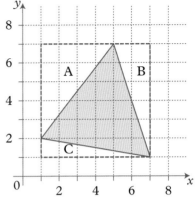

2 **a** Copy the diagram.

b Work out the areas of triangles A, B and C.

c Work out the area of the rectangle enclosed
by the broken lines.

d Hence work out the area of the shaded triangle.
Give the answer in square units.

For questions **3** to **7**, draw a pair of axes similar to those
in questions **1** and **2**. Plot the points in the order given
and find the area of the shape they enclose.

3 $(1, 4), (6, 8), (4, 1)$

4 $(1, 7), (8, 5), (4, 2)$

5 $(2, 4), (6, 1), (8, 7), (4, 8), (2, 4)$

6 $(1, 4), (5, 1), (7, 6), (4, 8), (1, 4)$

7 $(1, 6), (2, 2), (8, 6), (6, 8), (1, 6)$

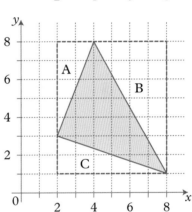

3.9.4 Areas of trapeziums and parallelograms

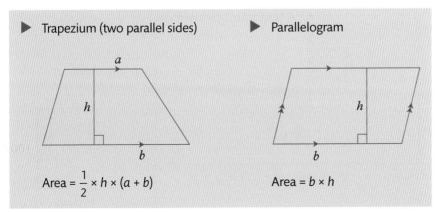

▶ Trapezium (two parallel sides)   ▶ Parallelogram

a

h

b

Area $= \dfrac{1}{2} \times h \times (a + b)$

h

b

Area $= b \times h$

> You will not have
> these formulae
> given to you in your
> examination.

Exercise 29 ❶

Find the area of each shape. All lengths are in cm.

1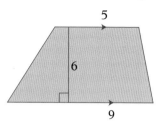

2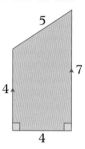

3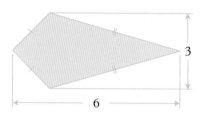

4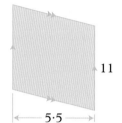

5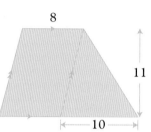

6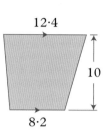

7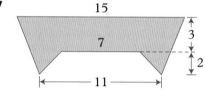

8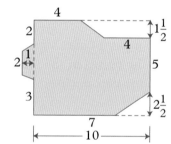

3.10 Perimeter

3.10.1 Perimeter of a shape

▶ The **perimeter** of a shape is the total length of its boundary.

EXAMPLE

Find the perimeters of these shapes.

a

8 cm

5 cm

b

4 cm

3 cm

5 cm

....

a Perimeter = 8 + 8 + 5 + 5

= 26 cm

b Perimeter = 3 + 4 + 5

= 12 cm

Exercise 30 ①

1 Find the perimeter of each shape. All lengths are in cm.

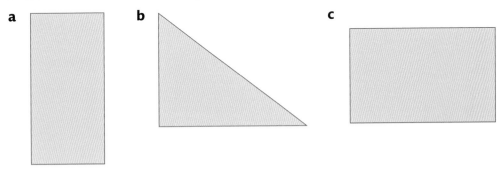

a 6, 4

b 13, 12, 5

c 11, 5, 9, 5, 8

d square 7

2 Measure the sides of these shapes with a ruler and work out the perimeter of each one.

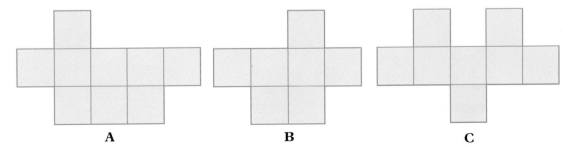

a

b

c

3 Here are three shapes made with centimetre squares.

A   B   C

a Find the perimeter of each shape.
b Draw a shape of your own design with a perimeter of 14 cm.
Ask another student to check your shape.

4 Use a ruler to find the perimeter of each picture.

a

b

c

5 The perimeter of a rectangular field is 160 m. The shorter side of the field is 30 m. How long is the longer side?

6 The perimeter of a rectangular garden is 56 m. The two longer sides are each 20 m long. How long are the shorter sides of the garden?

In questions **7** to **14**, find the perimeter of each shape.
All lengths are in cm.

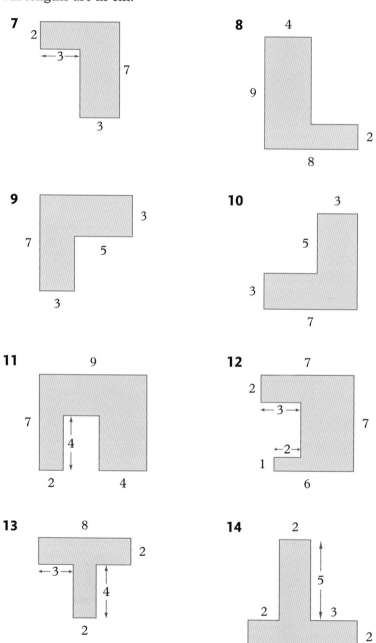

7

2

←3→

7

3

8

4

9

8

2

9

7

3

5

3

10

3

5

3

7

11

9

7

4

2   4

12

7

2

←3→

←2→

1

7

6

13

8

2

←3→

4

2

14

2

5

2

3

2

Test yourself

1 **a** A rectangle is 8 cm long and 5 cm wide.
    i Work out the area of this rectangle.
    ii Work out the perimeter of this rectangle.

   b A second rectangle has an area of 28 cm².
       Write down a possible pair of values for its
       length and width.

   c A third rectangle is half as long and half as
       wide as the **rectangle in part b**.
       What is the area of the third rectangle?

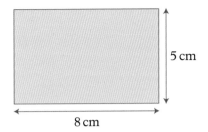

5 cm

8 cm

(*OCR, 2009*)

2 The diagram shows a rectangle.

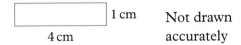

1 cm   Not drawn
accurately
4 cm

   a Four of these rectangles are put together as shown.

Not drawn
accurately

   Work out the shaded area.

   b The four rectangles are now put together to make
       this shape.

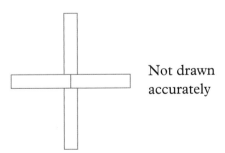

Not drawn
accurately

   Work out the perimeter of the shape.
   You **must** show your working.

(*AQA, 2012*)

3 *DAC*, *FCB* and *ABE* are straight lines.

Work out the size of the angle marked *x*.
You must give reasons for your answer.

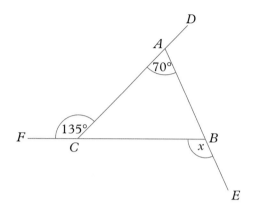

Diagram **NOT**
accurately drawn

Not to scale

(Edexcel, 2013)

4 *PQR* is a straight line.
PQ = *QS* = *QR*.
Angle *SPQ* = 25°.

Diagram **NOT**
accurately drawn

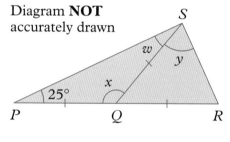

   a **i**   Write down the size of angle *w*.
      ii Work out the size of angle *x*.
   b Work out the size of angle *y*.

(Edexcel, 2008)

5 The diagram shows the distances, in kilometres, between some
towns, by road.

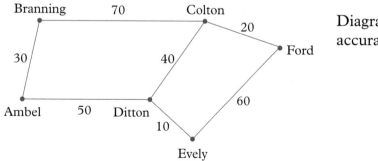

Diagram **NOT**
accurately drawn

Work out the shortest distance between Ambel and Ford by road.

(Edexcel, 2013)

6 a Work out 6^2.
  b In this cube each side has length 3 cm.
    What is the volume of the cube?

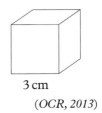

3 cm

(*OCR, 2013*)

7 Six shapes are made from centimetre squares.

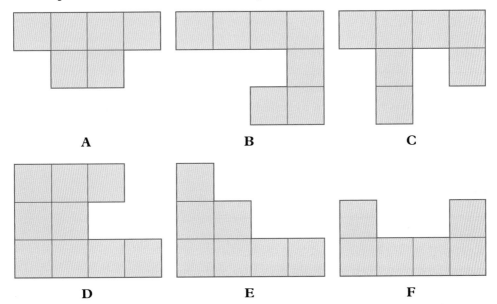

A   **B**   **C**

D   **E**   **F**

  a Copy and complete the following sentences.
    Shape A and Shape......... fit together to make a rectangle.
    Shape B and Shape D fit together to make a.........

  b Work out the area of Shape D.
    State the units of your answer.

(*AQA, 2013*)

8 Lines *AB* and *BC* are shown on the centimetre grid.
  a Write down the coordinates of point *A*.
  b *A*, *B* and *C* are three corners of a rectangle *ABCD*.
    Copy the grid and complete the rectangle.
  c Work out the perimeter of rectangle *ABCD*.

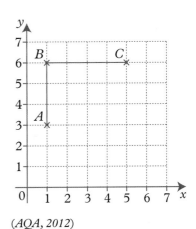

(*AQA, 2012*)

9 Jack is 1·28 metres tall.
He has a tent in the shape of a triangular prism.
The diagram shows the front view of the tent.

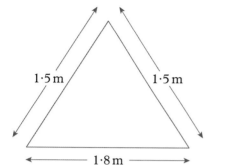

1·5 m   1·5 m

1·8 m

Not drawn
accurately

The base of the tent has been drawn to scale below.
Copy and complete the scale drawing to work out if Jack can
stand up in the middle of the tent. Show how you decide.

Scale: 1 cm represents 20 cm

_____

(AQA, 2013)

10 The diagrams show patterns made from grey tiles and white tiles.

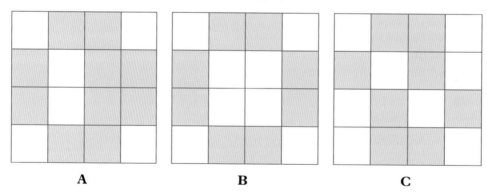

A   B   C

One of the patterns has exactly 1 line of symmetry.
a Write down the letter of this pattern.
b Write down the order of rotational symmetry of pattern C.

The diagram on the grid shows a tile in the shape of a hexagon.
c On triangular paper, show how the tile
will tessellate.
You should draw at least 6 tiles.

(Edexcel, 2013)

11 This L-shape has an area of 12 cm².
All corners are right angles.
All lengths are in centimetres.

Work out the value of x.

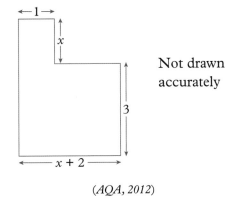

Not drawn
accurately

(*AQA, 2012*)

12 Look at this diagram.

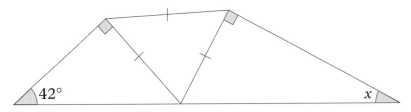

Work out the value of x.

(*MEI, Spec*)

13 a This is the plan and side elevation of a solid.

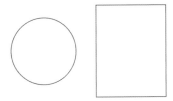

Plan   Side Elevation

What is the mathematical name of this solid?

b This is the plan and side elevation of a different solid.
They are drawn full size.

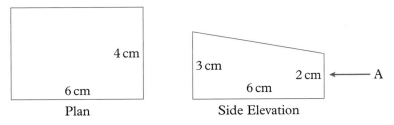

Draw accurately the front elevation of this solid, from
direction **A**, on square paper.

(*OCR, 2013*)

4 Statistics 1

In this unit you will:
- learn how to interpret and discuss graphs
- learn how to find the mean, median, mode and range for discrete, continuous and grouped data
- learn how to draw and interpret graphs, charts and diagrams
- learn how to compare sets of data.

Cyclists use statistics about their average speed, time and distance to measure their performance.

4.1 Interpreting graphs

4.1.1 Distance–time graphs

▶ A distance–time graph illustrates a journey by showing how far you have travelled during a period of time.

▶ A sloping line shows movement—the steeper the line, the faster the movement.

▶ The steepness of the line gives the speed of travel.

$$\text{Speed} = \frac{\text{Distance travelled}}{\text{Time taken}}$$

Speed is the rate of change of distance.

▶ A horizontal line shows there is no movement.

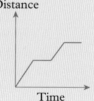

Distance

Time

'Rate of change' means how fast something is changing over time.

Exercise 1 ❶

Look at each graph and then answer the questions.

1 These graphs show the distance travelled from rest by an athlete, a cyclist and a car. In each case, find
   a how far they travelled in 5 seconds
   b how far they travelled in 10 seconds
   c their speed in metres per second.

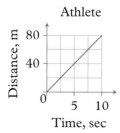

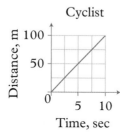

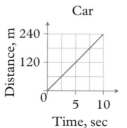

2 This graph shows a car journey from
Newbury to Tonbridge.

a Find the time when the car is

   i at Reigate

   ii at Basingstoke

   iii at Tonbridge.

b Where is the car

   i at 09:00

   ii at 09:45

   iii after travelling for two hours?

c Find the average speed of the car over
these three hours.

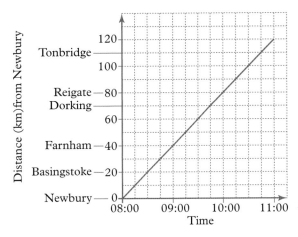

3 This graph shows a coach journey from
Bristol to Portsmouth.

a Find the time when the coach is

   i at Salisbury

   ii at Bath

   iii at Newton.

b Where is the coach

   i at 14:00

   ii at 15:45

   iii at 13:00

   iv at 16:15?

c How far is it from Bristol
to Westbury?

d How long does it take to travel from
Bristol to Westbury?

*e At what speed does the coach travel
from Bristol to Westbury?

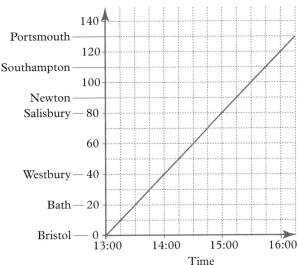

4 This graph shows a bicycle
journey from Royston to Harlow.

a Find the time when the bicycle is

   i at Harlow

   ii at Cambridge.

b Where is the bicycle

   i at 11:45

   ii at 12:45?

c How far is it from Royston
to Cambridge?

d How long does it take to
travel from Royston to Cambridge?

*e At what speed does the cyclist
travel from Royston to Cambridge?

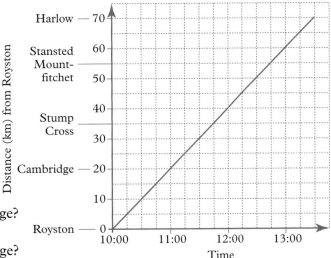

5 This graph shows a bus journey from Dover to Burgess Hill.

a When is the bus
  i at Throwley Forstal
  ii at Burgess Hill
  iii (more difficult) at Folkestone?

b Where is the bus
  i at 17:00
  ii at 17:30
  iii at 16:45?

c How far does the bus travel between 16:00 and 17:00?

*d At what speed does the bus travel from Dover to Sevenoaks?

e How far does the bus travel between 17:00 and 19:00?

*f At what speed does the bus travel from Sevenoaks to Burgess Hill?

6 Mike makes a journey partly on foot and partly on bicycle.

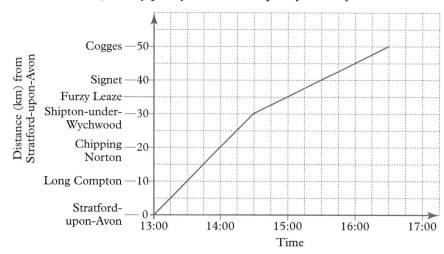

a When is Mike at
  i Signet   **ii** Cogges   **iii** Shipton-under-Wychwood?

b Where is he at
  i 14:00   **ii** 15:00   **iii** 13:30?

c How far does he travel between 13:00 and 14:30?

*d At what speed does he travel from Stratford-upon-Avon to Shipton-under-Wychwood?

e How far does he travel between 14:30 and 16:30?

*f At what speed does he complete the second part of the journey?

7 This graph shows Alex's journey cycling and running from Claydon to Bury St. Edmunds.

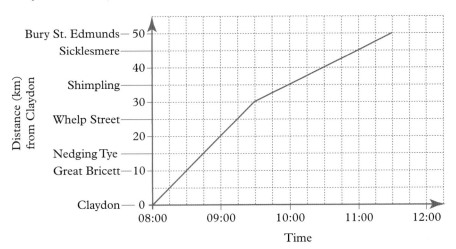

a When is Alex
  i at Nedging Tye **ii** at Whelp Street **iii** at Sicklesmere?
b Where is she
  i at 08:30 **ii** at 11:30 **iii** at 10:00?
c How far does she cycle between 08:00 and 09:30?
*d At what speed does she cycle from 08:00 until 09:30?
e How far does she run between 09:30 and 11:30?
*f At what speed does she run from 09:30 until 11:30?
*g What is Alex's **average** speed for the complete journey?

> Use the formula
> Average speed =
> $\dfrac{\text{Total distance}}{\text{Total time taken}}$

8 This graph shows a bus journey from York to Skipton and back again.
  a How long does the bus stop at Skipton?
*b At what speed does the bus travel
    i from York to Harrogate
    ii from Harrogate to Skipton
    iii from Skipton back to Harrogate?
  c When does the bus
    i return to York
    ii arrive at Skipton
    iii leave Harrogate?

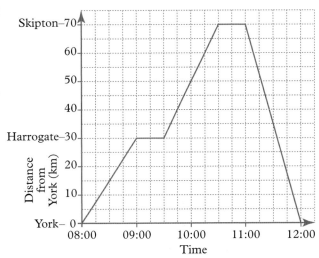

9 This graph shows a car journey from Albridge to Chington and back again.

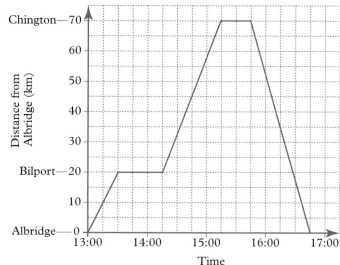

  a How long does the car stop at Bilport?
  b When does the car arrive at Chington?
  c When does the car leave Bilport?
  *d At what speed does the car travel
    i from Albridge to Bilport
    ii from Bilport to Chington
    iii from Chington back to Albridge?
  e At what time on the return journey is the car exactly halfway between Chington and Albridge?

10 The graph shows a return journey by car from Leeds to Scarborough.

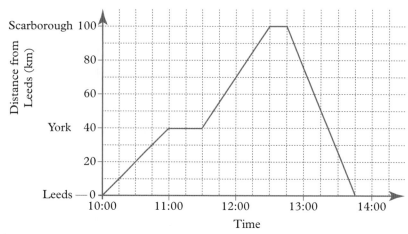

  a How far is it from Leeds to York?
  b How far is it from York to Scarborough?
  c At which two places does the car stop?
  d How long does the car stop at Scarborough?
  e When does the car
    i arrive in York   **ii** arrive back in Leeds?
  *f What is the speed of the car
    i from Leeds to York   **ii** from York to Scarborough
    iii from Scarborough to Leeds?

Speed =

$$\frac{\text{Distance travelled}}{\text{Time taken}}$$

11 Mr Berol and Mr Hale use the same road to travel between Aston and Borton.
This graph shows both journeys on one set of axes.

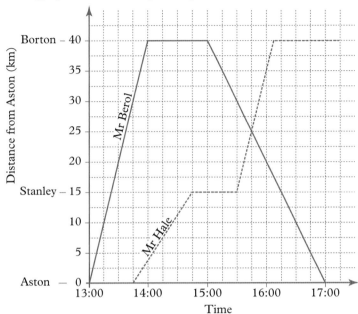

a At what time did
 i Mr Berol arrive in Borton **ii** Mr Hale leave Aston?

b i When did Mr Berol and Mr Hale pass each other?
 ii In which direction was Mr Berol travelling?

> Average speed =
>
> $\dfrac{\text{Total distance travelled}}{\text{Total time taken}}$

*c Find the speeds of
 i Mr Hale from Aston to Stanley **ii** Mr Berol from Aston to Borton
 iii Mr Hale from Stanley to Borton **iv** Mr Berol from Borton back to Aston.

d What is Mr Hale's average speed over his entire journey?

4.1.2 Velocity–time graphs

A velocity–time graph illustrates a journey by showing how the
velocity (or speed) changes with time.

> Velocity is the speed
> in a given direction.
> Velocity and speed
> are both measured
> in units such as
> m/s or km/h.

▶ A sloping line shows that your velocity (or speed) is changing. If it slopes
upwards, your speed is increasing. If it slopes downwards,
your speed is decreasing.

▶ The steepness of the line gives the acceleration.

$$\text{Acceleration} = \frac{\text{Change in speed}}{\text{Time taken}}$$

▶ Acceleration is the rate of change of velocity (or speed).

▶ A horizontal line shows that there is no change of speed. That is,
the speed is constant and there is no acceleration.

> Acceleration is
> measured in m/s^2
> or km/h^2.

Exercise 2 ①

1 These graphs show the speed of an athlete, a cyclist
   and a car as they set off from rest. In each case,
   find their acceleration. State why their acceleration is constant.

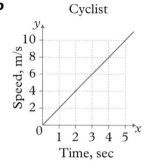

a Athlete **b** Cyclist **c** Car

2 **a** Find the acceleration of the three objects P, Q and R
      whose speeds vary with the time.
   b Why is your answer for R negative?
      What is happening to R?

P

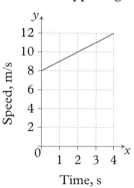

Q

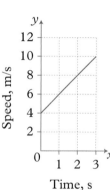

R

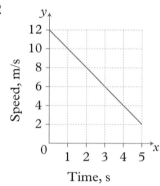

3 Two objects P and Q move and their speeds change over time.
   a At what times is P not moving?
   b At what times is Q not moving?
   c For how long do both of them travel at
      constant speed?
   d Copy and complete this table for the
      three legs of their journey, OA, AB and BC.

	Acceleration, m/s² given by		
	OA	**AB**	**BC**
P			
Q			

P

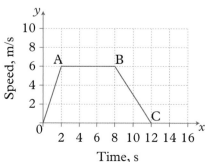

Q

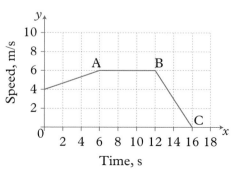

4 Draw axes, labelling the time axis in seconds from 0 to 10 and the speed axis in m/s from 0 to 10.
A bicycle travels at a constant speed of 8 m/s for 6 seconds, before slowing down gradually to rest over a further 4 seconds. Draw a graph of its speed against time, and find its acceleration in its final 4 seconds.

5 Draw axes, labelling the time axis in minutes from 0 to 10 and the speed axis in km/min from 0 to 2.
A train starts from rest and increases its speed steadily up to 2 km/min over 2 minutes. It then runs for a further 4 minutes at constant speed of 2 km/min, before gradually slowing down to rest in a final 4 minutes.
Draw a graph of speed against time on your axes, and find
  a its acceleration in its first two minutes
  b its acceleration in its final four minutes.

6 Draw axes, labelling the time axis in minutes from 0 to 20 and the speed axis in km/min from 0 to 2.
A train travels between two stations A and B. Starting from rest at A, it accelerates steadily for 2 minutes and reaches a speed of 1.5 km/min which it then holds steady for a further 12 minutes. Finally, it decelerates at a steady rate until it comes to rest 6 minutes later at section B.
Draw a graph of speed against time on your axes, and find
  a its acceleration in its first 2 minutes
  b its deceleration in its last 6 minutes.

4.1.3 Real-life graphs
You can use real-life graphs to find solutions to practical problems.

> **EXAMPLE**
>
> This graph converts miles into kilometres.
> Read off the axes to find
> 20 miles = 32 km
> 64 km = 40 miles.
>
>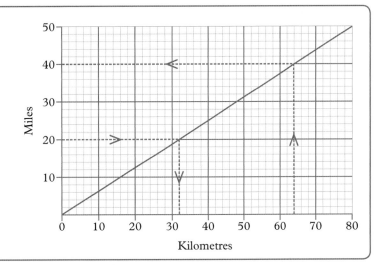

Exercise 3 ① ②

1 The graph converts pounds into euros.

   a Convert into pounds.
      i €4
      ii €3·30

   b Convert into euros.
      i £3
      ii £1·40

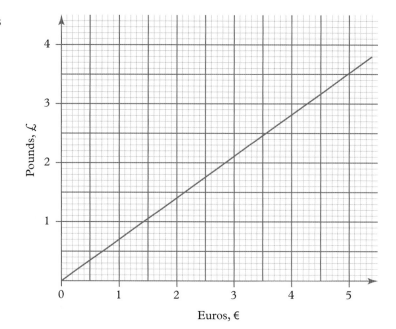

   c Henri, the highway robber, held up the Paris–Berlin stagecoach and got away with €50 000.
      How much was that in pounds?

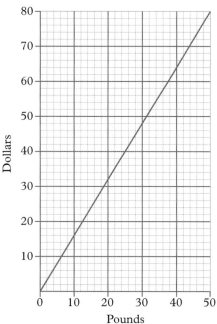

2 This graph converts US dollars to pounds.

   a Convert into pounds.
      i 40 dollars   **ii** 70 dollars
      iii 48 dollars   **iv** 64 dollars
      v 16 dollars   **vi** 24 dollars

   b Convert into dollars.
      i £20   **ii** £50   **iii** £6
      iv £16   **v** £36   **vi** £34

   c Before flying home from New York the Wilson family bought gifts for $72.
      How much did they spend in pounds?

3 The rupee is the currency used in India. This graph converts rupees into pounds.

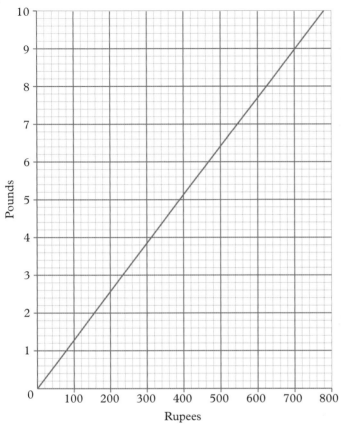

a Convert into pounds.

   i 280 rupees   **ii** 660 rupees   **iii** 420 rupees

   iv 380 rupees   **v** 600 rupees   **vi** 40 rupees

b Convert into rupees.

   i £8·00   **ii** £7·40   **iii** £4·60

   iv £2·00   **v** £10·00   **vi** £6·60

c On holiday in India, Stanley bought snake-charming lessons worth £9. How much did he pay in rupees?

4 This line graph shows the depth of water in a stream during one year.

a How deep was the stream in March?

b In which two months was the stream 8 cm deep?

c Which month saw the largest increase in depth?

Why do you think this happened?

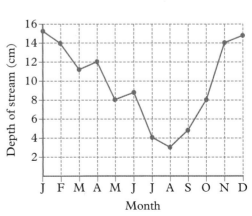

5 This line graph shows the average daily temperature in Stockholm, Sweden.

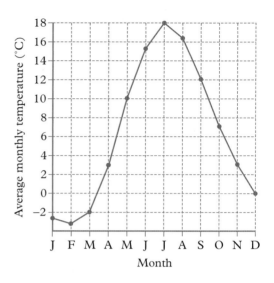

a What was the average temperature in June?

b In which month was the average temperature 7°C?

c In which two months was the average temperature 3°C?

d Which month saw the largest fall in temperature?

e The range is the difference between the highest and lowest temperatures. What was the range of temperature over the year?

*6 A car travels along a motorway and the amount of petrol in its tank is monitored as shown on the graph.

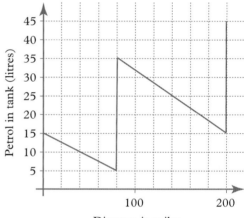

a How much petrol was bought at the first stop?

b What was the petrol consumption in miles per litre
   i   before the first stop
   ii between the two stops?

*7 Kendal Motors hires out vans.

KENDAL £35 MOTORS

plus 20 pence per mile (including VAT!)

a Copy and complete the table where x is the number of miles travelled and C is the total cost in pounds.

x	0	50	100	150	200	250	300
C	35			65			95

b Draw a graph of C against x, using scales of 2 cm for 50 miles on the x-axis and 1 cm for £10 on the C-axis.

c Use the graph to find the number of miles travelled when the total cost was £71.

*8 Jeff sets up his own business as a plumber.

24hr PLUMBING
07072 874561   call out £60
Plus £45 per hour   **NO VAT!**

    a Copy and complete the table where C stands for his total charge in pounds £ and h stands for the number of hours he works.

h	0	1	2	3
C		105		

    b Draw a graph with h across the page and C up the page. Use scales of 2 cm to 1 hour for h and 1 cm to £20 for C.

    c Use your graph to find how long Jeff worked if his charge was £127·50.

4.1.4   Sketch graphs

You can sketch a graph to see the shape of a general trend.
Sketch graphs have no detailed labels on the axes.

Exercise 4 ①

1 Which of the graphs A to D best fits the statement 'Unemployment is still rising but by less each month'.

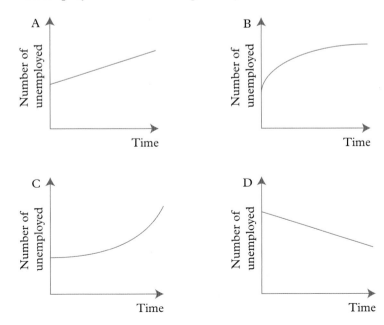

2 Match each of these statements to one of the graphs.
   a The temperature rises steadily.
   b The temperature begins to rise and then falls quickly.
   c The temperature rises more and more quickly.
   d The temperature stays the same and then falls quickly.
   e The temperature falls faster and faster.
   f The temperature rises, stays the same and then
      rises quickly.

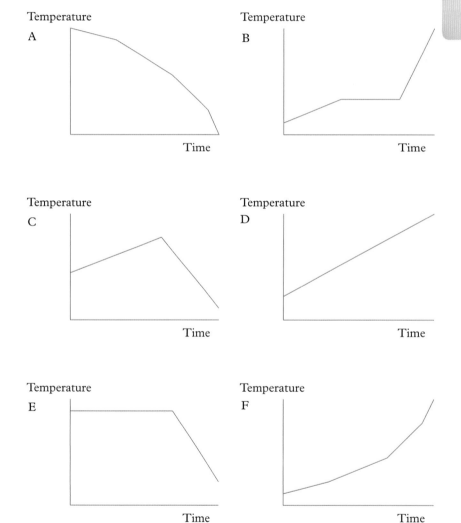

3 Which of the graphs A to D best fits the statement 'The price of oil was rising more rapidly in 2015 than at any time in the previous ten years'.

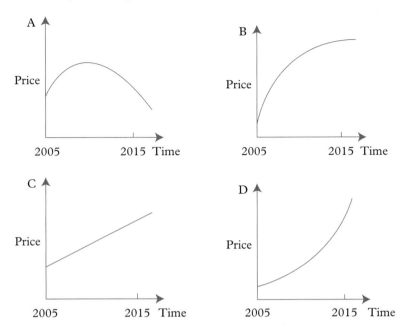

4 Which of the graphs A to D best fits each of the statements **a–d**?

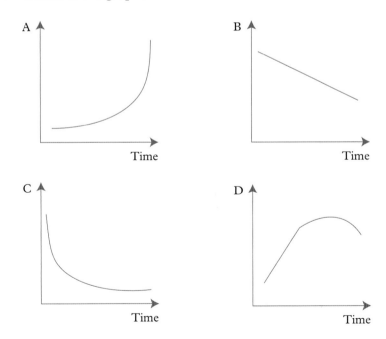

a The birthrate was falling but is now steady.
b Unemployment, which rose slowly until 2004, is now rising rapidly.
c Inflation, which has been rising steadily, is now beginning to fall.
d The price of gold has fallen steadily over the last year.

*5 The graph shows the motion of three cars A, B and C along the same road.
Answer these questions giving estimates where necessary.
  a Which car is in front after
    i 10 s **ii** 20 s?
  b When is B in the front?
  c Which car is going fastest after 5 s?
  d Which car starts slowly and then goes faster and faster?

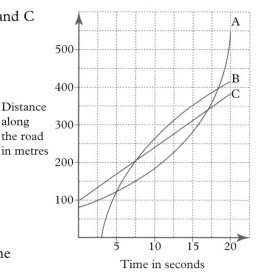

Distance along the road in metres

Time in seconds

*6 Water is poured at a constant rate into each of the containers A, B and C.
The graphs X, Y and Z show how the water level rises.
Decide which graph fits each container.

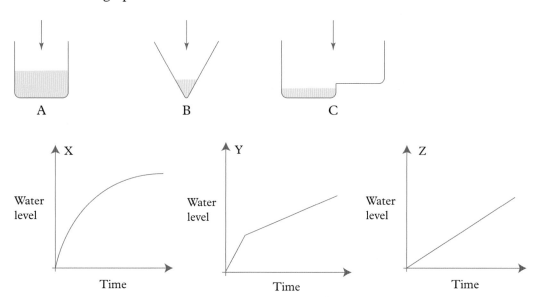

*7 The diagrams show three containers P, Q and R with water coming out as they empty.
Sketch three graphs similar to those in question **6** to show how the water level falls against time.

8 The graph below shows Linda's frustrating journey to work one day. After leaving home she was first caught behind a herd of cows being returned to their field. Next she had to slam on her brakes when a dog ran in front of the car. Then when she was finally on the motorway she got a puncture which she had to mend before she could continue.

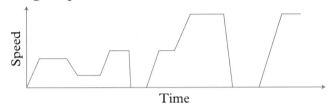

Copy the graph. Mark with an 'x', and label with the appropriate letter, the points where you think these events took place:

    A: leaving home

    B: cows finally going in their field

    C: the moment she first saw the dog run out

    D: the time at which she entered the motorway

    E: when she had finished mending the puncture.

4.2 Averages and spread

If you have a set of data, say exam marks or heights of people, and you are told to find the 'average', just what are you trying to find?

The answer is: a single number which can be used to represent the entire set of data.

The 'average' value can be found in three different ways: the **median**, the **mode** and the **mean**.

4.2.1 The median

> ▶ The data are arranged in order from the smallest to the largest; the middle number is then selected. This is called the **median**.

If there are two 'middle' numbers, the median is in the middle of these two numbers.

Median

Seven people are lined up in order of height. The median height is the height of the person in the middle.

EXAMPLE

Find the median of these sets of numbers
a 8   9   2   4   13   10   5
b 5   6   4   5   10   8

....

a Arrange the numbers in order of size.
   2   4   5   8   9   10   13   The median is **8**.
                  ↑

b Arrange the numbers in order of size.
   4   5   5   6   8   10   The median is **5·5**. This number is
              ↑   midway between 5 and 6.

Exercise 5 ①

1 Find the median for each set of numbers.
   a 5   6   7   3   11   9   14
   b 5   2   4   5   7
   c 18   4   17   10   11   16   15
   d 7   8   4   6   4   5   9   12
   e −1   4   −4   −2   5
   f 10   8   9   10   3   2   6   7   7
   g 2   −2   6   8   4   0   −1   5

2 The table shows the age, height and weight of seven students.

	Mike	Steve	Dora	Sam	Pat	Rayyan	Gary
Age (years)	16	15	17	15	16	15	16
Height (cm)	169	180	170	175	172	163	164
Weight (kg)	50	50	52	44	51	41	48

   a What is the median age?
   b What is the median height?
   c What is the median weight?

3 **a** Find the median of   4,   5,   7,   9,   1,   9,   2
    b **Use your answer** to find the median of
                   14,   15,   17,   19,   11,   19,   12

4 Sally throws a dice eight times and her father gives her 20p if the
   median score is more than 3.
   The dice scores are 6, 1, 2, 6, 4, 1, 3, 6.
   Find the median score and find out if Sally wins 20p.

5 **a** Write five numbers so that the median is 7.
  b Write seven numbers so that the median is 6.
  c Write six numbers so that the median is 8.

6 Twenty-one people were asked to estimate the
  weight of a stork (in kg). Their estimates were

7	30	16	30	17	21	9	24	18	9	25
21	14	8	21	12	7	10	27	20	11	

  What was the median estimate?

4.2.2 The mode

▶ The **mode** of a set of numbers is the number which occurs most often.

The mode is the easiest average to find. You can find it for
numerical data and for data which are not just numbers.
You could, for example, find the modal colour of T-shirts
worn by a group of children.

> **EXAMPLE**
>
> Find the mode of this set of data.
> 3 4 5 3 6 4 6 5 3 5 6 4 3 6 3
> ....
> The number 3 occurs most often. The mode is 3.

You can order data using a tally chart and frequency table to make
it easier to find the mode.

> **EXAMPLE**
>
> Find the mode of this set of data.
> 5 6 7 5 7 6 5 7 6 5
> ....
>
> | Number | Tally | Frequency |
> |:---:|:---:|:---:|
> | 5 | \|\|\|\| | ④ |
> | 6 | \|\|\| | 3 |
> | 7 | \|\|\| | 3 |
>
> The mode is 5, as it has the highest frequency.

5 is also called the
modal number.

Exercise 6 ①

1 Find the mode of each set of data.
   a 1 2 4 1 3 1 2 3
   b 3 4 5 6 6 3 6 5 4 6 6 3 4 5 6 4
   c 2 4 6 8 4 6 2 4 6 4 8 2 8 4 8 6
   d 2 2 2 2 3 3 3 3 4 4 4 4 4 5 5
   e red, red, blue, red, blue, white, white, red, blue, red

   f X, Y, Z, X, Y, Z, Z, Y, Z, X, Z

2 The frequency table shows the test results for a class of 30 students.

Mark	3	4	5	6	7	8
Frequency	2	5	4	7	6	6

What is the modal mark?

3 The temperature in a room is measured at midday
  every day for a month.
  The results are shown on the chart.
   a On how many days was the temperature
      17 °C?
   b What was the modal temperature?
   c How many days were there in that month?

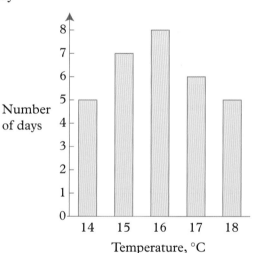

4 A dice was thrown 14 times.

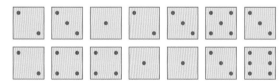

   a What was the modal score?
   b What was the median score?

5 a Write six numbers so that the mode is 2.
   b Write five numbers so that the median is 6 and the
      mode is 4.
   c Write seven numbers so that the median is 8 and the
      mode is 5.

6 The numbers of chicks in 32 nests were counted.

2	4	3	4	1	4	4	2
3	0	5	3	4	3	4	2
4	1	4	2	0	1	4	3
1	3	2	5	3	5	2	6

Make a tally and frequency chart and use it to find the modal number of chicks in a nest.

Number of chicks	Tally	Frequency
0	||	2
1		
2		
3		
4		
5		
6		

4.2.3 The mean

▶ To find the **mean**, add all the data and divide the total by the number of items.

EXAMPLE

a Find the mean of 3, 9, 6, 6, 8, 3, 9, 4
b Find the mean of 15, 7, 14, 17, 11, 16

....

a There are 8 numbers

$$\text{mean} = \frac{3 + 9 + 6 + 6 + 8 + 3 + 9 + 4}{8} = \frac{48}{8} = 6$$

b There are 6 numbers

$$\text{mean} = \frac{15 + 7 + 14 + 17 + 11 + 16}{6} = \frac{80}{6} = 13 \cdot 333 \dots$$

The mean = 13·3 correct to 1 decimal place (1 dp)

Exercise 7 ①

1 Find the mean of each set of data.
  a 3 6 9 5 8 11
  b 10 12 10 7 8 10 13
  c 3 3 6 7 1 4 5 5 2
  d 9 13 4 7 6
  e 1·3 0·6 1·4 3·7 0·4 1·6 0·9 0·5
  f 4 4 4 4 4 4 4 4 4 4 4 4 4 4

> You can use a calculator to find a mean. The button may look like this ($\bar{x}$).

2 In a test, the girls' marks were   8, 6, 7, 9, 5
      and the boys' marks were   6, 7, 9, 5, 8, 5.
Find the mean mark for the girls and the mean mark for the boys.

3 Mr and Mrs Amrit and their three children weigh 84 kg, 55 kg, 48 kg, 25 kg, and 9 kg.
Find the mean weight of the Amrit family.

4 **a** Write three numbers so that the mean is 5.
  b Write five numbers so that the mean is 8.
  c Write five numbers so that the mean is 6 and the median is 7.

5 Copy and complete.
  a The mean of 3, 5, 6 and ☐ is 6.
  b The mean of 7, 8, 4 and ☐ is 8.

6 **a** Calculate the mean of   2 5 7 7 4.
  b **Hence** find the mean of   32 35 37 37 34.
    (Try to do this **without** adding 32 + 35 + 37 + 37 + 34.)

7 Oliver's hobby is collecting crabs. As an experiment he decides to weigh each member of his collection as it crawls across a scale pan.
Their masses (in grams) were
5·2, 6·9, 2·7, 10·1, 3·6, 8·7, 2·7, 7·5
Find the mean mass of the crabs, correct to 1 decimal place.

8 The total mass of five men was 350 kg. Calculate the mean mass of the men.

9 The total height of 7 children is 1127 cm. Calculate the mean height of the children.

4.2.4 The range

The **range** is a measure of how spread out the data is. It is useful when comparing sets of data.

> ▶ The **range** is not an average. It is the difference between the largest value and the smallest value.
> For example: the range of the numbers 3, 7, 11, 12, 16 is 16 − 3 = 13.

Exercise 8 ①

1 The marks in a test were 2 2 3 5 7 7 7 10.
Copy and complete this sentence.
The range of the marks is 10 − ☐ = ☐

2 Find the range for each set of data.
  a 4 7 8 14 20 30
  b 1 2 2 2 4 40 50
  c 11 14 16 25 32 44 52

3 The temperature in a greenhouse was measured at midnight every day for a week. The results (in °C) were 1, −2, 4, 0, 5, 2, −1.
What was the range of the temperatures?

4 **a** Write six numbers so that the range is 10.
b Write five numbers so that the median is 7 and the range is 12.

5 There were 9 children at a party, including one up in the air. The mean age of the children was 15 and the range of their ages was 5.
Write each sentence below and write next to it whether it is **True**, **Possible** or **Impossible**.
a Every child was currently 15 years old.
b All the children were at least 13 years old.
c The oldest child was 5 years older than the youngest child.

6 There were six people living in a house. The **median** age of the people was 20 and the range of their ages was 3.
Write each sentence below and write next to it whether it is **True**, **Possible** or **Impossible**.
a Every person was either 19 or 20 years old.
b The oldest person in the house was 23 years old.
c All six people in the house could speak French.

7 The bar chart shows the scores for some children in a spelling test. The scores are collected together in groups.
a How many children took the test?
b What is the modal class?
c Explain why it is not possible to find the exact range for the scores in this test.

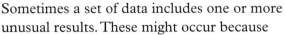

Sometimes a set of data includes one or more unusual results. These might occur because

● an error was made in collecting the data
● a rare event just happens to occur by chance.

A result which is distant from other results is called an **outlier**.
Whoever is analysing the data has to decide whether to include the outlier in the data set or whether to ignore it completely.

8 For each data set:
   i   identify any outliers
   ii include the outlier and find the range
   iii exclude the outlier and find the range
   iv say whether you think it is better to include or exclude the
          outlier from the data.
          Give the reasons.

   a The hours of sunshine each day in a week.
          2 3 4 6 7 7 15
   b The age of children invited to a birthday party.
          3 7 8 8 8 9 9
   c The number of passengers on a single-decker bus in the rush hour.
          29 36 38 40 42 42 67
   d The number of miles travelled by different cars on one litre of petrol.
          2 9 9 10 12 13 14 14
   e The number of brothers and sisters of each child in a class
          of nine children.
          0 0 1 1 1 2 2 3 9

Which is the 'best' average to use?

● The **median** is fairly easy to find and is not really affected by
  very large or very small values (outliers) that occur at the ends
  of the distribution.
  Look at these exam marks:
  20, 21, 21, 22, 23, 23, 25, 27, 27, 27, 29, 98, 98
                            ↑
  The median (25) gives a truer picture of the centre of the
  distribution than the mean (35·7).

● The **mode** of the exam marks is 27. It is easy to find and it takes
  away some of the effects of extreme values. However, it does
  have disadvantages, particularly in data which has two 'most
  popular' values, and it is not widely used.

● The **mean** takes account of all of the data and is the 'average'
  which most people first think of. It does, of course, take a little
  longer to calculate than either the mode or the median.

Exercise 9 Mixed questions ❶

1 Find the mean, median and mode of these sets of numbers.
   a 3   12 4   6   8   5   4
   b 7   21 2   17 3   13 7   4   9   7   9
   c 12 1   10 1   9   3   4   9   7   9
   d 8   0   3   3   1   7   4   1   4   4
   e Which is the best average to use in part **a**? Give your reason.

2 The temperature in °C on 17 days was:
1, 0, 2, 2, 0, 4, 1, 3, 2, 1, 2, 3, 4, 5, 4, 5, 5
  a What was the modal temperature?
  b What was the range?

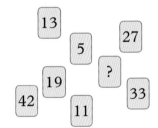

3 For this set of numbers, find the mean and the median.

> 2, 2, 3, 3, 3, 5, 101

Which average best describes the set of numbers?

4 Find
  a the range
  b the mode of these data.

4	2	5	4	5	12	4	1	3	3
3	5	5	3	4	5	2	7	4	2
4	12	1	7	12	1	3	8	10	4

5 The range for the eight numbers on these cards is 40.
Find the **two** possible values of the missing number.

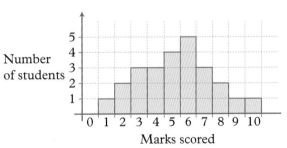

6 The mean weight of ten people in a lift is 70 kg.
The weight limit for the lift is 1000 kg. Roughly
how many more people can get into the lift?

7 The bar chart shows the marks scored
in a test. What is the modal mark?

8 Six boys have heights of 1·53 m, 1·49 m, 1·60 m, 1·65 m,
1·90 m and 1·43 m.
  a Find the mean height of the six boys.
  b Find the mean height of the remaining five boys when
    the shortest boy leaves.

9 In a maths test, the marks for the boys were 9, 7, 8, 7, 5 and
the marks for the girls were 6, 3, 9, 8, 2, 2.
  a Find the mean mark for the boys.
  b Find the mean mark for the girls.
  c Find the mean mark for the whole class.

10 Five different drinks have an average (mean) price of £1·50.
When a sixth drink is added the new average price is £1·60.
How much did the sixth drink cost?

11 Write five numbers so that
    the mean is 6
    the median is 5
    the mode is 4.

12 There were ten cowboys in a saloon. The mean age of the men was 25 and the range of their ages was 6.
Write each statement below and then write next to it whether it is 'true', 'possible' or 'false'.
  a The youngest man was 18 years old.
  b All the men were at least 20 years old.
  c The oldest person was 4 years older than the youngest.
  d Every man was between 20 and 26 years old.

13 These are the salaries of 5 employees in a small business.
Mr A: £22 500   Mr B: £17 900   Mr C: £21 400
Mr D: £22 500   Mr E : £155 300
  a Find the mean and the median of their salaries.
  b Which of these two averages does **not** give a fair 'average'? Explain why in one sentence.

14 A farmer has 32 cattle to sell.
The weights of the cattle in kg are

81	81	82	82	83	84	84	85
85	86	86	87	87	88	89	91
91	92	93	94	96	150	152	153
154	320	370	375	376	380	381	390

(Total weight = 5028 kg)

On the telephone to a potential buyer, the farmer describes the cattle and says the 'average' weight is 'over 157 kg'.
  a Find the mean weight and the median weight.
  b Which 'average' has the farmer used to describe his animals?
Does this average describe the cattle fairly?

4.2.5 Calculating the mean from a frequency table

EXAMPLE

The frequency table shows the weights of the eggs, to the nearest gram, bought in a supermarket.
Find the mean, median and modal weight.

Weight	58 g	59 g	60 g	61 g	62 g	63 g
Frequency	3	7	11	9	8	2

Mean weight of eggs

$$= \frac{(58 \times 3) + (59 \times 7) + (60 \times 11) + (61 \times 9) + (62 \times 8) + (63 \times 2)}{(3 + 7 + 11 + 9 + 8 + 2)} = \frac{2418}{40}$$

$= 60{\cdot}45 \text{ g}$

There are 40 eggs so the median weight is the number between the 20th and 21st numbers. You can see that both the 20th and 21st numbers are 60 g.
So, median weight = 60 g

The modal weight is the weight with the highest frequency.
The modal weight = 60 g

Exercise 10 ①

1 The frequency table shows the weights of the 40 apples sold in a shop.

Weight	70 g	80 g	90 g	100 g	110 g	120 g
Frequency	2	7	9	11	8	3

Calculate the mean weight of the apples.

2 The frequency table shows the price of a packet of butter in 30 different shops.

Price	49 p	50 p	51 p	52 p	53 p	54 p
Frequency	2	3	5	10	6	4

Calculate the mean price of a packet of butter.

3 A box contains 50 nails of different lengths as shown in the frequency table.

Length of nail	2 cm	3 cm	4 cm	5 cm	6 cm	7 cm
Frequency	4	7	9	12	10	8

Calculate the mean length of the nails.

4 The tables give the distribution of marks obtained by two classes in a test. For each class, find the mean, median and mode.

a Class 1

Mark	0	1	2	3	4	5	6
Frequency	3	5	8	9	5	7	3

b Class 2

Mark	15	16	17	18	19	20
Frequency	1	3	7	1	5	3

5 A teacher conducted a mental arithmetic test for 26 students and the marks out of 10 are shown in the table.

Mark	3	4	5	6	7	8	9	10
Frequency	6	3	1	2	0	5	5	4

a Find the mean, median and mode.

b The teacher congratulated the class saying that 'over three-quarters were above average'. Which average must the teacher be using?

*6 The table shows the number of goals scored in a series of football matches.

Number of goals	1	2	3
Number of matches	8	8	x

a If the modal number of goals is 3, find the smallest possible value of x.

b If the median number of goals is 2, find the largest possible value of x.

4.2.6 Data in groups

Results that are close together in value can be grouped in classes.
The data set is shown in a **grouped frequency table**.

EXAMPLE

The results of 51 students in a test are given in the frequency table.
Find the **a** mean **b** median **c** mode.

Mark	30–39	40–49	50–59	60–69
Frequency	7	14	21	9

....

a To find the mean, you approximate by saying each interval is represented by its midpoint. For the 30–39 interval, the midpoint is $(30 + 39) \div 2 = 34 \cdot 5$.

$$\text{Mean} = \frac{(34 \cdot 5 \times 7) + (44 \cdot 5 \times 14) + (54 \cdot 5 \times 21) + (64 \cdot 5 \times 9)}{7 + 14 + 21 + 9}$$

$$= 50 \cdot 774\,509\,8$$

$$= 51 \ (2 \text{ sf})$$

b The median is the 26th mark, which is in the interval 50–59. You cannot find the exact median.

c The **modal class** is 50–59. You cannot find an exact mode.

Don't forget that the mean is only an **estimate** because you do not have the raw data. You have made an assumption by using the midpoint of each interval.

Exercise 11 ①

1 The table gives the number of words in each sentence of a page in a book.
  a Copy and complete the table.
  b Work out an estimate for the mean number of words in a sentence.

Number of words	Frequency f	Midpoint x	fx
1–5	6	3	18
6–10	5	8	40
11–15	4		
16–20	2		
21–25	3		
Totals	20	—	

> You can often use a calculator for this calculation. The button may look like this $f(\bar{x})$.

2 The results of 24 students in a test are given in the table.

Mark	40–54	55–69	70–84	85–99
Frequency	5	8	7	4

  Find the midpoint of each group of marks and calculate an estimate of the mean mark.

3 The table shows the number of letters delivered to the 26 houses in a street.

Number of letters delivered	Number of houses (frequency)
0–2	10
3–4	8
5–7	5
8–12	3

  Calculate an estimate of the mean number of letters delivered per house.

4.3 Representing data

4.3.1 Types of data

- Data that can be placed into different non-numerical categories (colours of cars, names of students) is **categorical data**.
- Data that is numerical and found by counting (goals in a football match) is **discrete data**.
- Data that is numerical and found by measuring (the weight of athletes) is **continous data**.

Raw data can be collected from a whole **population** (people in a small village) using a census, or from a **sample** of population (people in a city) using a questionnaire. It can also be sampled by observation (colours of cars) or by experiment (weights of bags of potatoes). This kind of data is called **primary data**.

Secondary data is taken from other documents, records or the internet. It has been compiled by someone else.

Data can be gathered together into categories or classes and represented using tables, charts and graphs. Common methods use frequency tables, pictograms, bar charts, line charts, pie charts and lines graphs.

Exercise 12 ①

1 Would you collect data from the whole population or just a sample for these situations?

Population	**Data to collect**
a Members of your class in school	Their birthdays
b Trees in your school grounds	The distance around their trunks
c Trees in a forest	The distance around their trunks
d Passengers on a coach	The cost of their tickets
e Sheep on a hill farm	Their weights

2 Discuss and list the advantages and disadvantages of collecting data from
   a the whole population   **b** a sample of the population.

3 Is the data in this list categorical, discrete or continuous?
   a colour of socks worn by students in your class
   b the age of members of your family
   c runs in a cricket match
   d the score on a dice
   e the time to run a marathon
   f the height of basketball players
   g names given to new-born babies
   h the cost of an apple.

4 Is the data in each of these situations primary data or secondary data?
   a Male and female shoppers at a supermarket. Data collected by counting people as they enter.
   b Local house prices. Data collected by research on the internet and in local newspapers.
   c Amount of pocket money given to 12-year olds. Data collected by asking Year 8 students.
   d Prices of second hand cars. Data collected by visiting forecourts of garages.
   e Popularity of different airports with holidaymakers. Data collected by writing to airports for information on departures.

4.3.2 Pictograms

▶ In a pictogram you represent the frequency by a simple visual symbol that is repeated. A key is needed to say what the symbol stands for.

For example, this pictogram shows how many pizzas were sold on four days.

In the pictogram

  12 pizzas were sold on Monday
  10 pizzas were sold on Tuesday
   8 pizzas were sold on Wednesday
  17 pizzas were sold on Thursday.

The main problem with a pictogram is showing fractions of the symbol, which can sometimes only be approximate.

Key: ◯ represents 4 pizzas

Exercise 13 ①

1 The pictogram shows the money spent at a school tuck shop by four students.
  a Who spent most?
  b How much was spent altogether?
  c How would you show that someone spent 50p?

Key: £ represents £1

2 The pictogram shows the make of cars in a car park.
  a How many cars does the 🚗 represent?

Make	Number of cars	
Ford	4	🚗 🚗
Renault		🚗 🚗 🚗
Toyota	6	
Audi		🚗 🚗

  b Copy and complete the pictogram and draw a key.

3 The frequency table shows the number of letters posted to six houses one morning. Draw a pictogram to represent these data.

House 1	House 2	House 3	House 4	House 5	House 6
5	3	2	7	1	4

4.3.3 Pie charts

▶ In a pie chart all the data are represented by a circle and the sectors of the circle represent the different items.
▶ The angle of each sector is proportional to the frequency of that item.

Exercise 14

The questions in this exercise will remind you how to cancel fractions and how to find a fraction of a number. In questions **1** to **20**, simplify the fractions.

1 $\dfrac{20}{24}$ **2** $\dfrac{36}{48}$ **3** $\dfrac{45}{90}$ **4** $\dfrac{32}{40}$ **5** $\dfrac{21}{35}$

6 $\dfrac{15}{18}$ **7** $\dfrac{24}{30}$ **8** $\dfrac{120}{360}$ **9** $\dfrac{90}{360}$ **10** $\dfrac{150}{360}$

11 $\dfrac{180}{360}$ **12** $\dfrac{240}{360}$ **13** $\dfrac{300}{360}$ **14** $\dfrac{60}{360}$ **15** $\dfrac{45}{360}$

16 $\dfrac{40}{360}$ **17** $\dfrac{80}{360}$ **18** $\dfrac{210}{360}$ **19** $\dfrac{35}{360}$ **20** $\dfrac{54}{360}$

> Reminder:
> $\dfrac{15}{20} = \dfrac{3}{4}$
> Divide 15 and 20 by 5.

Work out these.

21 $\dfrac{1}{4}$ of 48   **22** $\dfrac{3}{4}$ of 60   **23** $\dfrac{2}{3}$ of 60

24 $\dfrac{2}{5}$ of 50   **25** $\dfrac{1}{8}$ of 88   **26** $\dfrac{3}{7}$ of 84

27 $\dfrac{2}{3}$ of 360°   **28** $\dfrac{5}{6}$ of 360°   **29** $\dfrac{3}{8}$ of 360°

30 $\dfrac{1}{12}$ of 360°   **31** $\dfrac{7}{10}$ of 360°   **32** $\dfrac{5}{12}$ of 360°

> Reminder:
> $\dfrac{3}{4}$ of 80
> Find $\dfrac{1}{4}$ of 80 and then multiply by 3.

EXAMPLE

The pie chart shows how 600 people are spending their summer holiday. How many people are
a camping   **b** touring   **c** going to the seaside?

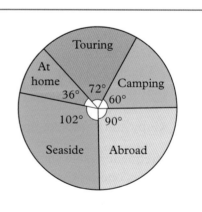

a Number of people camping $= \dfrac{60}{360} \times 600$

$\qquad\qquad\qquad\qquad\qquad = \dfrac{1}{6} \times 600$

$\qquad\qquad\qquad\qquad\qquad = 100$

b Number of people touring $= \dfrac{72}{360} \times 600$

$\qquad\qquad\qquad\qquad\qquad = \dfrac{1}{5} \times 600$

$\qquad\qquad\qquad\qquad\qquad = 120$

c Number of people at seaside $= \dfrac{102}{360} \times 600$

$\qquad\qquad\qquad\qquad\qquad\quad = 170$

Exercise 15 ①

1 The pie chart shows the country of origin of 120 passengers
   on board an aircraft.

   a Simplify these fractions.

   i $\dfrac{60}{360}$ **ii** $\dfrac{120}{360}$ **iii** $\dfrac{45}{360}$ **iv** $\dfrac{135}{360}$

   b What **fraction** of the passengers came from
   i France   **ii** USA
   iii Canada   **iv** UK?

   c Work out these.

   i $\dfrac{1}{6}$ of 120   **ii** $\dfrac{1}{3}$ of 120   **iii** $\dfrac{1}{8}$ of 120

   d How many passengers came from
   i France **ii** USA   **iii** Canada **iv** UK?

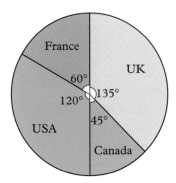

2 The colour of each of the 36 cars in a car park was
   noted and the results are shown in the pie chart.
   a Simplify these fractions.

   i $\dfrac{90}{360}$ **ii** $\dfrac{60}{360}$ **iii** $\dfrac{120}{360}$

   b What **fraction** of the cars were
   i red   **ii** green
   iii yellow   **iv** blue?

   c How many of the 36 cars were
   i red   **ii** green
   iii yellow   **iv** blue?

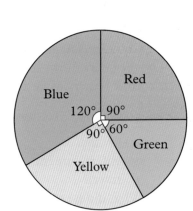

3 The pie chart shows how Simon spends £360 on
   various items in one week.
   a What **fraction** of his money did Simon spend on
   i food   **ii** fares
   iii rent   **iv** savings
   v entertainment **vi** clothes?

   b How much of the £360 did he spend on
   i food   **ii** fares
   iii rent   **iv** savings?

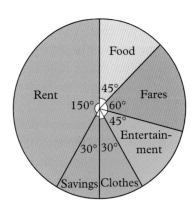

4 The total cost of a holiday was £900.
The pie chart shows how this cost was made up.

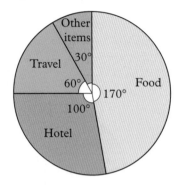

How much was spent on:
a food
b travel
c the hotel
d other items?

5 Mr Choudry had an income of £60 000.
The pie chart shows how he used the money.
 a How much did he spend on
  i food
  ii rent
  iii savings
  iv entertainment
  v travel?
 b What did he spend the most money on?
 c What did he spend the least money on?

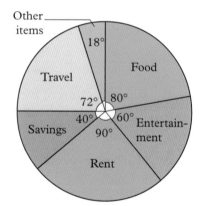

6 The pie chart shows how Sandra spends her
time in a maths lesson which lasts 60 minutes.
 a How much time does Sandra spend
  i getting ready to work
  ii talking
  iii sharpening a pencil?
 b Sandra spends 3 minutes working.
 What is the angle on the pie chart for
 the time spent working?

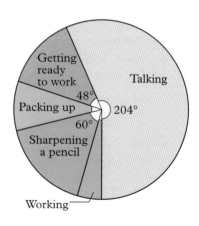

Exercise 16 ①

1 At the semi-final stage of the F.A. Cup, 72 neutral referees were asked to predict who they thought would win. Their answers were

Spurs	9	Chelsea	40
Manchester United	22	York City	1

a Work out

  i $\dfrac{9}{72}$ of 360°   **ii** $\dfrac{40}{72}$ of 360°

  iii $\dfrac{22}{72}$ of 360°   **iv** $\dfrac{1}{72}$ of 360°

b Draw an accurate pie chart to display the predictions of the 72 referees.

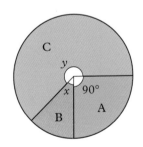

2 A survey was carried out to find what 400 students did at the end of Year 11.

> 120 went into the sixth form
> 160 went into employment
> 80 went to F.E. colleges
> 40 were unemployed.

a Simplify these fractions: $\dfrac{120}{400}; \dfrac{160}{400}; \dfrac{80}{400}; \dfrac{40}{400}$.

b Draw an accurate pie chart to show the information above.

3 In a survey on washing powder, 180 people were asked to state which brand they preferred.
45 chose Brand A.

If 30 people chose brand B and 105 chose Brand C, calculate the angles x and y.

4 A packet of breakfast cereal weighing 600 g contains four ingredients.
Calculate the angles on the pie chart and draw an accurate diagram.

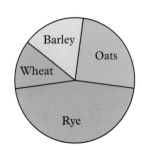

Oats	150 g
Barley	100 g
Wheat	75 g
Rye	275 g

5 The students at a school were asked to state their favourite colour. Here are the results.

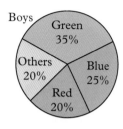

There were 100 boys   There were 60 girls

Tony says 'The same number of boys and girls chose red.'

Mel says 'More boys than girls chose blue.'

a Use both charts to explain whether or not Tony is right.

b Use both charts to explain whether or not Mel is right.

4.3.4 Bar charts and vertical line charts

On a **bar chart**, the vertical axis is always used to show the frequency. If you replace the columns by a line joining the midpoints of the tops of the columns, then you make a **frequency polygon**.

EXAMPLE

A shopper buys a bag of 40 pea pods. The pods are opened and the number of peas in each pod is counted with the results recorded in a frequency table.

Number of peas	3	4	5	6	7
Frequency	4	6	7	13	10

Show the data on a bar chart and frequency polygon.

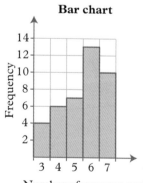 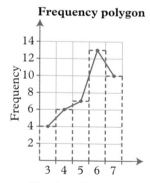

Note that the modal number of peas per pod is easy to find. It is the number that has the greatest frequency or the tallest column. In this case, the mode is 6.

Exercise 17 ❶

1 A survey of students in a class investigates the
   favourite meats they like to eat.

   a What is the mode of the data?

   b Which meat is the least favourite?

   c How many students are in the class?

   d What fraction of the students do not eat meat?

   e Use this chart to draw a frequency polygon of
        the data.

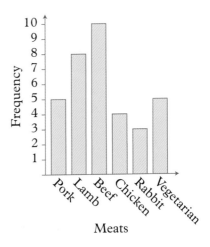

2 This table shows the number of mobile phones sold in a town over six months.

Month	January	February	March	April	May	June
Number sold	200	250	180	110	120	140

   a Draw a bar chart to display this data.

   b What is the total number of phones sold in this time?

   c What percentage of the total is sold during January to March?

   d Draw a frequency polygon of the data.

3 This bar chart gives the number of boys
   and girls in each year of a secondary
   school.

   a In which years are there more boys
        than girls?

   b How many girls are there altogether
        in the first three years of the school?

   c Which year below Year 12 has fewest
        students ?

   d If each Year 11 student on average
        sits six exams. What is the total
        number of exam entries for Year 11?

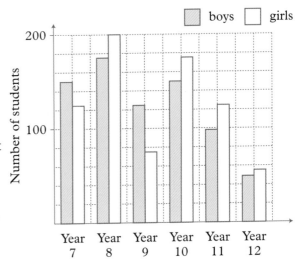

4 People going to work in a city used these types
of transport one morning in summer.

Transport	Car	Bus	Train	Cycle
Numbers (× 1000)	325	750	420	5

a Draw a bar chart and a frequency polygon of the data.
b What is the modal form of transport used by the
commuters?
c What percentage of people used the bus to go to work?

Instead of drawing columns, you can draw vertical lines to
represent the data and their frequencies. The height of the
vertical line is the same as the height of the corresponding
column. This is called a **vertical line chart**.
Here is the example of the shopper and the 40 pea pods.

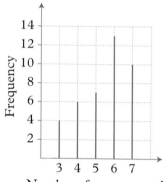

5 This vertical line chart shows the frequency of different
colours of hair in a sample of people in a hairdresser's shop.
a Copy and complete this table.

Hair colour	Blonde	Auburn	Brown	Black	Grey
Frequency					

b What is the modal colour?
c How many people are included in the sample?
d What percentage of the sample has blonde hair?

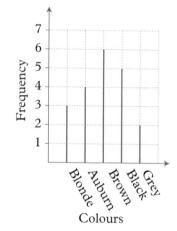

6 This table shows the number of passengers, excluding
the driver, in the cars driven into a city centre.

Number of passengers	0	1	2	3	4
Number of cars	30	20	5	3	2

a Draw a vertical line chart to show this data.
b How many cars are in this survey?
c What is the modal number of passengers?
d What percentage of cars carries no passengers?
e Should the local council encourage people to share cars?
What evidence can you give to support your answer?
Can you suggest what the council might do?

7 The sale of second-hand cars by Mr Rusty Wheeler over the last five years is shown in this table.

Year	2011	2012	2013	2014	2015
Number of cars	2600	2500	3200	3700	3000

   a Construct a vertical line chart to represent this data.
   b Write down the mode of the data.
   c What is the total number of cars sold?
   d What percentage of the total was sold in 2015?

4.3.5 Two-way tables

A school collects data about the reading ability of six-year-olds. Children are given a short passage and they 'pass' if they can read over three-quarters of the words. The results are given in this two-way table.

	Boys	Girls	Total
Can read	464	682	
Cannot read	317	388	
Total			

Useful information can be found from the table. Begin by working out the totals for each row and column. Find also the 'grand total' by adding together the totals in **either** the rows **or** columns.

	Boys	Girls	Total
Can read	464	682	1146
Cannot read	317	388	705
Total	781	1070	1851

↑
'Grand total'

EXAMPLE

Answer these questions using the above two-way table.
a What percentage of the boys can read?
b Similarly, what is the percentage of girls who can read?

....

a Out of 781 boys, 464 can read.
So percentage of boys who can read = $\dfrac{464}{781} \times 100\%$
$= 59 \cdot 4\%$ (1 dp)

b Out of 1070 girls, 682 can read.
So percentage of girls who can read = $\dfrac{682}{1070} \times 100\%$
$= 63 \cdot 7\%$ (1 dp)
So a slightly higher percentage of the girls can read.

Exercise 18 ①

Give percentages correct to one decimal place.

1 Here are ten shapes.

Copy and complete this two-way table to show the different shapes.

	Shaded	**Unshaded**
Squares		
Triangles		

2 The students in a class were asked to name their favourite sport.
Here are the results.

Boy	Football	Boy	Swimming	Boy	Football
Girl	Hockey	Boy	Football	Girl	Football
Girl	Football	Girl	Hockey	Girl	Swimming
Boy	Hockey	Girl	Swimming	Boy	Football
Boy	Football	Girl	Hockey	Boy	Hockey

a Record the results in a two-way table.
b How many boys were in the class?
c What percentage of the boys chose hockey?

3 This incomplete two-way table shows details of seven year-olds
who can/cannot ride a bicycle without stabilisers.

	Girls	**Boys**	**Total**
Can cycle	95		215
Cannot cycle		82	
Total			476

a Copy the table and work out the missing numbers.
b What percentage of the girls can cycle?
c What percentage of the boys can cycle?

4 Mrs Kotecha collected these data to help
assess the risk of various drivers who apply
for car insurance with her company. She
needs to know if men drivers are more or
less likely than women drivers to be involved
in motor accidents.

	Men	**Women**	**Total**
Had accident	75	88	
Had no accident	507	820	
Total			

a Copy the table and work out the totals for the rows and columns.
b What percentage of the men had accidents?
c What percentage of the women had accidents?
d What conclusions, if any, can you draw?

4.3.6 Times series

Some quantities, such as the price of houses, vary with time. A **time series** is a set of data collected at equal intervals over a period of time. The data can be represented as a **line graph**. Note that a line graph is made from a series of line segments rather from one continuous curve. The line graph is useful for showing whether there is a **trend** in the data.

EXAMPLE

This table shows the number of Christmas cards delivered to a block of flats over the ten days before Christmas Day.

a On which day were most cards delivered?

b How many cards were delivered in the last week?

c Is there any trend in the deliveries? Describe it.

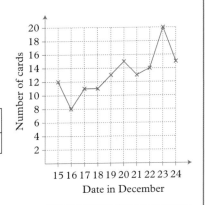

Date in December	15	16	17	18	19	20	21	22	23	24
Number of Cards	12	8	11	11	13	15	13	14	20	15

---

a December 23rd

b 11 + 13 + 15 + 13 + 14 + 20 + 15 = 101 cards

c Overall, there is an upward trend over the ten days.

Exercise 19 ②

1 The amount of money in Gary's bank account at the end of each month last year is shown on this line graph.

a At the end of which month was the balance in the account greatest?

b Describe the trend in the balances over the period January to July.

c In which two months did Gary spend most? Why do you think these drops in his balance happened?

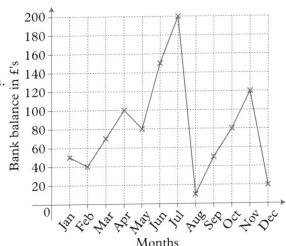

2 A patient is admitted to hospital and her temperature is measured every 4 hours. Normal body temperature is 98.4°F (on the Fahrenheit scale). Draw a line graph of the data.

Hours after admittance	0	4	8	12	16	20	24	28	32
Temperature, °F	104	105	102	100	101	102	99	98	99

a Her parents visited her two hours after she was admitted. What was her temperature then?

b How long did it take for her temperature to fall to normal?

c How many degrees above normal was her highest temperature?

d Describe the trend in her temperature over these 32 hours.

3 A school's weather station recorded these temperatures at 12 noon over one week of the summer term. Draw a line graph of the data.

Day	Mon	Tues	Wed	Thur	Fri	Sat	Sun
Temperature, °C	18	22	21	18	18	20	19

a What is the modal temperature during this week?

b Find the median temperature and calculate the mean temperature.

c Show the mean temperature on the graph by a horizontal dotted line.

d Describe the trend in the temperature.

4.3.7 Stem-and-leaf diagrams

▶ You can display data in a stem-and-leaf diagram.

Here are the marks of 20 girls in a science test.

54   42   61   47   24   43   55   62   30   27
28   43   54   46   25   32   49   73   50   45

Put the marks into groups 20–29, 30–39,… 70–79.
Choose the tens digit as the 'stem' and the units digit as the 'leaf'.

The first four marks are shown.

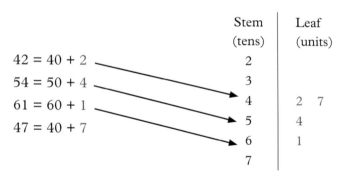

Key: 4|2 means 42

Here are the results … and then with the leaves in numerical order.

Stem	Leaf
2	4 7 8 5
3	0 2
4	2 7 3 3 6 9 5
5	4 5 4 0
6	1 2
7	3

Stem	Leaf
2	4 5 7 8
3	0 2
4	2 3 3 5 6 7 9
5	0 4 4 5
6	1 2
7	3

The diagram shows the shape of the distribution.
It is also easy to find the mode, the median and
the range from a stem-and-leaf diagram.

> The highest mark is
> 73 and the lowest
> mark is 24.
> The **range** of the
> marks is 73 − 24=49.

Back-to-back stem plots

Two sets of data can be compared using a **back-to-back stem plot**.
The two sets of leaves share the same stem.
Here are the marks of 20 boys who took the same science
test as the girls.

33	55	63	74	20	35	40	67	21	38
51	64	57	48	46	67	44	59	75	56

These marks are entered onto a back-to-back stem plot.

Boys	Stem	Girls
1 0	2	4 5 7 8
8 5 0	3	0 2
8 6 4 0	4	2 3 3 5 6 7 9
9 7 6 5 1	5	0 4 4 5
7 7 4 3	6	1 2
5 4	7	3

| It is helpful to have a key. | → | Key (boys) 1∣5 means 51. | | Key (girls) 2∣4 means 24. |

You can see that, overall, the boys achieved higher marks than
the girls in this test.

Exercise 20 ①

1 The marks of 24 students in a test were

41	23	35	15	40	39	47	29
52	54	45	27	28	36	48	51
59	65	42	32	46	53	66	38

Copy and complete the stem-and-leaf diagram.

Stem	Leaf
1	
2	3
3	
4	1
5	
6	

Key: 4 | 1 means 41

2 Draw a stem-and-leaf diagram for each set of data.

a 24 52 31 55 40 37 58 61 25 46
44 67 68 75 73 28 20 59 65 39

b 30 41 53 22 72 54 35 47
44 67 46 38 59 29 47 28

Stem	Leaf
2	
3	
4	
5	
6	
7	

3 Here is the stem-and-leaf diagram showing the masses, in kg, of some people in a lift.

a Write the range of the masses.

b How many people were in the lift?

c What is the median mass?

Key: 3 | 2 means 32 kg

Stem (tens)	Leaf (units)
3	2 5
4	1 1 3 7 8
5	0 2 5 8
6	4 8
7	1
8	2

4 In this question, the stem shows the units digit and the leaf shows the first digit after the decimal point.
Draw the stem-and-leaf diagram using these data.

2·4 3·1 5·2 4·7 1·4 6·2 4·5 3·3
4·0 6·3 3·7 6·7 4·6 4·9 5·1 5·5
1·8 3·8 4·5 2·4 5·8 3·3 4·6 2·8

a What is the median?

b What is the range?

Stem	Leaf
2	
3	
4	
5	
6	
7	

Key: 3 | 7 means 3·7

5 Here is a back-to-back stem plot showing the pulse rates of several people.

a How many men were tested?

b What was the median pulse rate for the women?

c Write a sentence to describe the main features of the data.

Men		Women
5 1	4	
7 4 2	5	3
8 2 0	6	2 1
5 2	7	4 4 5 8 9
2 6	8	2 5 7
4	9	2 8

Key (men)
1 | 4 means 41

Key (women)
5 | 3 means 35

4.4 Comparing sets of data

To compare two sets of data we need to write at least two things:
- Compare an average (mean, median or mode)
- Compare the range of each set of data (this measures how spread out the data is).

EXAMPLE

The weights of the players in two basketball teams are (in kg):

Team A: 58, 60, 60, 68, 81, 83, 94
Team B: 62, 66, 66, 72, 74, 76, 79, 81
Team A: median weight = 68 kg;
      range = 94 − 58 = 36 kg
Team B: median weight = 73 kg;
      range = 81 − 62 = 19 kg

Compare the weights of the players in Team A and Team B.

···

The median weight for Team A is lower than that for Team B and the range for Team A is much greater than that for Team B. The weights of the players in Team A are more spread out.

Exercise 21 ①

1 The six members of the Pearce family weigh 39 kg, 48 kg, 52 kg, 56 kg, 58 kg and 59 kg.
The five members of the Taylor family weigh 35 kg, 41 kg, 63 kg, 64 kg and 75 kg.
  a For each family work out the median weight and the range of the weights.
  b Write a sentence, similar to the one in the example, to compare the weights of the people in the two families.

2 Two marks for two classes in a test are shown in the stem-and-leaf diagrams.
  a Find the median and range for each class.
  b Write a sentence to compare the marks for the two classes.

Class 10 M		Class 10 S	
5	6 7	4	1 2 4
6	0 1	5	5 8 8
7	3 4 6	6	2
8	2 7	7	1 7
		8	0 8

Key: 5 | 6 means 56

Test yourself

1   The table shows the favourite drinks of 25 girls.

Drink	Milk	Tea	Coffee	Juice
Number of girls	9	7	4	5

  a   Copy and complete the bar chart to show this information.

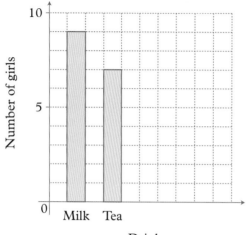

  b   The pictogram shows the favourite drinks of 25 boys.

Key ◹ = 2 boys

Milk

Tea

Coffee

Juice

How many **more** boys than girls prefer coffee?

(AQA, 2013)

2   Jim records how many text messages he receives each day for
ten days.
3   0   1   4   1   4   6   1   20   0
  a   Work out the median.
  b   Work out the mean.
  c   Which of these two averages better represents the data?
      Explain your answer.

(AQA, 2007)

3 Here are ten numbers.

7 6 8 4 5 9 7 3 6 7

a Work out the range.   **b** Work out the mean.

(Edexcel, 2008)

4 Dan leaves home at 0800.

He drives 60 miles from home in the first 90 minutes.

He stops for 30 minutes.

He then drives home at an average speed of 50 mph.

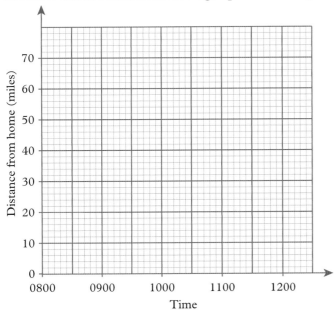

a Draw a distance-time graph to show Dan's journey.
Copy the axes above.

b A TV programme starts at 1130.
Does Dan get home in time for the start? Show how you decide.

(AQA, 2013)

5 The pie chart shows the sports played by 30 boys.

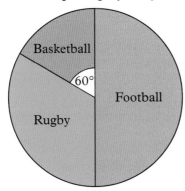

a How many boys play Football?

b How many boys play Rugby?

(AQA, 2013)

6 Here is a list of all the coins in Amira's purse.

£1	5p	20p	1p
20p	1p	10p	£1
20p	10p	£1	20p
10p	20p	20p	5p

Copy and complete the table for this information.

Coin	Tally	Frequency
£1		
50p		
20p		
10p		
5p		
2p		
1p		

(Edexcel, 2013)

7 The times, in minutes, taken by 15 women to complete a crossword are shown in this stem and leaf diagram.
  a Find the range of the women's times.
  b Find the median of the women's times.

	WOMEN
1	2 3 3 5 7 7 8 9
2	4 6 8 8
3	5 6 8

Key: 2 | 4 means 24 minutes

The women's times were then compared with the times of 15 men in this new stem and leaf diagram.

	WOMEN			MEN
1	2 3 3 5 7 7 8 9		1	2 4 4 6
2	4 6 8 8		2	1 3 3 3 9
3	5 6 8		3	7 8
4			4	3 4 5 6

Key: 2 | 4 means 24 minutes

  c The men say that they were quicker at completing the crossword.
    Do you agree? Give a reason for your answer.

(CCEA, 2013)

8 A group of students sat four separate tests, Test A, Test B, Test C and Test D, as part of their course.
Using the data from the marks scored in each of these tests the following sketch graphs were drawn. The same scales are used in each graph.

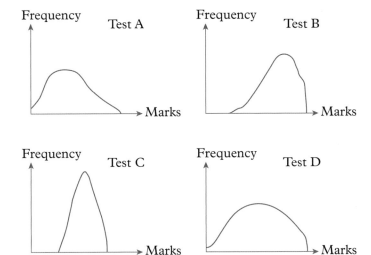

Copy and complete the following statements:

"Most of the students gained high marks in Test...........

"The range of students' marks was smallest in Test...........

"Most of the marks were low in Test...........

(WJEC, 2011)

9 **a** The lengths of Desmond's telephone calls, in minutes, are summarised in the table below.

Length of call (*t* minutes)	Number of calls		
$0 < t \le 10$	0		
$10 < t \le 20$	3		
$20 < t \le 30$	3		
$30 < t \le 40$	6		
$40 < t \le 50$	8		
$50 < t \le 60$	5		

Calculate an estimate of the mean length of Desmond's calls.

b The table below summarises the lengths, in minutes, of Harriet's calls in November and December.

	Mean	Range
November	34·2	67·4
December	39·7	43·8

i In which month were Harriet's calls longer on average? Explain how you decide.

ii In which month were the lengths of Harriet's calls more spread out? Explain how you decide.

(OCR, 2013)

5 Number 2

In this unit you will:
- revise written multiplication and division
- learn how to use fractions and percentages in calculations
- learn about ratio and proportion
- revise mental methods of calculation
- learn about rounding and estimating.

A map always needs a scale. The earliest large maps to scale in the U.K. were made in Scotland in 1747.

5.1 Long multiplication and division

5.1.1 Long multiplication

- To work out 327×53 use the fact that
  $327 \times 53 = (327 \times 50) + (327 \times 3)$.

  Set out the working like this   ... or like this

       327   327
        53 ×   53 ×
  16 350 → This is 327×50 981
      981 → This is 327×3 16 350
  17 331 → This is 327×53 17 331

  So $327 \times 53 = 17\,331$

Exercise 1 ❶

Work out these, without a calculator.

1 35×23	**2** 27×17	**3** 26×25	**4** 31×43
5 45×61	**6** 52×24	**7** 323×14	**8** 416×73
9 504×56	**10** 306×28	**11** 624×75	**12** 839×79
13 694×83	**14** 973×92	**15** 415×235	

5.1.2 Long division

With ordinary 'short' division, you divide and find remainders. The method for 'long' division is really the same but you set it out so that the remainders are easier to find.

EXAMPLE

Work out 736 ÷ 32

$$
\begin{array}{r}
23 \\
32\overline{)736} \\
-64\!\downarrow \\
\hline
96 \\
-96 \\
\hline
0
\end{array}
$$

32 into 73 goes 2 times
2 × 32 = 64
73 − 64 = 9 and 'bring down' 6
32 into 96 goes 3 times

Exercise 2 ❶

1 672 ÷ 21

2 425 ÷ 17

3 576 ÷ 32

4 247 ÷ 19

5 875 ÷ 25

6 574 ÷ 26

7 806 ÷ 34

8 748 ÷ 41

9 666 ÷ 24

10 707 ÷ 52

11 951 ÷ 27

12 806 ÷ 34

13 2917 ÷ 45

14 2735 ÷ 18

15 56 274 ÷ 19

EXAMPLE

Kieran can fit 23 matches into a box. How many boxes does he need for 575 matches?

$$
\begin{array}{r}
25 \\
23\overline{)575} \\
-46 \\
\hline
115 \\
-115 \\
\hline
0
\end{array}
$$

He needs 25 boxes.

Exercise 3 ❶

1 A shop owner buys 56 tins of paint at 84p each. How much does he spend altogether?

2 Eggs are packed eighteen to a box.

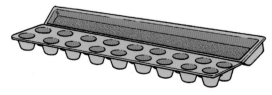 

How many boxes are needed for 828 eggs?

3 On average a man sells 146 newspapers a week. How many does he sell in a year?

4 Sally wants to buy as many 32p stamps as possible. She has £5 to buy them. How many can she buy and how much change is left?

5 How many 49-seater coaches will be needed for a school trip for a party of 366?

6 It costs £7905 to hire a plane for a day. A trip is organised for 93 people. How much does each person pay?

7 A lottery prize of £238 million was won by a syndicate of 17 people who shared the prize equally between them. How much did each person receive?

8 An office building has 24 windows on each of 8 floors. A window cleaner charges 42p for each window. How much is he paid for the whole building?

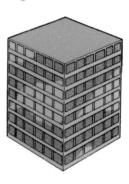

5.2 Fractions

5.2.1 What is a fraction?

▶ A fraction is a number that is less than a whole one. When a whole one is divided into equal parts, each of the parts is a fraction of the whole one.

2 equal parts

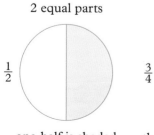

$\frac{1}{2}$

one-half is shaded

4 equal parts

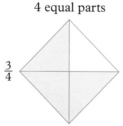

$\frac{3}{4}$

three-quarters is shaded

10 equal parts

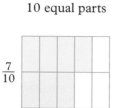

$\frac{7}{10}$

seven-tenths is shaded

Exercise 4 1

1 What fraction of each diagram is shaded?
   Write your answer in both words and figures.

a  **b**   **c** 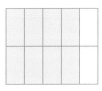 **d**

e  **f**   **g**   **h**

2 What fractions are shown by the arrows on each of these number lines?

a  **b** 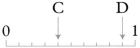 **c**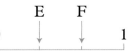

3 Draw three grids like this one and label them A, B, and C.

   In grid A, shade in $\frac{1}{4}$ of the squares.

   In grid B, shade in $\frac{1}{12}$ of the squares.

   In grid C, shade in $\frac{1}{3}$ of the squares.

5.2.2 Equivalent fractions

▶ **Equivalent fractions** are fractions that look different but are the same

For example

one-half $\frac{1}{2}$ = $\frac{2}{4}$ two-quarters

four-fifths $\frac{4}{5}$ = $\frac{8}{10}$ eight-tenths

▶ Changing a fraction into a simpler form is known as 'cancelling'.

This is also called simplifying.

Method: Find the highest number that divides exactly into both the numerator (top) and the denominator (bottom).

The fraction $\frac{15}{20}$ cancels to $\frac{3}{4}$.

$$\frac{15 \div 5}{20 \div 5} = \frac{3}{4} \qquad \frac{4 \div 4}{12 \div 4} = \frac{1}{3}$$

Exercise 5 ①

Use the fraction charts to copy and complete the equivalent fractions by putting the correct number in the box.

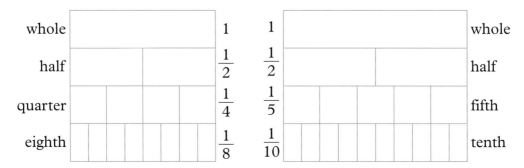

1 $\frac{1}{2} = \frac{\square}{4}$ **2** $\frac{1}{4} = \frac{\square}{8}$ **3** $\frac{1}{5} = \frac{\square}{10}$ **4** $\frac{3}{4} = \frac{\square}{8}$

5 $\frac{1}{2} = \frac{\square}{10}$ **6** $1 = \frac{\square}{4}$ **7** $\frac{3}{5} = \frac{\square}{10}$ **8** $\frac{4}{5} = \frac{\square}{10}$

9 Copy and complete these fraction chains.

     a $\frac{1}{2} = \frac{\square}{4} = \frac{\square}{8} = \frac{\square}{16}$ **b** $\frac{1}{2} = \frac{\square}{4} = \frac{\square}{6} = \frac{\square}{8} = \frac{\square}{10}$

Copy and cancel these fractions to their simplest form.

10 $\frac{9}{12}$ **11** $\frac{6}{24}$ **12** $\frac{8}{10}$ **13** $\frac{8}{20}$ **14** $\frac{9}{36}$

15 $\frac{8}{12}$ **16** $\frac{9}{15}$ **17** $\frac{6}{18}$ **18** $\frac{7}{21}$ **19** $\frac{32}{36}$

20 $\frac{24}{30}$ **21** $\frac{4}{12}$ **22** $\frac{4}{18}$ **23** $\frac{20}{30}$ **24** $\frac{14}{42}$

25 $\frac{6}{15}$ **26** $\frac{27}{45}$ **27** $\frac{56}{64}$ **28** $\frac{18}{30}$ **29** $\frac{28}{36}$

30 $\frac{18}{63}$ **31** $\frac{44}{55}$ **32** $\frac{24}{60}$ **33** $\frac{18}{72}$ **34** $\frac{75}{100}$

35 What fraction of £1 is 1p?

36 What fraction of £1 is 50p?

37 What fraction of one minute is 15 seconds?

38 What fraction of one hour is 10 minutes?

39 What fraction of £10 is £3?

40 What fraction of the months of the year begin with the letters J or A?

41 In a class of 30 children, 23 are right-handed.
What fraction are left-handed?

42 What fraction of the numbers from 1 to 30 contain the digit 9?

43 What fraction of these numbers are greater than ten?

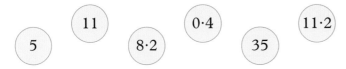

44 Sam and Sue are given a total of £120. Sam gets £75.
What fraction of the total does each of them get?

45 Express the first number as a fraction of the second number.
    a   20 as a fraction of 75   **b**   32 as a fraction of 80
    c   48 as a fraction of 96   **d**   48 as a fraction of 64
    e   180 as a fraction of 120   **f**   75 as a fraction of 60

46 June, July, August are the summer months. What proportion of the
year is summer?

47 A tin of biscuits has 32 chocolate and 24 plain ones.
What fraction of the tin is plain?

48 A history class has 21 boys and 14 girls. What fraction of
the class is girls? What fraction is boys?

49 What proportion of the whole numbers from 1 to 20 are prime numbers?

50 You have two scarves. The red scarf is 100 cm long and
the blue scarf is 80 cm long. The red scarf is one-and-a-quarter
times longer than the blue scarf because $\dfrac{100}{80} = \dfrac{5}{4} = 1\dfrac{1}{4}$.
You also have two belts. The black belt is 120 cm long and
the brown belt is 80 cm long. How many times longer is the
black belt than the blue belt?

51 Alan and Ben are brothers. Alan is 150 cm tall and Ben is 100 cm tall. How many times taller is Alan than Ben?

52 You invest £600 into a saving account and, one year later, the account is has £750 in it, How many times more have you now than you had at the start?

53 It is 60 miles from Aston to Weston. A diversion due to road works increases the journey to 80 miles.
   a How many times longer is the journey now?
   b What is the ratio of the direct distance to the longer diversion, in its simplest terms?

54 Jan has two pairs of socks. The short socks are 12 cm long. The long socks are 30 cm long.
   a How many times longer are the long socks than short socks?
   b What is ratio of the length of the long socks to the short socks, in its simplest terms?

55 Here are some number cards.

   a Use two cards to make a fraction which is equal to $\frac{1}{2}$.

   b Use three of the cards to make the smallest possible fraction.

*56 a Copy and complete. $\frac{2}{3} = \frac{\square}{12}$ $\frac{1}{2} = \frac{\square}{12}$ $\frac{3}{4} = \frac{\square}{12}$.

   b Hence write $\frac{2}{3}, \frac{1}{2}, \frac{3}{4}$ in order of size, smallest first.

*57 a Copy and complete $\frac{5}{6} = \frac{\square}{12}$, $\frac{2}{3} = \frac{\square}{12}$, $\frac{1}{4} = \frac{\square}{12}$.

   b Hence write $\frac{5}{6}, \frac{2}{3}, \frac{1}{4}$ in order of size, smallest first.

In questions **58** to **61**, write the fractions in order of size.

*58 $\frac{3}{5}, \frac{7}{10}, \frac{1}{2}$ ***59** $\frac{7}{12}, \frac{5}{6}, \frac{5}{4}$

*60 $\frac{1}{2}, \frac{3}{8}, \frac{3}{4}$ ***61** $\frac{7}{15}, \frac{1}{3}, \frac{2}{15}$

For harder fractions, use a calculator to change the fractions to decimals. Then write the fractions in order of size using their decimal equivalents.

*62 $\frac{3}{7}, \frac{4}{9}, \frac{3}{8}, \frac{2}{5}$ ***63** $\frac{7}{12}, \frac{2}{3}, \frac{8}{13}, \frac{11}{17}$

? 5.2.3 Proper and improper fractions

▶ A **proper fraction** is one in which the **numerator** (top number) is less than the **denominator** (bottom number).

$\dfrac{1}{2}$ ◄—numerator
$\phantom{\dfrac{1}{2}}$ ◄—denominator

The fractions $\dfrac{1}{2}, \dfrac{2}{3}, \dfrac{3}{4}$ and $\dfrac{99}{100}$ are all examples of proper fractions.

▶ An **improper fraction** is one in which the **numerator** is larger than the **denominator**. They are sometimes called 'top-heavy' fractions.

The fractions $\dfrac{3}{2}, \dfrac{4}{3}, \dfrac{8}{5}$ and $\dfrac{100}{33}$ are all examples of improper fractions.

▶ A **mixed number** contains both a whole number and a fraction. Improper fractions can be changed into mixed numbers and vice versa.

For example, $\dfrac{3}{2} = 1\dfrac{1}{2}$

EXAMPLE

a Change $\dfrac{3}{2}$ and $\dfrac{16}{3}$ into mixed numbers.

b Change $2\dfrac{1}{2}$ and $3\dfrac{5}{6}$ into improper fractions.

....

a $\dfrac{3}{2} = 1\dfrac{1}{2}$ 2 divides into 3 once, giving the whole number 1.

The remainder is 1 which is written as $\dfrac{1}{2}$.

$\dfrac{16}{3} = 5\dfrac{1}{3}$ 3 divides into 16 five times, giving the whole number 5.

The remainder is 1 which is written as $\dfrac{1}{3}$.

b $2\dfrac{1}{2} = \dfrac{5}{2}$ 2 times 2 gives 4 (4 halves).

Add 1 for the extra $\dfrac{1}{2}$ to 4 giving 5.

There are 5 halves altogether.

$3\dfrac{5}{6} = \dfrac{23}{6}$ 3 times 6 is 18 (18 sixths).

There are 5 sixths to add to these 18 sixths giving 23.

There are 23 sixths altogether.

Exercise 6 ①

Change these improper fractions to mixed numbers or whole numbers.

1 $\frac{7}{2}$	**2** $\frac{5}{3}$	**3** $\frac{7}{3}$	**4** $\frac{5}{4}$	**5** $\frac{8}{3}$
6 $\frac{8}{6}$	**7** $\frac{9}{3}$	**8** $\frac{9}{2}$	**9** $\frac{9}{4}$	**10** $\frac{10}{2}$
11 $\frac{10}{6}$	**12** $\frac{10}{7}$	**13** $\frac{13}{8}$	**14** $\frac{35}{15}$	**15** $\frac{42}{21}$
16 $\frac{120}{10}$	**17** $\frac{22}{7}$	**18** $\frac{15}{9}$	**19** $\frac{12}{5}$	**20** $\frac{150}{100}$

In questions **21** to **35**, change the mixed numbers to improper fractions.

21 $1\frac{1}{4}$	**22** $1\frac{1}{3}$	**23** $2\frac{1}{4}$	**24** $2\frac{2}{3}$	**25** $1\frac{7}{8}$
26 $1\frac{2}{3}$	**27** $3\frac{1}{7}$	**28** $2\frac{1}{6}$	**29** $4\frac{3}{4}$	**30** $7\frac{1}{2}$
31 $3\frac{5}{8}$	**32** $4\frac{2}{5}$	**33** $3\frac{2}{5}$	**34** $8\frac{1}{4}$	**35** $1\frac{3}{10}$

5.2.4 Fraction of a number

EXAMPLE

If four people share a lottery prize of £100 000 equally, what does each person receive?

Each person receives $\frac{1}{4}$ of £100 000.

That is £100 000 ÷ 4 = £25 000 each.

To find one **half** of a number, divide the number by **two**. To find one **third** of a number, divide the number by **three**.

Exercise 7 ①

1 Copy and complete this table.

	Fraction of quantity required	Divide the quantity by
a	$\frac{1}{2}$	2
b	$\frac{1}{3}$	
c	one-quarter	
d		10
e	$\frac{1}{5}$	

In questions **2** to **19**, copy and complete each calculation.

2 $\frac{1}{2}$ of £12 = ☐   **3** $\frac{1}{4}$ of £40 = ☐   **4** $\frac{1}{3}$ of 15 litres = ☐

5 $\frac{1}{4}$ of 20 kg = ☐   **6** $\frac{1}{2}$ of 48 cm = ☐   **7** $\frac{1}{5}$ of £20 = ☐

8 $\frac{1}{2}$ of 150 miles = ☐   **9** $\frac{1}{10}$ of £100 = ☐   **10** $\frac{1}{4}$ of 280 kg = ☐

11 $\frac{1}{5}$ of 30 litres = ☐   **12** $\frac{1}{8}$ of 24 kg = ☐   **13** $\frac{1}{100}$ of £2000 = ☐

14 $\frac{1}{11}$ of 66 kg = ☐   **15** $\frac{1}{7}$ of 490 eggs = ☐   **16** $\frac{1}{9}$ of 990 hens = ☐

17 $\frac{1}{8}$ of 888 miles = ☐   **18** $\frac{1}{5}$ of £60 = ☐   **19** $\frac{1}{8}$ of 320 pages = ☐

20 Here are calculations with letters.
Put the answers in order of size, smallest first.
Write the letters to make a word.

P	L	P
$\frac{1}{3}$ of 9	$\frac{1}{7}$ of 42	$\frac{1}{8}$ of 40

A	E
$\frac{1}{4}$ of 8	$\frac{1}{9}$ of 81

EXAMPLE

A petrol tank in a car holds 56 litres when full.
How much petrol is in the tank when it is $\frac{3}{4}$ full?

You need to work out $\frac{3}{4}$ of 56.

$\frac{1}{4}$ of 56 = 56 ÷ 4

     = 14

So $\frac{3}{4}$ of 56 = 3 × 14

      = 42

There are 42 litres in the tank when it is $\frac{3}{4}$ full.

Working:
$$\begin{array}{r} 1\,4 \\ 4\overline{)5^1 6} \end{array}$$

$\frac{3}{4}$ of 56 is 3 times as
much as $\frac{1}{4}$ of 56.

EXAMPLE

Work out

a $\frac{2}{5}$ of £55

b $\frac{3}{4}$ of 52 weeks

....

a $\frac{1}{5}$ of 55 = 55 ÷ 5

$\quad\quad = 11$

So $\frac{2}{5}$ of 55 = 2 × 11

$\quad\quad = 22$

$\frac{2}{5}$ of £55 = £22

b $\frac{1}{4}$ of 52 = 52 ÷ 4

$\quad\quad = 13$

So $\frac{3}{4}$ of 52 = 3 × 13

$\quad\quad = 39$

$\frac{3}{4}$ of 52 weeks = 39 weeks

Exercise 8 ①

Work out these amounts.

1 $\frac{2}{3}$ of £69

2 $\frac{3}{4}$ of £64

3 $\frac{3}{8}$ of 24 kg

4 $\frac{4}{5}$ of £65

5 $\frac{4}{7}$ of 84 miles

6 $\frac{5}{9}$ of 108 cm

7 $\frac{2}{3}$ of £216

8 $\frac{3}{4}$ of 20 kg

9 $\frac{7}{10}$ of 50 hens

10 $\frac{5}{8}$ of 480 cm

11 $\frac{3}{5}$ of 80 pence

12 $\frac{2}{3}$ of 600 miles

13 $\frac{4}{7}$ of 49p

14 $\frac{5}{8}$ of £4000

15 $\frac{3}{10}$ of 90p

16 $\frac{5}{9}$ of £144

17 $\frac{2}{11}$ of 88 kg

18 $\frac{3}{4}$ of 804 km

19 The total mark for a science test was 72. How many marks did Sandy get if she got $\frac{3}{4}$ of the full marks?

20 There are 450 apples on a tree and there are maggots in $\frac{2}{5}$ of them.
How many apples have maggots in them?

21 A DVD that cost £18 was sold on the Internet for $\frac{2}{9}$ of the original price. What was the selling price?

22 Mark has 216 houses on his paper round. One day his sister does $\frac{2}{3}$ of the houses. How many houses are left for Mark?

23 Here are calculations with letters. Put the answers in order of size, smallest first. Write the letters to make a word.

C	A	B
$\frac{2}{5}$ of 40	$\frac{1}{8}$ of 256	$\frac{2}{3}$ of 18

E	M	K	H
$\frac{1}{4}$ of 60	$\frac{2}{9}$ of 180	$\frac{3}{7}$ of 42	$\frac{5}{8}$ of 48

5.2.5 Adding and subtracting fractions

▶ You can add or subtract fractions when they have the same denominator.

For fractions with different denominators, first change them into equivalent fractions with the same denominator.

EXAMPLE

Find   **a** $\frac{3}{8} + \frac{1}{8}$ **b** $\frac{1}{2} + \frac{1}{5}$ **c** $\frac{3}{4} - \frac{1}{3}$

a $\frac{3}{8} + \frac{1}{8} = \frac{4}{8} = \frac{1}{2}$

Notice that you add the numerators but not the denominators!

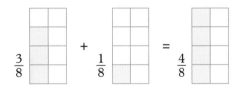

b $\frac{1}{2} + \frac{1}{5} =$

$\frac{5}{10} + \frac{2}{10} = \frac{7}{10}$

The L.C.M. of 2 and 5 is 10.

c $\frac{3}{4} - \frac{1}{3} =$

$\frac{9}{12} - \frac{4}{12} = \frac{5}{12}$

The L.C.M. of 4 and 3 is 12.

Exercise 9 ❶

1 Work out

a $\dfrac{5}{7} + \dfrac{1}{7}$ b $\dfrac{4}{9} + \dfrac{1}{9}$ c $\dfrac{7}{10} - \dfrac{1}{10}$ d $\dfrac{3}{5} + \dfrac{2}{5}$

e $\dfrac{1}{8} + \dfrac{4}{8}$ f $\dfrac{2}{11} + \dfrac{3}{11}$ g $\dfrac{5}{12} - \dfrac{1}{12}$ h $\dfrac{7}{15} - \dfrac{3}{15}$

2 Copy and complete

a $\dfrac{1}{4} + \dfrac{1}{8} = \dfrac{\square}{8} + \dfrac{1}{8} =$ b $\dfrac{1}{2} + \dfrac{2}{5} = \dfrac{\square}{10} + \dfrac{\square}{10} =$ c $\dfrac{2}{5} + \dfrac{1}{3} = \dfrac{\square}{15} + \dfrac{\square}{15} =$

d $\dfrac{1}{2} + \dfrac{1}{5} + \dfrac{\square}{10} + \dfrac{\square}{10} =$ e $\dfrac{3}{4} + \dfrac{1}{8} = \dfrac{\square}{8} + \dfrac{1}{8} =$ f $\dfrac{1}{2} + \dfrac{1}{6} = \dfrac{\square}{6} + \dfrac{1}{6} =$

3 Work out

a $\dfrac{1}{2} + \dfrac{1}{3}$ b $\dfrac{1}{4} + \dfrac{1}{6}$ c $\dfrac{1}{3} + \dfrac{1}{4}$ d $\dfrac{1}{4} + \dfrac{1}{5}$

e $\dfrac{2}{5} + \dfrac{2}{3}$ f $\dfrac{1}{4} + \dfrac{2}{5}$ g $\dfrac{1}{2} + \dfrac{2}{7}$ h $1\dfrac{1}{4} + 2\dfrac{1}{2}$

4 Work out

a $\dfrac{2}{3} - \dfrac{1}{2}$ b $\dfrac{3}{4} - \dfrac{1}{3}$ c $\dfrac{4}{5} - \dfrac{1}{2}$

d $\dfrac{2}{3} - \dfrac{1}{4}$ e $\dfrac{1}{2} - \dfrac{2}{5}$ f $\dfrac{4}{5} - \dfrac{2}{3}$

5 Work out the perimeter of this photo.

$\dfrac{1}{4}$ m

$\dfrac{1}{5}$ m

> Perimeter = sum of the lengths of the sides

6 Add together $\dfrac{1}{4}$, $\dfrac{1}{3}$ and $\dfrac{1}{6}$.

*7 Of the cars which failed MoTs

$\dfrac{1}{4}$ failed on brakes

$\dfrac{1}{3}$ failed on steering

$\dfrac{1}{6}$ failed on lights

the rest failed on worn tyres.

What fraction of the cars failed on worn tyres?

? ***8** In this equation, all the asterisks stand for the same number. What is the number?

$$\left[\frac{\star}{\star} - \frac{\star}{6} = \frac{\star}{36} \right]$$

Try different numbers until it works.

*9 The shaded border round the rectangle is $\frac{1}{16}$ inch wide. Find the outside length and width.

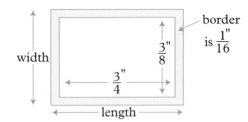

border is $\frac{1"}{16}$

width

$\frac{3"}{8}$

$\frac{3"}{4}$

length

5.2.6 Multiplying and dividing fractions

EXAMPLE

Work out

a $\frac{2}{5} \times 6$

b $2\frac{2}{5} \times \frac{1}{5}$

a $\frac{2}{5} \times 6 = \frac{2 \times 6}{5}$

$= \frac{12}{5} = 2\frac{2}{5}$

b $2\frac{2}{5} \times \frac{1}{5}$

$= \frac{12}{5} \times \frac{1}{5}$

$= \frac{12 \times 1}{5 \times 5} = \frac{12}{25}$

Change $2\frac{2}{5}$ to an improper fraction.

EXAMPLE

Work out

a $\frac{5}{11} \div \frac{1}{2}$

b $\frac{3}{8} \div 4$

c $\frac{3}{5} \times 4$

The reciprocal of $\frac{1}{2}$ is 2.

a $\frac{5}{11} \div \frac{1}{2}$

$= \frac{5}{11} \times \frac{2}{1} = \frac{10}{11}$

b $\frac{3}{8} \div 4$

$= \frac{3}{8} \times \frac{1}{4}$

$= \frac{3}{32}$

c $\frac{3}{5} \times 4$

$= \frac{3}{5} \times \frac{4}{1}$

$= \frac{12}{5}$

$= 2\frac{2}{5}$

Invert $\frac{1}{2}$ and then multiply.

Invert $\frac{4}{1}$ and then multiply.

Write 4 as $\frac{4}{1}$.

Exercise 10 1

1 Work out

    a $\dfrac{2}{3} \times \dfrac{1}{5}$ **b** $\dfrac{3}{4} \times \dfrac{1}{7}$ **c** $\dfrac{4}{5} \times \dfrac{1}{3}$ **d** $\dfrac{5}{6} \times \dfrac{1}{7}$

    e $\dfrac{5}{9} \times \dfrac{1}{3}$ **f** $\dfrac{4}{11} \times \dfrac{1}{6}$ **g** $\dfrac{7}{8} \times \dfrac{1}{4}$ **h** $\dfrac{8}{9} \times \dfrac{1}{4}$

2 **a** $\dfrac{1}{2} \times \dfrac{4}{5}$ **b** $\dfrac{1}{3} \times \dfrac{6}{7}$ **c** $\dfrac{2}{3} \times \dfrac{1}{4}$ **d** $\dfrac{7}{4} \times \dfrac{4}{9}$

    e $\dfrac{9}{10} \times \dfrac{4}{3}$ **f** $\dfrac{5}{12} \times \dfrac{6}{7}$ **g** $\dfrac{3}{4} \times \dfrac{12}{13}$ **h** $1\dfrac{1}{2} \times \dfrac{2}{3}$

3 **a** $2\dfrac{1}{4} \times \dfrac{1}{3}$ **b** $1\dfrac{3}{4} \times \dfrac{1}{2}$ **c** $2\dfrac{1}{2} \times \dfrac{3}{10}$ **d** $3\dfrac{1}{2} \times \dfrac{1}{5}$

    e $\dfrac{3}{4} \times 2$ **f** $\dfrac{3}{5} \times 2$ **g** $5 \times \dfrac{7}{10}$ **h** $8 \times \dfrac{3}{4}$

4 **a** $\dfrac{3}{4} \div \dfrac{1}{2}$ **b** $\dfrac{3}{5} \div \dfrac{1}{3}$ **c** $\dfrac{5}{6} \div \dfrac{1}{4}$ **d** $\dfrac{2}{3} \div \dfrac{1}{4}$

    e $\dfrac{5}{6} \div 3$ **f** $\dfrac{5}{7} \div 3$ **g** $\dfrac{3}{5} \div 2$ **h** $\dfrac{4}{7} \div 5$

5 Copy and complete this multiplication square.

$\times$	$\dfrac{2}{3}$		
$\dfrac{1}{2}$		$\dfrac{3}{8}$	
		$\dfrac{3}{16}$	
$\dfrac{2}{5}$			$\dfrac{2}{25}$

6 Work out one-half of one-third of £320.

7 A rubber ball is dropped from a height of 300 cm. After each bounce, the ball rises to $\dfrac{4}{5}$ of its previous height. How high will it rise after the second bounce?

8 Callum spends his income in three ways:

    a $\dfrac{2}{5}$ of his income goes on clothes.

    b $\dfrac{2}{3}$ of what is left goes on phone credit, food and transport.

    c he saves the rest for a new guitar.

    What fraction of his income does he save?

9 Copy and fill in the missing numbers so that the answer is always $\frac{3}{8}$.

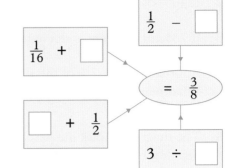

10 This question involves adding, subtracting, multiplying and dividing.

a $\frac{2}{3} \times 3$ **b** $\frac{5}{6} + \frac{1}{2}$ **c** $\frac{4}{7} \div \frac{1}{3}$ **d** $\frac{7}{10} - \frac{2}{5}$

*11 Work out

$$\frac{1}{3} \times \frac{2}{4} \times \frac{3}{5} \times \frac{4}{6} \times \frac{5}{7} \times \frac{6}{8} \times \frac{7}{9}$$

5.3 Percentages

5.3.1 What is a percentage?

▶ Percentages are equivalent to fractions with a denominator (bottom number) equal to 100.

So 21% means $\frac{21}{100}$, 63% means $\frac{63}{100}$ and so on.

● Some percentages are used often and you should learn them by heart.

$10\% = \frac{10}{100} = \frac{1}{10}$ $30\% = \frac{30}{100} = \frac{3}{10}$ $20\% = \frac{20}{100} = \frac{1}{5}$

$40\% = \frac{2}{5}$ $60\% = \frac{3}{5}$ $80\% = \frac{4}{5}$ $25\% = \frac{1}{4}$

$75\% = \frac{3}{4}$ $33\frac{1}{3}\% = \frac{1}{3}$ $66\frac{2}{3}\% = \frac{2}{3}$ $50\% = \frac{1}{2}$

Exercise 11 ①

1 Draw each square and write underneath it
   a what fraction is shaded   **b** what percentage is shaded.

i 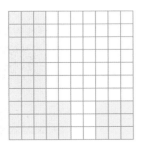   **ii** 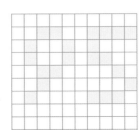   **iii**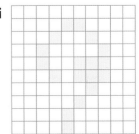

2 If 40% of a square is shaded, what percentage of the square is not shaded?

3 If 73% of a square is shaded, what percentage of the square is not shaded?

4 Approximately 67% of the Earth's surface is covered with water.
What percentage of the Earth's surface is land?

5 The diagram shows the percentage of people who took part in activities offered at a sports centre on a Friday night.

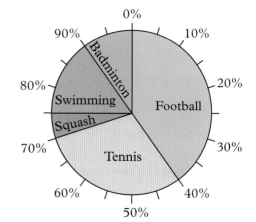

  a What percentage went swimming?
  b What percentage played squash?
  c What percentage played a racket sport?
  d What percentage did not play football?
  e What percentage played activities involving a ball?

6 Copy these and fill in the squares.

  a $30\% = \dfrac{\square}{100} = \dfrac{\square}{10}$ **b** $\dfrac{3}{4} = \dfrac{\square}{100} = \square\%$ **c** $\dfrac{1}{10} = \square\%$

  d $1\% = \dfrac{\square}{100}$ **e** $80\% = \dfrac{\square}{\square}$ **f** $\dfrac{1}{3} = \square\%$

7 For each shape, write
  a what fraction is shaded   **b** what percentage is shaded.

A   B   C   D

E   F   G   H

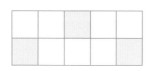

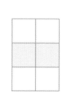

8 These pictures show how much petrol is in a car.
E is Empty and F is Full.
What percentage of a full tank is in each car?

a E ⊢———————————⊣ F

b E ⊢———————————⊣ F

c E ⊢———————————⊣ F

d E ⊢———————————⊣ F

9 Write the percentage that could be used in each sentence.

   a   Three-quarters of the students skated to school.
   b   Three out of five workers voted for a strike.
   c   Lela got 15 out of 20 in the maths test.
   d   One in three cats prefer 'Kattomeat'.
   e   Half of the customers at a computer store thought
       that prices were going down.
   f   One in four mothers think children are too tidy
       at home.

10 Draw three diagrams of your own design, like those
in question **7**, and shade in
   a   30%   **b**   75%   **c** $66\frac{2}{3}\%$

5.3.2 Changing fractions to percentages

▶   To change a fraction to a percentage you multiply by 100%.

EXAMPLE

a Change $\frac{7}{20}$ to a percentage.   **b** Change $\frac{3}{8}$ to a percentage.

a $\dfrac{7}{20} \times \dfrac{100\%}{1} = \dfrac{7}{\overset{}{\underset{1}{20}}} \times \dfrac{\overset{5}{100\%}}{1} = 35\%$ **b** $\dfrac{3}{8} \times \dfrac{100\%}{1} = \dfrac{300\%}{8} = 37{\cdot}5\%$

Multiplying by 100% means multiplying by one whole, so you
are not changing the value of the fraction.

Exercise 12 ❶

Change to percentages

1 $\dfrac{3}{4}$ **2** $\dfrac{2}{5}$ **3** $\dfrac{3}{8}$ **4** $\dfrac{9}{10}$ **5** $\dfrac{17}{20}$

6 $\dfrac{5}{20}$ **7** $\dfrac{17}{25}$ **8** $\dfrac{3}{20}$ **9** $\dfrac{49}{50}$ **10** $\dfrac{7}{100}$

11 $\dfrac{15}{60}$ **12** $\dfrac{16}{50}$ **13** $\dfrac{27}{40}$ **14** $\dfrac{1}{8}$ **15** $\dfrac{235}{1000}$

16 16 marks out of 25.   **17** 18 marks out of 20.

18 12 marks out of 30.   **19** £45 out of £200

20 £17 out of £50   **21** £93 out of £100

22 a In a test Ann obtained 11 marks out of 25. What is her percentage mark?

    b In a second test she obtained 13 marks out of 20. What is her percentage mark?

23 A motorist has to drive a distance of 400 km. After an hour he has driven 84 km.
What percentage of his journey has he now completed?

24 In a survey, 1600 people were asked if they preferred ITV1 or BBC1.
    800 chose ITV1
    640 chose BBC1
    160 had no television set.
    a What percentage chose ITV1?
    b What percentage chose BBC1?
    c What percentage had no television?

25 Of the people in a room, 11 are men and 14 are women.
    a How many people are in the room?
    b What percentage of the people are men?
    c What percentage of the people are women?

26 Three girls in different classes all had maths tests on the same day.
    Jane scored 27 out of 50.
    Susan scored 28 out of 40.
    Jackie scored 39 out of 75.
Work out their marks as percentages and put them in order with the highest first.

5.3.3 Working out percentages

> **EXAMPLE**
>
> **a** Work out 22% of £40.   **b** Work out 16% of £85.
>
> ....
>
> $\dfrac{22}{100} \times \dfrac{40}{1} = \dfrac{880}{100}$ (Alternative method)
>
> Since $16\% = \dfrac{16}{100}$, you can replace 16% by 0·16
>
> 22% of £40 = £8·80   So 16% of £85 = 0·16 × 85 = £13·60

Exercise 13 ①

Work out

1 20% of £60	**2** 10% of £80	**3** 5% of £200
4 6% of £50	**5** 4% of £60	**6** 30% of £80
7 9% of £500	**8** 18% of £400	**9** 61% of £400
10 12% of £80	**11** 6% of $700	**12** 11% of $800
13 5% of 160 kg	**14** 20% of 60 kg	**15** 68% of 400 g
16 15% of 300 m	**17** 2% of 2000 km	**18** 71% of $1000
19 26% of 19 kg	**20** 1% of 6000 g	**21** 8·5% of €2400

22 Ayesha earns £25 for doing a paper round. How much **extra** does she earn when she gets a 20% rise?

23 Full marks in a maths test is 80. How many marks did Tim get if he got 60%?

24 There are 240 children at a school and 65% of them walk to school. How many children is that?

25 The normal price of a garden hoe is £15.
In a sale, prices are reduced by 20%.
Find the sale price of the hoe.

26 Sam buys a car for £4200.
How much does he pay each month?

EXAMPLE

A coat originally cost £24. Calculate the new price after a 5% reduction.

Price reduction $= 5\%$ of £24

$$= \frac{5}{100} \times \frac{24}{1} = £1{\cdot}20$$

New price of coat $= £24 - £1{\cdot}20$

$$= £22{\cdot}80$$

There is another way of finding the reduced price of the coat. The price, after a 5% reduction, will be 95% of the original price.

So, new price of coat $= 95\%$ of £24

$$= 0{\cdot}95 \times 24$$

$$= £22{\cdot}80$$

> For a 5% **increase** you multiply by 1·05.

EXAMPLE

A CD originally cost £11·60. Calculate the new price after a 7% increase.

New price $= 107\%$ of £11·60

$$= 1{\cdot}07 \times 11{\cdot}6$$

$$= £12{\cdot}41 \text{ to the nearest penny.}$$

Exercise 14 ①

1 Increase a price of £60 by 5%.
2 Increase a price of £800 by 8%.

3 Increase a price of £82·50 by 6%.
4 Increase a price of £65 by 60%.

5 Reduce a price of £2000 by 2%.
6 Increase a price of £440 by 80%.

7 Increase a price of £66 by 100%.
8 Reduce a price of £91·50 by 50%.

9 Increase a price of £88·24 by 25%.
10 Reduce a price of £63 by $33\frac{1}{3}\%$.

11 Increase a price of £8·50 by 46%.
12 Increase a price of £240 by 11%.

13 Increase a price of £5·75 by 20%.
14 Reduce a price of £8500 by 4%.

15 Increase a price of £11·20 by 15%.
16 Reduce a price of £88 by 10%.

In the remaining questions, give the answers to the nearest penny.

17 Increase a price of £28·20 by 13%.
18 Increase a price of £8·55 by 4%.

19 Reduce a price of £9·60 by 7%.
20 Increase a price of £12·80 by 11%.

Exercise 15 ❶

1 In a closing-down sale a shop reduces all its prices by 20%.
Find the sale price of a coat which previously cost £44.

2 The price of a car was £5400 but it is increased by 6%.
What is the new price?

3 The price of a sideboard was £245 but, because the
sideboard is scratched, the price is reduced by 30%. What is
the new price?

4 A hi-fi shop offers a 7% discount for cash. How much does a
cash-paying customer pay for an amplifier advertised at £95?

5 The insurance premium for a car is normally £90. With a
'no-claim bonus' the premium is reduced by 35%. What is
the reduced premium?

6 Myxomatosis kills 92% of a colony of 300 rabbits. How
many rabbits survive?

7 The population of a town increased by 32% between 1945
and 2015. If there were 45 000 people in 1945, what was
the 2015 population?

8 A restaurant adds a 12% service charge onto the basic
price of meals. How much do I pay for a meal with a basic
price of £18·50?

*9 A new-born baby weighs 3·1 kg. Her weight increases by 8%
over the next fortnight. What does she weigh then?

*10 A large snake normally weighs 12·2 kg.
After swallowing a rat, the weight of the
snake is increased by 7%. How
much does it weigh after dinner?

*11 At the beginning of the year a car is valued at £3250.
During the year its value falls by 15%. How much is it
worth at the end of the year?

*12 The area of a square is 400 cm².
  a The area is increased by 20%. Find the new area.
  b The new area is then decreased by 20%.
    Find the final area.

400 cm²

Exercise 16 ①

VAT stands for Value Added Tax. It is added to the price of products before they are sold.

Find the total bill.

1 2 hammers at £5·30 each
50 screws at 25p for 10
5 bulbs at 38p each
1 tape measure at £1·15
VAT at 20% is added to the total cost.

2 5 litres of oil at 85p per litre
3 spanners at £1·25 each
2 manuals at £4·30 each
200 bolts at 90p for 10
VAT at 20% is added to the total cost.

3 12 rolls of wallpaper at £3·70 per roll
3 packets of paste at £0·55 per packet
2 brushes at £2·40 each
1 step ladder at £15·50
VAT at 20% is added to the total cost.

4 5 golf clubs at £12·45 each
48 golf balls at £15 per dozen
100 tees at 1p each
1 bag at £21·50
1 umbrella at £12·99
VAT at 20% is added to the total cost.

5.3.4 Percentage change

Price changes are sometimes easier to understand when expressed as a percentage of the original price. For example, if the price of a car goes up from £7000 to £7070, this is only a 1% increase. If the price of a jacket went up from £100 to £170 this would be a 70% increase! In both cases the actual increase is the same: £70.

▶ Percentage increase $= \dfrac{\text{actual increase}}{\text{original value}} \times \dfrac{100}{1}$

▶ Percentage decrease $= \dfrac{\text{actual decrease}}{\text{original value}} \times \dfrac{100}{1}$

EXAMPLE

The price of a car is increased from £6400 to £6800. What percentage change is that?

Percentage increase $= \dfrac{400}{6400} \times \dfrac{100\%}{1} = \dfrac{40000\%}{6400} = 6\cdot25\%$

Exercise 17 1

In questions **1** to **10**, calculate the percentage increase.

	Original price	Final price
1	£50	£54
2	£80	£88
3	£180	£225
4	£100	£102
5	£75	£78
6	£400	£410
7	€5000	€6000
8	£210	£315
9	£600	£690
10	$4000	$7200

In questions **11** to **20**, calculate the percentage decrease.

	Original price	Final price
11	£800	£600
12	£50	£40
13	£120	£105
14	£420	£280
15	£6000	£1200
16	€880	€836
17	$15000	$14100
18	$7·50	$6·00
19	£8·20	£7·79
20	£16000	£15600

If the final selling price is greater than the original cost price, the seller has made a profit. If it is less, they have made a loss. You can find this profit or loss as a percentage of the original price the seller paid (cost price).

EXAMPLE

John buys a coat for £16 and sells it again for £20. What is his percentage profit?

$$\text{Percentage profit} = \frac{20 - 16}{16} \times \frac{100\%}{0} = \frac{4}{6} \times \frac{100\%}{1} = 25\%$$

$$\text{Percentage profit} = \frac{\text{actual profit}}{\text{cost price}} \times \frac{100\%}{1} \text{ or percentage loss} = \frac{\text{actual loss}}{\text{cost price}} \times \frac{100\%}{1}$$

Exercise 18 1

Find the percentage profit or the percentage loss using one of these formulae. Give the answers correct to one decimal place.

	Cost price	Selling price
1	£11	£15
2	£21	£25
3	£36	£43
4	£41	£50
5	£411	£461
6	£5·32	£5·82
7	£6·14	£7·00
8	£2·13	£2·50
9	£6·11	£8·11
10	£18·15	£20

	Cost price	Selling price
11	£20	£18·47
12	£17	£11
13	£13	£9
14	£211	£200
15	£8·15	£7
16	£2·62	£3
17	£1·52	£1·81
18	$13·50	$13·98
19	$3·05	$4·00
20	$1705	$1816

Exercise 19 ①

1   The number of people employed by a firm increased from
    250 to 280. Calculate the percentage increase in the
    workforce.

2   During the first four weeks of his life, Samuel's weight
    increases from 3000 g to 3870 g. Calculate the percentage
    increase in his weight.

3   After cleaning out the drains, Peter's clothes went down in
    value from £80 to £56. Calculate the percentage decrease in
    the value of his clothes.

4   When cold, an iron rod is 200 cm long. After it is heated, the
    length of the rod increases to 200·5 cm. Calculate the
    percentage increase in the length.

5   A man buys a car for £4000 and sells it for £4600. Calculate
    his percentage profit.

6   A shopkeeper buys jumpers for £6·20 and sells them for
    £9·99. Calculate the percentage profit, correct to one
    decimal place.

7   A grocer buys bananas at 20p per pound but, after the fruit
    are spoiled, he has to sell them at only 17p per pound.
    Calculate the percentage loss.

5.3.5   Inflation

Usually the price of items such as bread, furniture and
clothes will increase every year. This is called **inflation**.

Suppose a fridge cost £180 in 2015 and the annual rate of inflation
was 2%.

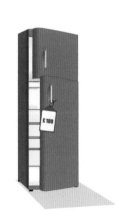

In 2016 the increase in price = 2% of £180 = £3·60
So the price of the fridge in 2016 would be £183·60.

Alternatively, take the initial cost of the fridge in 2015 as 100%.
One year later, its cost is 102% of the initial cost.
So the price of the fridge in 2016 = 102% of £180 = 1·02 × 180
                                                  = £183·60.

The annual rate of inflation is an average of the percentage rise of the price of goods that people buy.

The rate of inflation is important because it is used to calculate the increase in pensions for retired people and it is sometimes used as a guide for the pay increases of people like nurses or teachers who are employed by the government.

If the price of goods **decreases** from year to year, this is called **deflation**.

Exercise 20 ①

1   Today's cost of items **a** to **e** are given in this table. Find their cost after one year with the given inflation rates.

	a	b	c	d	e
Today's cost	£280	£350	£350	£410	£410
Annual inflation rate	2%	5%	8%	1%	4%

2   The annual rate of inflation stays steady at 2% for 3 years.
Today's price of a freezer for the kitchen is £200.
Find its price after   **a** one year   **b** two years   **c** three years.

3   A car costs £9560 today. Its price rises in line with inflation which is 2% for the first year and then 3% for the following year.
Find the price of the car after one year and after two years.

5.3.6   Reverse percentages

EXAMPLE

After an increase of 10%, the price of a ticket on a rollercoaster is £3·96. What was the price before the increase?

A common mistake here is to work out 10% of £3.96. This is wrong because the increase is 10% of the old price, not 10% of the new price.

110% of old price = £3·96

$$1\% \text{ of old price} = \frac{3 \cdot 96}{110}$$

$$100\% \text{ of old price} = \frac{3 \cdot 96}{110} \times 100$$

The old price = £3·60

Exercise 21 ①

1 After an increase of 5% the price of a fridge is £441.
Find the price of the fridge before the increase.

2 After a 4% pay rise, the Sarah's salary was £18 200.
What was her salary before the increase?

3 Find the missing prices.

	Item	Old price	New price	% change
a	Jacket	?	£55	10% increase
b	Dress	?	£212	6% increase
c	CD player	?	£56.16	4% increase
d	TV	?	£195	30% increase
e	Car	?	£3960	65% increase

4 Between 2008 and 2016 the population of an island fell by 4%.
The population in 2016 was 201600. Find the population in 2008.

5 After being ill for 3 months, a man's weight went down by 12%.
Find his original weight if he weighed 74·8 kg after the illness.

6 Find the missing prices.

	Object	Old price	New price	% change
a	Football	£9·50	?	3% increase
b	Radio	?	£34·88	9% increase
c	Roller blades	£52	?	15% increase
d	Golf club	?	£41·40	8% increase

7 The diagram shows two rectangles.
The width and height of rectangle B are both 20%
greater than the width and height of rectangle A.

Use the figures given to find the width and height
of rectangle A.

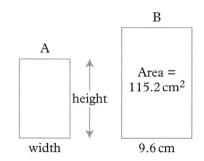

8 Here is the label on a carton of yoghurt.
The figure on the right is smudged out.
Work out what the figure should be.

30% *extra* FREE

455g *for the price of*

9 After a severe storm, the cost of staying at a beach
resort for a week went down by 20% to £1240.
Find the original cost before the storm.

10 Between 2010 and 2015 the population of a town fell by 6%.
The population in 2015 was 186,120. Find the population in 2010.

Exercise 22 ①

1 A supermarket has a special offer on its own brand
breakfast cereal. Find the missing number.

2 In a restaurant, VAT at 20% is added to the cost of a
meal. The total cost of a meal is £44·22 including
VAT. Find the cost of the meal before the VAT was added.

40% *extra* FREE

for the price of 330g

3 Compared to 2014, the number of schools in France in 2015
increased by 3%. There were 1065 more schools in 2015. How
many schools were there in 2014?

4 A horse lost 4% of its body weight after an illness. He now
weighs 307·2 kg. How much did it weigh before the illness?

5 When heated, the volume of a pan increased by 2% to
420 cm^3. What was its volume when cold?

6 After scoring 44 goals in a season, a footballer's pay
went up by 30% to £188,500 per week.
What was he paid before the pay rise?

7 Find the missing prices.

	Item	Old price	New price	% change
a	Calculator	?	£ 11·34	8% increase
b	Football	?	£ 16·80	12% increase
c	Train ticket	?	£ 22·88	4% increase
d	TV	?	£ 169·20	6% increase
e	Necklace	?	£ 40·80	15% decrease
f	Watch	?	£ 8.40	20% decrease

8 The cost of dog food for Mrs Brand's dog went up to £3·68 after a 14% increase. How much did it cost before the increase?

5.3.7 Compound interest

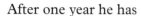

> **EXAMPLE**
>
> A bank pays a fixed interest of 10% on money in savings accounts. A man puts £1000 in the bank.
>
> After one year he has
> $$1000 + 10\% \text{ of } 1000 = 1000 + 100 = £1100$$
> After two year he has
> $$1100 + 10\% \text{ of } 1100 = £1210$$
> After three year he has
> $$1210 + 10\% \text{ of } 1210 = £1331$$
> [Check that this is $1·1^3 \times 1000$]
>
> In general after n years, the money in the bank will be £$(1·1^n \times 1000)$.

Exercise 23 ②

1 £2000 is invested at 10% per year compound interest. How much money will there be after 2 years?

2 A bank pays interest of 5%. Kim puts £5000 in the bank. How much has she after **a** one year, **b** three years?

3 An office worker is paid £20000 a year per year. His pay is increased by 5% each year. What will he be paid in three year's time?

4 A car loses 20% of its value every year. Steve bought it for £4000. How much will it be worth after:
    a 1 year?   **b** 3 years?

5 Assuming an average inflation rate of 2% per annum, work out the probable cost of the following items in 3 years.
    a car £4500   **b** T.V. £280   **c** house £150000

6 Compound interest is paid at 8% per annum. For how many years must £200 be left in the scheme to total over £500?

7 A savings scheme pays 6% compound interest per annum. If Jack puts £400 in the scheme calculate how much will be in the account after
    a 1 year   **b** 2 years   **c** 10 years?

8 The rate of inflation is 5% per annum and a supermarket bill is £98. How much would the same shopping cost in 10 years time?

9 What sum of money invested at 4% compound interest per annum will grow to £146 after 5 years?

10 A car goes down in value by 15% of its previous year's value. If Louise bought a new car for £16 500, how much will it be worth after
  a 1 year   **b** 5 years?

5.4 Ratio

5.4.1 What is a ratio?

▶ You use a ratio to compare the sizes of two or more quantities.

● Suppose in a classroom there are 9 girls and 12 boys.
The ratio of girls to boys is

girls : boys = 9 : 12

You can simplify the ratio by dividing both numbers by a common factor, 3.

So girls : boys = 3 : 4.

You can use the ratio of girls to boys to find what fraction of the class are girls and what fraction are boys.
Add 3 + 4 to give 7.

$\frac{3}{7}$ of the class are girls and $\frac{4}{7}$ of the class are boys.

▶ You must use the same units in both parts of a ratio.

For example, you write the ratio of 22 cm to 1 m as 22 cm to 100 cm.
Then divide both sides by 2 to get 22 : 100 = 11 : 50.

Exercise 24 ①

1 Express each ratio in its simplest form.
  a 9 : 6   **b** 15 : 25   **c** 10 : 40   **d** 30 : 12   **e** 48 : 44
  f 18 to 24   **g** 40 to 25   **h** 21 to 49   **i** 60 : 42   **j** 16 : 22
  k 9 : 6 : 12 **l** 40 : 5 : 15 **m** 12 : 10 : 8 **n** 18 : 12 : 18

2 In a box there are 9 pencils and 12 pens. Find the ratio of pencils to pens. Give your answer in its simplest form.

3 In a hall there are 36 chairs and 9 tables. Find the ratio of chairs to table. Give your answer in its simplest form.

4 Express each ratio in its simplest form. Remember to use the same units in both parts.
   a 8 kg to 20 kg   **b** 20p to £1   **c** 20 cm to 1 m
   d 20 minutes to 1 hour   **e** 2 kg to 500 g   **f** 400 m to 2 km
   g 5 mm to 5 cm   **h** 25p to £1·50   **i** 40 cm to 1·2 m

5 In a minibus there are 5 women and 3 men.
   What fraction of the people are men?

6 In a room there are 36 plastic chairs and 9 wooden chairs.
   What fraction of the chairs are wooden?

7 In a class the ratio of girls to boys is 3 : 2.
   What fraction of the class are girls?

8 In a shop the ratio of apples to pears to oranges is 3 : 4 : 5.
   What fraction of the total fruit are
   a apples   **b** pears   **c** oranges?

9 In an office there are twice as many men as women.
   What fraction of the people in the office are men?

5.4.2 Dividing quantities in a given ratio

> **EXAMPLE**
>
> Share £60 in the ratio 2 : 3.
>
> Total number of shares = 2 + 3 = 5
> So one share = £60 ÷ 5 = £12
> Two shares = 2 × £12 = £24 and three shares 3 × £12 = £36.
> The two amounts are £24 and £36.

Check by addition
£24 + £36 = £60

Exercise 25 ①

1 Share £30 in the ratio 1 : 2.

2 Share £60 in the ratio 3 : 1.

3 Divide 880 g of food between a cat and a dog in the ratio 3 : 5.

4 Divide $1080 between Sam and Chris in the ratio 4 : 5.

5 Share 126 litres of petrol between Steven and Dave in the ratio 2 : 5.

6 Share £60 in the ratio 1 : 2 : 3.

7 Alan, Brian and Dawn divided £560 between them in the
ratio 2 : 1 : 5. How much did Brian receive?

8 A sum of £120 is divided in the ratio 3 : 4 : 5. What is the largest share?

9 At an election, 7800 people voted Labour, Conservative or
Liberal Democrat in the ratio 4 : 3 : 5. How many people voted
Liberal Democrat?

10 Find the ratio of shaded area to unshaded area for each diagram.

a   **b**   **c**

11 Using the diagrams in question 10, find the ratio of shaded area
to the **total** area of each diagram.

12 Dileep has £27. He spends £18 of his money on a book.
  a What is the ratio of the cost of the book to how much he
     had initially?
  b What fraction of his money did he spend on the book?

13 Sandy's bedroom has a floor area of 12 m². The total bedroom
space in the house is 48 m².
  a What is the ratio of her bedroom area to the total bedroom area?
  b What fraction is her bedroom area of the total bedroom area?

14 You have £1350 and you put £450 of it into a saving account.
What fraction have you put into the saving account? What is the
ratio of the money in the saving account to the total sum of
money that you have?

15 The local council has £375,000 to spend on its roads. Of this
money, it spends £125,000 on new tarmac. What fraction of the
total money is spent on new tarmac? What is the ratio of money
spent on tarmac to the total amount spent on roads?

EXAMPLE

In a class, the ratio of boys to girls is 3 : 4.
If there are 12 boys, how many girls are there?

....

       Boys : Girls = 3 : 4
Multiply both parts by 4 as 3 × 4 = 12.
       Boys : Girls = 12 : 16
So there are 12 boys and 16 girls.

Exercise 26 ①

1 In a room, the ratio of boys to girls is 3 : 2. If there are 12 boys, how many girls are there?

2 In a room, the ratio of men to women is 4 : 1. If there are 20 men, how many women are there?

3 In a box, the ratio of nails to screws is 5 : 3. If there are 15 nails, how many screws are there?

4 An alloy consists of copper, zinc and tin in the ratio 1 : 3 : 4. If there is 10 g of copper in the alloy, find the weights of zinc and tin.

5 In a shop, the ratio of oranges to apples is 2 : 5. If there are 60 apples, how many oranges are there?

6 A recipe for 5 people calls for 1·5 kg of meat. How much meat is required if the recipe is adapted to feed 8 people?

> Find how much meat is needed for one person.

7 A cake for 6 people requires 4 eggs. How many eggs are needed to make a cake big enough for 9 people?

8 A photocopier enlarges the original in the ratio 2 : 3. The height of a tree is 12 cm on the original. How tall is the tree on the enlarged copy?

12 cm

original

enlarged copy

9 A photocopier enlarges copies in the ratio 4 : 5. The length of the headline 'BRIDGE COLLAPSES' is 18 cm on the original. How long is the headline on the enlarged copy?

—

10 A photocopier **reduces** in the ratio 5 : 3. The height of a church spire is 20 cm on the original. How high is the church spire on the reduced copy?

11 A cake weighing 550 g has three ingredients: flour, sugar and butter. There is twice as much flour as sugar and one and a half times as much sugar as butter. How much flour is there?

12 If $\frac{5}{8}$ of the children in a school are boys, what is the ratio of boys to girls?

13 A man and a woman share a bingo prize of £1000 between them in the ratio 1 : 4. The woman shares her part between herself, her mother and her dog in the ratio 2 : 1 : 1.

How much does her dog receive?

14 The number of pages in a newspaper is increased from 36 to 54. The price is increased in the same ratio. If the old price was 28p, what will the new price be?

*15 Two friends bought a house for £220 000. Sam paid £140 000 and Joe paid the rest. Three years later they sold the house for £275 000. How much should Sam receive from the sale?

*16 Concrete is made from 1 part cement, 2 parts sand and 5 parts aggregate (by volume). How much cement is needed to make 2 m³ of concrete?

5.4.3 More applications
Exercise 27 ①

1 Use this graph to find how many Swiss francs (SF) you get for
  a £2   b £3·20
  c £4·80   d £2·60

Now use the graph to change these francs back to pounds.
  e 10 SF   f 8·50 SF
  g 6·50 SF   h 11·50 SF

2 a Use the graph to find how many francs you get for £1.
  b A business man changes £3400 into Swiss francs and receives 3400 × k francs. What is the value of k and how many francs is £3400 worth?

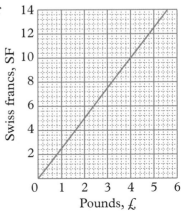

3 You visit the USA and the exchange rate is £1 = 1·65 US dollars.

    a You arrive in the USA and change £650 into dollars. How many dollars do you get?

    b When you come back to the UK, you change $120 back into UK £ at the same rate of exchange. How many pounds do you get?

    c Why might these answers are not be the actual sums you receive?

4 a £5 is changed into 6·20 euros. How many euros is £1 worth?

    b £8 is exchanged into 8 × k euros. Write the value of k as a decimal and calculate how many euros £8 is worth.

5 A weight of 7 kg converts (approximately) to 15 British pounds (lbs). So, a weight of 12 kg converts (approximately) to 12 × k lbs.

    a Write the value of k as a fraction.

    b Calculate how many pounds (lbs) are there in 12 kg to the nearest lb

6 Ravi Gupta changes pounds £ into rupees four times on his visit to India. The exchange rate is the same each time. Copy and complete this table.

Pounds, £	120	200	70	
Rupees	11000			1500

7 A DIY shop sells 12 nuts and bolts for 84p. A hardware shop sells 15 identical nuts and bolts for 90p. Find which shop offers the better buy.

8 A supermarket sells 3·5 litres of milk for £1·66 and 2·5 litres of the same milk for £1·22. Which size of milk is the better buy?

9 Packs of 4 tins of beans are on 'special offer' at £1·44 for a pack. Packs of 6 tins of the same beans are sold for £2·13 per pack. Which pack is the better buy?

10 Bank A sells 15 US dollars for £9·49. Bank B sells 25 US dollars for £14·75. Which bank offers the better buy for a visitor who travels

    a to the USA from the UK   **b** to the UK from the USA?

11 Pink paint is mixed from 2 litres of red paint and 3 litres of white paint. If you want 20 litres of pink paint, how much red paint and how much white paint do you need?

12 Dark red paint is mixed from 5 litres of bright red paint and 2 litres of black paint. How many litres of bright red paint and black paint do you have to mix to make 28 litres of dark red paint?

13 You make brown paint by mixing red and green paint in the ratio 2 : 3. If you have a 10 litre tin of green paint and a huge tub of red paint, how many litres of brown paint can you make?

14 Apples are bagged so that the ratio of ripe to the unripe apples is 3 : 2 in each bag. If each bag has 15 apples in it, how many ripe and how many unripe apples are there in each bag?

15 An orange drink is made by mixing water and concentrated orange juice in the ratio of 5 : 1 by volume.

   a What fraction of the drink is water?

   b If I have 3 litres of concentrate juice, how much water do I need to make this mixture?

   c I want to make 24 litres of the drink. How much concentrated juice do I need?

16 Dilute sulphuric acid is made by mixing water and concentrated acid. A typical car battery contains dilute acid which is 70% water.

   a What is the ratio of water to concentrated acid in a car battery?

   b How much concentrated acid is needed to make 10 litres of car-battery acid?

*17 Elderflower cordial is diluted to make a drink with a ratio of cordial : water of 1 : 10.

   a How much cordial is there in a 5.5 litre carton of the drink?

   b If a manufacturer of 5.5 litre cartons increases the amount of water by 10%, what fraction of the new mixture is cordial?

5.4.4 Map scales and ratio

The map below is drawn to a scale of 1 : 50 000.
1 cm on the map represents 50 000 cm on the land.

> You always draw an accurate map to a scale.

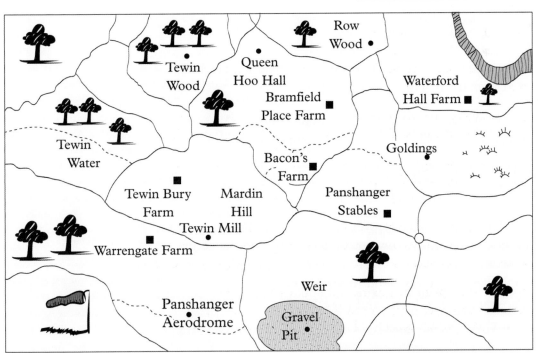

> **EXAMPLE**
>
> On a map of scale 1 : 25 000, two towns appear 10 cm apart.
> What is the actual distance between the towns in km?
>
> ··
>
> 1 cm on map = 25 000 cm on land
> 10 cm on map = 250 000 cm on land
> 250 000 cm = 2500 m
>   = 2·5 km
> The towns are 2·5 km apart.

Exercise 28 ①

1 The scale of a map is 1 : 1000. Find the actual length in metres represented on the map by 20 cm.

2 The scale of a map is 1 : 10 000. Find the actual length in metres represented on the map by 5 cm.

3 Copy and complete the table.

	Map scale	Length on map	Actual length on land
a	1 : 10 000	10 cm	__km
b	1 : 2000	10 cm	__m
c	1 : 25 000	4 cm	__km
d	1 : 10 000	6 cm	__km

4 Find the actual distance in metres between two points which are 6·3 cm apart on a map whose scale is 1 : 1000.

5 On a map of scale 1 : 300 000, the distance between York and Harrogate is 8 cm. What is the actual distance in km?

6 A builder's plan is drawn to a scale of 1 cm to 10 m. How long is a road which is 12 cm on the plan?

*7 The map on page 237 is drawn to a scale of 1 : 50 000.
Make your own measurements with a ruler to find the actual distance in km between
   a Goldings and Tewin Wood (marked ●)
   b Panshanger Aerodrome and Row Wood
   c Gravel Pit and Queen Hoo Hall.

EXAMPLE

The distance between two towns is 18 km.
How far apart will they be on a map of scale 1 : 50 000?

....

18 km = 1 800 000 cm

1 800 000 cm on land = $\dfrac{1}{50000}$ × 1 800 000 = 36 cm on map

Distance between towns on map = 36 cm

Exercise 29 ①

1 The distance between two towns is 15 km. How far apart will they be on a map of scale 1 : 10 000?

2 The distance between two points is 25 km. How far apart will they be on a map of scale 1 : 20 000?

3 The length of a road is 2·8 km. How long will the road be on a map of scale 1 : 10 000?

4 The length of a reservoir is 5·9 km. How long will it be on a map of scale 1 : 100 000?

5 Copy and complete the table.

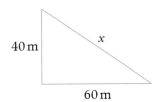

	Map scale	Actual length on land	Length on map
a	1 : 20 000	12 km	__cm
b	1 : 10 000	8·4 km	__cm
c	1 : 50 000	28 km	__cm
d	1 : 40 000	56 km	__cm
e	1 : 5000	5 km	__cm

6 The scale of a drawing is 1 cm to 10 m. The length of a wall is 25 m. What length will the wall be on the drawing?

Exercise 30 ①

In questions **1** and **2**, draw an accurate scale drawing of each shape using the scale shown.

1

12 m

6 m

Use 1 cm for every 3 m

2

40 m

x

60 m

Use 1 cm to every 10 m. Measure and write down the length of the side marked x (in cm).

3 This is a plan of Mr. Cooper's house and gardens. It has been drawn to a scale of 1 cm for every 2 m.

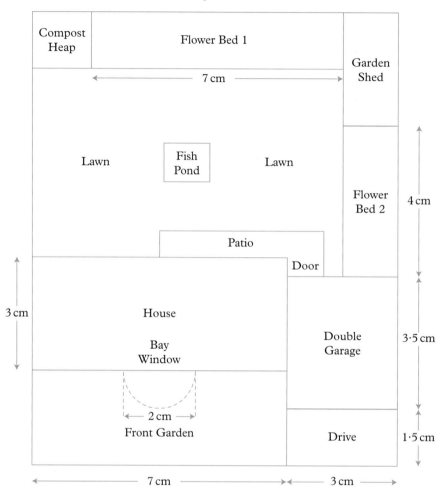

a How wide is:
   i the front garden
   ii the drive
   iii the bay window?
b How long is flower bed 2?
c How wide is flower bed 1?
d If the fish pond is 4 m wide, what size should it be on the plan?
e Measure carefully the width of the patio on the plan.
   How many *metres* wide is the real patio?
f What is the real *area* of the double garage?

5.5 Proportion

5.5.1 Direct proportion

When you buy petrol, the more you buy, the more money you have to pay. So if 2·2 litres costs 297p, then 4·4 litres costs 594p – which is twice as much.

The price you pay, £y, is **directly proportional** to the amount bought, x litres.

To show that two quantities are proportional, you use the symbol ∝. In this example, we can write *price ∝ amount*

or, in symbols, $\qquad\qquad\qquad y \quad ∝ \quad x.$

A graph of y against x is a straight line.

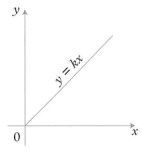

You say 'y varies directly as x' or 'y is proportional to x'.

> ▶ When y and x are **directly proportional**, the graph of y against x is a straight line which passes through origin.

EXAMPLE

If 12 calculators cost £54, find the cost of 17 calculators.

a First method

Find the cost of just one item.

12 calculators cost £54

1 calculator costs £54 ÷ 12 = £4·50

17 calculators cost £4·50 × 17
$\qquad\qquad\qquad = £76·50$

b Second method

Multiply by a fraction to get the answer in one step.

The answer is **more than** £54, so we need to multiply by a fraction which is **more than** 1, using the numbers 12 and 17.

17 calculators cost £54 × $\dfrac{17}{12}$ = £76·50

Exercise 31 ① ②

1 The variables in these tables are directly proportional.
Copy and complete the table. On square grid paper,
draw the graph of each table and label it.

a

Weight of sweets, g	1	2	3	4	5	10
Cost, pence	2					

b

Distance cycled, km	1	2	4	8	10	20
Time taken, min	4					

c

Loaves of bread	1	2	3	5	10	12
Cost, £	0.60					

2 If 5 books cost £15, find the cost of 8.

3 If 7 apples cost 63p, find the cost of 12.

4 If 4 batteries cost 180p, find the cost of 7.

5 Toy cars cost £3·36 for 8. Find the cost of
    a   3 toy cars   **b**   10 toy cars.

6 Crisps cost £1·32 for 12 packets.
Find the cost of
    a   20 packets   **b**   200 packets.

7 Stair carpet costs £78 for 12 m.
Find the cost of 15 m.

8 The total weight of 7 ceramic tiles is 1750 g. What do 11 tiles weigh?

9 A machine fills 2000 bottles in 10 minutes. How many bottles
will it fill in 7 minutes?

10 The total contents of 8 cartons of fruit drink is 12 litres. How much
fruit drink is there is 3 cartons?

11 Find the cost of 15 cakes if 9 cakes cost £2·07.

12 Find the cost of 7 screws if 20 screws cost £4·60.

13 How much would 7 cauliflowers cost, if 10 cauliflowers cost £5·70?

14 A machine takes 20 seconds to make 8 coins. How long does
it take to make 50 coins?

15 A plane flies 50 km in 15 minutes. Howlong will it take to fly 300 km?

Exercise 32 1 2

1 The pairs of variables, x and y, are directly proportional.
Copy and complete this table for each equation in x and y.
On square grid paper, draw the graph of each table and label
it with its equation.

x	1	2	3	5	10
y					

  a $y = 2x$　　　　**b** $y = 3x$　　　　**c** $y = 4x$

2 If 8 pencils cost 56p, how many can you buy for 70p?

3 If you can buy 6 pineapples for £3·12, how many can you
buy for £5·20?

4 If 20 m² of carpet costs £150, what area of carpet can you
buy for £90?

5 Tins of cat food cost £2·88 for 12.
How many tins can I buy for £2·16?

6 Twenty men produce 500 articles in 6 days. How many articles
would 4 men produce in 6 days?

7 Forty people take 8 days to produce 400 articles. How many
articles would 16 people produce in 8 days?

8 12 people produce 600 components in 12 hours.
How many components would 9 people produce in 12 hours?

9 7 cycles cost £623.
  a What is the cost of 3 cycles?
  b How many cycles can be bought for £979?

10 11 CDs cost £93·50.
  a What is the cost of 15 CDs?
  b How many CDs can be bought for £17?

11 $2\frac{1}{2}$ m of metal tube cost £1·40.

  Find the cost of
  a 4 m　　　　　　　　**b** $7\frac{1}{2}$ m.

12 $3\frac{1}{4}$ kg of sweets costs £2·60.

  Find the cost of
  a 2 kg　　　　　　　　**b** $4\frac{1}{2}$ kg.

13 A car travels 210 km on 30 litres of petrol. How much petrol is needed for a journey of 245 km?

14 A light aircraft flies 375 km on 150 litres of fuel. How much fuel is needed for a journey of 500 km?

15 A tank travels 140 miles on 40 gallons of fuel. How much fuel is needed for a journey of 245 miles?

5.5.2 Indirect proportion

If you travel 120 miles on a motorway at 30 miles per hour, you take 4 hours.
If you travel the same distance at 60 miles per hour, you take only 2 hours.
The faster you go, the less time you take.
In this case if you double your speed, you halve the time taken.

The time, y hours, taken for a journey is **inversely proportional** to your speed, x mph.

In this example, we can write $time \propto \dfrac{1}{speed}$.

or, in symbols, $$y \propto \frac{1}{x}.$$

A graph of y against x is a curved line showing y decreasing as x increases.

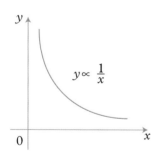

You say 'y varies inversely as x' or 'y is inversely proportional to x'.

▶ When y and x are **inversely proportional**, the graph of y against x is a curved line where y decreases as x increases.

The equation of the curve is $y = \dfrac{k}{x}$ where k is a constant.

EXAMPLE

The time taken for a liner to sail to a foreign port is 48 hours at a speed of 20 km/h.
Find the time taken if its increases its speed to 30 km/h.

a   First method

Find the time at a(very slow) speed of 1 km/h.

At 20 km/h, it takes 48 hours.
At 1 km/h it takes 20 times longer.
It takes $48 \times 20 = 960$ h

At 30 km/h, it is 30 times quicker.
It takes $\dfrac{960}{30} = 32$ hours.

b   Second method

Multiply by a fraction to get the answer in one step.

The answer is **less than** 48 hours because it goes quicker, so we need to multiply by a fraction which is **less than** 1, using the numbers 20 and 30.

At 30 km/h, it takes $48 \times \dfrac{20}{30} = 32$ hours

Exercise 33 ① ②

1   The variables in these tables are inversely proportional. Copy and complete the tables.
    On square grid paper, draw the graph of each table and label it.

   a

Time taken, h	1	2	3	4	5	10
Average speed, km/h	60					

   b

Number of people employed	1	2	3	5	10	12
Time taken for the job, h	120					

   c

Number of lottery winners	1	2	4	5	6	8
Prize money, £	24,000					

   d   What do you notice if you multiply the number pairs in each of these tables?

2   Six boys can paint a fence in 4 days. How long would 8 boys take to paint the same fence?

3   Water from 4 taps fill a water tank in 12 minutes. If one tap is closed, how long would it take the remaining 3 taps to fill in the same tank?

4   A bag of oats will feed 6 horses for 10 days. How many days would the same bag feed 4 horses?

5 A fast train travelling at 80 km/h takes 3 hours per journey. How long will it take a slower train travelling at 60 km/h?

6 A shelf can be filled with 18 books if each of them is 2 cm thick. If these books are replaced by 12 thicker books, what will be the average thickness of these new books?

7 Stephanie buys enough material to make 8 curtains, each of which is 3 metres wide. How wide would the curtains have been if she used the same material for 6 curtains?

8 If 9 men built a wall in 20 days, how long would 12 men have taken?

9 I buy stamps with 18 stamps in each of 12 rows. How many rows would there be if I bought the same number with only 8 stamps per row?

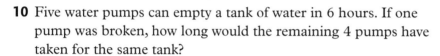

10 Five water pumps can empty a tank of water in 6 hours. If one pump was broken, how long would the remaining 4 pumps have taken for the same tank?

11 Tom can afford to buy 12 footballs if they cost £10·53 each. However if Tom decided to buy 13 smaller footballs and spend the same sum of money, what is the cost of a smaller football?

12 For each of these tables, $y \propto \dfrac{1}{x}$.

Copy the table and find the missing values of y.
Find the equation connecting x and y.

a

x	1	2	3	6	12
y	12				

b

x	2	4	6	8	12
y	24				

13 Which of these equations indicate that $y \propto x$ and which indicate that $y \propto \dfrac{1}{x}$?

a $y = 4x$ **b** $y = 6x$ **c** $y = \dfrac{12}{x}$ **d** $y = \dfrac{15}{x}$

e $y = \dfrac{1}{2}x$ **f** $y = \dfrac{8}{x}$ **g** $y = \dfrac{1}{4}x$ **h** $x \times y = 24$

14 Sketch the graphs of these equations. This means draw the general shape of the graph without numbering the axes.
Say if the graphs indicate direct proportion or inverse proportion.

a $y = 3x$ **b** $y = \dfrac{20}{x}$ **c** $y = \dfrac{3}{4}x$ **d** $x \times y = 15$

5.6 Mental arithmetic

Ideally these questions should be read out by a teacher or another student and you should not be reading them. Each of the 30 questions should be repeated once. The answer, and only the answer, should be written down.

Each test, including the recording of results, should take about 30 minutes.

If you do not have anyone to read out the questions for you, try to do the test without writing down any detailed working.

Test 1

1 Find the cost in pounds of ten books at 35 pence each.

2 Add together £4·20 and 75 pence.

3 What number divided by six gives an answer of four?

4 I spend £1·60 and pay with £2. My change consists of three coins. What are they?

5 Find the difference between $13\frac{1}{2}$ and 20.

6 Write one centimetre as a fraction of one metre.

7 How many ten pence coins are there in a pile worth £5?

8 Ten per cent of the students in a school play hockey, 15% play basketball and the rest play football. What percentage play football?

9 In a room of 20 people, three-quarters were women. What was the number of women?

10 Four lemons costing eleven pence each are bought with a one pound coin. What is the change?

11 Write the number that is thirteen less than one hundred.

12 What number is ten times as big as 0·6?

13 A hockey pitch measures 20 metres by 40 metres. Find the distance around the pitch.

14 What is twenty multiplied by ten?

15 The side of a square is four metres. What is the area of the square?

16 How many 2p coins are worth the same as ten 5p coins?

17 What number must be added to $1\frac{1}{4}$ to make $2\frac{1}{2}$?

18 Three books cost six pounds. How much will five books cost?

19 A rubber costs 20 pence. How many can be bought for £2?

20 What number is a hundred times as big as 0·15?

21 How many millimetres are there in 5 cm?

22 Find the mean of 12 and 20.

23 In the morning, the temperature is minus three degrees Celsius. What will be the temperature after it rises eleven degrees?

24 A certain number multiplied by itself gives 81 as the answer. What is half of that number?

25 The difference between two numbers is 15. One of the numbers is 90. What is the other?

26 How many half-litre glasses can be filled from a vessel containing ten litres of water?

27 How much will a dozen oranges cost at 20 pence each?

28 On a coach, forty-one out of fifty people are men. What percentage is this?

29 A prize of £400 000 is shared equally between one hundred people. How much does each person receive?

30 How many quarters make up two whole ones?

Test 2

1 What are 48 twos?

2 How many fives are there in ninety-five?

3 What is 6:30 a.m. on the 24-hour clock?

4 Add together £2·25 and 50 pence.

5 I go shopping with £2·80 and buy a magazine for ninety pence. How much money have I left?.

6 Two angles of a triangle are 65° and 20°. What is the third angle?

7 Write in figures the number 'five million, eighteen thousand and one'.

8 How many 20 pence biros can be bought for £3?

9 Work out 1% of £600.

10 A packet of 10 small cakes costs 35 pence. How much does each cake cost?

11 Add ten to 9 fives.

12 A packet of flour weighing 2400 grams is divided into three equal parts. How heavy is each part?

13 Add together 7, 23 and 44.

14 A car does 7 miles per litre of petrol. How far does the car travel on 40 litres of petrol?

15 How many twenty pence coins are needed to make eight pounds?

16 A certain butterfly lives for just 96 hours. How many days is this?

17 What number is 25 more than 37?

18 Find the average of 2, 5 and 8.

19 Pears cost eleven pence each. How many can I buy for sixty pence?

20 How many minutes are there in eight hours?

21 Write seven-tenths as a decimal number.

22 One-third of a number is six. What is the number?

23 Write one-fifth as a decimal.

24 Which is the larger: 0·7 or 0·071?

25 If a woman earns £8·40 per hour, how much does she earn in ten hours?

26 A car costing £2500 is reduced by £45. What is the new price?

27 How many half-kilogram packets of sugar can be filled from a large bowl containing 32 kilograms?

28 My daily paper costs 60 pence and I buy the paper six days a week. What is my weekly bill?

29 A car journey of 110 miles took two hours. What was the average speed of the car?

30 How many days will there be in February 2011?

Test 3

1 What number is fifteen more than fifty-five?

2 What is a tenth of 2400?

3 What is twenty times forty-five?

4 Write in figures the number ten thousand, seven hundred and five.

5 A play lasting $2\frac{1}{4}$ hours starts at half-past eight. When does it finish?

6 What number is fifty-five less than 300?

7 What is one-quarter of twenty-eight?

8 A book costs £1·95. How much change do I receive from a five pound note?

9 What is half of half of sixty?

10 What four coins make 61 pence?

11 Work out $\frac{1}{2}$ plus $\frac{1}{4}$ and give the answer as a decimal.

12 A box holds 16 cans. How many boxes are needed for 80 cans?

13 If the 25th of December is a Tuesday, what day of the week is the first of January?

14 By how much is two kilos more than 500 g?

15 Write fifteen thousand and fifty pence in pounds and pence.

16 The sides of a square field measure 160 metres. Find the total distance around the field.

17 A quarter of my wages is taken in tax. What percentage have I got left?

18 A bingo prize of £150 000 is shared equally between five people. How much does each person receive?

19 Ice creams cost twenty-four pence each. How many can I buy with one pound?

20 A bag contains 22 five-pence coins. How much is in the bag?

21 What is two point three multiplied by ten?

22 A wine merchant puts 100 bottles in crates of 12. How many crates does he need?

23 Add together 73 and 18.

24 What is 5% of £120?

25 Peaches cost twenty pence each. How much do I pay for seven peaches?

26 A toy costs 54 pence. Find the change from a five-pound note.

27 A boy goes to and from school by bus and a ticket costs 33 pence each way. How much does he spend in a five-day week?

28 In your purse, you have two ten-pound notes, three five-pound notes and seven one-pound coins. How much have you got altogether?

29 What is double seventeen?

30 Fifty per cent of a number is thirty-two. What is the number?

Test 4

1 What is the change from a £10 note for goods costing £1·95?

2 Add 12 to 7 nines.

3 How many 20 pence coins are needed to make £5?

4 A pile of 100 sheets of paper is 10 cm thick. How thick is each sheet?

5 Lemons cost 7 pence each or 60 pence a dozen. How much is saved by buying a dozen instead of 12 separate lemons?

6 How many weeks are there in two years?

7 What is 1% of £40?

8 How much more than £92 is £110?

9 My watch reads five past 6. It is 15 minutes fast. What is the correct time?

10 How many degrees are there in three right angles?

11 I am facing south-west and the wind is hitting me on my back. What direction is the wind coming from?

12 If eight per cent of students of a school are absent, what percentage of students are present?

13 I go shopping with £5 and buy 3 items at 25 pence each. How much money have I left?

14 From one thousand and seven, take away nine.

15 If I can cycle a mile in 3 minutes, how many miles can I cycle in one hour?

16 How many millimetres are there in 20 cm?

17 A metal rod 90 cm long is cut into four equal parts. How long is one part?

18 Find the cost of fifteen items at 5 pence each.

19 A 2 pence coin is about 2mm thick. How many coins are in a pile which is 2 cm high?

20 Add up the first four odd numbers.

21 Add up the first four even numbers.

22 My daily paper costs 25 pence. I pay for it with a £10 note. What change do I receive?

*23 A film starts at 8:50 p.m. and finishes at 9:15 p.m. How long is the film?

24 School finishes at twenty to four. What is that on the 24-hour clock?

25 What is the sum of the numbers 1, 2, 3, 4 and 5?

26 What is 10% of £7?

27 How many 2 pence coins are needed to make £4?

28 35% of a class prefer BBC1 and 30% prefer ITV1. What percentage prefer the other channels?

29 How many minutes is it between 6:20 p.m. and 8:00 p.m.?

30 What is the cost of 1000 books at £2·50 each?

Test 5

1 What are eight twenties?

2 What number is nineteen more than eighty-seven?

3 Write in figures the number six-thousand and eleven.

4 What is the cost of six items at thirty-five pence each?

5 What is the sum of sixty-three and twenty-nine?

6 How many sevens are there in eighty-four?

7 A pair of shorts costs £8·99. How much change do you get from a £10 note?

8 Subtract forty-five centimeters from two metres, giving your answer in metres as a decimal number.

9 If you have three thousand and eleven pennies, how much do you have in pounds and pence?

10 The perimeter of a square is sixteen centimetres. What is the length of one side of the square?

11 How many metres are there in 1·5 kilometres?

12 What is fifty per cent of fifty pounds?

13 How many sides has a heptagon?

14 I think of a number, double it and the answer is five. What was the number I thought of?

15 When playing darts, you score double twenty and treble eight. What is your total score?

16 How many inches are there in one foot?

17 A petrol pump delivers 1 litre in 5 seconds. How many litres will it deliver in one minute?

18 If seventy-seven per cent of the students in a school are right-handed, what percentage are left-handed?

19 How much more than 119 is 150?

20 What is the cube root of 27?

21 What number is eight squared?

22 A rectangular lawn is 7 yards wide and 12 yards long. What area does it cover?

23 How many centimetres are there in 1 km?

24 A ship was due at noon on Tuesday, but it arrived at 15:00 on Wednesday. How many hours late was it?

25 A litre of wine fills 9 glasses. How many litre bottles are needed to fill 50 glasses?

26 Work out 25% of £40.

27 How many grams are there in half a kilogram?

28 How many seconds are there in 3 minutes?

29 Write any multiple of nine.

30 If the eighth of May is a Monday, what day of the week is the thirteenth of May?

Test 6

1 What is the angle between the hands of a clock at two o'clock?

2 What is a half of a half of 60?

3 In a test, Paul got 15 out of 20. What percentage is that?

4 Work out $2 \times 20 \times 200$.

5 Two friends share a bill for £52. How much does each person pay?

6 What number is halfway between six and sixteen?

7 A television programme starts at five minutes to seven and lasts thirty-five minutes. At what time does the programme finish?

8 What is $\frac{1}{5}$ as a percentage?

9 What is twenty-fifteen in twelve hour clock time?

10 Work out 0·1 plus 0·01.

11 What number should you subtract from fifty-one to get the answer twenty-four?

12 What is ten per cent of £25?

13 A car has a 1795 cc engine. What is that approximately in litres?

14 Write the number two thousand, one hundred and four in figures.

15 Work out 75% of £20.

16 How many minutes are there in $2\frac{1}{2}$ hours?

17 A pie chart has a red sector representing 10% of the whole chart. What is the angle of the sector?

18 How many five-pence coins are needed to make £3?

19 What is the reflex angle between clock hands showing three o'clock?

20 A rectangular pane of glass is 3 feet long and 2 feet wide. Glass costs £1·50 per square foot. How much will the pane cost?

21 A car journey of 150 miles took 3 hours. What was the average speed?

22 Add 218 to 32.

23 Pencils cost 10 pence each. How many can I buy with £2·50?

24 Write the next prime number after 31.

25 A ruler costs 32 pence. What is the total cost of three rulers?

26 A salesman receives commission of 2% on sales. How much commission does he receive when he sells a computer for £1000?

27 How many edges does a cube have?

28 What four coins make seventy-six pence?

29 One angle in an isosceles triangle is one hundred and ten degrees. How large is each of the other two angles?

30 What number is one hundred times bigger than nought point two?

Test 7

1 Two angles of a triangle are 42° and 56°. What is the third angle?

2 Telephone charges are increased by 10%. What is the new charge for a call which previously cost 60p?

3 Write the number two thousand, one hundred and seven in figures.

4 What is nine hundred and fifty-eight to the nearest hundred?

5 In a survey, three-quarters of people like football. What percentage of people like football?

6 What decimal number is twenty-three divided by one hundred?

7 How many twenty-pence coins make three pounds?

8 How many twelve-pence pencils can you buy for one pound?

9 One-third of a number is eight. What is the number?

10 Write nine-tenths as a decimal number.

11 What is the name of the quadrilateral which has only one pair of parallel sides?

12 What number is 10 less than ninety thousand?

13 A sphere is a prism. True or false?

14 What is the probability of scoring less than six on a fair dice?

15 I think of a number, divide it by three and the answer is seven. What number did I think of?

16 What is the area of a rectangle nine metres by seven metres?

17 How many quarters are there in one and a half?

18 How many millimetres are there in one metre?

19 How many hours of recording time are there on a two hundred and forty minute DVD?

20 What number is squared to produce eighty-one?

21 A packet of peanuts costs 65 pence. I buy two packets and pay with a ten-pound note. Find the change.

22 What is a half of a half of 22?

23 I bought three kilograms of flour and I used four hundred and fifty grams. How many grams of flour do I have left?

24 A bingo prize of two hundred thousand pounds is shared equally between five people. How much does each person receive?

25 What is the angle between the hands of a clock at 4 o'clock?

26 Write down a sensible estimate for eleven multiplied by ninety-nine.

27 How many 24p stamps can I buy for £2?

28 A milk crate has space for 24 bottles. How many crates are needed for 100 bottles?

29 Write nought point two five as a fraction.

30 What is two hundred and ten divided by one hundred?

Test 8

1 $39 + 22$
2 $60 - 21$
3 20% of 50
4 $0.2 + 0.62$
5 20×6
6 $200 - 145$
7 £5 − £1·20
8 9×7
9 $14 + 140$
10 50×22
11 5% of 300
12 Half of 330
13 $600 - 245$
14 $2.4 + 1.7$
15 $200 \div 5$
16 8×25
17 60p + £1·50
18 $82 - 63$
19 $7 + 8 + 9$
20 $2 \times 3 \times 4$

Test 9

1 65×2
2 $84 - 7$
3 $0.7 + 0.3$
4 23×100
5 7^2
6 £10 − 50p
7 50% of 684
8 $1 - 0.2$
9 25×12
10 $8 + 9 + 10$
11 $210 \div 7$
12 100×100
13 $5.5 + 1.5$
14 $(2 + 3)^2$
15 $240 \div 6$
16 $8 - 2.5$
17 Half of 38·4
18 $400 \div 50$
19 15% of 300
20 $18 + 81$

Test 10

1 $7 + 77$
2 $330 - 295$
3 $3 (8 - 2)^2$
4 $37 + 63$
5 30×7
6 25×200
7 12×7
8 $5^2 - 5$
9 $76 + 14$
10 5% of 440
11 $500 - 85$
12 $0.6 + 0.4$
13 10^3
14 $425 - 198$
15 200×8
16 $420 \div 7$
17 $4^2 + 7$
18 $9 - 0.2$
19 $2.6 + 2.6$
20 $2600 \div 100$

Test 11

1 25% of 880
2 $400 \div 20$
3 8×7
4 $3.5 + 0.35$
5 1% of 20 000
6 $301 - 102$
7 $(3 + 8)^2$
8 Half of 630
9 1000×0.5
10 $3 \times 4 \times 5$
11 25×16
12 $54 \div 9$
13 30×60
14 $22 + 23 + 24$
15 $6^2 - 2^2$
16 $1100 - 999$
17 $11 - 0.3$
18 $200 \div 200$
19 60×60
20 $4 \times 5 \times 6$

5.7 Rounding

A car travels a distance of 158 miles in $3\frac{1}{2}$ hours. What is its average speed?

$$\text{Speed} = \frac{\text{Distance}}{\text{Time}} = \frac{158}{3 \cdot 5}$$

> You have met this formula in Chapter 4.

On a calculator, the answer is 45·142 857 14 mph.
It is not sensible to give all these figures in the answer.
The distance and time may not be all that accurate. It would be reasonable to give the answer as '45 mph'. This is called 'rounding'.

> ▶ You can approximate in several ways:
> • You can round to **the nearest whole number**
> • You can round to one or more **significant figures**
> • You can round to one or more **decimal places**.

5.7.1 Rounding to the nearest whole number, ten, hundred or thousand

> ▶ To round a number to the nearest whole number, look at the first digit after the decimal point to see if it is **'five or more'**.
> If that first digit is **'five or more'** you round **up**. Otherwise you round **down**.

Here are some examples.
a 13·82 = 14 to the nearest whole number
b 6·2 = 6 to the nearest whole number
c 211·54 = 200 to the nearest whole hundred
d 45·68 = 50 to the nearest ten

Exercise 34 ①

Round these numbers to the nearest whole number.

1 18·32	**2** 22·8	**3** 41·51
4 3·24	**5** 224·9	**6** 36·11
7 8·07	**8** 56·52	**9** 3·911

Round these numbers to the nearest hundred.

10 2314   **11** 187·2   **12** 111·1

13 2464   **14** 2048   **15** 712·89

16 62·66   **17** 5888   **18** 5742·2

19 154·1   **20** 65 540

Work these out on a calculator and then give the answer to the nearest whole number.

21 $8·2 \times 1·7$ **22** $11·3 \times 11·4$ **23** $8·06 \times ·19$

24 $6·9 \times 1·5$ **25** $0·71 \times 5·2$ **26** $16·4 \times 2$

27 $8·05 \div 5$ **28** $11·2 \div 3$ **29** $6·6 \times 6·6$

30 $27 \div 4·7$ **31** $10 \div 0·42$ **32** $624 \div 11$

33 $7084 \div 211$ **34** $6168 \div 217$ **35** $18·2^2$

36 $6·09^2$ **37** $6·84 + 11·471$ **38** $19 - 7·364$

39 $22·1 + 0·724$ **40** $16·3 - 7·82$

5.7.2 Rounding to one significant figure

▶ To round a number to one significant figure, look at the place value of the first non-zero digit and then round to this place value.

Here are some examples.

a 41 is 40 to one significant figure

b 278 is 300 to one significant figure

c 4562 is 5000 to one significant figure

d 84 796 is 80 000 to one significant figure

> Look at the red digit to see if it is 'five or more.'

Exercise 35 ①

Round these numbers to one significant figure.

1 214   **2** 378   **3** 4911   **4** 6684

5 8209   **6** 4592   **7** 376   **8** 29

9 42   **10** 196   **11** 417   **12** 4211

13 685   **14** 701   **15** 6666   **16** 28

17 4192   **18** 16 234   **19** 8523   **20** 672

Work these out on a calculator and then give the answer to one significant figure.

21 41×11 **22** 7×229 **23** 82×83 **24** 17×5

25 $3540 \div 15$ **26** $1426 \div 23$ **27** $682 \div 31$ **28** $1760 \div 32$

29 $4714 + 525$ **30** $6024 - 4111$ **31** $378 + 5972$ **32** $84 + 871 + 246$

33 The newspaper article contains several numbers in bold type.

   a For each number, decide whether or not to replace the number with an **approximate** value to an appropriate degree of accuracy. ('appropriate' means 'sensible')

   b Some of the numbers should **not** be replaced. State which these are.

> The Olympic swimming pool in Beijing contained **1493.2** m³ of water at a temperature of **23.41**°C. The crowd of **2108** cheered as Marisa won the **100** m butterfly in a new World Record time of **58.23** seconds. Altogether there were about **5173** swimmers taking part in the swimming events. The next Games take place in **2012** in ...

5.7.3 Rounding to two or three significant figures

> **EXAMPLE**
>
> Write these numbers correct to three significant figures (3 sf).
>
>   **a** 2·6582   **b** 0·5142
>
>   **c** 84 660   **d** 0·040 31
>
> ··
>
> In each case, look at the fourth significant figure to see if it is 'five or more'. Here you count figures from the left starting from the first **non-zero** figure.
>
>   **a** 2·658**2** = 2·66 to 3 sf   **b** 0·514**2** = 0·514 to 3 sf   Ignore the zero before the ·5
>
>   **c** 84 6**6**0 = 84 700 to 3 sf   **d** 0·040 3**1** = 0·0403 to 3 sf

Exercise 36 ①

In questions **1** to **8**, write the numbers correct to three significant figures.

  1 2·3462   **2** 0·814 38   **3** 26·241   **4** 35·55

  5 112·74   **6** 210·82   **7** 0·8254   **8** 0·031 162

In questions **9** to **16**, write the numbers correct to two significant figures.

  9 5·894   **10** 1·232   **11** 0·5456   **12** 0·7163

13 0·1443   **14** 1·831   **15** 24·83   **16** 31·37

In questions **17** to **24**, write the numbers correct to three significant figures.

17 486·72   **18** 500·36   **19** 2·8888   **20** 3·1125

21 0·071542   **22** 3·0405   **23** 2463·5   **24** 488852

In questions **25** to **36**, write the numbers to the degree of accuracy indicated.

25 0·5126 (3 sf)   **26** 5·821 (2 sf)   **27** 65·89 (2 sf)

28 587·55 (3 sf)   **29** 0·581 (1 sf)   **30** 0·0713 (1 sf)

31 5·8354 (3 sf)   **32** 87·84 (2 sf)   **33** 2482 (2 sf)

34 52 666 (3 sf)   **35** 0·0058 (1 sf)   **36** 6568 (1 sf)

> sf is shorthand for significant figures.

5.7.4 Rounding to one, two or three decimal places

EXAMPLE

Write these numbers correct to two decimal places (2 dp).

a   8·358   **b**   0·0328   **c**   74·355

In each case, look at the third decimal place to see if it is 'five or more'. Here you count figures after the decimal point.

a   8·**3**5**8** = 8·34 to 2 dp   **b**   0·0**3**2**8** = 0·03 to 2 dp   **c**   74·**3**5**5** = 74·36 to 2 dp

Exercise 37 ①

In questions **1** to **8**, write the numbers correct to two decimal places (2 dp).

> dp is shorthand for decimal places.

1 5·381   **2** 11·0482   **3** 0·414   **4** 0·3666

5 8·015   **6** 87·044   **7** 9·0062   **8** 0·0724

In questions **9** to **16**, write the numbers correct to one decimal place.

9 8·424   **10** 0·7413   **11** 0·382   **12** 0·095

13 6·083   **14** 19·53   **15** 8·111   **16** 7·071

In questions **17** to **28**, write the numbers to the degree of accuracy indicated.

17 8·155 (2 dp)   **18** 3·042 (1 dp)   **19** 0·5454 (3 dp)

20 0·005 55 (4 dp)   **21** 0·7071 (2 dp)   **22** 6·8271 (2 dp)

23 0·8413 (1 dp)   **24** 19·646 (2 dp)   **25** 0·071 35 (4 dp)

26 60·051 (1 dp)   **27** 7·30 (1 dp)   **28** 5·424 (2 dp)

29 Use a ruler to measure the dimensions of the rectangles.

    a Write the length and width in cm correct to one dp.

    b Work out the area of each rectangle and give the answer in cm² correct to one dp.

i

ii

Exercise 38 ❶

Write the answers to the degree of accuracy indicated.

1 0.153×3.74	(2 dp)		**2** $18.09 \div 5.24$	(3 sf)
3 184×2.342	(3 sf)		**4** $17.2 \div 0.89$	(1 dp)
5 $58 \div 261$	(2 sf)		**6** 88.8×44.4	(1 dp)
7 $(8.4 - 1.32) \times 7.5$	(2 sf)		**8** $(121 + 3758) \div 211$	(3 sf)
9 $(1.24 - 1.144) \times 0.61$	(3 dp)		**10** $1 \div 0.935$	(1 dp)
11 78.3524^2	(3 sf)		**12** $(18.25 - 6.941)^2$	(2 dp)
13 $9.245^2 - 65.2$	(1 dp)		**14** $(2 - 0.666) - 0.028$	(3 sf)
15 8.433	(1 dp)		**16** $0.924^2 - 0.835^2$	(2 dp)

5.8 Estimating

5.8.1 Estimating answers

In some circumstances it is not realistic to work out the exact answer to a problem. You can give an estimate for the answer. For example, a builder does not know **exactly** how many bricks a new garage will need. He may estimate that he needs 2500 bricks and place an order for that number. In practice, he may need only 2237.

Estimate the answers to these calculations.

  a $9.7 \times 3.1 \approx 10 \times 3$ About 30.

  b $81.4 \times 98.2 \approx 80 \times 100$ About 8000.

  c $19.2 \times 49.1 \approx 20 \times 50$ About 1000.

  d $102.7 \div 19.6 \approx 100 \div 20$ About 5.

▶   Estimates can also help you check that your exact answer is correct.

Exercise 39 ①

Write each question and decide (by estimating) which answer is closest to the exact answer. Do not do the calculations exactly.
Try to make the estimates mentally without writing any working.

	Question	Answer A	Answer B	Answer C
1	$7 \cdot 79 \div 1 \cdot 9$	8	4	2
2	$27 \cdot 03 \div 5 \cdot 1$	5	0·5	9
3	$59 \cdot 78 \div 9 \cdot 8$	12	3	6
4	$58 \cdot 4 \times 102$	600	6000	2400
5	$6 \cdot 8 \times 11 \cdot 4$	19	280	80
6	$97 \times 1 \cdot 08$	100	50	1000
7	$972 \times 20 \cdot 2$	2000	20 000	9000
8	$7 \cdot 1 \times 103$	70	700	7000
9	$18 \cdot 9 \times 21$	400	60	200
10	$1 \cdot 078 \div 0 \cdot 98$	6	10	1
11	$1250 \cdot 5 \div 6 \cdot 1$	20	200	60
12	$20 \cdot 48 \div 3 \cdot 2$	6	12	3
13	$25 \cdot 11 \div 3 \cdot 1$	8	15	20
14	$216 \div 0 \cdot 9$	50	24	240
15	$19 \cdot 2 + 0 \cdot 41$	23	8	20
16	$207 + 18 \cdot 34$	25	230	1200
17	$68 \cdot 2 - 1 \cdot 38$	100	48	66
18	$7 - 0 \cdot 64$	6	1·5	0·5
19	$974 \times 0 \cdot 11$	9	100	500
20	$551 \cdot 1 \div 11$	7	50	5000
21	$207 \cdot 1 + 11 \cdot 65$	300	20	200
22	$664 \times 0 \cdot 51$	150	300	800
23	$(5 \cdot 6 - 0 \cdot 21) \times 39$	400	200	20
24	$\dfrac{17 \cdot 5 \times 42}{2 \cdot 5}$	300	500	100
25	$(906 + 4 \cdot 1) \times 0 \cdot 31$	600	300	30
26	$\dfrac{543 + 472}{18 \cdot 1 + 10 \cdot 9}$	60	30	100
27	$18 \cdot 9 \times 21 \cdot 4$	200	400	4000
28	$5 \cdot 14 \times 5 \cdot 99$	15	10	30
29	$811 \times 11 \cdot 72$	8000	4000	800
30	99×98	1 million	100 000	10 000
31	$1 \cdot 09 \times 29 \cdot 6$	20	30	60
32	$81 413 \times 10 \cdot 96$	8 million	1 million	800 000
33	$601 \div 3 \cdot 92$	50	100	150

Exercise 40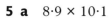

1 For a wedding the caterers provide food at £39·75 per head. When Janet married John there were 207 guests at the wedding. Estimate the total cost of the food.

2 985 people share the cost of hiring an ice rink. About how much does each person pay if the total cost is £6017?

3 A home-made birthday card costs £3·95. Estimate the total cost of 107 cards.

4 The rent for a flat is £104 per week. Estimate the total spent on rent in one year.

In questions **5** and **6**, there are six calculations and six answers. Write each calculation and insert the correct answer from the list given. Use estimation.

5 a 8·9 × 10·1   **b** 7·98 ÷ 1·9   **c** 112 × 3·2

    d 11·6 + 47·2   **e** 2·82 ÷ 9·4   **f** 262 ÷ 100

Answers: | 2·62 | 58·8 | 0·3 | 89·89 | 358·4 | 4·2 |

6 a 49·5 ÷ 11   **b** 21 × 22   **c** 9·1 × 104

    d 86 − 8·2   **e** 2·4 ÷ 12   **f** 651 ÷ 31

Answers: | 21 | 946·4 | 0·2 | 4·5 | 462 | 77·8 |

In questions **7** to **14**, estimate which answer is closest to the actual answer.

7 The height of a double-decker bus

A	B	C
3 m	6 m	10 m

8 The height of the tallest player in the Olympic basketball competition

A	B	C
18 m	3·0 m	2·2 m

9 The mass of a £1 coin

A	B	C
1 g	10 g	100 g

10 The volume of your classroom

A	B	C
20 cubic metres	200 cubic metres	2000 cubic metres

11 The top speed of a Grand Prix racing car

A	B	C
600 km/h	80 km/h	300 km/h

12 The number of times your heart beats in one day (24 h)

A	B	C
10 000	100 000	1 000 000

The heart of an average person beats 72 times per minute.

13 The thickness of one page in this book

A	B	C
0·01 cm	0·001 cm	0·0001 cm

14 The number of cars in a traffic jam 10 km long on a 3-lane motorway. (Assume each car takes up 10 m of road.)

A	B	C
3000	30 000	300 000

15 A rectangular floor measures 28 metres by 43 metres. It costs £6·95 per square metre to treat the floor for damp. Estimate the cost of this job, showing all your working.

*16 A shopkeeper wants to buy

28 windows at £97   each
63 lights   at £4·95   each
32 shelves   at £19·99 each.

Do a rough calculation to find the total cost.

5.8.2   Checking answers

Here are five calculations followed by appropriate checks that use inverse operations.

a   22·5 ÷ 5 = 4·5   check   4·5 × 5
b   29·5 − 1·47 = 28·03   check   28·03 + 1·47
c   78·5 × 20 = 1570   check   1570 ÷ 20
d $\sqrt{11}$ = 3·316 62   check   3·316 62^2
e   14·7+28·1+17·4+9·9   check   9·9 + 17·4 + 28·1 + 14·7 (add in reverse order)

Exercise 41 ①

1   Do these and check using inverse operations.

a   83·5 × 20 = ☐   check   ☐ ÷ 20
b   104 − 13·2 = ☐   check   ☐ + 13·2
c   228·2 ÷ 7 = ☐   check   ☐ × 7
d $\sqrt{28}$ = ☐   check   ☐^{2}
e   11·5 + 2·7 + 9·8 + 20·7   check   20·7 + 9·8 + 2·7 + 11·5

2 **a** Will the answer to 64 × 0·8 be larger or smaller than 64?

b Will the answer to 21040 ÷ 7 be larger or smaller than 210?

c Will the answer to 17·4 × 0·9 be larger or smaller than 17·4?

3 These cards, **a** to **f**, show the answers found by six students. Some are correct but some are clearly impossible or highly unlikely. Decide which answers are 'OK' and which are 'impossible' or 'highly unlikely'.

a The height of the three boxes on this trolley is 110 cm

b Time taken to walk 1 mile to school is 21 minutes

c The height of the classroom door is 220 cm

d The number of schools in the UK is 500 000

e The mean value of the numbers 32, 35, 31, 36, 32 is 37·8

f One per cent of the UK population is 60 000 people.

5.9 Solving numerical problems 2

Exercise 42 ①

1 Four dozen bags of grain weigh 2016 kg. How much does each bag weigh?

2 An office building has twelve floors and each floor has twenty windows.
A window cleaner charges 50p per window.
How much will he charge to clean all the windows in the building?

3 Answer 'always true' or 'sometimes false' for each statement.

a $3 × n = 3 + n$ **b** $n × n = n^2$ **c** $m + n = n + m$

d $3n − n = 3$ **e** $a − b = b − a$ **f** $n + n^2 = n^3$

4 A rectangular wheat field is 200m by 400 m. One hectare is $10\,000\,m^2$ and each hectare produces 3 tonnes of wheat. How much wheat is produced in this field?

5 A powerful computer is hired out at a rate of 50p per minute. How much will it cost to hire the computer from 06:30 to 18:00?

6 An old sailor keeps all of his savings in gold. Altogether the gold weighs ten pounds. The price of gold is \$520 an ounce. How much is his gold worth?

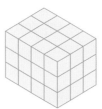 1 pound = 16 ounces

7 This packet of sugar cubes costs 60p.

How much would you have to pay for this packet?

8 A wall measuring 3 m by 2 m is to be covered with square tiles of side 10 cm. If tiles cost £ 3·40 for a packet of ten, how much will the tiles cost to cover the wall?

9 Draw the next member of each sequence.

a

b

c

Exercise 43 ①

1 The table shows the results of a test given to 60 students.

Mark	0	1	2	3	4	5
Number of students	1	4	10	12	15	18

   a How many students scored fewer than 3 marks?
   b Find the percentage of students who scored
     i 2 marks   **ii** 5 marks.

2 The thirteenth number in the sequence
1, 3, 9, 27, . . . is 531 441.
   a What is the position-to-term rule for the sequence?
   b What is the twelfth number?
   c What is the fourteenth number?

3 Six sacks of corn will feed 80 hens for 12 days.
Copy and complete these statements.

   a 18 sacks of corn will feed 80 hens for ____ days.
   b 6 sacks of corn will feed 40 hens for ____ days.
   c 60 sacks of corn will feed 40 hens for ____ days.

4 Copy each sequence and write the next two numbers.

    a 2, 8, 20, _____, _____   **b** 1, 8, 27, _____, _____

    c 144, 121, 100, _____, _____   **d** $\frac{3}{4}, \frac{4}{5}, \frac{5}{6},$ _____, _____

5 What is the smaller angle between the hands of a clock at four o'clock?

6 Write the coordinates of the points which make up the '3.' You must give the points in the correct order starting at point(1, 2).

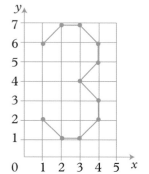

7 I think of a number. If I add 7 and then multiply the result by 10, the answer is 90. What number was I thinking of?

8 How many 31p stamps can be bought for £2 and how much change will there be?

9 How many 10mm pieces of string can be cut from a string of length 1 metre?

10 Here are some function machines. Copy them and fill in the missing numbers.

    a 30 → $\boxed{\times 3}$ → ?   **b** 0·1 → $\boxed{\times 10}$ → ?   **c** 207 → $\boxed{\div 9}$ → ?

    d ? → $\boxed{+18}$ → 62   **e** ? → $\boxed{\div 7}$ → 11   **f** ? → $\boxed{\times 6}$ → 666

Exercise 44 ①

1 A slimmer's calorie guide shows how many calories are contained in various foods.

Bread 1·2 calories per g
Cheese 2·5 calories per g
Meat 1·6 calories per g
Butter 6 calories per g

Calculate the number of calories in this meal:
50 g bread, 40 g cheese, 100 g meat, 15 g butter.

2 Write these as single numbers.

    a 8^2 **b** 14   **c** 10^2

    d 3×10^3 **e** 25   **f** 3^4

3 Here is a row of numbers

    1 2 3 4 5 6 7 8 9 10 11 12 13 14 15 16 17 18

    a Find **two** numbers next to each other which add up to 27.

    b Find **three** numbers next to each other which add up to 24.

    c Find **four** numbers next to each other which add up to 46.

4 Find the area of the shaded shape.

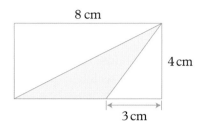

8 cm

4 cm

3 cm

5 Write the number that is 10 more than

   a 263   **b** 7447   **c** 74·5   **d** 295

6 Write the next number in each sequence. Then, for each sequence,
write either the term-to-term rule or theposition-to-term rule.

   a   2   9   16   23   ?

   b   2   4   8   16   ?

   c   80   40   20   10   ?

7 A group of four adults are planning
a holiday in France. The box shows
the ferry costs for the return journey.

| Adult | £25 |
| Car | £62 |

   The distance they travel around France is
estimated at 2000 km and petrol costs 1·4 euros
per litre. The car travels 10 km on one litre of petrol.
Find the total cost of the ferry and the petrol in euros,
given that £1 = 1·20 euros.

8 A journey by boat takes 2 hours 47 minutes. How long will
it take at half the speed?

9 Ten posts are equally spaced in a straight line. It is 450m from the
first to the tenth post. What is the distance between successive posts?

10 Find the smallest whole number that is exactly divisible
by all the numbers 1 to 5 inclusive.

Exercise 45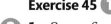

1 Seven fig rolls together weigh 560 g. A calorie guide
shows that 10 g of fig roll contains 52 calories.
  a How much does one fig roll weigh?
  b How many calories are there in 1 g of fig roll?
  c How many calories are there in one fig roll?

2 Find each of these mystery numbers.
  a I am an odd number and a prime number. I am
     a factor of 14.
  b I am a two-digit multiple of 50.
  c I am one less than a prime number which is even.
  d I am odd, greater than one and a factor of both 20 and 30.

3 To the nearest whole number, 5·84 is 6, 7·781 is 8 and
16·23 is 16.
  a Use these approximate values to obtain an approximate
     result for $\dfrac{5 \cdot 84 \times 16 \cdot 23}{7 \cdot 781}$.
  b Use the same approach to obtain approximate results for

  i $\dfrac{15 \cdot 72 \times 9 \cdot 78}{20 \cdot 24}$ **ii** $\dfrac{23 \cdot 85 \times 9 \cdot 892}{4 \cdot 867}$

4 King Richard is given three coins which
look identical, but in fact one of them is an
overweight fake.
Describe how he could discover the fake
using an ordinary balance and only **one**
weighing operation.

5 A light aircraft flies 375 km on 150 litres of fuel. How much
fuel does it need for a journey of 500 km?

6 The head teacher of a school discovers an oil well in the school
playground. As is the custom in such cases, he receives all the money
from the oil. The oil comes out of the well at a rate of £15 for every
minute of the day and night. How much does the head teacher
receive in a 24-hour day?

7 A map uses a scale of 1 to 100 000. Calculate the actual length,
in km, of a canal which is 5·4 cm long on the map.

8 Copy and complete this multiplication square.

×	6	3		
		15		35
			36	
2				
			32	56

9 I think of a number. If I subtract 4, and then divide the result by 4, the answer is 3. What number was I thinking of?

10 Copy the 9 dots. Try to draw five straight lines which pass through all of the 9 dots, without taking your pen from the paper and without going over any line twice.

- •   •   •

- •   •   •

- •   •   •

> The lines can go outside the pattern of dots.

Exercise 46

1 A car travels 35 m in 0·7 seconds. How far does it travel in
  a 0·1 seconds **b** 1 second **c** 2 minutes?

2 Twelve calculators cost £102. How many calculators can you buy for £76·50?

3 The outline of a 50p coin is shown.

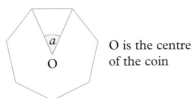

O is the centre of the coin

Calculate the size of the angle, a, to the nearest $\dfrac{1}{10}$ of a degree.

4 The diagram shows the map of a
farm which grows four different
crops in the regions shown.

Each square represents one acre.
a What is the total area of the farm?
b What area is used for crop A?
c What percentage of the farm is used for crop C?
d What percentage of the farm is used for crop D?

5 A farm uses 40 litres of heating oil every day and each
litre costs 56 p. How much is spent on oil for the 365 days
of a whole year?

6 Adam worked 7 hours per day from Monday to Friday
and 4 hours overtime on Saturday. The rate of pay from
Monday to Friday is £9 per hour and the overtime rate is
time and a half. How much did he earn during the week?

7 An examination is marked out of a total of 120 marks.
How many marks did Alan get if he scores 65% of the marks?

8 A shopkeeper buys coffee at £3·65 per kg and sells it at 95p
per 100 g. How much profit does he make per kg?

9 Five 2s can be used to make 25 as shown.

$$22 + 2 + \frac{2}{2} = 25$$

a Use four 9s to make 100   **b** Use three 6s to make 7

c Use three 5s to make 60   **d** Use five 5s to make 61

e Use four 7s to make 1   **f** Use three 8s to make 11

10 Copy and find the missing digits in these calculations.

a

	2	
+	5	4
		7

b

	1	7
+		6
	6	

c

	5		2
+	1	3	
		1	8

d

	4		4
+		5	
	8	2	4

e

	8	
−		4
	5	2

f

	8		2
−		5	
	2	3	2

Test yourself

 You may **not** use a calculator in questions **1** to **12**.

1 **a** What percentage of pupils in Class A are girls when $\frac{3}{5}$ of the pupils are boys?

  b Class B has 25 pupils of which 44% are girls.
    How many girls are there in this class?

  c Can you tell whether or not there are more girls in Class B than in Class A? You must give a reason for your answer.

(WJEC, 2011)

2 Here is a list of ingredients for making 18 mince pies.

> **Ingredients for 18 mince pies**
> 225 g of butter
> 350 g of flour
> 100 g of sugar
> 280 g of mincemeat
> 1 egg

Elaine wants to make 45 mince pies.
Elaine has
  1 kg of butter
  1 kg of flour
  500 g of sugar
  600 g of mincemeat
  6 eggs
Does Elaine have enough of each ingredient to make 45 mince pies?
You must show clearly how you got your answer.

(Edexcel, 2013)

3 **a** Cheryl scored 60 marks out of 80 in a test.
    Express Cheryl's score as a percentage.

  b Share £300 in the ratio 5 : 7.

  c Calculate 75% of £562.80.

(WJEC, 2011)

4 Colin, Dave and Emma share some money.
Colin gets $\frac{3}{10}$ of the money.
Emma and Dave share the rest of the money in the ratio 3 : 2
What is Dave's share of the money?

(Edexcel, 2013)

5 Naomi used $3\frac{3}{4}$ cans of paint for the living room and $1\frac{2}{3}$ cans for the kitchen. Show how to work out the total amount of paint used without using a calculator.

(CCEA, 2013)

6 The table shows Ann's marks in two tests.

Test	Mark
1	60 out of 80
2	70 out of 100

In which test did Ann do better?
You **must** show your working.

(AQA, 2007)

7 Work out
   a $2655 \div 49$

   b 417×28

   c   40% of 150

(OCR, 2004)

8 A school hall costs £200 to hire.
Parents get a 10% discount.
30 parents share the cost equally.

How many does each parent pay?

(AQA, 2013)

9 A book store displays the following offer.

> **BUY 3 AND GET THE CHEAPEST FREE**

Karl wants to buy seven books costing £5·99, £7·50, £7·50, £7·99, £7·99, £10 and £10. Karl wants to pay as little as possible for the books.

Show whether this offer is as good for Karl as another offer, which would give him 25% off the total cost of the seven books.

(WJEC, 2012)

10 Naomi wins £1200 in a lottery.

She spends $\frac{1}{4}$ of the money on a laptop and she gives $\frac{1}{10}$ of the money to her parents. How much money does she have left?

(OCR, 2013)

11 Calculate.

$$\frac{6 \cdot 3^2 - 3 \cdot 7}{5 \cdot 8}$$

Write your answer correct to 2 decimal places.

(OCR, 2013)

12 Robert and his family are going on holiday to France.
A bank gives Robert this chart to help him to
change between pounds (£) and euros (€).

Robert changes £600 into euros (€).
a How many euros should Robert get?

In France, a laptop costs €540.
In England, the same laptop costs £460.
b Work out the difference between the cost of
the laptop in France and the cost of the laptop
in England.
You must show clearly how you got your answer.

(Edexcel, 2013)

pounds (£)		euros (€)
1	=	1·2
2	=	2·4
5	=	6·0
10	=	12·0
20	=	24·0
50	=	60·0
100	=	120·0

13 Ben bought a car for £12 000.
Each year the value of the car depreciated by 10%.
Work out the value of the car two years after he bought it.

(Edexcel, 2003)

You may use a calculator in questions **14** to **22**.

14 A sports competition was held in London and Paris. The prize money
in London was £140000 and the prize money in Paris was €160000
(£1 = €1.15)

Which city paid out the most prize money?
You must show working.

(CCEA, 2013)

15 Tyre pressure for bicycles is measured in pounds
per square inch (PSI) or Bars.
Here is a conversion graph for PSI and Bars.
a Use the graph to convert 40 PSI to Bars.
b The tyre pressure for a racing bicycle is 100 PSI.
Work out this pressure in Bars.
You **must** show your working.

(AQA, 2012)

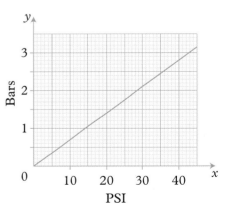

16 Audrey, Becks and Clare invest money in a business
in the ratio 3:4:5
Audrey invests £5400
How much do Becks and Clare each invest?

(CCEA, 2013)

17 a Lucinda is making mushroom soup.

> Mushroom Soup
> Serves 4 people
>
> 500 g mushrooms
> 1 litre chicken stock
> 100g butter
> 3 tablespoons flour
> 4 tablespoons cream

She needs to make enough soup to serve 6 people.
i How much butter does she need?
ii Mushrooms are sold in 300 g packs.
How many packs does Lucinda need to buy?

b The soup takes 15 minutes to prepare and 20 minutes to cook.
Lucinda wants to serve the soup at 1315.
At what time should she start preparing the soup?

c Jo has 22 litres of hot chocolate to pour into mugs.
The mugs are cylinders with an internal diameter of 9cm and an
internal height of 12 cm. Each mug is filled to 1 cm from the top.
How many mugs can Jo fill?

(OCR, 2013)

18 In 2011, Greenmeadows Tennis Club had 25 members and in
2012 it had 31 members. Calculate the percentage increase in
the number of members.

(OCR, 2013)

19 A garage sells British cars and foreign cars.
The ratio of the number of British cars sold to the number of
foreign cars sold is 2 : 7
The garage sells 45 cars in one week.
a Work out the number of British cars the garage sold that week.

A car tyre costs £80 plus VAT at $17\frac{1}{2}\%$.

b Work out the total cost of the tyre.

(Edexcel, 2008)

20 A company offers its workers a choice on how much their salary
will increase next year. Each worker can receive either a £500
increase or a 2% increase on their present salary.

Janet is currently on a salary of £24 000 per year.

Which option should Janet choose? You **must show the calculations**
that support your answer.

(WJEC, 2012)

21 a Which of these fractions is nearest in value to $\frac{1}{4}$?

$$\frac{3}{10} \quad \frac{7}{20} \quad \frac{4}{15} \quad \frac{17}{60}$$

Show your working.

b Write down the meaning of $0.5\dot{7}$

(CCEA, 2013)

22 a The timetable below shows the times of the Talyllyn Railway from Tywyn Wharf to Nant Gwernol during one day.

Tywyn Wharf	10 30	11 40	12 15	13 20	14 30	15 05	16 10
Rhydyronen	10 42	11 52	12 27	13 32	14 42	15 17	16 22
Dolgoch Falls	11 01	12 11	12 48	13 53	15 01	15 38	16 43
Abergynolwyn	11 15	12 25	13 00	14 05	15 15	15 50	16 55
Nant Gwernol	11 23	12 33	13 08	14 13	15 23	15 58	17 18

i How many trains leave Tywyn Wharf during the afternoon?

ii At what time does the 11 40 train from Tywyn Wharf arrive at Nant Gwernol?

iii How long, in minutes, is the journey on the 11 40 train from Tywyn Wharf to Nant Gwernol?

b The return timetable from Nant Gwernol to Tywyn Wharf is shown below.

Nant Gwernol	11 35	12 45	13 20	14 25	15 35	16 10	17 27
Abergynolwyn	11 41	12 51	13 26	14 31	15 41	16 16	17 33
Dolgoch Falls	12 24	13 29	14 06	15 14	16 19	16 56	17 49
Rhydyronen	12 39	13 44	14 21	15 29	16 34	17 11	18 04
Tywyn Wharf	12 55	14 00	14 40	15 45	16 50	17 27	18 20

Sam catches the 11 40 train from Tywyn Wharf to Nant Gwernol. He must arrive back at Tywyn Wharf no later than 17 00. What is the longest time Sam could spend in Nant Gwernol?

c Mr and Mrs Townley have two children aged 4 and 10. They are going to travel on the railway from Tywyn Wharf to Nant Gwernol and then back to Tywyn Wharf. The costs of travel are shown below.

	Return
Adult	£13·00
Child (5–15 years)	£3·00
Family (2 adults + 1 child over 5)	£28·50
Under 5 years travel FREE.	

How much cheaper is it to buy a family ticket than to buy individual tickets?

(OCR, 2013)

6 Algebra 2

In this unit you will:

- revise using brackets in equations
- learn how to change the subject of a formula
- learn how to display and interpret inequalities
- revise straight-line graphs and learn about the equation of a straight line
- learn how to solve simultaneous equations
- learn how to plot and solve quadratic functions.

You can model the path of a ball through the air with an equation.

6.1 Factors

▶ You can factorise expressions by taking out **common factors**.

EXAMPLE

Factorise these expressions. **a** $12a - 15b$
b $3x^2 - 2x$
c $2xy + 6y^2$

....

a 3 divides exactly into 12 and 15, so 3 is a **common factor**.
$12a - 15b = 3(4a - 5b)$
b x is a common factor.
$3x^2 - 2x = x(3x - 2)$
c 2 and y are both common factors.
$2xy + 6y^2 = 2y(x + 3y)$

Exercise 1 ①

In questions **1** to **10**, copy and complete the statements.

1 $6x + 4y = 2(3x + \square)$

2 $9x + 12y = 3(\square + 4y)$

3 $10a + 4b = 2(5a + \square)$

4 $4x + 12y = 4(\square + \square)$

5 $10a + 15b = 5(\square + \square)$

6 $18x - 24y = 6(3x - \square)$

7 $8u - 28v = \square(\square - 7v)$

8 $15s + 25t = \square(3s + \square)$

9 $24m + 40n = \square(3m + \square)$

10 $27c - 72d = \square(\square - 8d)$

In questions **11** to **31**, factorise the expressions.

11 $20a + 8b$ **12** $30x - 24y$ **13** $27c - 33d$

14 $35u + 49v$ **15** $12s - 32t$ **16** $40x - 16t$

17 $24x + 84y$ **18** $12x + 8y + 16z$ **19** $12a - 6b + 9c$

20 $10x - 20y + 25z$ **21** $20a - 12b - 28c$ **22** $48m + 8n - 24x$

23 $42x + 49y - 21z$ **24** $6x^2 + 15y^2$ **25** $20x^2 - 15y^2$

26 $7a^2 + 28b^2$ **27** $27a + 63b - 36c$ **28** $12x^2 + 24xy + 18y^2$

29 $64p - 72q - 40r$ **30** $36x - 60y + 96z$ **31** $9x + 6xy - 3x^2$

Factorise these expressions.

32 $x^2 - 5x$ **33** $2x^2 - 3x$ **34** $7x^2 + x$

35 $y^2 + 4y$ **36** $2x^2 + 8x$ **37** $4y^2 - 4y$

38 $p^2 - 2p$ **39** $6a^2 + 2a$ **40** $2ab - a$

41 $3xy + 2x$ **42** $3t + 9t^2$ **43** $4 - 8x^2$

44 $5x - 10x^3$ **45** $4\pi r^2 + \pi rh$ **46** $\pi r^2 + 2\pi r$

6.2   Changing the subject of a formula

▶   You can rearrange any formula to have a different symbol before the equals sign. This is called 'changing the subject' of the formula.

You need to remember to do the same operation to both sides at each step to keep the formula correct.

6.2.1   Simple formulae

EXAMPLE

Make x the subject in these formulae.

a $ax - p = t$ **b** $y(x + y) = v^2$

....

a $ax - p = t$ **b** $y(x + y) = v^2$

       $ax = t + p$ add p $yx + y^2 = v^2$ expand the bracket

       $x = \dfrac{t + p}{a}$ divide by a $yx = v^2 - y^2$ subtract y^2

                                                     $x = \dfrac{v^2 - y^2}{y}$ divide by y

Exercise 2 ①

Make x the subject.

1 $x + b = e$ **2** $x - t = m$ **3** $x - f = a + b$

4 $x + h = A + B$ **5** $x + t = y + t$ **6** $a + x = b$

7 $k + x = m$ **8** $v + x = w + y$ **9** $ax = b$

10 $hx = m$ **11** $mx = a + b$ **12** $kx = c - d$

13 $vx = e + n$ **14** $3x = y + z$ **15** $xp = r$

16 $xm = h - m$ **17** $ax + t = a$ **18** $mx - e = k$

19 $ux - h = m$ **20** $ex + q = t$ **21** $kx - u^2 = v^2$

22 $gx + t^2 = s^2$ **23** $xa + k = m^2$ **24** $xm - v = m$

25 $a + bx = c$ **26** $t + sx = y$ **27** $y + cx = z$

28 $a + hx = 2a$ **29** $mx - b = b$ **30** $kx + ab = cd$

31 $a(x - b) = c$ **32** $c(x - d) = e$ **33** $m(x + m) = n^2$

34 $k(x - a) = t$ **35** $h(x - h) = k$ **36** $m(x + b) = n$

37 $a(x - a) = a^2$ **38** $c(a + x) = d$ **39** $m(b + x) = e$

> In questions **31** to **39**, multiply out the brackets first.

6.2.2   Formulae involving fractions

EXAMPLE

Make x the subject in these formulae.

a $\dfrac{x}{a} = p$

b $\dfrac{m}{x} = t$

....

a $\dfrac{x}{a} = p$ multiply by a

  $x = ap$

b $\dfrac{m}{x} = t$

  $m = xt$ multiply by x

  $\dfrac{m}{t} = x$ divide by t

Exercise 3 ①

Make x the subject.

1 $\dfrac{x}{t} = m$ **2** $\dfrac{x}{e} = n$ **3** $\dfrac{x}{p} = a$

4 $am = \dfrac{x}{t}$ **5** $bc = \dfrac{x}{a}$ **6** $e = \dfrac{x}{y^2}$

7 $\dfrac{x}{a} = b + c$ **8** $\dfrac{x}{t} = c - d$ **9** $\dfrac{x}{m} = s + t$

10 $\dfrac{x}{k} = h + i$ **11** $\dfrac{x}{b} = \dfrac{a}{c}$ **12** $\dfrac{x}{m} = \dfrac{z}{y}$

13 $\dfrac{x}{h} = \dfrac{c}{d}$

14 $\dfrac{m}{n} = \dfrac{x}{e}$

15 $\dfrac{b}{e} = \dfrac{x}{h}$

16 $\dfrac{x}{a+b} = c$

17 $\dfrac{x}{h+k} = m$

18 $\dfrac{x}{u} = \dfrac{m}{y}$

19 $\dfrac{x}{h-k} = t$

20 $\dfrac{x}{a+b} = z+t$

21 $t = \dfrac{e}{x}$

22 $a = \dfrac{e}{x}$

23 $m = \dfrac{h}{x}$

24 $\dfrac{a}{b} = \dfrac{c}{x}$

> When x is on the bottom, multiply both sides by x.

6.2.3 Formulae with negative x-terms

EXAMPLE

Make x the subject of these formulae.

a $t - x = a^2$

b $h - bx = m$

---

a $t - x = a^2$

$\quad\quad t = a^2 + x$

$\quad t - a^2 = x$

$\quad$ or $\quad x = t - a^2$

b $h - bx = m$

$\quad\quad h = m + bx$

$\quad h - m = bx$

$\quad \dfrac{h - m}{b} = x$

Exercise 4 ①

Make x the subject.

1 $a - x = y$

2 $h - x = m$

3 $z - x = q$

4 $v = b - x$

5 $m = k - x$

6 $h - cx = d$

7 $y - mx = c$

8 $k - ex = h$

9 $a^2 - bx = d$

10 $m^2 - tx = n^2$

11 $v^2 - ax = w$

12 $y - x = y^2$

13 $k - t^2x = m$

14 $e = b - cx$

15 $z = h - gx$

16 $a + b = c - dx$

17 $y^2 = v^2 - kx$

18 $h = d - fx$

19 $a(b - x) = c$

20 $h(m - x) = n$

Exercise 5 ①

1 A formula for calculating velocity is $v = u + at$.
  a Rearrange the formula to express a in terms of v, u and t.
  b Calculate a when $v = 20$, $u = 4$, $t = 8$.

> Make a the subject.

2 $P = \dfrac{mk}{y}$

  a Express k in terms of P, m and y.

  b Express y in terms of P, m and k.

3 A formula for calculating the cost of repair bills, £R, is $R = \dfrac{n - d}{p}$.

  a Express n in terms of R, p and d.
  b Calculate n when $R = 400$, $p = 3$ and $d = 55$.

Exercise 6 ①

This exercise is different, as it contains a mixture of questions from the last four exercises.

Make the letter in square brackets the subject of the formula.

1 $ax - d = h$ [x]　　**2** $zy + k = m$ [y]　　**3** $d(y + e) = f$ [y]

4 $m(a + k) = d$ [k]　　**5** $a + bm = c$ [m]　　**6** $ae = b$ [e]

7 $yt = z + a$ [t]　　**8** $ax - c = e$ [x]　　**9** $my - n = b$ [y]

10 $a(z + a) = b$ [z]　　**11** $\dfrac{a}{x} = d$ [x]　　**12** $\dfrac{k}{m} = t$ [k]

13 $\dfrac{u}{m} = n$ [u]　　**14** $\dfrac{y}{x} = d$ [x]　　**15** $\dfrac{a}{m} = t$ [m]

16 $\dfrac{d}{g} = n$ [g]　　**17** $\dfrac{t}{k} = (a + b)$ [t]　　**18** $y = \dfrac{v}{e}$ [e]

19 $c = \dfrac{m}{y}$ [y]　　**20** $\dfrac{a}{m} = b$ [a]　　**21** $g(m + a) = b$ [m]

22 $h(h + g) = x^2$ [g]　　**23** $y - t = z$ [t]　　**24** $me = c$ [e]

25 $a(y + x) = t$ [x]　　**26** $uv - t^2 = y^2$ [v]　　**27** $pk + t = c$ [k]

28 $k - w = m$ [w]　　**29** $b - an = c$ [n]　　**30** $m(a + y) = c$ [y]

31 $pq - x = ab$ [x]　　**32** $a^2 - bk = t$ [k]　　**33** $v^2z = w$ [z]

34 $c = t - u$ [u]　　**35** $xc + t = 2t$ [c]　　**36** $m(n + w) = k$ [w]

37 $v - mx = t$ [m]　　**38** $c = a(y + b)$ [y]　　**39** $m(a - c) = e$ [c]

40 $ba = ce$ [a]　　**41** $\dfrac{a}{p} = q$ [p]　　**42** $\dfrac{a}{n} = e$ [n]

6.3　Inequalities

6.3.1　Symbols

▶ There are four inequality symbols.
$x < 4$ means 'x is **less than** 4'
$y > 7$ means 'y is **greater than** 7'
$z \le 10$ means 'z is **less than or equal to** 10'
$t \ge -3$ means 't is **greater than or equal to** -3'

▶ If there are two symbols in one statement, look at each part separately.

For example, if n is an **integer** and $3 < n \le 7$, then n has to lie between 3 and 7. It can equal 7 but not equal 3.
So n can be 4, 5, 6 or 7 only.

An integer is a whole number.

EXAMPLE

Show on a number line the given range of values of x.
a $x > 1$ **b** $x \leq -2$ **c** $1 \leq x < 4$

. . . .

a $x > 1$

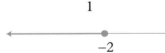

Use an open circle to show that 1 is not included.

b $x \leq -2$

Use a filled in circle to show that -2 is included.

c $1 \leq x < 4$

Exercise 7 ②

1 Write each statement with either > or < in the box.
     a 3 ☐ 7   **b** 0 ☐ −2   **c** 3·1 ☐ 3·01
     d −3 ☐ −5   **e** 100 m ☐ 1 m   **f** 1 kg ☐ 1 lb

2 Write the inequality shown. Use x for the variable.

     a   2   **b**   5   **c**   100

     d   −2   2   **e**   −6   **f**   3   8

3 Draw number lines to show these inequalities.
     a $x \geq 7$ **b** $x < 2\cdot5$ **c** $1 < x < 7$ **d** $0 \leq x \leq 4$ **e** $-1 < x \leq 5$

4 Write an inequality for each statement.
     a You must be at least 16 to get married.
        [Use A for age.]
     b Vitamin J1 is not recommended for people over 70 or for children 3 years or under.
     c To braise a rabbit, the oven temperature should be between 150 °C and 175 °C.
        [Use T for temperature.]
     d Applicants for training as paratroopers must be at least 1·75 m tall.
        [Use h for height.]

5 Answer 'true' or 'false'.
     a n is an integer and $1 < n \leq 4$, so n can be 2, 3 or 4 only.
     b x is an integer and $2 \leq x < 5$, so x can be 2, 3 or 4 only.
     c p is an integer and $p \geq 10$, so p can be 10, 11, 12, 13…

6.3.2 Solving inequalities

Follow the same procedure that you use for solving equations except that, when you multiply or divide by a **negative** number, the inequality is **reversed**.

For example, $4 > -2$ but, after multiplying by -2, you get $-8 < 4$

It is best to avoid dividing by a negative number.

> You are aiming for just an x on one side of the inequality.

EXAMPLE

a $2x - 1 > 5$

$\quad 2x > 5 + 1$ add 1

$\quad x > \dfrac{6}{2}$ divide by 2

$\quad x > 3$

b $x + 1 \leq 3x$

$\quad 1 \leq 3x - x$ subtract x from both sides

$\quad 1 \leq 2x$

$\quad \dfrac{1}{2} \leq x$ or $x \geq \dfrac{1}{2}$ divide by 2

Exercise 8 ②

Solve these inequalities.

1 $x - 3 > 10$ **2** $x + 1 < 0$ **3** $5 > x - 7$

4 $2x + 1 \leq 6$ **5** $3x - 4 > 5$ **6** $10 \leq 2x - 6$

7 $5x < x + 1$ **8** $2x \geq x - 3$ **9** $4 + x < -4$

10 $3x + 1 < 2x + 5$ **11** $2(x + 1) > x - 7$ **12** $7 < 15 - x$

13 $9 > 12 - x$ **14** $4 - 2x \leq 2$ **15** $3(x - 1) < 2(1 - x)$

16 $7 - 3x < 0$ **17** $\dfrac{x}{3} < -1$ **18** $\dfrac{2x}{5} > 3$

19 $2x > 0$ **20** $\dfrac{x}{4} < 0$

*21 The height of the picture has to be greater than the width. Find the range of possible values of x.

$2(x + 1)$

$x + 7$

In questions **22** to **27**, solve the two inequalities separately.

*22 $10 \leq 2x \leq x + 9$ ***23** $x < 3x + 2 < 2x + 6$

*24 $10 \leq 2x - 1 \leq x + 5$ ***25** $3 < 3x - 1 < 2x + 7$

*26 $x - 10 < 2(x - 1) < x$ ***27** $4x + 1 < 8x < 3(x + 2)$

Exercise 9

1 The area of the rectangle must be greater than the area of the triangle. Find the range of possible values of x.

For questions **2** to **8**, list the solutions which satisfy the given conditions.

2 $3a + 1 < 20$ a is a positive integer.

3 $b - 1 \geq 6$ b is a prime number less than 20.

4 $1 < z < 50$ z is a square number.

5 $2x > -10$ x is a negative integer.

6 $x + 1 < 2x < x + 13$ x is an integer.

7 $0 \leq 2z - 3 \leq z + 8$ z is a prime number.

8 $\dfrac{a}{2} + 10 > a$ a is a positive even number.

9 Given that $4x > 1$ and $\dfrac{x}{3} \leq 1\dfrac{1}{3}$, list the possible integer values of x.

10 State the smallest integer n for which $4n > 19$.

11 Given that $-4 \leq a \leq 3$ and $-5 \leq b \leq 4$, find
   a   the largest possible value of a^2
   b   the smallest possible value of ab
   c   the largest possible value of ab
   d   the value of b if $b^2 = 25$.

 12 For any shape of triangle ABC, copy and complete the statement
AB + BC $\square$ AC
by writing <, > or = inside the box.

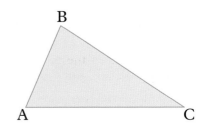

13 Find a simple fraction r such that $\dfrac{1}{3} < r < \dfrac{2}{3}$.

14 Find the largest prime number p such that $p^2 < 400$.

15 Find the integer n such that $n < \sqrt{300} < n + 1$.

16 If $f(x) = 2x - 1$ and $g(x) = 10 - x$, for what values of x is $f(x) > g(x)$?

*17 If $2^r > 100$, what is the smallest integer value of r?

*18 Given $\left(\dfrac{1}{3}\right)^x < \dfrac{1}{200}$, what is the smallest integer value of x?

*19 Find the smallest integer value of x which satisfies $x^x > 10\,000$.

*20 What integer values of x satisfy
$100 < 5^x < 10\,000$?

6.4 Straight-line graphs

6.4.1 Horizontal and vertical lines

- The points P, Q, R and S have coordinates (4, 4), (4, 3), (4, 2) and (4, 1). They all lie on a straight line. Since the x-coordinate of all the points is 4, you say the **equation** of the line is $x = 4$.

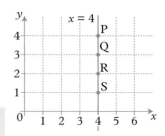

▶ All vertical lines have equations x = a number

- The points A, B, C and D have coordinates (1, 3), (2, 3), (3, 3) and (4, 3). They all lie on a straight line. Since the y-coordinate of all the points is 3, you say the **equation** of the line is $y = 3$.

▶ All horizontal lines have equations y = a number

Exercise 10 ①

1 Write the equations for the lines marked A, B and C.

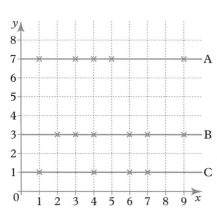

2 Write the equations for the lines marked P, Q and R.

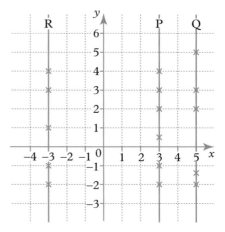

3 **a** Draw axes with values of x and y as in question **1**.
b Draw the lines $y = 3$ and $x = 2$. At what point do they cross?
c Draw the lines $y = 1$ and $x = 5$. At what point do they cross?
d Draw the lines $x = 8$ and $y = 6$. At what point do they cross?

4 In the diagram, E and N lie on the line $y = 1$. B and K lie on the line $x = 5$. Find the equation of the line passing through these points.

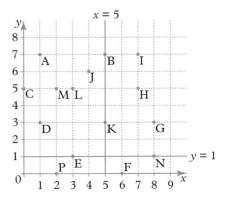

a A and D   **e** L and E

b A, B and I   **f** D, K and G

c M and P   **g** C, M, L and H

d I and H   **h** P and F

6.4.2 Relating *x* and *y*

● The sloping line passes through these points (1, 1), (2, 2), (3, 3), (4, 4), (5, 5). For each point, the y-coordinate is equal to the x-coordinate. The equation of the line is $y = x$ or $x = y$.

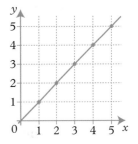

● This line passes through (0, 1), (1, 2), (2, 3), (3, 4), (4, 5).

For each point, the y-coordinate is one more than the x-coordinate. The equation of the line is $y = x + 1$.

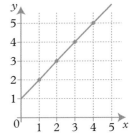

▶ Equations of the form **$y = mx + c$** always give straight-line graphs.

6.4.3 Drawing graphs

● Think about the graph $y = 2x - 3$. There are two operations.

| multiply by 2 | subtract 3 |

You multiply x before you subtract 3.
A function machine shows this.

$$x \longrightarrow \boxed{\times 2} \longrightarrow \boxed{-3} \longrightarrow y$$

So when $x = 2$, $y = 2 \times 2 - 3 = 1$. On the graph, you plot (2, 1).
    when $x = 3$, $y = 2 \times 3 - 3 = 3$. On the graph, you plot (3, 3).
    when $x = 5$, $y = 2 \times 5 - 3 = 7$. On the graph, you plot (5, 7).

Here is the graph of $y = 2x - 3$.

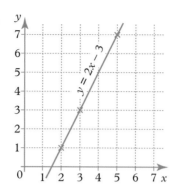

- Think about the graph of $y = 3(x + 5)$.

  You do the operation in brackets first.

  The function machine is

  $x \rightarrow \boxed{+5} \rightarrow \boxed{\times 3} \rightarrow y$

  When $x = 1$, $y = (1 + 5) \times 3 = 18$. You plot $(1, 18)$.

  $\quad\quad\;\; x = 2$, $y = (2 + 5) \times 3 = 21$. You plot $(2, 21)$.

  $\quad\quad\;\; x = 6$, $y = (6 + 5) \times 3 = 33$. You plot $(6, 33)$.

Exercise 11 ①

For each question, copy and complete the table using the function machine. Then draw the graph using the scale of 1 cm = 1 unit on both axes.

1 $y = 2x + 3$ for x from 0 to 6.

$x \rightarrow \boxed{\times 2} \rightarrow \boxed{+3} \rightarrow y$

x	0	1	2	3	4	5	6
y	3				11		
Coordinates					(4, 11)		

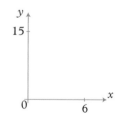

2 $y = x + 3$ for x from 0 to 7.

$x \rightarrow \boxed{+3} \rightarrow y$

x	0	1	2	3	4	5	6	7
y			5					
Coordinates			(2, 5)					

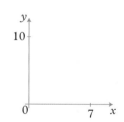

3 $y = 2x$ for x from 0 to 5.

$x \rightarrow \boxed{\times 2} \rightarrow y$

x	0	1	2	3	4	5
y		2				
Coordinates		(1, 2)				

4 $y = 2x + 4$ for x from 0 to 5.

$x \rightarrow \boxed{\times 2} \rightarrow \boxed{+4} \rightarrow y$

x	0	1	2	3	4	5
y			8			

5 $y = 2x - 1$ for x from 0 to 5.

$x \rightarrow \boxed{\times 2} \rightarrow \boxed{-1} \rightarrow y$

x	0	1	2	3	4	5
y	−1					

6 $y = 3x + 1$ for x from 0 to 3.

7 $y = 2(x + 1)$ for x from 0 to 5.

$x \rightarrow \boxed{+1} \rightarrow \boxed{\times 2} \rightarrow y$

8 $y = 3(x + 2)$ for x from 0 to 4.

$x \rightarrow \boxed{+2} \rightarrow \boxed{\times 3} \rightarrow y$

9 $y = \dfrac{x}{2}$ for x from 0 to 7.

$x \rightarrow \boxed{\div 2} \rightarrow y$

x	0	1	2	3	4	5	6	7
y						$2\frac{1}{2}$		

10 $y = 5 + x$ for $0 \le x \le 5$

11 $y = 6 - x$ for $0 \le x \le 6$

12 $y = 2x - 3$ for $0 \le x \le 5$

Exercise 12 ①

In this exercise, the x-values may be positive or negative.
For each question, make a table and then draw the graph.

1 $y = 2x + 1$ for x from −1 to 4.

$x \rightarrow \boxed{\times 2} \rightarrow \boxed{+1} \rightarrow y$

x	−1	0	1	2	3	4
y	−1	1				

2 $y = 3x + 2$ for x from -2 to 3.

x	−2	−1	0	1	2	3
y	−4		2			

3 $y = x - 3$ for x from -2 to 5.

4 $y = x + 6$ for x from -3 to 3.

5 $y = 2x - 1$ for x from -3 to 3.

6 $y = 4x + 1$ for x from -2 to 2.

*7 $y = 5 - x$ for x from 0 to 5.

$x \longrightarrow$ subtract from 5 $\longrightarrow y$

*8 $y = 10 - 2x$ for x from 0 to 6.

*9 $y = 4 - 2x$ for x from 0 to 3.

$x \longrightarrow$ × 2 $\longrightarrow$ subtract from 4 $\longrightarrow y$

*10 $y = 2(6 - x)$ for x from 0 to 6.

*11 $y = 2(x + 1)$ for x from -2 to 4.

12 $y = 3(x - 1)$ for x from -2 to 2.

*13 a Draw the graph of $y = 2x + 1$ for $0 \le x \le 4$.
    b Draw the graph of $y = 8$.
    c Write the coordinates of the point where the two lines meet.

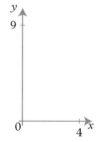

*14 a Draw the graph of $y = x$ for $0 \le x \le 6$.
    b Draw the graph of $y = 1$.
    c Draw the graph of $x = 5$.
    d Shade in the triangle formed by the three lines.

*15 Draw the graph of $x + y = 6$ for $0 \le x \le 6$.

*16 a Draw, on the same page, the graphs of $x + y = 8$ and
      $y = x$ for $0 \le x \le 8$.
    b Write the coordinates of the point where the two lines meet.

6.5 Equations of lines in the form *y = mx + c*

6.5.1 Gradient

▶ The **gradient** of a straight line is a measure of how steep it is.

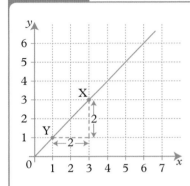

Gradient of line XY $= \dfrac{3-1}{3-1} = \dfrac{2}{2}$

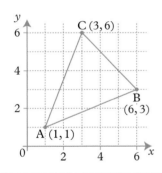

Gradient of line AB $= \dfrac{3-1}{6-1} = \dfrac{2}{5}$

Gradient of line AC $= \dfrac{6-1}{3-1} = \dfrac{5}{2}$

Gradient of line BC $= \dfrac{6-3}{3-6} = -1$

▶ A line which slopes **upwards** to the right has a **positive** gradient.

▶ A line which slopes **downwards** to the right has a **negative** gradient.

▶ Gradient $= \dfrac{\text{difference in } y\text{-coordinates}}{\text{difference in } x\text{-coordinates}}$

Exercise 13 ①

1 Find the gradients of AB, BC, AC.

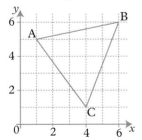

2 Find the gradients of PQ, PR, QR.

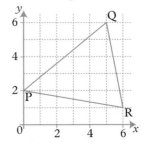

3 Find the gradients of the lines joining these pairs of points.

    a $(5, 2) \rightarrow (7, 8)$ **b** $(-1, 3) \rightarrow (1, 6)$

    c $(1, 4) \rightarrow (5, 6)$ **d** $(4, 10) \rightarrow (0, 20)$

4 Find the value of a if the line joining the points $(3a, 4)$ and $(a, 2)$ has a gradient of 1.

5 **a** Write the gradient of the line joining the points $(6, n)$ and $(3, 4)$ in terms of n.

    b Find the value of n if the line is parallel to the x-axis.

6.5.2 The form $y = mx + c$

Here are two straight lines.

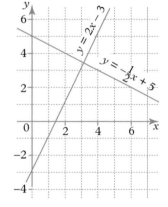

For $y = 2x - 3$, the gradient is 2 and the y-intercept is -3.

For $y = -\dfrac{1}{2}x + 5$, the gradient is $-\dfrac{1}{2}$ and the y-intercept is 5.

These two lines illustrate a general rule.

> ▶ When the equation of a straight line is in the form $y = mx + c$, the gradient of the line is m and the intercept on the y-axis is c.

An intercept is where a line crosses an axis.

EXAMPLE

Draw the line $y = 2x + 3$ on a sketch graph.

The line $y = 2x + 3$ has a gradient of 2 and cuts the y-axis at $(0, 3)$

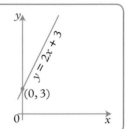

The word 'sketch' means that you do not plot a series of points but simply show the position and slope of the line.

Exercise 14 ②

Write the equations of these lines in the form $y = mx + c$.

a

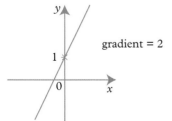

gradient = 2

b

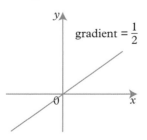

gradient = $\dfrac{1}{2}$

c

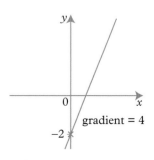

gradient = 4

d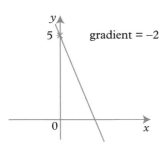
gradient = −2

e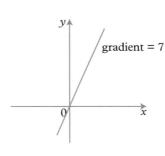
gradient = 7

f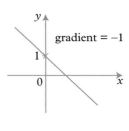
gradient = −1

g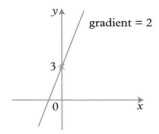
gradient = 2

h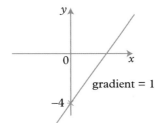
gradient = 1

i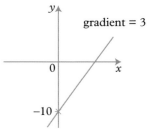
gradient = 3

2 Here are the equations of four lines.

$y = 3x - 7$ $y = 2x - 7$ $y = -2x$ $y = 3x$

Which two lines are parallel?

> Parallel lines have the same gradient.

3 Write the equation of the line with
  a gradient 4 and y-intercept −3
  b gradient −3 and y-intercept 5
  c gradient $\frac{1}{3}$ and y-intercept −2.

4 Write the equation of any line which is parallel to the line
  $y = 5x - 11$.

EXAMPLE

Draw the line $x + 2y - 6 = 0$ on a sketch graph.

First rearrange the equation to make y the subject.
$x + 2y - 6 = 0$
$\qquad 2y = -x + 6$
$\qquad 2y = -\frac{1}{2}x + 3$

The line has a gradient of $-\frac{1}{2}$ and cuts the y-axis at $(0, 3)$.

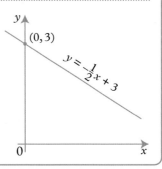

Exercise 15 ②

In questions **1** to **20**, write
a the gradient of the line
b the intercept on the *y*-axis.
Then draw a small sketch graph of each line.

1 $y = x + 3$

2 $y = x - 2$

3 $y = 2x + 1$

4 $y = 2x - 5$

5 $y = 3x + 4$

6 $y = \dfrac{1}{2}x + 6$

7 $y = 3x - 2$

8 $y = 2x$

9 $y = \dfrac{1}{4}x - 4$

10 $y = -x + 3$

11 $y = 6 - 2x$

12 $y = 2 - x$

13 $y + 2x = 3$

14 $3x + y + 4 = 0$

15 $2y - x = 6$

16 $3y + x - 9 = 0$

17 $4x - y = 5$

18 $3x - 2y = 8$

19 $10x - y = 0$

20 $y - 4 = 0$

> You may need
> to rearrange the
> equations into the
> form $y = mx + c$.

*21 Find the equations of the lines A and B.

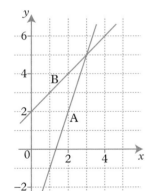

*22 Find the equations of the lines C and D.

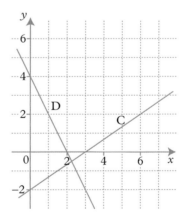

6.5.3   Rates of change

When a quantity changes over time at a constant rate, its graph
against time is a straight line.
The gradient of the straight line gives its rate of change.

EXAMPLE

The temperature of a liquid changes over time as shown in this graph.
Find the rate of change of temperature.

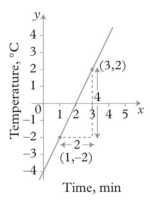

The temperature is increasing over time.
Draw a right-angled triangle on the line.

The rate of increase of temperature = the gradient of the line

$$= \frac{\text{difference in } y\text{-coordinates}}{\text{difference in } x\text{-coordinates}}$$

$$= \frac{2 - (-2)}{3 - 1} = \frac{4}{2} = 2\,°\text{C}\,/\,\text{min}$$

Exercise 16 ②

1 In a science experiment the temperature of some water is controlled so that it rises steadily over a period of 5 minutes, as shown in the diagram.
What is the rate of change of the temperature?

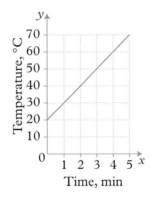

2 A car starts a motorway journey with a full tank of 40 litres.
It uses this petrol at a steady rate over a period of 5 hours as shown.
What is the rate at which it uses petrol?

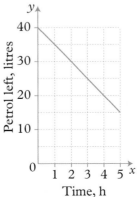

3 A balloon rises steadily after release from ground level. Use the graph to find its rate of change of height.

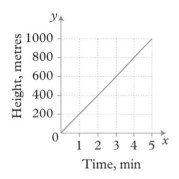

4 The temperature of a classroom in winter changes during the morning.

Time	8 am	9 am	10 am	11 am	12 noon
Temperature, °C	−2	4	10	16	22

Draw a graph of temperature against time and find the rate of increase of temperature in °C/hour.

5 The depth, d metres, of water in a tank increases steadily over time, t hours, such that $d = 2t + 1$.

a Copy and complete this table.

t, hours	0	1	2	3	4
d, metres					

b Draw the graph of d against t and find the rate of increase of the depth of the water.

6 A submarine dives slowly. The top of its tower is at a height, h metres, above the surface of the sea such that $h = 4 - 3t$, where t is the time in minutes after the start of the dive.

a Copy and complete this table.

t, min	0	1	2	3	4
h, metres					

b What is the rate of change of the height, h.
Can you find its value without drawing a graph?

6.6 Simultaneous equations using graphs

6.6.1 Using graphs

EXAMPLE

Louise and Philip are two children.
Louise is 5 years older than Philip.
The sum of their ages is 12 years.
How old is each child?

Let Louise be x years old and Philip be y years old.

So you can say $x + y = 12$ sum = 12
and $x - y = 5$ difference = 5

Draw, on the same grid, the graphs of $x + y = 12$
                                  and $x - y = 5$.

$x + y = 12$ goes through $(0, 12)$, $(2, 10)$, $(6, 6)$, $(12, 0)$.
$x - y = 5$ goes through $(5, 0)$, $(7, 2)$, $(10, 5)$.

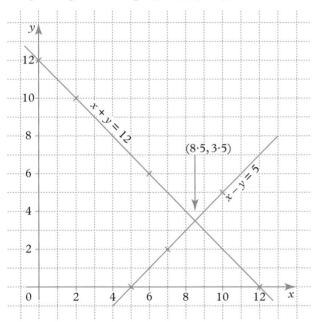

The point $(8 \cdot 5, 3 \cdot 5)$ lies on both lines.
It is where the lines cross.

So $x = 8 \cdot 5$, $y = 3 \cdot 5$ are the solutions of the **simultaneous**
equations $x + y = 12$, $x - y = 5$.

So Louise is $8\frac{1}{2}$ years old and Philip is $3\frac{1}{2}$ years old.

There are many values of x and y that fit the equation $x + y = 12$, and many values that fit the equation $x - y = 5$, **but** only the values $x = 8 \cdot 5$ and $y = 3 \cdot 5$ fit **both** equations.

Exercise 17 ②

1 Use the graphs to solve these simultaneous equations.

a $x + y = 10$
$y - 2x = 1$

b $2x + 5y = 17$
$y - 2x = 1$

c $x + y = 10$
$2x + 5y = 17$

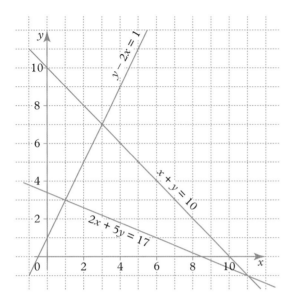

In questions **2** to **6**, solve the simultaneous equations by drawing graphs. Use a scale of 1 cm to 1 unit on both axes.

2 $x + y = 6$
$2x + y = 8$
Draw axes with x and y from 0 to 8.

3 $x + 2y = 8$
$3x + y = 9$
Draw axes with x and y from 0 to 9.

4 $x + 3y = 6$
$x - y = 2$
Draw axes with x from 0 to 8 and y from −2 to 4.

5 $5x + y = 10$
$x - y = -4$
Draw axes with x from −4 to 4 and y from 0 to 10.

6 $a + 2b = 11$
$2a + b = 13$
Here the unknowns are a and b. Draw the a-axis across the page from 0 to 13 and the b-axis up the page, also from 0 to 13.

7 There are four lines drawn here. Write the solutions to these pairs of equations.

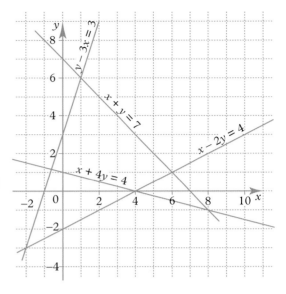

a $x - 2y = 4$
  $x + 4y = 4$

b $x + y = 7$
  $y - 3x = 3$

c $y - 3x = 3$
  $x - 2y = 4$

d $x + 4y = 4$
  $x + y = 7$

e $x + 4y = 4$
  $y - 3x = 3$ (For **e**, give x and y correct to 1 dp.)

6.7 Quadratic expressions and equations

6.7.1 Expanding brackets

You can already expand a bracket such as $4(2x + 3)$ by multiplying by 4. So $4(2x + 3) = 4 \times 2x + 4 \times 3 = 8x + 12$

To expand two brackets such as $(x + 3)(x + 2)$, use the area of a rectangle to help.

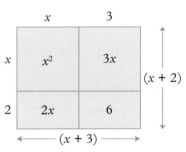

Total area $= (x + 3)(x + 2)$
$= x^2 + 2x + 3x + 6$
$= x^2 + 5x + 6$

After a little practice, it is possible to do without the diagram.

EXAMPLE

Expand $(3x - 2)(2x - 1)$

$(3x - 2)(2x - 1) = 3x(2x - 1) - 2(2x - 1)$ Multiply $(2x - 1)$ by $3x$ and then by -2
$= 6x^2 - 3x - 4x + 2$
$= 6x^2 - 7x + 2$

Exercise 18 ②

Remove the brackets and simplify.

1 $(x + 1)(x + 3)$

2 $(x + 3)(x + 2)$

3 $(y + 4)(y + 5)$

4 $(x + 3)(x + 4)$

5 $(x + 5)(x + 2)$

6 $(x + 8)(x + 2)$

7 $(a + 7)(a + 5)$

8 $(z + 9)(z + 2)$

9 $(x + 3)(x + 3)$

10 $(k + 11)(k + 11)$

11 $(2x + 1)(x + 3)$

12 $(3x + 4)(x + 2)$

13 $(2y + 3)(y + 1)$

14 $(7y + 1)(7y + 1)$

15 $(3x + 2)(3x + 2)$

16 $(5 + x)(4 + x)$

17 $(x + 4)^2$

18 $(x + 2)^2$

19 $(x + 1)^2$

20 $(2x + 1)^2$

21 $(y + 5)^2$

22 $(3y + 1)^2$

23 $3(x + 2)^2$

> Remember that
> $(x + 4)^2$ means
> $(x + 4)(x + 4)$

*24 a If n is an integer, explain why $(2n - 1)$ is an odd number.

  b Write the next odd number after $(2n - 1)$ and show that the sum of these two numbers is a multiple of 4.

6.7.2 Factorising quadratic expressions

We have seen how to expand a pair of brackets such as $(x + 2)(x - 4)$ to give $x^2 - 2x - 8$.

The reverse of this process is called factorising.

EXAMPLE

Factorise $x^2 + 6x + 8$.

a Find two numbers which multiply to give 8 and add up to 6. [i.e. 4 and 2]

b Put these numbers into brackets.
So $x^2 + 6x + 8 = (x + 4)(x + 2)$

> Check by expanding
> $(x + 4)(x + 2)$

EXAMPLE

Factorise   **a** $x^2 + 2x - 15$
     b $x^2 - 6x + 8$

a Two numbers which multiply to give -15 and add up to $+2$ are -3 and 5. So $x^2 + 2x - 15 = (x - 3)(x + 5)$

b Two numbers which multiply to give $+8$ and add up to -6 are -2 and -4.
So $x^2 - 6x + 8 = (x - 2)(x - 4)$

Exercise 19 ②

Copy and complete

1 $x^2 + 7x + 10$
   $= (x + 2)(x + \square)$

2 $x^2 + 7x + 12$
   $= (x + 3)(x + \square)$

3 $x^2 + 8x + 15$
   $= (x + 3)(x + \square)$

Factorise these quadratic expressions.

4 $x^2 + 10x + 21$ **5** $x^2 + 8x + 12$ **6** $x^2 + 12x + 35$

7 $x^2 + 11x + 24$ **8** $x^2 + 10x + 25$ **9** $x^2 + 15x + 36$

10 $a^2 - 3a - 10$ **11** $a^2 - a - 12$ **12** $a^2 + a - 6$

13 $x^2 - 2x - 35$ **14** $x^2 - 5x - 24$ **15** $x^2 - 6x + 8$

Factorise these:

16 $y^2 - 5y + 6$ **17** $x^2 - 8x + 15$ **18** $a^2 - a - 6$

19 $a^2 + 14a + 45$ **20** $b^2 - 4b - 21$ **21** $x^2 - 8x + 16$

22 $y^2 + 2y + 1$ **23** $y^2 - 3y - 28$ **24** $x^2 - x - 20$

6.7.3 The difference of two squares

Expanding the brackets
$$(x - 3)(x + 3) = x(x + 3) - 3(x + 3)$$
$$= x^2 + 3x - 3x - 9$$
$$= x^2 - 9$$

gives the answer as the difference of two squares, x^2 and 3^2.

So, in reverse, factorising $x^2 - 9 = (x - 3)(x + 3)$.

▶ $x^2 - y^2 = (x - y)(x + y)$.

Remember this result.

EXAMPLE

Factorise

a $y^2 - 16$ **b** $4a^2 - b^2$

....

a $y^2 - 16 = (y - 4)(y + 4)$ **b** $4a^2 - b^2 = (2a - b)(2a + b)$

Exercise 20 ②

Factorise these expressions.

1 $y^2 - a^2$ **2** $m^2 - n^2$ **3** $x^2 - t^2$

4 $y^2 - 1$ **5** $x^2 - 9$ **6** $a^2 - 25$

7 $x^2 - \dfrac{1}{4}$ **8** $x^2 - \dfrac{1}{9}$ **9** $4x^2 - y^2$

10 $a^2 - 4b^2$ **11** $25x^2 - 4y^2$ **12** $9x^2 - 16y^2$

13 $4x^2 - z^2$ **14** $x^3 - x$ **15** $a(a^2 - b^2)$

16 $4x^3 - x$ **17** $4a^2 - 100$ **18** $y(y^2 - 9)$

19   Find the exact value of $100\,003^2 - 99\,997^2$.

20 Find the exact value of $1\,500\,002^2 - 1\,499\,998^2$.

*21 Rewrite 9991 as the difference of two squares. Use your answer to find the prime factors of 9991.

6.7.4 Methods of solving quadratic equations

Quadratic equations always have an x^2-term, and often an x-term as well as a number term. They generally have two different solutions.

Solution by factors

Consider the equation $a \times b = 0$, where a and b are numbers.

- The product $a \times b$ can only be zero if either a or b (or both) is equal to zero.

> **EXAMPLE**
>
> Solve the equation $x^2 + x - 12 = 0$.
>
> Factorising, $(x - 3)(x + 4) = 0$
>   either $x - 3 = 0$ or $x + 4 = 0$
>   so $x = 3$ or $x = -4$

Exercise 21 ②

Solve these equations.

1 $x^2 + 7x + 12 = 0$	**2** $x^2 + 7x + 10 = 0$	**3** $x^2 + 2x - 15 = 0$
4 $x^2 + x - 6 = 0$	**5** $x^2 - 8x + 12 = 0$	**6** $x^2 + 10x + 21 = 0$
7 $x^2 - 5x + 6 = 0$	**8** $x^2 - 4x - 5 = 0$	**9** $x^2 + 5x - 14 = 0$
10 $y^2 - 15y + 56 = 0$	**11** $12y^2 - 16y + 5 = 0$	**12** $y^2 + 2y - 63 = 0$
13 $x^2 + 2x + 1 = 0$	**14** $x^2 - 6x + 9 = 0$	**15** $x^2 + 10x + 25 = 0$
16 $x^2 - 14x + 49 = 0$	**17** $z^2 - 8z - 65 = 0$	**18** $y^2 - 2y + 1 = 0$

Quadratics with only two terms

> **EXAMPLE**
>
> **a** Solve the equation $x^2 - 7x = 0$.   **b** Solve the equation $x^2 - 100 = 0$.
>
> **a** Factorising, $x(x - 7) = 0$ **b** Rearranging, $x^2 = 100$
>   Either $x = 0$ or $x - 7 = 0$ Take square root,
> $x = 7$ $x = 10$ or -10
>   The solution are $x = 0$ and $x = 7$.
>
> > Alternatively use the difference of squares, that is $(x - 10)(x + 10) = 0$ so either $x = 10$ or -10

Exercise 22 ❷

Solve these equations.

1 $x^2 - 3x = 0$ **2** $x^2 + 7x = 0$ **3** $2x^2 - 2x = 0$ **4** $3x^2 - x = 0$

5 $x^2 - 16 = 0$ **6** $x^2 - 49 = 0$ **7** $4x^2 - 1 = 0$ **8** $9x^2 - 4 = 0$

9 $6y^2 + 9y = 0$ **10** $6a^2 - 9a = 0$ **11** $10x^2 - 55x = 0$ **12** $16x^2 - 1 = 0$

13 $y^2 - \frac{1}{4} = 0$ **14** $56x^2 - 35x = 0$ **15** $36x^2 - 3x = 0$ **16** $x^2 = 6x$

17 $x^2 = 11x$ **18** $2x^2 = 3x$ **19** $x^2 = x$ **20** $4x = x^2$

21 Find the length AB in terms of x. ABCD is a rectangle.

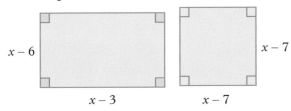

22 The area of the rectangle is $24\,\text{m}^2$ greater than the area of the square. Find x.

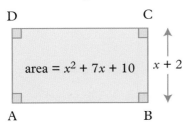

6.7.5 Using quadratic equations to solve problems

EXAMPLE

A patio is 5 metres longer than it is wide. Its area is $24\,\text{m}^2$.
Find its length.

Let the width $= x$
So the length $= x + 5$
The area is given by $x(x + 5) = 24$
Expand the bracket $x^2 + 5x = 24$
Rearrange $x^2 + 5x - 24 = 0$
Factorise $(x + 8)(x - 3) = 0$
                     $x = -8 \text{ or } 3$
x cannot be negative, so $x = 3$ and $x + 5 = 8$.
The patio is $8\,\text{m}$ long.

Exercise 23 ②

Solve by forming a quadratic equation.

1 Two numbers, which differ by 3, have a product of 88. Find them.

> Call the numbers x and $x + 3$.

2 The product of two consecutive odd numbers is 143. Find the numbers.

> If the first odd number is x, what is the next odd number?

3 The height of a photo exceeds the width by 7 cm. If the area is 60 cm², find the height of the photo.

4 The length of a rectangle exceeds the width by 2 cm. If the diagonal is 10 cm long, find the width of the rectangle.

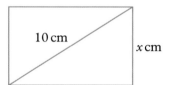

5 The area of the rectangle exceeds the area of the square by 11 m². Find x.

$x + 7$ m $x + 4$ m

$x + 3$ m $x + 4$ m

*6 Three consecutive integers are written as x, $x + 1$, $x + 2$. The square of the largest number is 45 less than the sum of the squares of the other numbers. Find the three numbers.

*7 The area of rectangle A is twice the area of rectangle B.

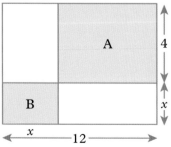

Find x.

6.8 Sketching graphs

It is helpful to know the general shape of some of the more common curves.

▶ **Quadratic curves** have an x^2 as the highest power of x.

e.g. $y = 3x^2 + x - 5$ and $y = 8 + 3x - x^2$

a when the x^2 term is positive
the curve is U shaped.

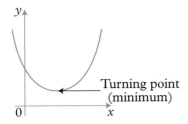

Turning point
(minimum)

b when the x^2 term is negative the
curve is an inverted U-shape.

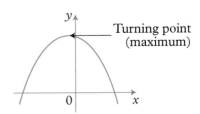

Turning point
(maximum)

▶ **Cubic curves** have an x^3 term as the highest power of x.

e.g. $y = x^3 - 4x + 5$ and $y = 6x - 2x^3$

a when the x^3 term is positive, the curve can be like one of the
two shown below. Notice that as x gets large, so does y.

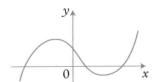

b when the x^3 term is negative, the curve can be like one of the two
shown below. Notice that as x gets large, y is large but negative.

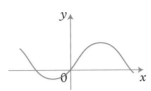

▶ **Reciprocal functions** have a term like $\frac{1}{x}$.

e.g. $y = \dfrac{15}{x}$, $y = \dfrac{10}{x} - 3$

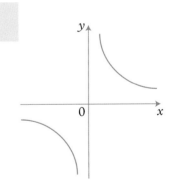

The curve has a break at $x = 0$.
The x-axis and y-axis are called **asymptotes** to the curve.
The curve gets very near but never actually touches the
asymptotes.

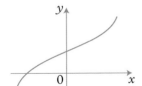
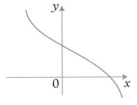

Exercise 24 ❷

1 What sort of curves are these? Give as much information as you can.
(e.g. 'quadratic, x^2 term positive')

a
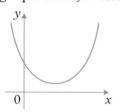

b
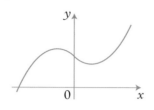

c
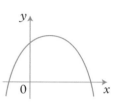

d
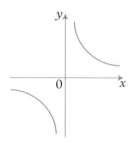

e
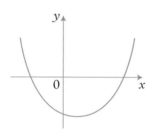

f
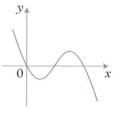

2 Sketch the general shape of the following curves.

    a $y = 2x^2 - x + 3$ **b** $y = x^3 + 1$

    c $y = 10x - x^2$ **d** $y = 2x^3 + x - 7$

    e $y = 9 - x^2$ **f** $y = \dfrac{20}{x}$

3 Sketch the graph of $y = (x - 2)(x - 5)$.
Start by finding the two points where the graph cuts the
x-axis (i.e. where $y = 0$). These are the **roots** of the equation.

4 Match each graph with one of the equations from the list below.

    $y = x(x - 2)$ $y = \dfrac{8}{x}$ $y = x^3$ $y = 2 + x - x^2$

    $y = (x - 2)(x - 4)$ $y = x^3 - 4x$

A
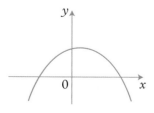

B
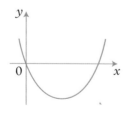

C
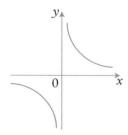

D
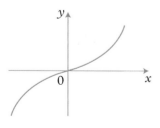

E
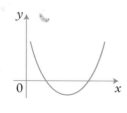

F
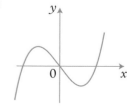

5 Sketch the graph of $y = (x - 2)(x + 4)$. Start by finding the two points where the graph cuts the x-axis (i.e. where $y = 0$)

6 Sketch the graph of

a $y = (x - 1)(x - 5)$ **b** $y = x^3$ **c** $y = \dfrac{12}{x}$

d $y = 2x + 1$ **e** $y = (3 - x)(3 + x)$ **f** $y = x^2 + 1$

6.9   Graphs of quadratic functions

A quadratic function has an x^2-term and usually an x-term and a number term. For example $y = x^2 + 2x$, $y = x^2 + x + 3$ and $y = x^2 - 7$ are all quadratic functions.

6.9.1   Table of values

Constructing a table of values helps you to calculate the values of y. Have a row in the table for each term in the equation.

EXAMPLE

Work out a table of values for the graph of $y = x^2 + 1$ for x from -3 to $+3$.

When $x = -3$,
$x^2 = (-3)^2 = 9$.

x	−3	−2	1	0	1	2	3
x²	9	4	1	0	1	4	9
1	1	1	1	1	1	1	1
y	10	5	2	1	2	5	10

Add these two rows to get y.

EXAMPLE

Work out a table of values for the graph of $y = x^2 + 2x$ for values of x from -3 to $+3$.

Be careful with negative numbers: when $x = -3$, $2x = 2 \times (-3)$ $= -6$.

x	3	−2	−1	0	1	2	3
x²	9	4	1	0	1	4	9
2x	6	−4	−2	0	2	4	6
y	3	0	1	0	3	8	15

Add these two rows to get y.

Exercise 25 2

Copy and complete the tables of values. Include as many rows as
there are terms in each equation.

1 $y = x^2 + 4$

x	−3	−2	−1	0	1	2	3
x²							
+4	4	4	4	4			
y							

2 $y = x^2 − 7$

x	−3	−2	−1	0	1	2	3
x²	9						
−7	−7	−7					
y							

3 $y = x^2 + 3x$

x	−4	−3	−2	1	0	1	2
x²	16	9	4				
+3x	−12	−9	−6				
y	4	0	−2				

4 $y = x^2 + 5x$

x	−4	−3	−2	−1	0	1	2
x²	16						
+5x	−20						
y	−4						

5 $y = x^2 − 2x$

x	−3	−2	−1	0	1	2	3
x²							9
−2x	6						−6
y							3

6 $y = x^2 − 4x$

x	−2	−1	0	1	2	3	4
x²							
−4x	6						
y							

7 $y = x^2 + 2x + 3$

x	−3	−2	−1	0	1	2	3
x²	9	4					
+2x	−6	−4					
+3	3	3	3	3			
y	−6	−3					

8 $y = x^2 + 3x − 2$

x	−4	−3	−2	−1	0	1	2
x²							
+3x							
−2	−2	−2					
y							

9 $y = x^2 + 4x − 5$

x	−3	−2	−1	0	1	2	3
x²							
+4x							
−5							
y							

10 $y = x^2 − 2x + 6$

x	−3	−2	−1	0	1	2	3
x²							
−2x							
+6							
y							

In questions **11** to **18**, draw and complete a table of values for the values of x given.

11 $y = x^2 + 4x$ $\qquad$ x from -3 to $+3$.

12 $y = x^2 - 6x$ $\qquad$ x from -4 to $+2$.

13 $y = x^2 + 8$ $\qquad$ x from -3 to $+3$.

14 $y = x^2 + 3x + 1$ $\qquad$ x from -4 to $+2$.

15 $y = x^2 - 5x + 3$ $\qquad$ x from -3 to $+3$.

16 $y = x^2 - 3x - 5$ $\qquad$ x from -2 to $+4$.

17 $y = x^2 - 3$ $\qquad$ x from -5 to $+1$.

18 $y = 2x^2 + 1$ $\qquad$ x from -3 to $+3$.

6.9.2 Curved graphs

EXAMPLE

Draw the graph of $y = x^2 + x - 2$ for values of x from -3 to $+3$.
Use your graph to write the y-intercept, the turning point and the roots of the function.

a

x	−3	−2	−1	0	1	2	3
x²	9	4	1	0	1	4	9
+x	−3	−2	−1	0	1	2	3
−2	−2	−2	−2	−2	−2	−2	−2
y	4	0	−2	−2	0	4	10

b Plot the x- and y-values from the table.

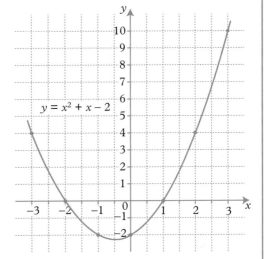

c The curve cuts the y-axis at the point $(0, -2)$.
The y-intercept is -2.

The turning point is a minimum at $\left(-\dfrac{1}{2}, -2\dfrac{1}{4}\right)$.

The roots are the x-values of the points where the curve cuts the x-axis. The roots are $x = -2$ and $x = 1$.

Exercise 26 ❷

For each question, make a table of values and then draw the graph.
Suggested scales: 2 cm to 1 unit on the x-axis and 1 cm to 1 unit on the y-axis.

For each graph, write the y-intercept, the turning point and the roots
of the function. Say if the turning point is a maximum or a minimum.

1 $y = x^2 + 2$ for x from -3 to $+3$.

x	−3	−2	1	0	1	2	3
x²	9	4	1	0	1		
+2	2	2	2				
y	11	6	3				

2 $y = x^2 + 5$ for x from -3 to $+3$.

3 $y = x^2 - 4$ for x from -3 to $+3$.

4 $y = x^2 - 8$ for x from -3 to $+3$.

5 $y = x^2 + 2x$ for x from -4 to $+2$.

x	−4	−3	−2	−1	0	1	2
x²	16	9					4
+2x	8	−6					4
y	8	3					8

6 $y = x^2 + 4x$ for x from -5 to $+1$.

7 $y = x^2 + 4x - 1$ for x from -2 to $+4$.

8 $y = x^2 + 2x - 5$ for x from -4 to $+2$.

9 $y = x^2 + 3x + 1$ for x from -4 to $+2$.

6.9.3 Graphical solution of equations

You can find an approximate solution from an accurately drawn graph
for a wide range of equations, many of which are impossible to solve
exactly by other methods.

EXAMPLE

Draw the graphs of $y = x^2 - 2x$ and $y = 3$.
Hence find approximate solutions to the equations
a $x^2 - 2x = 0$ and **b** $x^2 - 2x = 3$

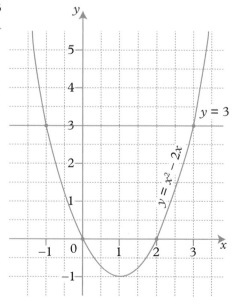

The solutions to the equation are
the x-values at the two points of
intersection.

a Look where the curve meets the line
$y = 0$ (the x-axis).
There are two solutions: $x = 0$ and $x = 2$.

b Look where the curve meets the line
$y = 3$.
There are two solutions: $x = -1$ and
$x = 3$.

Exercise 27 ②

1 The diagram shows the graphs of
$y = x^2 - 2x - 3$, $y = 3$ and $y = -2$.

Use the graphs to find approximate
solutions to these equations.

a $x^2 - 2x - 3 = 3$

b $x^2 - 2x - 3 = -2$

a $x^2 - 2x - 3 = 0$

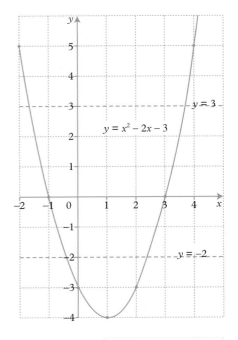

In questions **2** to **6**, use a scale of 2 cm to 1 unit for
x and 1 cm to 1 unit for y.

2 Draw the graph of the function $y = x^2 - 2x$ for
$-1 \leq x \leq 4$.
Hence find approximate solutions of the equations
a $x^2 - 2x = 1$　　　　**b** $x^2 - 2x = 0$

3 Draw the graph of the function $y = x^2 - 3x + 5$
for $-1 \leq x \leq 5$.
Hence find approximate solutions of the equations
a $x^2 - 3x + 5 = 5$　　　　**b** $x^2 - 3x + 5 = 8$

> This means use
> x-values from −1
> up to 4.

4 Draw the graph of $y = x^2 - 2x + 2$ for $-2 \leq x \leq 4$.
By drawing other graphs, solve the equations
a $x^2 - 2x + 2 = 8$　　　　**b** $x^2 - 2x + 2 = 3$

5 Draw the graph of $y = x^2 - 7x$ for $0 \leq x \leq 7$.
a Use the graph to find approximate solutions of the equation $x^2 - 7x = -3$.
b Explain why the equation $x^2 - 7x = -14$ does not have a solution.

6 a Draw axes with x from −6 to 4 and y from −20 to 10.
Use a scale of 1 cm to 1 unit for x and 2 cm to
5 units for y.
b Draw the graphs of these equations.

$y = x^2 + 2x - 15$

$y = x$　　　　$y = 0$

$y = -5$　　　　$y = -19$

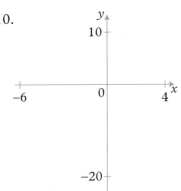

c Hence solve the equations

i $x^2 + 2x - 15 = -5$

ii $x^2 + 2x - 15 = 0$

iii $x^2 + 2x - 15 = -19$

iv $x^2 + 2x - 15 = x$

7 a Draw axes with x from -4 to 4 and y from -6 to 10.
Use a scale of $1\,cm$ to 1 unit for both axes.

b Draw the graphs of these equations.

$y = x^2 - 6$

$y = 4 - x$

$y = -3$

$y = 4$

$y = 0$

$y = x$

c Hence solve the equations

i $x^2 - 6 = 4$ **ii** $x^2 - 6 = 0$

iii $x^2 - 6 = -3$ **iv** $x^2 - 6 = x$

v $x^2 - 6 = 4 - x$

Solve these problems using either an algebraic or graphical method.

8 A garage is x metres wide. It is 5 metres longer than it is wide.
The area of the garage floor is $24\,m^2$.
Write an equation in x and solve it to find the dimensions of
the garage.

9 The width of a room is 4 metres less than its length. If the area
of carpet needed to cover the floor is $32\,m^2$, find the length of
the room.

10 The square of a number is added to the double of the number to
give a total of 24. Find the number.

6.10   Simultaneous equations – algebraic solutions

If we have two equations involving the same two unknowns (usually
x and y), then we say that we have two **simultaneous equations**.
To solve the simultaneous equations, we have to find the numbers
that x and y stand for.

6.10.1 Elimination method

EXAMPLE

Solve the simultaneous equations:

$6x + 9y = 57$ ①   Label the equation ① and ②

$6x + 2y = 22$ ②

Subtract ① – ②: $7y = 35$

                     $y = 5$ We have 'eliminated' x

Chose one of the original equations.

            $6x + 9y = 57$

We know that $y = 5$, so substitute for y

               $6x + 45 = 57$

                $6x = 57 - 45$

                $6x = 12$

                  $x = 2$

So $x = 2$ and $y = 5$

EXAMPLE

Solve the simultaneous equations

         $3x - y = 3$ ①

           $x + y = 5$ ②

Add ① + ②: $4x = 8$ We have 'eliminated' y.

            $x = 2$

Put $x = 2$ in equation ①: $3 \times 2 - y = 3$

                        $6 - y = 3$

                           $y = 3$

The solution is $x = 2$, $y = 3$.

Exercise 28 ②

Solve the simultaneous equation by adding or subtracting these equations.

1 $2x + 7y = 24$ **2** $4x + 7y = 37$
     $2x + 5y = 20$ $4x + 3y = 25$

3 $7x + 4y = 41$ **4** $5x + 6y = 34$
     $7x + 2y = 31$ $5x + 3y = 22$

5 $5x + 4y = 39$ **6** $3x + 6y = 78$
      $5x + y = 36$ $3x + 2y = 38$

7 $x + 7y = 54$ **8** $8x + 7y = 30$
      $x + 2y = 19$ $8x + y = 18$

9 $2x + 10y = 40$ **10** $3x + 6y = 51$
      $2x + 2y = 24$ $x + 6y = 33$

Directed numbers revision

At this stage it is helpful to revise adding / subtracting positive and negative numbers.

Exercise 29 ①

Work out

1 $13 + (-9)$	**2** $10 - (-2)$	**3** $-3 - 4$	**4** $6 - 11$
5 $-2x - (+3x)$	**6** $-5y - (-9y)$	**7** $6x + (-3x)$	**8** $8 - (+8)$
9 $-4y + (-4y)$	**10** $8x - (5x)$	**11** $-2 - (-2)$	**12** $11x - (-2x)$
13 $8 - (-8)$	**14** $-12 + 19$	**15** $3x - (-4x)$	**16** $-3 + (-7)$

Often you have to make the same multiple of x (or y) before you can eliminate them. You do this by multiplying the equations.

EXAMPLE

Solve the equations $\qquad 2x + 3y = 5 \qquad\qquad \ldots [1]$
$\qquad\qquad\qquad\qquad\qquad 5x + 2y = -16 \qquad\quad \ldots [2]$

$[1] \times 5 \qquad\qquad 10x + 15y = 25 \qquad\quad \ldots [3]$
$[2] \times 2 \qquad\qquad 10x - 4y = -32 \qquad\quad \ldots [4]$

Both equations now have $10x$. Eliminate them by subtraction.

$[3] - [4] \qquad\qquad 15y - (-4y) = 25 - (-32)$
$\qquad\qquad\qquad\qquad\qquad 19y = 57$
$\qquad\qquad\qquad\qquad\qquad y = 3$

Substitute for y in $[1] \quad 2x + 3 \times 3 = 5$
$\qquad\qquad\qquad\qquad\qquad 2x = 5 - 9 = -4$
$\qquad\qquad\qquad\qquad\qquad x = -2$

The solution is $x = -2$, $y = 3$.

> Remember to check your solution.

Exercise 30 ②

Use the elimination method to solve these pairs of simultaneous equations.

1 $2x + 5y = 24$	**2** $5x + 2y = 13$	**3** $3x + y = 11$	**4** $x + 2y = 17$
$\quad 4x + 3y = 20$	$\quad 2x + 6y = 26$	$\quad 9x + 2y = 28$	$\quad 8x + 3y = 45$
5 $3x + 2y = 19$	**6** $2a + 3b = 9$	**7** $2x + 7y = 17$	**8** $5x + 3y = 23$
$\quad x + 8y = 21$	$\quad 4a + b = 13$	$\quad 5x + 3y = -1$	$\quad 2x + 4y = 12$
9 $3x + 2y = 11$	**10** $3x + 2y = 7$	**11** $x - 2y = -4$	**12** $5x - 7y = 27$
$\quad 2x - y = -3$	$\quad 2x - 3y = -4$	$\quad 3x + y = 9$	$\quad 3x - 4y = 16$
13 $x + 3y = 7$	**14** $3a - b = 9$	**15** $2x - y = 5$	**16** $3x - y = 17$
$\quad 2y - x = 3$	$\quad 2a + 2b = 14$	$\quad x + y = 7$	$\quad 3x + y = 13$

6.10.2 Substitution method

EXAMPLE

Solve the simultaneous equations

$$3x - 2y = 0 \qquad [1]$$
$$2x + y = 7 \qquad [2]$$

a Label the equations so that the working is made clear.
b In this case, write y in terms of x from equation [2].
c Substitute the expression for y in equation [1] and solve to find x.
d Find y from equation [2] using the value of x.

$$2x + y = 7$$
$$y = 7 - 2x$$

Substituting for y in [1]:

$$3x - 2(7 - 2x) = 0$$
$$3x - 14 + 4x = 0$$
$$7x = 14$$
$$x = 2$$

Substituting for x in [2]

$$2 \times 2 + y = 7$$
$$y = 3$$

The solution is $x = 2, \quad y = 3$.

Exercise 31 ②

Use the substitution method to solve these pairs of simultaneous equations.

1 $2x + y = 5$
   $x + 3y = 5$

2 $x + 2y = 8$
   $2x + 3y = 14$

3 $3x + y = 10$
   $x - y = 2$

4 $2x + y = -3$
   $x - y = 2$

5 $4x + y = 14$
   $x + 5y = 13$

6 $x + 2y = 1$
   $2x + 3y = 4$

7 $2x + y = 5$
   $3x - 2y = 4$

8 $2x + y = 13$
   $5x - 4y = 13$

9 $7x + 2y = 19$
   $x - y = 4$

10 $b - a = -5$
    $a + b = -1$

11 $a + 4b = 6$
    $8b - a = -3$

12 $a + b = 4$
    $2a + b = 5$

13 $3m = 2n - 6\frac{1}{2}$
    $4m + n = 6$

14 $2w + 3x - 13 = 0$
    $x + 5w - 13 = 0$

15 $x + 2(y - 6) = 0$
    $3x + 4y = 30$

16 $2x = 4 + z$
    $6x - 5z = 18$

17 $3m - n = 5$
    $2m + 5n = 9$

18 $5c - d - 11 = 0$
    $4d + 3c = 2$

6.10.3 Solving problems using simultaneous equations

As with linear equations, solving problems involves four steps.

a Call the two unknown numbers x and y.

b Write the problems in the form of two equations.

c Solve the equations and give the answers in words.

d Check *your* solution using the problem and not your equations.

EXAMPLE

A motorist buys 24 litres of petrol and 5 litres of oil for £26·75, while another motorist buys 18 litres of petrol and 10 litres of oil for £31.

Find the cost of 1 litre of petrol and 1 litre of oil.

i Let the cost of 1 litre of petrol be x pence and the cost of 1 litre of oil be y pence.

ii $24x + 5y = 2675$ ...[1]
$18x + 10y = 3100$ ...[2]

iii Solve the equations to get the solutions $x = 75$, $y = 175$
1 litre of petrol costs 75 pence.
1 litre of oil costs 175 pence.

iv Check $24 \times 75 + 5 \times 175 = 2675p = £26.75$
$18 \times 75 + 10 \times 175 = 3100p = £31$

> Notice that, in the equations, you change the units from pounds to pence.

Exercise 32 ②

Solve each problem by forming a pair of simultaneous equations.

> Let the numbers be x and y.

1 Find two numbers with a sum of 15 and a difference of 4.

2 Twice the larger number added to three times the smaller number gives 21. Find the numbers, if the difference between them is 3.

3 Twice one number plus the order number add up to 12. The sum of the two numbers is 7. Find the numbers.

4 Double the larger number plus three times the smaller number makes 31. The difference between the numbers is 3.
Find the numbers.

5 Here is a puzzle. The ? and * stand for numbers which you have to find. The totals for the rows and columns are given.
Write two equations involving ? and * and solve them to find the values of ? and *.

?	*	?	*	36
?	*	*	?	36
*	?	*	*	33
?	*	?	*	36
39	33	36	33	

6 Angle x is 9° greater than angle y.
Find the angles of the triangle.

7 The cost of 2 cups and 3 plates is £18.
The cost of 5 cups and 1 plate is £19.
Let the cost of a cup be c and the cost of a plate be p.
Write two simultaneous equations and solve them to find c and p.

8 Shoppers can buy either two television and three DVD players for £1750 or four televisions and one DVD player for £1250.
Find the cost of one of each.

9 A pigeon can lay either white or brown eggs. Three white eggs and two brown eggs weigh 13 ounces, while five white eggs and four brown eggs weigh 24 ounces. Find the weight of a brown egg, b, and of a white egg, w.

10 The wage bill for five builders and six joiners is £1340, while the bill for eight builders and three joiners is £1220. Find the wage of one builder and one joiner.

11 The line $y = mx + c$ passes through the points (2, 5) and (4, 13). Find m and c.

> For the point (2, 5),
> $5 = m \times 2 + c.$
> $(y = mx + c)$

12 A bag contains forty coins, all of them either 2p or 5p coins. If the value of the money in the bag is £1·55, find the number of each kind. Let the number of 2p coins be t and the number of 5p coins be f.

13 A slot machine takes only 10p and 50p coins and contains a total of twenty-one coins altogether. If the value of the coins is £4·90, find the number of coins of each value.

14 Thirty tickets were sold for a concert, some at 60p and the rest at £1. If the total raised was £22, how many had the cheaper tickets?

Test Yourself

1 Factorise $3y + y^2$

(Edexcel, 2003)

2 a Solve this inequality.

$$3x - 4 \le 8$$

b Represent your solution on a number line.

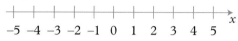

(OCR, 2013)

3 a Simplify fully.

i $p + 7p - 5p$

ii $3x + 4y - 4 + 5x - y$

b Use the formula $B = \dfrac{n}{5}$ to find B when $n = 45$.

c Use the formula $K = 2g - 3h$ to find K when $g = 7$ and $h = 4$.

(OCR, 2013)

4 a Show the inequality $x > -2$ on a number line.

$$-4 \; -3 \; -2 \; -1 \; 0 \; 1 \; 2 \; 3 \; 4 \quad x$$

b Solve the inequality $3x + 5 \le 11$

(AQA, 2013)

5 a Copy and complete this table for $y = 4x - 2$.

x	−2	−1	0	1	2
y		−6	−2		6

b Draw the graph of $y = 4x - 2$ on a grid with x-axes from −3 to 3 and y-axes from 7 to −11.

c Use your graph to find the value of x when $y = 4$

(OCR, 2013)

6 $p = 5$

$r = 2$

a Work out the value of $4p + 3r$

n is an even number.

b What type of number is $n + 1$?

(Edexcel, 2008)

7 **a** Simplify $4a + 3c - 2a + c$

  b $S = \dfrac{1}{2} at^2$

  Find the value of S when $t = 3$ and $a = \dfrac{1}{4}$

  c Factorise $x^2 - 5x$

  d Solve $7x - 19 = 3(x - 3)$

(Edexcel, 2008)

8 Draw the graph of $y = 3 - 2x$ on a grid with x from -3 to 3 and y from -6 to 8.

(CCEA, 2013)

9 **a** Lizzie has a part-time job putting leaflets into envelopes.
  She earns £30 a day for filling **up to** 90 envelopes.
  She earns 20p for every **extra** envelope she fills after 90.
  i Complete this table showing how much she can earn.

Number of envelopes filled	60	70	80	90	100	110	120	130	140
Earnings £		30		30				38	

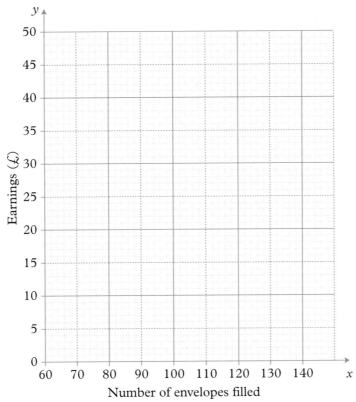

  ii Plot the pairs of values on a copy of the grid and join them using straight lines.

b Alec also has a job filling envelopes. He earns 30p for **every** envelope he fills.
  i On your grid draw the straight line graph to show Alec's earnings for filling from 60 to 140 envelopes. Label this line A.
  ii One day Alec and Lizzie find they have both earned the same amount of money and filled the same number of envelopes. How many envelopes did they each fill?

(OCR, 2013)

10 Line A is drawn on the grid.

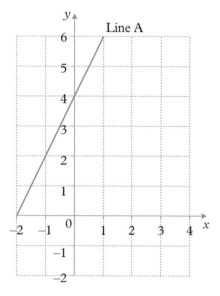

a Write down the coordinates of the point where line A crosses the *y*-axis.

b The equation of line A is $y = 2x + 4$. Write down the gradient of line A.

c Write down the equation of the line that is parallel to line A and that passes through the point $(0, 1)$

(OCR, 2013)

11 a Simplify $3x - 2y - x + 6y$

b Expand $4(3t - 1)$

c Solve $\frac{y}{4} = 8$

d Solve $3p - 2 = 16$

(CCEA, 2013)

12 a Complete the table of values for $y = x^2 + x - 3$

x	−4	−3	−2	−1	0	1	2
y	9	3	−1	−3	−3		3

b Draw the graph of $y = x^2 + x - 3$ for values of x from −4 to 2. Use values of y from 10 to −4.

c Use your graph to solve the equation $x^2 + x - 3 = 0$

(OCR, 2012)

13 a Complete the table for $y = x^2 - 2$.

x	−3	−2	−1	0	1	2	3
y	7			−2			7

b Draw the graph of $y = x^2 - 2$. Use x values from −3 to 3 and y values from 8 to −4.

c Use your graph to solve the equation $x^2 - 2 = 0$.

(OCR, 2012)

7 Geometry 2

In this unit you will:

- revise transformations of shapes
- learn about vectors
- learn about circles, including their area, perimeter, arc length and sector area
- revise volume and surface area calculations
- learn how to use bearings
- learn about interior and exterior angles of polygons
- learn about loci
- learn how to use Pythagoras' theorem.
- learn about trigonometry, including sine, cosine and tangent.

You can use formulae to work out the properties of spheres, including curved surface area and volume.

7.1 Transforming shapes

7.1.1 Reflection

▶ When you **reflect** an object in a line, all points map to equidistant points on the opposite side of the line.

▶ The line joining pairs of object and image points is perpendicular to the mirror line.

- In a reflection, the object and its image are **congruent** because they are the same size and shape.

EXAMPLE

a A′B′C′D′ is the image of ABCD after reflection in the broken line (the mirror line).

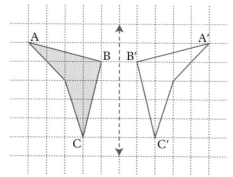

b Δ2 is the image of Δ1 after reflection in the diagonal mirror line.

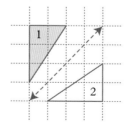

Point A′ is the reflection of point A.

Exercise 1 ①

On square grid paper, draw the object and its image after
reflection in the broken line.

1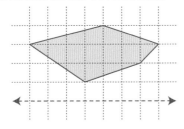

2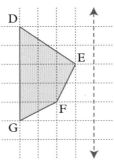

3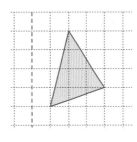

4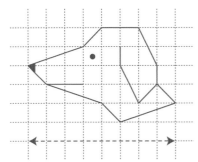

5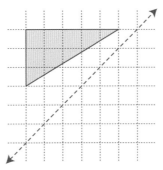

6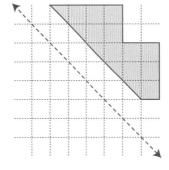

7 Copy the diagram.
   a Reflect Δ1 in the line AB.
      Label the image Δ2.
   b Reflect Δ2 in the line CD.
      Label the image Δ3.

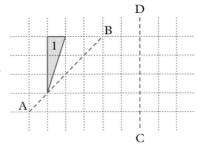

Exercise 2 ①

1 Copy the diagram.
Draw the image of the shape after reflection in
   a the x-axis. Label it 1
   b the y-axis. Label it 2
   c the line $x = 3$. Label it 3.

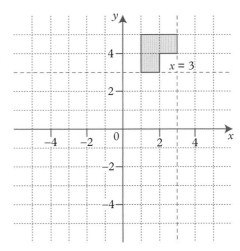

For questions **2** to **5**, draw a pair of axes so that both x and y can take values from -7 to $+7$.

2 **a** Plot and label P(7, 5), Q(7, 2), R(5, 2).

    b Draw the lines $y = -1$, $x = 1$ and $y = x$.
Use dotted lines.

    c Draw the image of $\triangle$PQR after reflection in

      i the line $y = -1$. Label it $\triangle$1

      ii the line $x = 1$. Label it $\triangle$2

      iii the line $y = x$. Label it $\triangle$3.

    d Write the coordinates of the image of point P in each case.

3 **a** Plot and label L(7, −7), M(7, −1), N(5, −1).

    b Draw the lines $y = x$ and $y = -x$.
Use dotted lines.

    c Draw the image of $\triangle$LMN after reflection in

      i the x-axis. Label it $\triangle$1

      ii the line $y = x$. Label it $\triangle$2

      iii the line $y = -x$. Label it $\triangle$3.

    d Write the coordinates of the image of point L in each case.

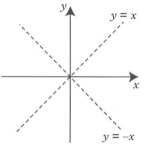

4 **a** Draw the line $x + y = 7$. (It passes through (0, 7) and (7, 0).)

    b Draw $\triangle$1 at (−3, −1), (−1, −1), (−1, −4).

    c Reflect $\triangle$1 in the y-axis on to $\triangle$2.

    d Reflect $\triangle$2 in the x-axis on to $\triangle$3.

    e Reflect $\triangle$3 in the line $x + y = 7$ on to $\triangle$4.

    f Reflect $\triangle$4 in the y-axis on to$\triangle$5.

    g Write the coordinates of $\triangle$5.

5 **a** Draw the lines $y = 2$, $x = -1$ and $y = x$.

    b Draw $\triangle$1 at (1, −3), (−3, −3), (−3, −5).

    c Reflect $\triangle$1 in the line $y = x$ on to $\triangle$2.

    d Reflect $\triangle$2 in the line $y = 2$ on to $\triangle$3.

    e Reflect $\triangle$3 in the line $x = -1$ on to $\triangle$4.

    f Reflect $\triangle$4 in the line $y = x$ on to $\triangle$5.

    g Write the coordinates of $\triangle$5.

6 Find the equation of the mirror line for these reflections.

   a   Δ1 on to Δ2

   b   Δ1 on to Δ3

   c   Δ1 on to Δ4

   d   Δ1 on to Δ5.

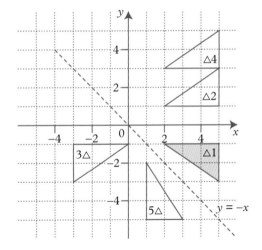

7.1.2 Rotation

▶ You need three properties to describe a **rotation**:

1   the centre

2   the angle (either in degrees or as a fraction of a turn)

3   the direction (clockwise or anticlockwise).

ΔA′B′C′ is the image of ΔABC after a 90° clockwise rotation about centre O.

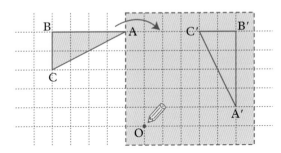

Draw ΔABC on tracing paper and then put the tip of your pencil on O. Turn the tracing paper 90° clockwise about O. The tracing paper now shows the position of ΔA′B′C′.

Exercise 3 ①

Draw the object and its image after the rotation given.

Take O as the centre of rotation in each case.

1

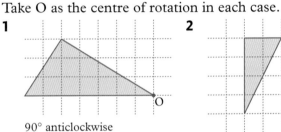

90° anticlockwise

2

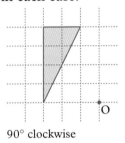

90° clockwise

3

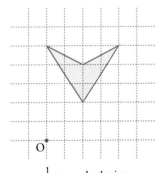

a $\frac{1}{4}$ turn clockwise

4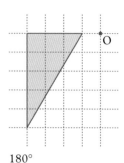

180°

5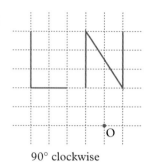

90° clockwise

6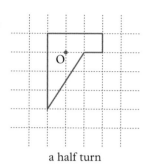

a half turn

7 The shape below has been rotated about several different centres to form this pattern.

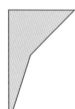

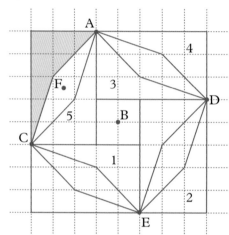

Describe the rotation that takes the shaded shape on to shape 1, shape 2, shape 3, shape 4 and shape 5. For each rotation, give the centre (A, B, C, D, E or F), the angle and the direction of the rotation. For example, 'centre C, 90°, clockwise'.

Exercise 4 ①

1 Copy the diagram.

  a Rotate △ABC 90° clockwise about (0, 0). Label it △1.

  b Rotate △DEF 180° clockwise about (0, 0). Label it △2.

  c Rotate △GHI 90° clockwise about (0, 0). Label it △3.

For questions **2** and **3**, draw a pair of axes with values

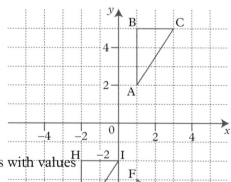

of x and y from -7 to $+7$.

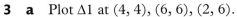

2 a Plot Δ1 at $(2, 3)$, $(6, 3)$, $(3, 6)$.

  b Rotate Δ1 90° clockwise about $(2, 1)$ on to Δ2.

  c Rotate Δ2 180° about $(0, 0)$ on to Δ3.

  d Rotate Δ3 90° anticlockwise about $(1, 1)$ on to Δ4.

  e Write the coordinates of Δ4.

3 a Plot Δ1 at $(4, 4)$, $(6, 6)$, $(2, 6)$.

  b Rotate Δ1 90° anticlockwise about $(6, 0)$ on to Δ2.

  c Rotate Δ2 90° anticlockwise about $(-3, -4)$ on to Δ3.

  d Rotate Δ3 90° clockwise about $(-3, 2)$ on to Δ4.

  e Write the coordinates of Δ4.

7.1.3 Finding the centre of a rotation

Exercise 5 ❶

In questions **1** to **3,** copy the diagram exactly on square grid paper. Then use tracing paper to find the center of the rotation that takes the shaded shape on to the unshaded shape. Mark the centre of rotation with a cross.

1

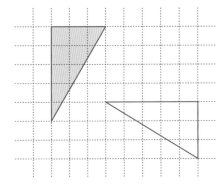

2

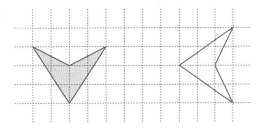

3

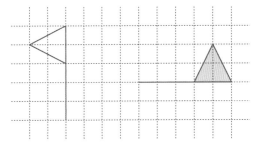

4 Copy the diagram.
Find the coordinates of the centres of
these rotations.

a Δ1 → Δ2
b Δ1 → Δ3
c Δ1 → Δ4
d Δ1 → Δ5

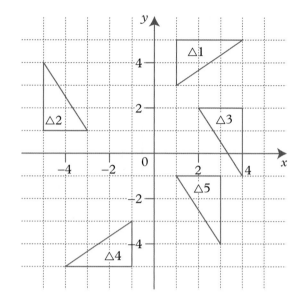

For questions **5** and **6** draw a pair of axes with values of
x and y from −7 to +7.

5 a Plot and label these triangles.
Δ1: (3, 4), (7, 4), (3, 7)
Δ2: (3, 2), (6, 2), (3,−2)
Δ3: (−7, −4), (−3, −4), (−3, −7)
Δ4: (−2, 1), (−5, 1), (−2, 5)
Δ5: (2, −3), (5, −3), (2, −7)

b Find the coordinates of the centres of these rotations.
i Δ1 → Δ2   **ii** Δ1 → Δ3
iii Δ1 → Δ4   **iv** Δ1 → Δ5

6 a Plot and label these triangles.
Δ1: (−4, −3), (−4, −7), (−6, −7)
Δ2: (−3, 4), (−7, 4), (−7, 6)
Δ3: (−2, 1), (2, 1), (2, −1)
Δ4: (0, 7), (4, 7), (4, 5)
Δ5: (2, −3), (4, −3), (2, −7)

b Find the coordinates of the centres of these rotations.
i Δ1 → Δ2   **ii** Δ1 → Δ3
iii Δ1 → Δ4   **iv** Δ1 → Δ5

7.1.4 Enlargement

A

B

C

Photo A has been enlarged to give photos B and C. Notice that the shape of the face is exactly the same in all the pictures

Photo A measures 22 mm by 27 mm
Photo B measures 44 mm by 54 mm
Photo C measures 66 mm by 81 mm

From A to B, both the width and the height have been multiplied by 2.
You say B is an enlargement of A with a **scale factor** of 2.
Similarly C is an enlargement of A with a scale factor of 3.
Also C is an enlargement of B with a scale factor of $1\frac{1}{2}$.

You can find the scale factor of an enlargement by dividing corresponding lengths on two pictures.

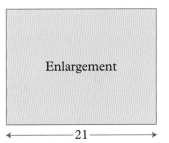

In this enlargement, the scale factor is $\frac{21}{14} = \frac{3}{2} = 1\frac{1}{2}$.

Exercise 6 ① ②

1 This picture is to be enlarged to fit exactly in a frame.
Which of these frames will the picture fit?
Write 'yes' or 'no'.
 a 100 mm by 7 m
 b 110 mm by 70 mm
 c 150 mm by 105 mm
 d 500 mm by 300 mm.

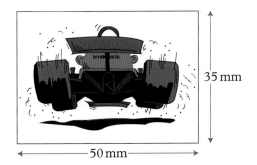

2 This picture is to be enlarged so that it fits exactly into the frame. Find the length x.

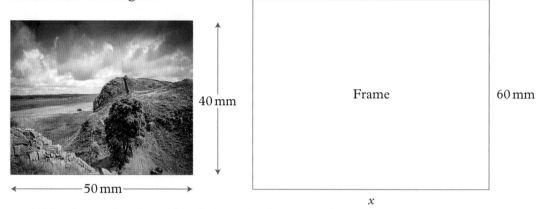

3 a This picture is enlarged to fit into the frame on the right. Calculate y.

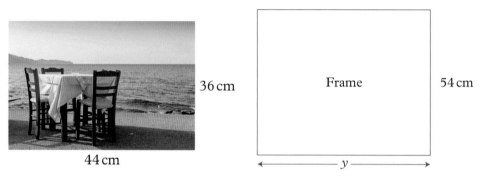

b The same picture is reduced to fit into this frame. Calculate z.

4 Janine has started to draw a two-times enlargement of a house using the squares.

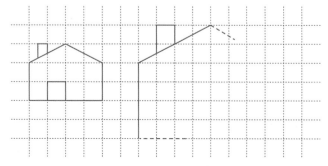

Draw the complete enlargement (use square grid paper).

5 Draw a three-times enlargement of this figure on square grid paper.

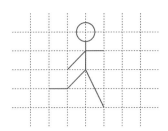

6 Draw an enlargement of this shape on square grid paper with a scale factor of 2.
Measure the angles a and b on the shape and its image. Write the correct version of each sentence.

  a 'In an enlargement, the angles in a shape are changed/unchanged.'

  b 'In an enlargement, the object and the image are congruent/not congruent.'

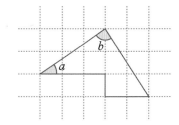

7 This diagram shows an arrowhead and its enlargement. Notice that lines drawn through corresponding points (A, A′ or B, B′) all go through one point O.
This point is called the **centre of enlargement**.
Copy and complete by inserting the correct scale factor.
OA′ = ___ × OA
OB′ = ___ × OB.

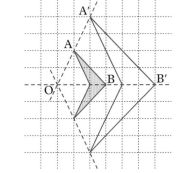

8 Copy this shape and its enlargement onto square grid paper. Draw construction lines to find the centre of enlargement.

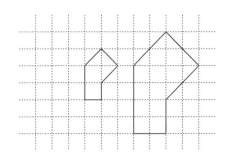

9 Copy the diagram, leaving space on the left for construction lines.

  a Mark the centre of enlargement for

  i Δ1 → Δ2

  ii Δ1 → Δ3

  iii Δ2 → Δ3

  b Write the scale factor for the enlargement Δ2 → Δ3.

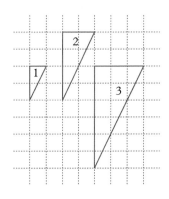

7.1.5 Centre of enlargement

▶ For a mathematical description of an **enlargement** you
need two properties:

1 the scale factor **2** the centre of enlargement.

The triangle ABC is enlarged on to triangle A'B'C' with a scale
factor of 3 and centre O.

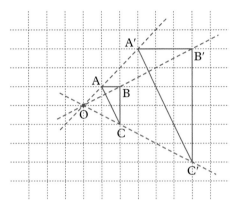

Note: OA' = 3 × OA   OB' = 3 × OB   OC' = 3 × OC

> These lengths are measured from the centre of enlargement.

and   A' B' = 3 × AB   B' C' = 3 × BC   C'A' = 3 × CA

> These lengths are measured on the two triangles.

Exercise 7 ❶ ❷

Copy each diagram and draw an enlargement using the centre O
and the scale factor given.

1

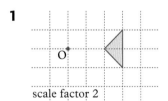

scale factor 2

2

scale factor 3

3

scale factor 2

4

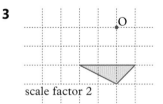

scale factor 3

5

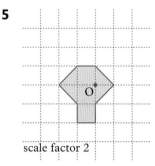

scale factor 2

6

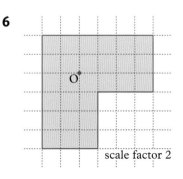

scale factor 2

7 a Copy the diagram.

b Draw the image of Δ1 after enlargement with scale factor 3, centre (0, 0). Label the image Δ4.

c Draw the image of Δ2 after enlargement with scale factor 2, centre (−1, 3). Label the image Δ5.

d Draw the image of Δ3 after enlargement with scale factor 2, centre (−1, −5). Label the image Δ6.

e Write the coordinates of the 'pointed ends' of Δ4, Δ5 and Δ6. (The 'pointed end' is the vertex of the triangle with the smallest angle.)

f The smallest angle in Δ1 is 27°. What is the size of the smallest angle in the enlargement of Δ1?

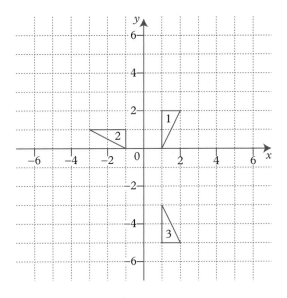

For questions **8** and **9** draw a pair of axes with values from −7 to +7.

8 a Plot and label the triangles
Δ1: (5, 5), (5, 7), (4, 7)   Δ2: (−6, −5), (−3, −5), (−3, −4)
Δ3: (1, −4), (1, −6), (2, −6)

b Draw the image of Δ1 after enlargement with scale factor 2, centre (7, 7). Label the image Δ4.

c Draw the image of Δ2 after enlargement with scale factor 3, centre (−6, −7). Label the image Δ5.

d Draw the image of Δ3 after enlargement with scale factor 2, centre (−1, −5). Label the image Δ6.

e Write the coordinates of the 'pointed ends' of Δ4, Δ5 and Δ6.

9 a Plot and label the triangles
Δ1: (5, 3), (5, 6), (4, 6)
Δ2: (4, −3), (1, −3), (1, −2)
Δ3: (−4, −7), (−7, −7), (−7, −6)

b Draw the image of Δ1 after enlargement with scale factor 2, centre (7, 7). Label the image Δ4.

c Draw the image of Δ2 after enlargement with scale factor 3, centre (5, −4). Label the image Δ5.

d Draw the image of Δ3 after enlargement with scale factor 4, centre (−7, −7). Label the image Δ6.

e Write the coordinates of the 'pointed ends' of Δ4, Δ5 and Δ6.

7.1.6 Enlargements with fractional scale factors

▶ Even though a shape is smaller, mathematicians still call it an enlargement with a **fractional scale factor**.

The unshaded shape is the image of the shaded shape after an enlargement with scale factor $\frac{1}{2}$, centre O.

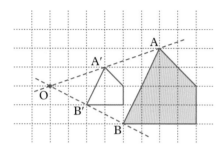

Note that $OA' = \frac{1}{2} \times OA$

$$OB' = \frac{1}{2} \times OB$$

$$A'B' = \frac{1}{2} \times AB$$

Exercise 8 ②

Copy each diagram and draw an enlargement using the centre O and the scale factor given.

1

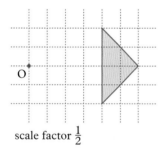

scale factor $\frac{1}{2}$

2

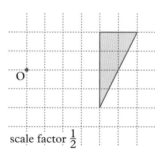

scale factor $\frac{1}{2}$

3

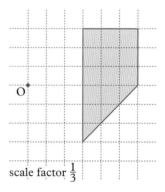

scale factor $\frac{1}{3}$

4 a Plot and label the triangles
Δ1: (7, 6), (1, 6), (1, 3)
Δ2: (7, −1), (7, −7), (3, −7)
Δ3: (−5, 7), (−5, 1), (−7, 1)

b Draw Δ4, the image of Δ1 after an enlargement with scale factor $\frac{1}{3}$, centre (−2, 0).

c Draw Δ5, the image of Δ2 after an enlargement with scale factor $\frac{1}{2}$, centre (−5, −7).

d Draw Δ6, the image of Δ3 after an enlargement with scale factor $\frac{1}{2}$, centre (−7, −5).

7.1.7 The effect of enlargement on perimeter, area and volume

● Here are two rectangles. Rectangle B is an enlargement of rectangle A with a scale factor of 2.

 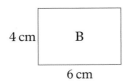

2 cm | A |
3 cm

4 cm | B |
6 cm

Perimeter = 10 cm

Area   = 6 cm²

Perimeter = 20 cm

Area   = 24 cm²

Note that   the perimeter of B = 2 × the perimeter of A.

the area of B = 4 × the area of A.

The scale factor of the enlargement is 2.
The area of B is *not* twice the area of A.

● Here are two cubes. The edges of cube Q are twice the length of the edges of cube P.

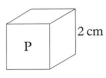 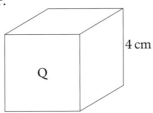

2 cm

P

4 cm

Q

Volume = 8 cm³

Volume = 64 cm³

Notice that the **volume** of cube Q is *not* just twice the volume of cube P. The volume of Q = 8 × the volume of P.

> ▶ Enlargement does not have the same effect on area and volume as it does on length.

7.2 Vectors

7.2.1 Translation

> ▶ A **translation** is simply a 'shift', for example '3 units to the right' or '4 units down'. There is no turning or reflection and the object stays the same size.

a Δ1 is mapped on to Δ2 by the translation with vector $\begin{pmatrix} 4 \\ 2 \end{pmatrix}$

b Δ2 is mapped on to Δ3 by the translation with vector $\begin{pmatrix} 2 \\ -3 \end{pmatrix}$

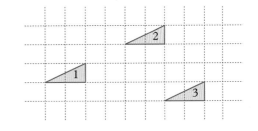

c Δ3 is mapped on to Δ2 by the translation with vector $\begin{pmatrix} -2 \\ 3 \end{pmatrix}$

> To find the vector, just look at one vertex on the shape and its image.

▶ In a vector, the top number gives the number of units across (positive to the right) and the bottom number gives the number of units up/down (positive upwards).

So $\begin{pmatrix} 4 \\ 2 \end{pmatrix}$ is 4 right 2 up $\begin{pmatrix} -2 \\ 3 \end{pmatrix}$ is 2 left 3 up $\begin{pmatrix} 3 \\ 0 \end{pmatrix}$ is 3 to the right

Exercise 9 ①

1 Look at the diagram.
Write the vector for each of these translations.

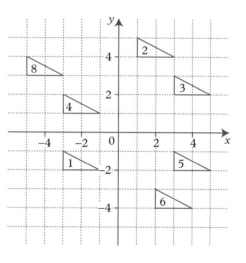

a Δ1 → Δ2   **b** Δ1 → Δ3

c Δ1 → Δ4   **d** Δ1 → Δ5

e Δ1 → Δ6   **f** Δ6 → Δ5

g Δ1 → Δ8   **h** Δ2 → Δ3

i Δ2 → Δ4   **j** Δ2 → Δ5

k Δ2 → Δ6   **l** Δ2 → Δ8

m Δ3 → Δ5   **n** Δ8 → Δ2

7.2.2 Equal vectors

▶ Two vectors are equal if they have the same length **and** the same direction.

The actual position of the vector on the diagram or in space is of no consequence.

You can write a vector as a bold letter, for example, **x**. In your own work, you can show vectors by writing a squiggly line under the letter, for example a̰. On a coordinate grid such as the one in Exercise 9, you can show a vector as a column vector.

Equal vectors have identical column vectors.

7.2.3 Addition of vectors

Vectors **a** and **b** are represented by line segments. They can be added by using the 'nose-to-tail' method to give a single equivalent vector.

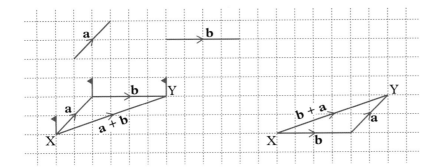

The 'tail' of vector **b** is joined to the 'nose' of vector **a.**

Alternatively the tail of **a** can be joined to the 'nose' of vector **b.**

In both cases the vector $\overrightarrow{XY}$ has the same length and direction and therefore **a + b = b + a.**

In the first diagram the flag is moved by translation **a** and then translation **b.** The translation **a + b** is the equivalent or **resultant** translation.

7.2.4 Multiplication by a scalar

A scalar quantity has magnitude but no direction (for example, mass, volume, temperature). Ordinary numbers are scalars.

$$\overrightarrow{\underset{\textbf{x}}{\qquad}} \quad \overrightarrow{\underset{2\textbf{x}}{\qquad\qquad}} \qquad\qquad \overrightarrow{\underset{\textbf{x}}{\qquad}} \quad \overleftarrow{\underset{-3\textbf{x}}{\qquad\qquad}}$$

When vector **x** is multiplied by 2, the result is 2**x.**

When **x** is multiplied by −3 the result is −3**x.**

> **EXAMPLE**
>
> The diagram shows vectors **a** and **b.** Draw a diagram to show $\overrightarrow{OP}$ and $\overrightarrow{OQ}$ such that
>
> $\overrightarrow{OP} = 3\textbf{a} + \textbf{b} \quad \overrightarrow{OQ} = -2\textbf{a} - 3\textbf{b}$
>
>
>
>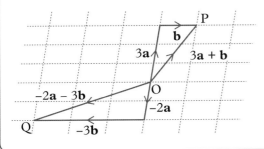

(1) A negative sign reverses the direction of the vector.
(2) The result **a − b** is **a + −b**. So, subtracting **b** is equivalent to adding the negative of **b**.

Exercise 10 ❷

In questions **1** to **15**, use this diagram to describe the vectors given in terms of **c** and **d** where **c** = $\overrightarrow{QN}$ and **d** = $\overrightarrow{QR}$.
For example, $\overrightarrow{QS}$ = 2**d**, $\overrightarrow{TD}$ = **c** + **d**.

K	J	I	H	G	F	E
L	M	N	A	B	C	D
O	P	Q	R	S	T	U
V	W	X	Y	Z		

1 $\overrightarrow{AB}$ **2** $\overrightarrow{SG}$ **3** $\overrightarrow{VK}$ **4** $\overrightarrow{KH}$ **5** $\overrightarrow{OT}$

6 $\overrightarrow{WJ}$ **7** $\overrightarrow{FH}$ **8** $\overrightarrow{FT}$ **9** $\overrightarrow{KV}$ **10** $\overrightarrow{NQ}$

11 $\overrightarrow{OM}$ **12** $\overrightarrow{SD}$ **13** $\overrightarrow{PI}$ **14** $\overrightarrow{YG}$ **15** $\overrightarrow{OI}$

In questions **16** to **21**, use the same diagram to find these vectors in terms of the capital letters, starting from Q each time. For example,
3**d** = $\overrightarrow{QT}$, **c** + **d** = $\overrightarrow{QA}$.

16 2**c**   **17** 4**d**   **18** 2**c** + **d**

19 2**d** + **c**   **20** 3**d** + 2**c**   **21** 2**c** – **d**

In questions **22** and **23**, use this diagram.
$\overrightarrow{LM}$ = **a**, $\overrightarrow{LQ}$ = **b**.

22 Write these vectors in terms of **a** and **b**.

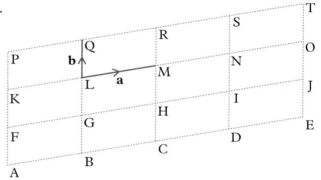

  a $\overrightarrow{GN}$ **b** $\overrightarrow{CO}$

  c $\overrightarrow{TN}$ **d** $\overrightarrow{FT}$

  e $\overrightarrow{KC}$ **f** $\overrightarrow{CJ}$

23 From your answers to question **22**, find the vector which is
  a parallel to $\overrightarrow{LR}$
  b 'opposite' to $\overrightarrow{LR}$
  c parallel to $\overrightarrow{CJ}$ with twice the magnitude
  d parallel to the vector (**a** – **b**).

In questions **24** to **27**, write each vector in terms of **a**, **b**, or **a** and **b**.

24 a $\overrightarrow{BA}$
   b $\overrightarrow{AC}$
   c $\overrightarrow{DB}$
   d $\overrightarrow{AD}$

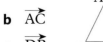

25 a $\overrightarrow{ZX}$
   b $\overrightarrow{YW}$
   c $\overrightarrow{XY}$
   d $\overrightarrow{XZ}$

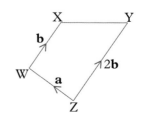

26 a $\overrightarrow{MK}$
b $\overrightarrow{NL}$
c $\overrightarrow{NK}$
d $\overrightarrow{KN}$

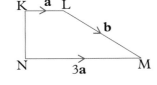

27 a $\overrightarrow{FE}$
b $\overrightarrow{BC}$
c $\overrightarrow{FC}$
d $\overrightarrow{DA}$

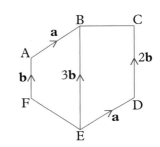

28 The points A, B and C lie on a straight line and the vector
$\overrightarrow{AB}$ is **a + 2b.**
Which of these vectors is possible for $\overrightarrow{AC}$?

a 3**a** + 6**b**   **b** 4**a** + 4**b**   **c** **a** – 2**b**   **d** 5**a** + 10**b**

29 Find three pairs of parallel vectors from those below.

a + 3b	**a – b**	**6a – 3b**	**2a + 6b**	**3a – 3b**	**2a – b**	**a + b**
A	B	C	D	E	F	G

7.3 The circle

7.3.1 Parts of a circle

▶ All points on a circle are an equal distance from the centre.
▶ A line from the centre to any point on the circle is a **radius**.
▶ A line across the circle through the centre is a **diameter**.
▶ The **circumference** is the perimeter of the circle.

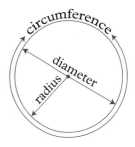

Exercise 11 ❶ ❷

1 Write the radius and the diameter of each circle.

a

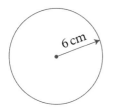

b

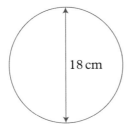

c

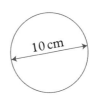

d

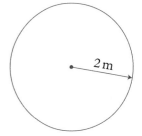

2 Find the radius and diameter by measuring with a ruler.

3 Find the radius and diameter by measuring with a ruler.

4 The diagrams show a regular pentagon ABCDE and a regular hexagon ABCDEF.
What angles would there be at the centre of a regular octagon?

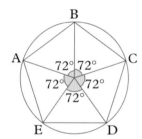

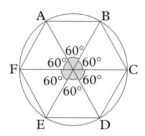

5 Copy and complete each sentence by choosing the correct word from the list.
   a The line AB is a _____.
   b The line CD is a _____.
   c The angle OPC is _____.
   d The shaded region above AB is a _____.
   e The curved line PQ is an _____.
   f The shaded region OPQ is a _____.

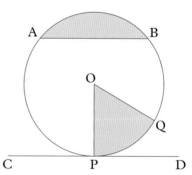

arc	chord	sector	segment	tangent	ninety degrees

7.3.2 Circumference of a circle

▶ The circumference of a circle is given by the formula $C = \pi d$ or $C = 2\pi r$

Learn this important formula.

EXAMPLE

Find the circumference.

a

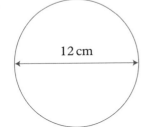
12 cm

b

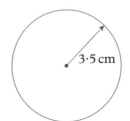
3·5 cm

a $C = \pi d$
$C = \pi \times 12$
$C = 37{\cdot}7\,\text{cm}$ (to 3 sf)

b $C = \pi \times 7$
$C = 22{\cdot}0\,\text{cm}$ (to 3 sf)

π is a special number. Its value is 3·142 approximately.

Use the π button on your calculator.

In part **b,** you need the **diameter**. The diameter is twice the radius, so d = 7 cm.

Note: Sometimes it is convenient to give the answer in a form involving π.
In this example, you could write '$C = 12\pi\,\text{cm}$' and '$C = 7\pi\,\text{cm}$'.

Exercise 12 ①

Find the circumference of the circles in questions **1** to **8**.
Use the π button on a calculator or take $\pi = 3{\cdot}142$.
Give the answers correct to 3 significant figures.

1
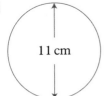
11 cm

2
8 cm

3
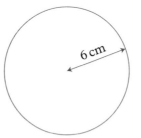
6 cm

4
5 cm

5

5 cm

6
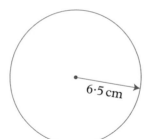
6·5 cm

7

1 cm

8
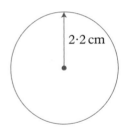
2·2 cm

In questions **9** to **12**, find the circumference of each circle and write your answer as a multiple of π.

9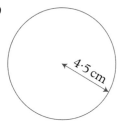

10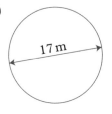
17 m

11
7·1 m

12
23 m

4·5 cm

13 A 10p coin has a diameter of 2·4 cm and a 5p coin has a diameter of 1·6 cm.
How much longer, to the nearest mm, is the circumference of the 10p coin?

14 A circular pond has a diameter of 2·7 m.
Calculate the length of the perimeter of the pond. Write your answer as a multiple of π.

 15 A running track has two semicircular ends of radius 34 m and two straights of 93·2 m as shown.

Calculate the total distance around the track to the nearest metre.

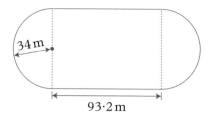
34 m
93·2 m

*16 A fly, perched on the tip of the minute hand of a grandfather clock, is 14·4 cm from the centre of the clock face. How far does the fly move between 12:00 and 13:00?

*17 A penny-farthing bicycle is shown. In a journey, the front wheel rotates completely 156 times.
How far does the bicycle travel?

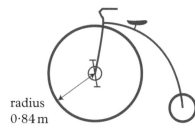
radius
0·84 m

*18 How many complete revolutions does a cycle wheel of diameter 60 cm make in travelling 400 m?

*19 Circle A has radius 5 cm and circle B has diameter 13 cm.
Find the circumference of each circle, leaving π in your answer.

*20 The diagram shows a framework for a target, consisting of 2 circles of wire and 6 straight pieces of wire. The radius of the outer circle is 30 cm and the radius of the inner circle is 15 cm.

Calculate the total length of wire needed for the whole framework.

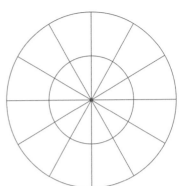

7.3.3 Area of a circle

▶ The area of a circle of radius *r* is given by the formula $A = \pi r^2$

Learn this important formula.

EXAMPLE

Find the area of this circle.

$A = \pi r^2$ can be written as

$A = \pi \times r \times r$

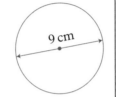

9 cm

In this circle, $r = 4{\cdot}5$ cm

Area of circle $= \pi \times 4{\cdot}5^2$

$= 63{\cdot}6 \text{ cm}^2$ (to 3 sf)

Remember the formula is $\pi(r^2)$ **not** $(\pi r)^2$.

On a calculator, work out the answer like this:

Note that on many calculators you need to press SHIFT before the π button.

| 4·5 | × | 4·5 | × | π | = |

or, on a modern calculator, like this:

| 4·5 | x^2 | × | π | = |

Exercise 13 ①

In questions **1** to **8**, find the area of the circle. Use the π button on a calculator or use π = 3·14. Give the answers correct to three significant figures.

1

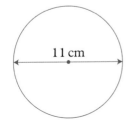

11 cm

2

5 cm

3

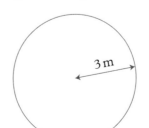

3 m

4

7 m

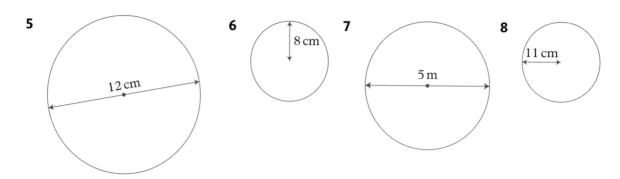

5 12 cm

6 8 cm

7 5 m

8 11 cm

In questions **9** to **16**, remember to add the correct units to your answer.

9 A spinner of radius 7·5 cm is divided into six equal sectors.
Calculate the area of each sector.

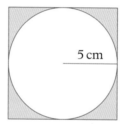

10 A circular swimming pool of diameter 12·6m is to be covered
by a plastic sheet to keep out leaves and insects.
Work out the area of the pool.

11 A circle of radius 5 cm is inscribed inside
a square as shown. Find the area shaded.

5 cm

? **12** A large circular lawn is sprayed with weedkiller.
Each square metre of grass requires 2 g of weedkiller.
How much weedkiller is needed for a lawn of radius 27 m?

? **13** Discs of radius 4 cm are cut from a rectangular plastic sheet
of length 84 cm and width 24 cm.
How many complete discs can be cut out?
Find the area of the sheet wasted.

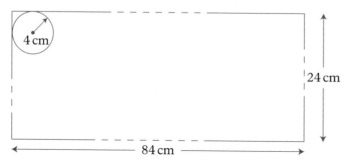

4 cm

24 cm

84 cm

14 A circular pond of radius 6 m is surrounded by a path of width 1 m. Find the area of the path.

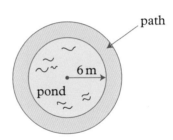

path

6 m

pond

In questions **15** and **16**, find the shaded area.

15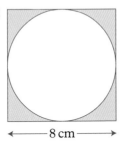

8 cm

16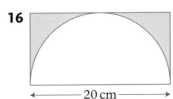

20 cm

7.3.4 More complicated shapes

EXAMPLE

For this shape find
a the perimeter
b the area.

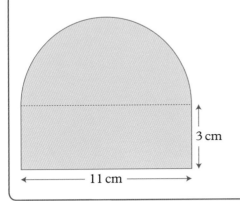

3 cm

11 cm

a Perimeter $= \frac{1}{2} \times \pi \times 11 + 11 + 3 + 3$

$= 34 \cdot 3$ cm (3 sf)

b Area $= \frac{1}{2} \times \pi \times 5 \cdot 5^2 + (11 \times 3)$

$= 80 \cdot 5$ cm^2 (3 sf)

Exercise 14 ❶ ❷

Use the π button on a calculator or take π = 3·14. Give the answers correct to 3 sf. For each shape, find the perimeter.

1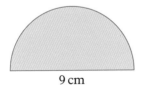

9 cm

2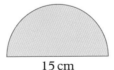

15 cm

3

8 m

4 3·2 cm

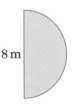

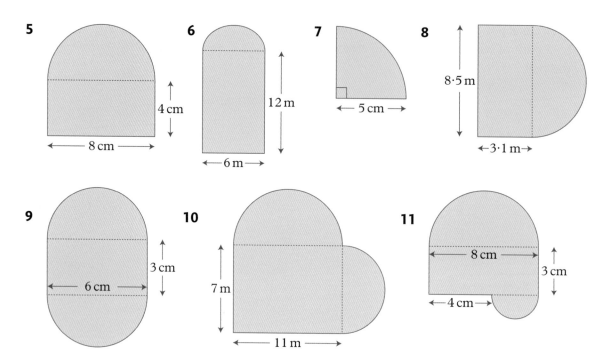

5 4 cm 8 cm

6 12 m 6 m

7 5 cm

8 8·5 m 3·1 m

9 3 cm 6 cm

10 7 m 11 m

11 8 cm 3 cm 4 cm

7.3.5 Arcs and sectors

Arc length

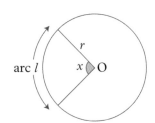

▶ Arc length, $\ell = \dfrac{x}{360} \times 2\pi r$

The arc length is a fraction of the whole circumference depending on the angle at the centre of the circle.

Sector area

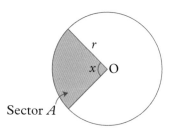

▶ Sector area, $A = \dfrac{x}{360} \times \pi r^2$

The sector area is a fraction of the whole area depending on the angle at the centre of the circle.

EXAMPLE

Find the area of the shaded sector.

7 cm
130°

$\text{Area of sector} = \dfrac{130}{360} \times \pi \times 7 \times 7 = 55.6 \text{ cm}^2 (\text{3sf.})$

Exercise 15 ❷

1 Find the length of the minor arc AB.

a

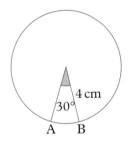

b

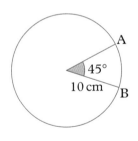

c

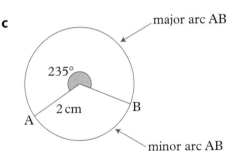

2 A pendulum of length 55 cm swings through an angle of 16°.
Through what distance does the tip of the pendulum swing?

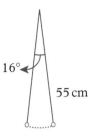

3 Find the area of the shaded sector.

a

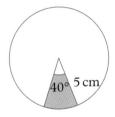

b

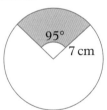

c

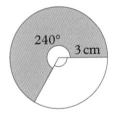

4 This question refers to the diagram.
Find the quantities marked with a ✓.
Leave your answers in terms of π.

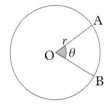

	r	θ	arc AB	area of sector AOB
a	8 cm	45°	✓	
b	10 cm	70°		✓
c	5·8 cm	115°	✓	
d	28 cm	22°		✓
e	65 cm	107°		✓

5 A fly, perched on the tip of the minute hand of a grandfather
clock, is 12.1 cm from the centre of the clock face.
How far does the fly move between 12:00 and 12:15?

6 Find the shaded area in each diagram.

a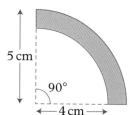

b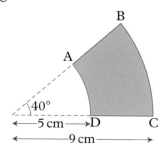

7 Find the length of the perimeter of the shaded area in question **6b**.

8 A sector is cut from a round cheese as shown.
Calculate the volume of the piece of cheese.

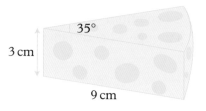

9 The length of the minor arc AB of a circle, centre O, is 2π cm
and the length of the major arc is 22π cm. Find
   a the radius of the circle
   b the acute angle AOB.

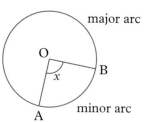

10 A wheel of radius 10 cm is turning at a rate of 5 revolutions
per minute. Calculate
   a the angle through which the wheel turns in 1 second
   b the distance moved by a point on the rim in 2 seconds.

[1 revolution = 360°]

EXAMPLE

The diagram shows an arc of length 10 cm on a circle of radius r.
Find r.

$$\text{Arc length} = \frac{42}{360} \times 2 \times \pi \times r$$

$$\therefore \frac{42}{360} \times 2 \times \pi \times r = 10$$

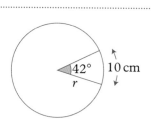

$$r = \frac{10 \times 360}{42 \times 2 \times \pi}$$

$$r = 13{\cdot}6 \text{ cm (3 sf)}$$

Exercise 16 ②

1 Calculate the area of the coloured sector.

a

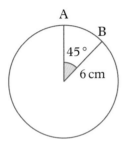

4 cm

90°

b

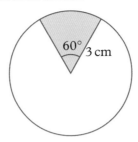

60° 3 cm

c

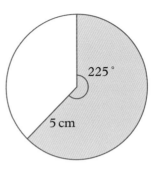

225°

5 cm

2 Calculate the length of the arc AB in each diagram.

a

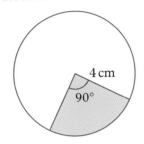

A

B

45°

6 cm

b

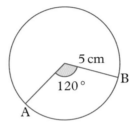

5 cm

120°

B

A

c

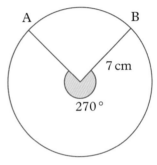

A   B

7 cm

270°

On these questions cover extended topics.

3 In the diagram the arc length is *l* and the sector area is *A*.

  a Find *r*, when *x* = 55° and *l* = 6 cm

  b Find *r*, when *x* = 72° and *l* = 8 cm

  c Find *A*, when *x* = 60° and *r* = 5 cm

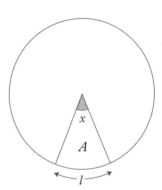

x

A

l

4 Show that arc PQ = 2π cm.

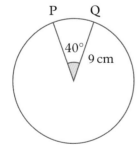

P   Q

40°

9 cm

5 Find the angle *x* so that the arc length is equal to the radius. The angle you have found is called a radian. Your calculator may have a ⎡RAD⎤ mode, where ⎡RAD⎤ stands for radian.

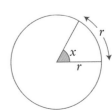

r

x

r

7.4 Volume and surface area

7.4.1 Cuboids

▶ The amount of space which an object occupies is called its **volume**.

A cube with edges 1 cm long has a volume of one cubic centimetre, which you write as $1\,cm^3$.

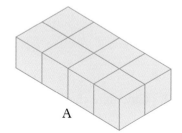

A

B

- A **cuboid** is a 3-D object whose faces are all rectangles. Shapes A and B are cuboids.

  Do not confuse a cuboid with a **cube**, where all the edges are the same length.

  A cube is a special kind of cuboid.

▶ The volume of a cuboid is given by
  Volume = length × width × height

For cuboid A, volume = $4 \times 2 \times 1 = 8\,cm^3$
For cuboid B, volume = $2 \times 3 \times 2 = 12\,cm^3$

- **Faces, edges and vertices**
  The **faces** of the cuboid are the flat surfaces on the shape. There are 6 faces on a cuboid.

  The **edges** of the cuboid are the lines where two faces meet. There are 12 edges on a cuboid.

  The **vertices** of the cuboid are where the edges meet at a point. There are 8 vertices on a cuboid.

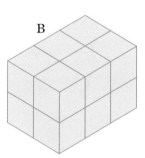

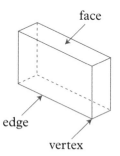

▶ The **surface area** of a shape is the sum of the area of all its faces.

EXAMPLE

a Find the volume and surface area of this cuboid.

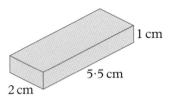

b Find the volume and surface area of cube of side 5 cm.

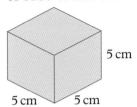

a Volume = length × width × height
$$= 5·5 × 2 × 1$$
$$= 11 \text{ cm}^3$$
Surface area of end = 2 × 1 = 2 cm²
side = 5·5 × 1 = 5·5 cm²
top = 5·5 × 2 = 11 cm²
Total surface area = 2(2 + 5·5 + 11)
$$= 37 \text{ cm}^2$$

b Volume = 5 × 5 × 5
$$= 125 \text{ cm}^3$$
Surface area of one face
$$= 5 × 5 = 25 \text{ cm}^2$$
Total surface area = 6 × 25
$$= 150 \text{ cm}^2$$

Exercise 17 ①

In questions **1** to **6**, work out the volume and surface area of each cuboid.
All lengths are in cm.

1

2

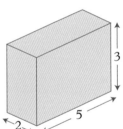

3

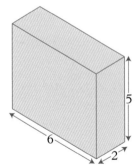

4

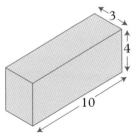

5

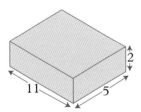

6

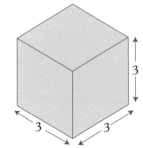

7 This large cube is cut into lots of identical small cubes as shown.
Calculate the volume of each small cube.

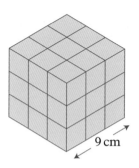

9 cm

For questions **8** to **13**, draw the solid on isometric paper and write the volume of the object. All the objects are made from centimetre cubes.

8

9

10

11

12

13

14 The diagram shows an empty swimming pool. Water is pumped into the pool at a rate of 1 m³ per minute. How long will it take to fill the pool?

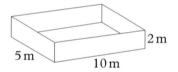

2 m

5 m

10 m

15 The diagram shows the net for a cube.
Calculate the volume of the cube.

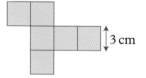

3 cm

16 Find the missing side x in each case.
All lengths are in cm.

a

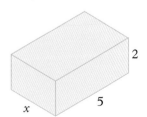

2

5

x

Volume = 30 cm³

b

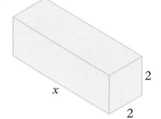

2

x

2

Volume = 16 cm³

c

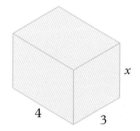

x

4

3

Volume = 24 cm³

17 A cylindrical tin of volume $120\,\text{cm}^3$ is filled with sand from a rectangular box. How many times can the tin be filled if the dimensions of the box are $50\,\text{cm}$ by $40\,\text{cm}$ by $20\,\text{cm}$?

18 The diagram shows a cube of volume $1\,\text{m}^3$.
How many cubic centimetres (cm^3) are there in $1\,\text{m}^3$?

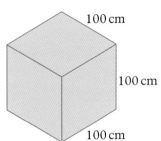

100 cm

100 cm

100 cm

*19 Mr Gibson decided to build a garage and began by calculating the number of bricks required. The garage was to be $6\,\text{m}$ by $4\,\text{m}$ and $2{\cdot}5\,\text{m}$ in height. Each brick measures $22\,\text{cm}$ by $10\,\text{cm}$ by $7\,\text{cm}$. Mr Gibson estimated that he would need about $40\,000$ bricks. Is this a reasonable estimate?

7.4.2 Volume of a prism and a cylinder

▶ A prism is an object with a uniform cross-section.

▶ A cylinder is a prism with a circular cross-section.

l

Area $= A$

Volume of prism
$=$ area of cross-section $\times$ length
$= A \times l$

h

radius $= r$
height $= h$
Volume of cylinder $=$ area of cross-section $\times$ height
Volume $= \pi r^2 h$

EXAMPLE

a Find the volume of the prism.

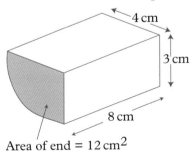

4 cm

3 cm

8 cm

Area of end $= 12\,\text{cm}^2$

b Find the volume of the cylinder.

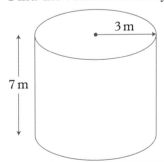

3 m

7 m

a Volume $= A \times l$
     $= 12 \times 8$
     $= 96\,\text{cm}^3$

b Volume $= \pi \times 3^2 \times 7$
     $= 198\,\text{m}^3$ (3 sf)

Exercise 18 ① ②

1 Calculate the volume of each prism. All lengths are in cm.

a Area of end = 15 cm²

10

b Area of end = 5 cm²

12

c
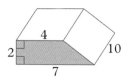
8 3 1 3 2

d
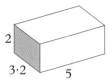
2 3·2 5

e
3 9 7

f
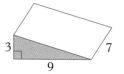
2 4 7 10

2 A cylindrical bar has a cross-sectional area of 12 cm² and a length of two **metres**. Calculate the volume of the bar in cm³.

3 The diagram represents a building.
  a Calculate the area of the shaded end.
  b Calculate the volume of the building.

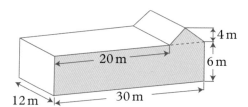
4 m 20 m 6 m 12 m 30 m

4 Calculate the volume of each cylinder.

a
2 cm
5 cm

b
3 cm
4 cm

c
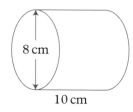
8 cm 10 cm

5 Calculate the volumes of these cylinders.
  a $r = 4$ cm, $h = 10$ cm **b** $r = 11$ m, $h = 2$ m

6 A gas cylinder has diameter 18 cm and length 40 cm. Calculate the capacity of the cylinder, correct to the nearest litre.

7 A solid cylinder of radius 5 cm and length 15 cm is made from material of density 6 g/cm³. Calculate the mass of the cylinder.

8 Cylinders are cut along axes of symmetry to form these objects. Find the volume of each object.

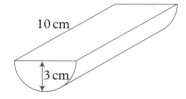
10 cm 3 cm

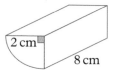
2 cm 8 cm

9 A rectangular block has dimensions $20\,cm \times 7\,cm \times 7\,cm$. Find the volume of the largest solid cylinder which can be cut from this block.

*10 The two solid cylinders shown have the same mass. Calculate the density, $x\,g/cm^3$, of cylinder B.

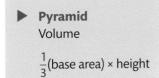

*11 Water flows through a circular pipe of internal diameter $3\,cm$ at a speed of $10\,cm/s$. If the pipe is full, how many litres of water issue from the pipe in one minute?

7.4.3 Volume of pyramids, spheres and cones

▶ **Pyramid**
Volume

$\dfrac{1}{3}$(base area) × height

▶ **Sphere**
Volume

$\dfrac{4}{3}\pi r^3$

▶ **Cone**
Volume

$= \dfrac{1}{3}\pi r^2 h$

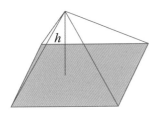

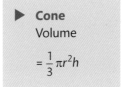

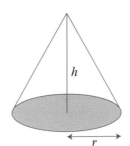

EXAMPLE

a Calculate the volume of a cone of radius $2\,cm$ and height $5\,cm$.

b A solid sphere of radius $4\,cm$ is made of metal of density $8\,g/cm^3$. Calculate the mass of the sphere.

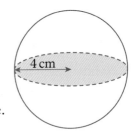

---

a Volume $= \dfrac{1}{3} \times \pi \times 2^2 \times 5$

$= \dfrac{\pi \times 20}{3}$

$= 20{\cdot}9\,cm^3$ (3 sf)

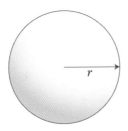

b Volume $= \dfrac{4}{3} \times \pi \times 4^3\,cm^3$

Mass $= \dfrac{4}{3} \times \pi \times 4^3 \times 8$

$= 2140\,g$ (3 sf)

Exercise 19 **2**

In questions **1** to **6**, find the volume of each object. All lengths are in cm.

1
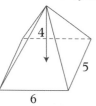

2

3
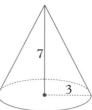

4
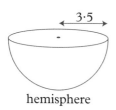

hemisphere

5
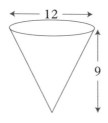

6
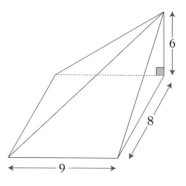

7 A solid sphere is made of metal of density $9 \, \text{g/cm}^3$.
Calculate the mass of the sphere if the radius is 5 cm.

8 Find the volume of a hemisphere of radius 5 cm.

9 Calculate the volume of a cone of height and radius 7 m.

10 A cone is attached to a hemisphere of radius 4 cm.
If the total length of the object is 10 cm
 a find the height of the cone
 b find the total volume of the object.

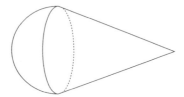

11 Find the volume of a pyramid of height 5 m and base area $12 \, \text{m}^2$.

12 A single drop of oil is a sphere of radius 3 mm. Find its volume.

*13 A toy consists of a cylinder of diameter 6 cm
'sandwiched' between a hemisphere and a cone
of the same diameter. If the cone is of height 8 cm
and the cylinder is of height 10 cm, find the total
volume of the toy.

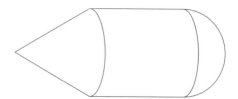

*14 Water is flowing into an inverted cone, of diameter and
height 30 cm, at a rate of 4 litres per minute.
How long, in seconds, will it take to fill the cone?

7.5 Surface areas of cylinders, cones and spheres

This section is about the surface areas of the **curved** parts of cylinders, spheres and cones. The areas of the plane faces are easier to find.

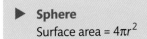

▶ Cylinder	▶ Sphere	▶ Cone
Area of each end = πr^2	Surface area = $4\pi r^2$	Area of base = πr^2
Curved surface area = $2\pi rh$		Curved surface area = πrl, where l is the slant height

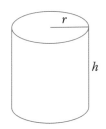

EXAMPLE

Calculate the **total** surface area of a solid cylinder of radius 3 cm and height 8 cm.

Curved surface area = $2\pi rh$
$$= 2 \times \pi \times 3 \times 8$$
$$= 48\pi \text{ cm}^2$$
Area of two ends $= 2 \times \pi r^2$
$$= 2 \times \pi \times 3^2$$
$$= 18\pi \text{ cm}^2$$
Total surface area $= (48\pi + 18\pi) \text{ cm}^2$
$$= 66\pi \text{ cm}^2$$
$$= 207 \text{ cm}^2 \text{ (3 sf)}$$

3 cm

8 cm

Exercise 20 ❷

1 Work out the **curved** surface area of these objects.
   Leave π in your answers. All lengths are in cm.

a
$r = 3$

b
$r = 4$
$h = 5$

c
$l = 10$
$r = 6$

d
$r = 0.7$
$h = 1$

2 Work out the **total** surface area of these objects.
Leave π in your answers. All lengths are in cm.

a
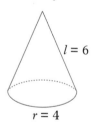
$l = 6$
$r = 4$

b
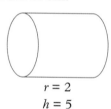
$r = 2$
$h = 5$

c
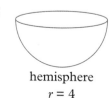
hemisphere
$r = 4$

d
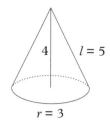
4 $l = 5$
$r = 3$

3 A solid cylinder of height 10 cm and radius 4 cm is to be plated with material costing £11 per cm². Find the cost of the plating.

4 A tin of paint covers a surface area of 60 m² and costs £30. Find the cost of painting the outside surface of a cylindrical gas holder of height 30 m and radius 18 m. The top of the gas holder is a flat circle.

5 Calculate the total surface area of the combined cone/cylinder/hemisphere.

Hint: *l* = 13

12 cm   10 cm   5 cm
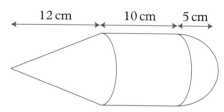

6 Find the surface area of
  a a football of diameter 22 cm
  b a tennis ball of radius 3·5 cm.

7 A cone has a slant height of 10 cm and a diameter of 8 cm. Find its surface area in terms of π if
  a the cone is hollow
  b the cone is solid.

7.6 Bearings and scale drawing

7.6.1 Bearings

Bearings are used where there are no roads to guide the way. Ships, aircraft and mountaineers use bearings to work out where they are.

▶ Bearings are measured **clockwise** from north.

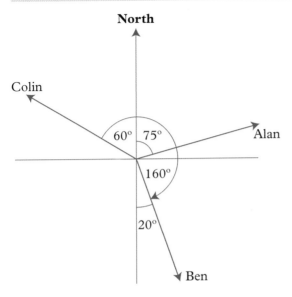

Alan is walking on a bearing of 075°.
Ben is walking on a bearing of 160°.
Colin is walking on a bearing of 300°.

> Notice that you put 3 digits in a bearing – the direction 75° east of north is 075°.

Exercise 21 ❶

The diagrams show the directions in which several people are travelling. Work out the bearing for each person.

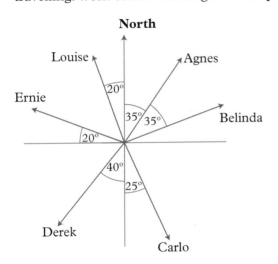

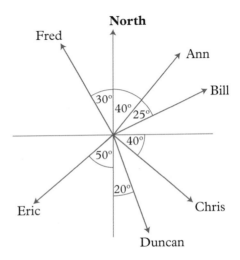

Exercise 22 ❶

The map of North America shows six radar tracking stations, A, B, C, D, E, F.
Use a protractor to measure these bearings.

1 From A, measure
the bearing of
  a F
  b B
  c C.

2 From C, measure
the bearing of
  a E
  b B
  c D.

3 From F, measure
the bearing of
  a D
  b A.

4 From B, measure
the bearing of
  a A
  b E
  c C.

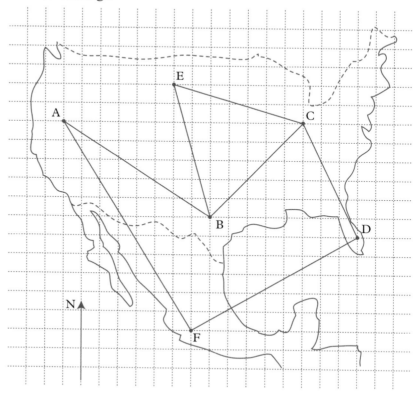

7.6.2 Scale drawing

You can solve problems involving compass directions using a scale drawing.

EXAMPLE

A ship sails 10 km north, then changes course to south-east, and sails a further 8 km.
How far is the ship from its starting point?

Use a scale of 1 cm to 1 km.
  a Mark a starting point S and draw a vertical line. This is
    the north line.
  b Mark a point A, 10 cm from S.
  c Draw a line at 45° to SA, as shown, and mark a point
    F, which is 8 cm from A.
  d Measure the distance SF.

$$SF = 7 \cdot 1 \, cm$$

The ship is 7·1 km from its starting point.
(An answer between 7·0 km and 7·2 km would be acceptable.)

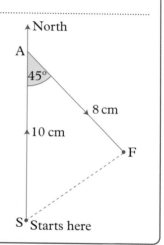

Exercise 23 ①

Use a scale of 1 cm to represent 1 km.

1   A ship sails 8 km due east and then a further
    6 km due south. Find the distance of the ship
    from its starting point.

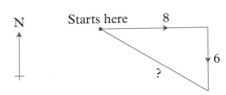

2   A ship sails 7 km due west and then a further 4 km
    north-east. Find the distance of the ship from its
    starting point.

3   A ship sails 8 km due north from point P and then a
    further 7 km on a bearing 080°, as in the diagram
    (which is not drawn to scale).
    How far is the ship now from its starting point?

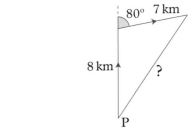

4   A ship sails 9 km from point A on a bearing 090° and
    then a further 6 km on a bearing 050°,
    as shown in the diagram. How far is the
    ship now from its starting point?

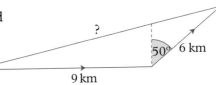

5   A ship sails 6 km on a bearing 160° from the harbour, H,
    and then a further 10 km on a bearing 240°, as shown.
    a   How far is the ship from its starting point?
    b   On what bearing must the ship sail so that it returns
        to its starting point?

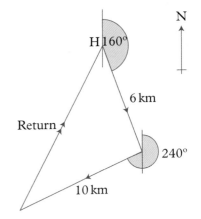

Exercise 24 ①

Draw the points P and Q in the middle of a clean page of square grid paper.
Mark the points A, B, C, D and E accurately, using the information given.

1   A is on a bearing of 040° from P and 015° from Q.

2   B is on a bearing of 076° from P and 067° from Q.

3   C is on a bearing of 114° from P and 127° from Q.

4   D is on a bearing of 325° from P and 308° from Q.

5   E is on a bearing of 180° from P and 208° from Q.

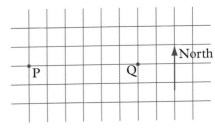

7.7 Angles in polygons

7.7.1 Polygons

▶ A polygon is a flat shape with 3 or more straight sides.

A quadrilateral is a polygon with 4 sides. The table shows the names of the more common polygons.

Name	Number of sides
Quadrilateral	4
Pentagon	5
Hexagon	6
Octagon	8

▶ A **regular** polygon has all its sides of equal length and all its angles are equal.

Here is a regular hexagon.

Exercise 25 ❶

1 **a** Draw a sketch of a pentagon.
 b On your sketch, draw all the diagonals of the pentagon.

2 **a** Draw a hexagon and draw all the diagonals.
 b How many are there?

3 What is the name for a regular polygon with
 a three sides
 b four sides?

4 **a** Draw a pentagon with two right angles.
 b Draw a pentagon with one pair of parallel sides.

7.7.2 Sum of the interior angles of a polygon

▶ You already know that the sum of the angles in a triangle is 180° and the sum of the angles in a quadrilateral is 360°.

● You can divide a pentagon into 3 triangles and a hexagon into 4 triangles. The sum of the angles in each shape is equal to the sum of the angles in the triangles that make the shape.

Pentagon, 5 sides

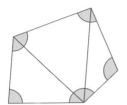

There are 3 triangles.
Sum of the interior
angles = 3 × 180°
= 540°
= (5 − 2) × 180°

Hexagon, 6 sides

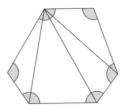

There are 4 triangles.
Sum of the interior
angles = 4 × 180°
= 720°
= (6 − 2) × 180°

Notice that
3 = 5 - 2
4 = 6 - 2
Number of triangles =
Number of sides – 2

▶ The sum of the interior angles of a polygon with *n* sides is
(*n* – 2) × 180°

In a **regular** pentagon, all the angles are the same.
As there are five equal angles, each angle = 540° ÷ 5
= 108°

▶ Each interior angle in a **regular** polygon with *n* sides is
$\frac{(n-2) \times 180°}{n}$.

Exercise 26 ①

1 Find the angles marked with letters.

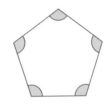

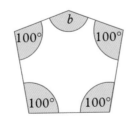

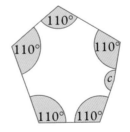

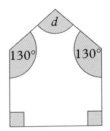

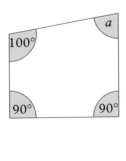

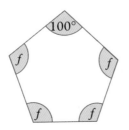

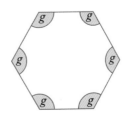

2 The diagram shows a regular pentagon.
   a Find the angle DCE.
   b Show that EC is parallel to AB.

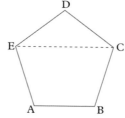

3 Use this formula to find the sum of the interior angles in:
   a an octagon (8 sides)
   b a decagon (10 sides)

> Sum of interior angles of a polygon with n sides is $(n - 2) \times 180°$

4 **a** Work out the sum of the interior angles in a polygon with 20 sides.
   b What is the size of each interior angle in a **regular** polygon with 20 sides?

5 A **regular** dodecagon has 12 sides.
   a Calculate the size of each interior angle, i.
   b Use your answer to find the size of each exterior angle, e.

7.7.3 Exterior angles of a polygon

> ▶ The exterior angle of a polygon is the angle between an extended side and the adjacent side of the polygon.

p, q, r, s and t are all exterior angles.

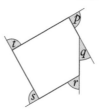

If you fit all the exterior angles together, you can see that the sum of the angles is 360°.
This is true for any polygon.

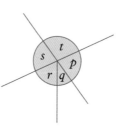

> ▶ The sum of the exterior angles of any polygon = 360°.
> ▶ In a **regular** polygon, all exterior angles are equal.
> ▶ For a **regular** polygon with n sides, each exterior angle = $\dfrac{360°}{n}$.

So, for a regular pentagon, each exterior angle = $\dfrac{360°}{5}$ = 72°

And, for a regular hexagon, each exterior angle = $\dfrac{360°}{6}$ = 60°

EXAMPLE

The diagram shows a **regular** octagon.
a Calculate the size of each exterior angle (marked e).
b Calculate the size of each interior angle (marked i).

....

a There are 8 exterior angles and the sum of these
angles is 360°.
They are equal angles because it is a regular polygon.

So angle $e = \dfrac{360°}{8} = 45°$

b $e + i = 180°$ (angles on a straight line)
So $i = 135°$

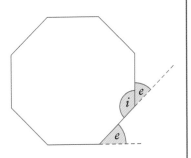

Exercise 27 ❶

1 Copy and complete: 'The sum of the exterior angles of any polygon is_____'.

2 Here is a regular pentagon with its exterior
angles marked. How big is each exterior angle?

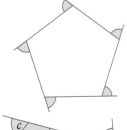

3 Look at the polygon in the diagram.
a Calculate each exterior angle.
b Check that the total of the
exterior angles is 360°.

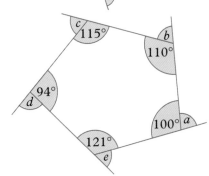

4 The diagram shows a regular decagon.
a Calculate the angle a.
b Calculate the interior angle of
a regular decagon.

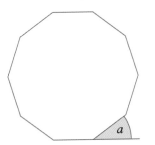

5 Find the exterior angle of a regular polygon with
a 9 sides **b** 18 sides **c** 60 sides.

6 Find the angles marked with letters.

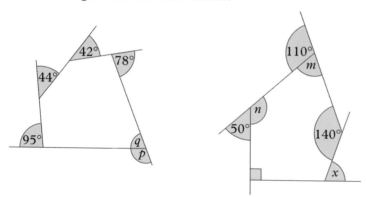

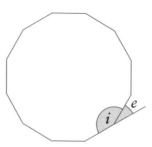

7 A regular dodecagon has 12 sides.

  a Calculate the size of each exterior angle, e.

  b Use your answer to find the size of each interior angle, i.

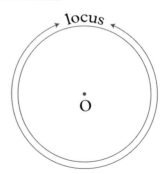

8 Each exterior angle of a regular polygon is 15°. How many sides has the polygon?

9 Each exterior angle of a regular polygon is 18°. How many sides has the polygon?

10 Each interior angle of a regular polygon is 140°. How many sides has the polygon?

7.8 Locus and constructions

> The word **locus** describes the position of points which obey a certain rule. The locus can be the path traced out by a moving point.

Circle

The locus of points which are equidistant from a fixed point O is drawn here. It is a circle with centre O.

Perpendicular bisector

With centres A and B, draw two arcs. Join the points
where these arcs intersect. The perpendicular bisector is
the broken line.
Note that the perpendicular bisector is the locus of
points which are equidistant from points A and B.

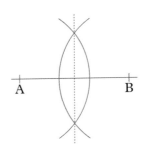

Perpendicular from point P to a line

P.

P.

line

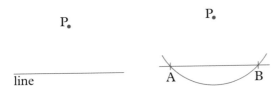

With centre P draw an arc to cut
the line at A and B.

Construct the perpendicular
bisector of AB.

This can be used to find the shortest distance from a point to a line.
Note that point M is the midpoint of AB. The distance PM is the
shortest distance from the point P to the line AB.

Angle bisector

With centre A draw arc PQ. With centres at P and Q draw two more
arcs that intersect at R. Join R to A. The dotted line AR is the angle
bisector. Note that the angle bisector is the locus of points which are
equidistant from the lines AP and AQ.

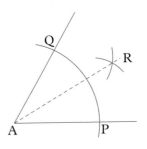

Exercise 28 ②

Look at the previous constructions. Then do questions **1** to **6**
using only a pencil, a ruler and a pair of compasses.

1   Draw an angle of about 60°. Construct the bisector of the angle.

2   Draw an angle of about 130°. Construct the bisector of the angle.

3   Construct an equilateral triangle of side 6 cm.

4   Use the construction in question 3 to construct an angle of 30°.

5   Draw a circle of radius 3 cm. Keep your compasses at the same
setting and draw 6 equal arcs around the circle as shown.
Join the arcs to construct a regular hexagon of side 3 cm.

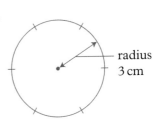

radius
3 cm

6   Construct a regular hexagon of side 4 cm.

Exercise 29 ❷

1 Draw the locus of a point P which moves so that it is always 3 cm from a fixed point X.

2 Mark two points P and Q which are 10 cm apart.
Draw the locus of points which are equidistant from P and Q.

3 Draw two lines AB and AC of length 8 cm, where ∠BAC = 40°.
Draw the locus of points which are equidistant from AB and AC.

4 A sphere rolls along a surface from A to B. Copy each diagram and sketch the locus of the centre of the sphere as it moves from A to B in each case.

a　　　　　　**b**　　　　　　**c**

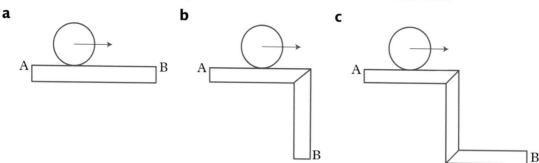

5 A rectangular slab ABCD is rotated around corner B from position 1 to position 2.
Draw a diagram, on square grid paper, to show
a the locus of corner A
b the locus of corner C.

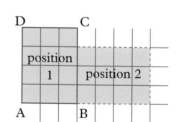

6 The diagram shows a section of coastline with a lighthouse L and coastguard C.
A sinking ship sends a distress signal.
The ship appears to be up to 40 km from L and up to 20 km from C.
Copy the diagram and show the region in which the ship could be.

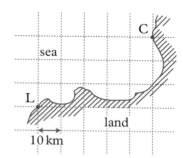

7 a Draw the triangle LMN full size.
b Draw the locus of the points which are
　i equidistant from L and N
　ii equidistant from LN and LM
　iii 4 cm from M.
　(Draw the three loci in different colours.)

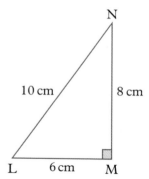

8 Draw a circle of radius 3 cm and a straight line that cuts the circle twice. Shade the area that is inside the smaller segment of the circle.

*9 Draw a line AB of length 6 cm. Draw the locus of a point P so that angle ABP = 90°.

*10 Mr Gibson's garden has a fence on two sides and trees at two corners.
He wants to build a sand pit so that it is
a equidistant from the 2 fences
b equidistant from the 2 trees.
Make a scale drawing (1 cm = 1 m) and mark where Mr Gibson can put the sand pit.

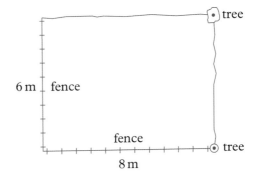

7.9 Pythagoras' theorem

Pythagoras (569–500 BC) was one of the first great mathematical names in Greek antiquity. He settled in southern Italy and formed a mysterious brotherhood with his students who were bound by an oath not to reveal the secrets of numbers and who exercised great influence. They laid the foundations of arithmetic through geometry and were among the first mathematicians to develop the idea of proof.

7.9.1 The theorem

▶ In a right-angled triangle, the square on the hypotenuse is equal to the sum of the squares on the other two sides.

$$c^2 = a^2 + b^2$$

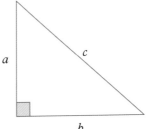

7.9.2 Finding the longest side (the hypotenuse)

EXAMPLE

a Find the length x.

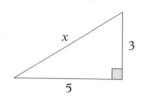

All measurements are in cm.

b Find the length y.

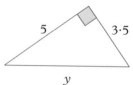

$x^2 = 3^2 + 5^2$
$x^2 = 9 + 25$
$x^2 = 34$
$x = \sqrt{34}$
$x = 5 \cdot 83$ (2 dp)

The hypotenuse is on its own in the equation.

$y^2 = 5^2 + 3 \cdot 5^2$
$y^2 = 25 + 12 \cdot 25$
$y^2 = 37 \cdot 25$
$y = \sqrt{37 \cdot 25}$
$y = 6 \cdot 10$ (2 dp)

Exercise 30 ②

Give your answers correct to 2 dp, where necessary. The units are all in cm.

1 Find x.

a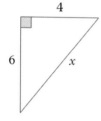

b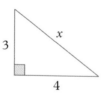

c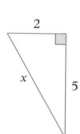

d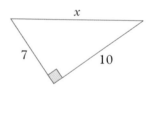

e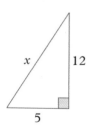

f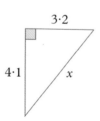

g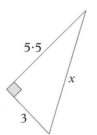

h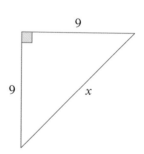

2 Find the length of a diagonal of a square of side 8 cm.

8 cm

3 **a** Find the length of a diagonal of a rectangle of length 7 cm and width 4·5 cm.

  b Find the length of a diagonal of a rectangle of length 10 cm and width 7·2 cm.

4 A ladder leans against a vertical wall. The foot of the ladder is 2 m from the bottom of the wall and the ladder reaches 6 m up the wall.
How long is the ladder?

5 Shruti walks 100 m north and then 55 m due east. How far is she directly now from her starting point?

7.9.3 Finding a shorter side

EXAMPLE

a Find the length x.

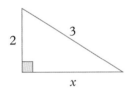

All measurements are in cm.

b Find the length y.

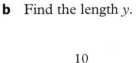

$x^2 + 2^2 = 3^2$
$x^2 + 4 \ = 9$
$x^2 = 5$
$x = \sqrt{5}$
$x = 2 \cdot 24$ (2 dp)

The hypotenuse is on its own in the equation.

$y^2 + 4^2 = 10^2$
$y^2 + 16 = 100$
$y^2 = 84$
$y = \sqrt{84}$
$y = 9 \cdot 17$ (2 dp)

Exercise 31 ②

1 Find y. Give your answers to 2 dp where necessary. All units are in cm.

a

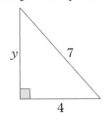

b

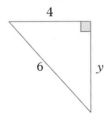

c

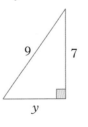

d

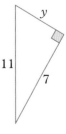

e

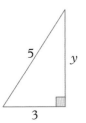

f

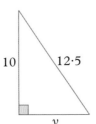

g

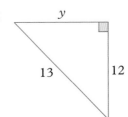

h

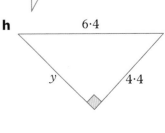

2 A 4 m ladder rests against a vertical wall with
its foot 2 m from the wall.
How far up the wall does the ladder reach?

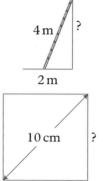

3 The diagonals of a square are of length 10 cm.
How long is each side of the square?

Exercise 32 Mixed questions ②

In questions **1** to **8**, find x. All the lengths are in cm.

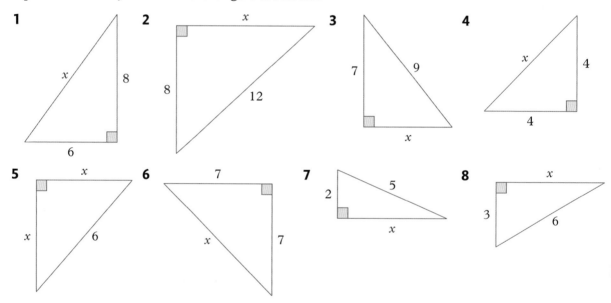

1

2

3

4

5

6

7

8

9 Find the length of a diagonal of a rectangle of length 9 cm and width 4 cm.

10 In the diagram, A is the point(1, 2)
and B is the point(6, 4).

Work out the length AB.
(First find the lengths of AN
and BN.)

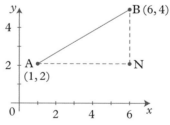

11 An isosceles triangle has sides 10 cm,
10 cm and 4 cm.
Find the height of the triangle.

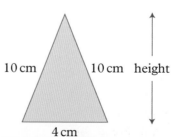

12 A ladder rests against a vertical wall with its foot 1·5 m from the wall.
The ladder reaches 4 m up the wall. How long is the ladder?

13 A ship sails 20 km due north and then 35 km due east.
How far is it directly from its starting point?

14 The square and the rectangle have diagonals of equal length. Find x.

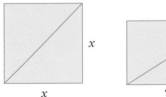 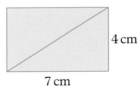

Exercise 33 ②

1 a Find the height of the triangle, h.
 b Find the area of the triangle ABC.

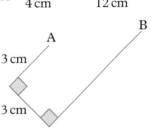

2 A thin wire of length 18 cm is bent
in the shape shown.
Calculate the distance from A to B.

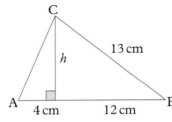

3 A paint tin is a cylinder of radius 12 cm and height 22 cm.
Leonardo, the painter, drops his stirring stick into the tin
and it disappears. Work out the maximum length of the stick.

4 Find the length of line segments:
 a AB **b** CA

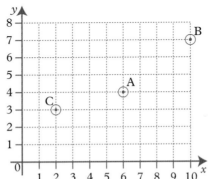

5 A man is building a garage and needs
to check that two of the walls are built
at 90° to each other.
The walls are of length 4 m and 56 m.
What should be the length across the
diagonal AB if the walls are at 90°?

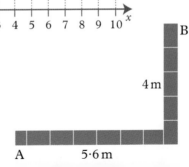

Questions **6** to **11** are more difficult. You cannot find x immediately.

In questions **6** to **11**, find x.

6

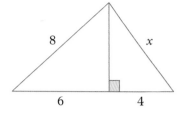

7

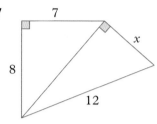

8

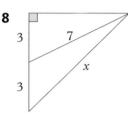

9

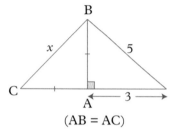

(AB = AC)

10

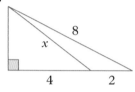

11

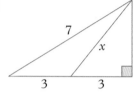

7.10 Trigonometry

You use trigonometry to calculate sides and angles in triangles. The triangle must have a right angle.

The side opposite the right angle is called the **hypotenuse** (H). It is the longest side.

The side opposite the marked angle is called the **opposite** (O).

The other side is called the **adjacent** (A).

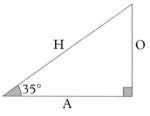

Consider two triangles, one of which is an enlargement of the other.
It is clear that, for the angle 30°, the ratio

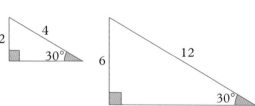

$$\frac{\text{opposite}}{\text{hypotenuse}} = \frac{6}{12} = \frac{2}{4} = \frac{1}{2}$$

This is the same for both triangles as the triangles are similar.

7.10.1 Sine, cosine, tangent

There are three important ratios for angle x.

> You need to learn these for use in your examination.

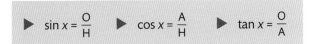

▶ $\sin x = \dfrac{O}{H}$ ▶ $\cos x = \dfrac{A}{H}$ ▶ $\tan x = \dfrac{O}{A}$

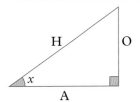

Exercise 34 ②

Draw each triangle and label O, A and H in relation to the angle marked. Write each of these as a fraction. Question **1** is done for you.

1 $\sin w = \dfrac{O}{H} = \dfrac{3}{5}$

2 $\sin x$	**3** $\sin y$	**4** $\sin z$
5 $\cos w$	**6** $\cos x$	**7** $\cos y$
8 $\tan w$	**9** $\tan x$	**10** $\tan y$
11 $\cos z$	**12** $\tan z$	

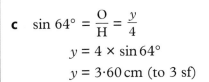

7.10.2 Finding the length of a side

EXAMPLE

a Find the length l.   **b** Find the length x.   **c** Find the length y.

a $\cos 32° = \dfrac{A}{H} = \dfrac{l}{10}$

     $l = 10 \times \cos 32°$

     $l = 8{\cdot}48\,\text{cm}$ (to 3 sf)

> You have 'A' and 'H' so use cos.

b $\tan 38° = \dfrac{O}{A} = \dfrac{x}{7}$

     $x = 7 \times \tan 38°$

     $x = 5{\cdot}47\,\text{cm}$ (to 3 sf)

> You have 'O' and 'A' so use tan.

c $\sin 64° = \dfrac{O}{H} = \dfrac{y}{4}$

     $y = 4 \times \sin 64°$

     $y = 3{\cdot}60\,\text{cm}$ (to 3 sf)

> You have 'O' and 'H' so use sin.

Exercise 35 ②

Find the lengths marked with letters. All lengths are in cm.
Give your answers correct to 3 sf.

Use 'sin' in
questions **1, 2, 3**.

1
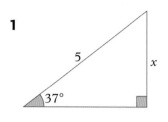

2
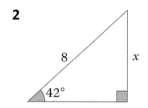

3
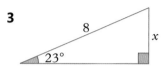

4
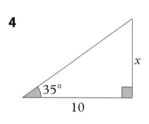

5
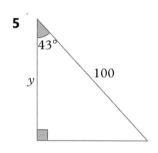

6
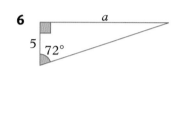

7
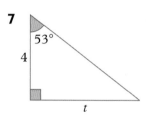

8
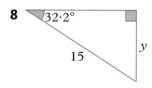

9
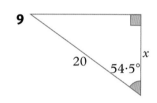

10
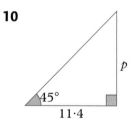

11
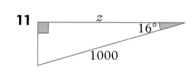

12
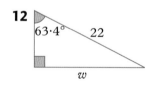

13
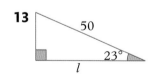

14
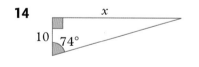

15
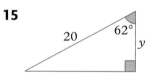

16
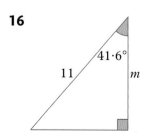

17
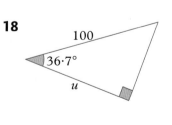

18
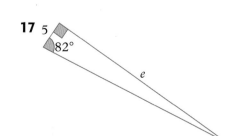

7.10.3 Finding the hypotenuse

EXAMPLE

Find the length x.

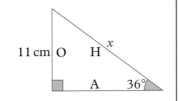

$\sin 36° = \dfrac{O}{H} = \dfrac{11}{x}$

$x \sin 36° = 11$ Multiply by x

$x = \dfrac{11}{\sin 36°} = 18.7\,\text{cm}$ (to 3 sf)

> You have 'O' and 'H' so use sin.

Exercise 36 ②

This exercise is more difficult. Find the lengths marked with letters.

1

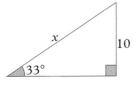

2

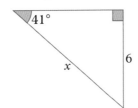

3

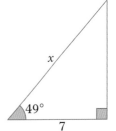

4

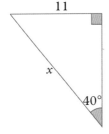

5

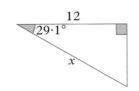

6

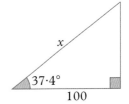

7 You can use the triangles in the diagram to find sin, cos and tan of 30°, 45° and 60°. For example, $\sin 30° = \dfrac{1}{2}$ and $\tan 60° = \sqrt{3}$. Copy and complete this table.

> You need to learn these exact values for sin, cos, tan.

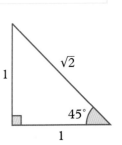

	30°	45°	60°	120°	135°	150°
sin	$\dfrac{1}{2}$					
cos						
tan			$\sqrt{3}$			

7.10.4 Finding angles

EXAMPLE

Find the angle x.

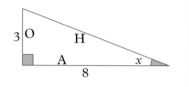

$$\tan x = \frac{O}{A} = \frac{3}{8}$$

$\tan x = 0{\cdot}375$

$\quad x = 20{\cdot}6°$ (to 1 dp)

On a calculator:

| 3 | ÷ | 8 | = | INV | tan |

You have 'O' and
'A' so use tan.

Exercise 37 ❷

Find the angles marked with letters.
Give the answers correct to 1 dp.

1

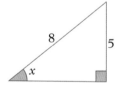

2

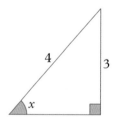

3

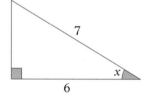

4

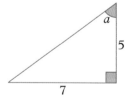

5

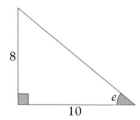

6

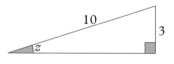

7

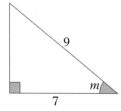

8

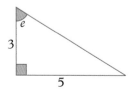

9

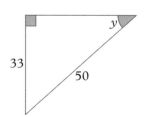

10

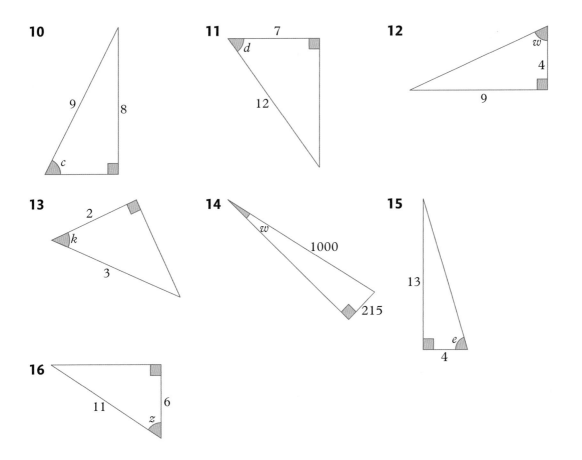

11 7

d

12

12 *w*

4

9

13 2

k

3

14 *w*

1000

215

15 13

e

4

16 6

11

z

7.10.5 Angles of elevation and depression and bearings revision

Angle of elevation

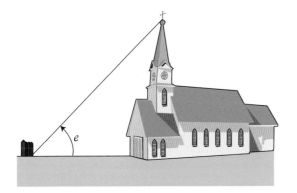

e is the angle of elevation of the steeple from the gate.

Look **up** from the horizontal.

Angle of depression

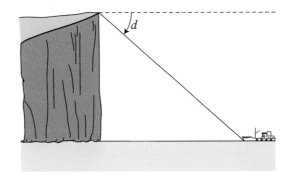

d is the angle of depression of the boat from the cliff top.

Look **down** from the horizontal.

Bearings revision

▶ Bearings are measured **clockwise** from north.

a

North

B

60° 70° A

300°

b N

P 140°

N

40°

Q

320°

Always use three
digits in a bearing.

Ship A sails on a bearing 070°.
Ship B sails on a bearing 300°.

The bearing of Q from P is 140°.
The bearing of P from Q is 320°

Exercise 38 ②

Begin each question by drawing a large clear diagram.

1 A ladder of length 4 m rests against a vertical wall so that the
base of the ladder is 1·5 m from the wall.
Calculate the angle between the ladder and the ground.

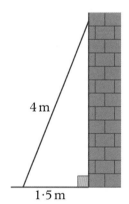

4 m

1·5 m

2 A ladder of length 4 m rests against a vertical wall so that the
angle between the ladder and the ground is 66°. How far up
the wall does the ladder reach?

3 From a distance of 20 m the angle of elevation to the
top of a tower is 35°. How high is the tower?

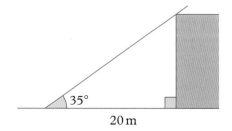

35°

20 m

4 A point G is 40 m away from a building, which is 15 m high.
What is the angle of elevation to the top of the building from G?

5 A boy is flying a kite from a string of length 60 m.
If the string is taut and makes an angle of 71° with the
horizontal, what is the height of the kite? Ignore the
height of the boy.

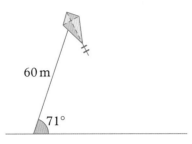

6 A straight tunnel is 80 m long and slopes downwards at an angle of
11° to the horizontal. Find the vertical drop in travelling from the
top to the bottom of the tunnel.

7 AB is a chord of a circle of radius 5 cm and centre O.
The perpendicular bisector of AB passes through
O and also bisects the angle AOB. If ∠AOB = 100°
calculate the length of the chord AB.

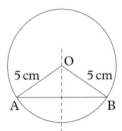

8 A ship is due south of a lighthouse L. It sails on
a bearing of 055° for a distance of 80 km until it
is due east of the lighthouse.
How far is it now from the lighthouse?

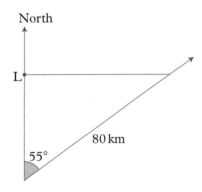

9 A ship is due south of a lighthouse. It sails on a bearing of 071°
for a distance of 200 km until it is due east of the lighthouse.
How far is it now from the lighthouse?

10 An is isosceles triangle has sides of length 8 cm, 8 cm and 5 cm.
Find the angle between the two equal sides.

11 The angles of an isosceles triangle are 66°, 66° and 48°.
If the shortest side of the triangle is 8·4 cm, find the length
of one of the two equal sides.

*12 From the top of a tower of height 75 m, a guard sees two prisoners, both due east of him.

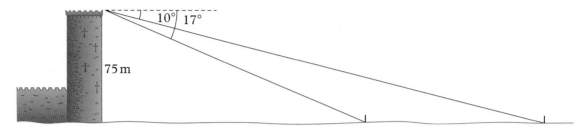

If the angles of depression of the two prisoners are 10° and 17°, calculate the distance between them.

Test yourself

1 Here is a cube.

    a How many vertices does a cube have?

    b Draw a net of a cube on a squared grid.

The diagram shows a cube of side 3 cm.

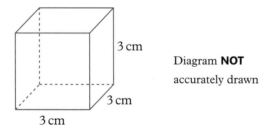

3 cm

Diagram **NOT** accurately drawn

3 cm

3 cm

    c Work out the total surface area of this cube.

(Edexcel, 2013)

2 a A factory manufactures cuboids, each measuring 8·5 cm by 6·6 cm by 3·7 cm.
    Calculate the volume of one of these cuboids, giving the units of your answer.

  b The cuboids are packed into rectangular boxes.
    Each box is 85 cm long, 66 cm wide and 37 cm high.
    What is the maximum number of cuboids that can be packed in each box?

(WJEC, 2011)

3 The diagram shows a container in the shape of a cuboid.

   a Work out the volume of the container.
State the units of your answer.

   b Ben wants to paint the four outside
walls and the top of the container.
One tin of paint covers $6\,m^2$.
How many tins of paint does Ben need?
You **must** show your working.

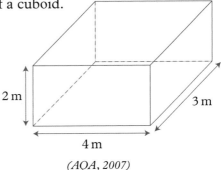

2 m   3 m

4 m

(AQA, 2007)

4 The radius, r, of the cylinder is $10\,cm$.
The height, h, is $4\,cm$.

The volume, V, of a cylinder is $V = \pi r^2 h$

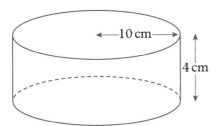

—10 cm—

4 cm

Work out the volume of the cylinder.
Use $\pi = 3 \cdot 1$

(AQA, 2013)

5 This scale drawing shows the area where Edward is staying.

  • Hotel   N

Scale: 1 cm represents 4 km

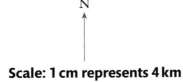

• Drosier Car Hire depot

   a Measure the bearing of the car hire depot from the hotel.

   b What is the real distance from the car hire depot to the hotel?

   c Edward is going to visit a cathedral.
The cathedral is $30\,km$ on a bearing of $285°$ from the hotel.

      Copy the drawing and mark the position of the cathedral on the diagram.

(OCR, 2013)

6 The diagram shows the position of Elaine's house, H, and her position, X, on a walk. The scale of the diagram is 1 cm represents 2 km.

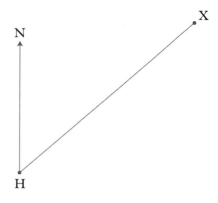

a Measure and write the bearing and distance, in km, of X from H.

Elaine then walks to a position, Y, which is 15 km from H and on a bearing of 260° from H.

b Copy the diagram and mark the position of Y.

(OCR, Spec)

7 The diagram shows a cylinder with a height of 5 cm and a diameter of 16 cm.

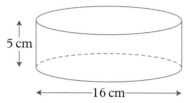

Not drawn accurately

Calculate the volume of the cylinder.
Give your answer in terms of π.
State the units of your answer.

(AQA, 2008)

8 This quadrilateral is drawn on a centimetre square grid.

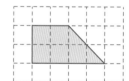

a i What is the mathematical name of the quadrilateral? Choose from the words in the box.

Kite	trapezium	parallelogram	rhombus

ii Work out the area of the quadrilateral.

b On a copy of the diagram below, reflect the quadrilateral in the line **m**.

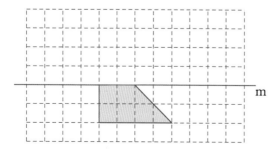

c On a copy of the diagram below, reflect the quadrilateral in the line **n**.

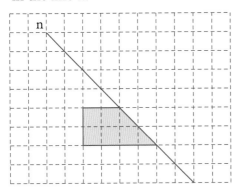

(OCR, 2013)

9 **a** Draw on a copy of this shape any lines of symmetry.

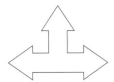

b **i** Copy and complete the following: This shape has rotational symmetry of order ————.

ii Shade 3 more triangles on a copy of the shape below so that the pattern will have rotational symmetry of order 3.

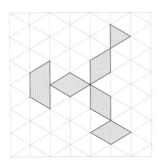

(CCEA, 2013)

10 Enlarge a copy of this shape using a scale factor of 2

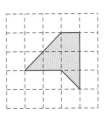

(CCEA, 2013)

11 A circle has a radius of 6 cm.
A square has a side of length 12 cm.

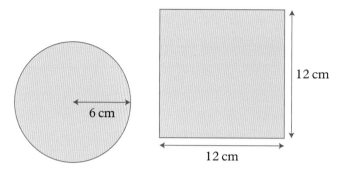

Work out the difference between the area of the circle and
the area of the square.
Give your answer correct to one decimal place.

(Edexcel, 2008)

12 a Calculate the area of the trapezium.

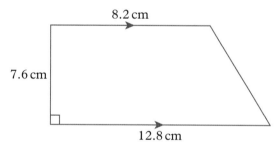

Diagram not drawn to scale

b Calculate the area of a semicircle with a diameter of 44·8 cm.

(WJEC, 2011)

13 A circle has a radius of 11·7 cm.
Work out the circumference of the circle.

(CCEA, 2013)

14 *In this question, use a pair of compasses and a ruler.*
Leave in all your construction lines.

Triangle ABC has sides AB = 8·5 cm, AC = 7·3 cm and BC = 6·8 cm.

Copy and complete the accurate drawing of triangle ABC.
Side AB has been drawn for you.

A   8.5 cm   B

(OCR, 2012)

15 Dionne cuts six identical circles from a rectangle of fabric to make mats.

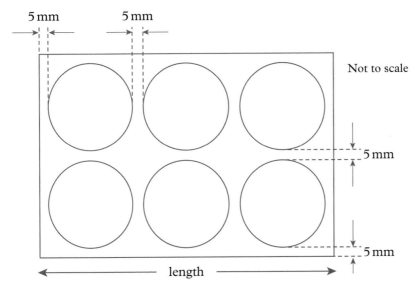

5 mm   5 mm

Not to scale

5 mm

5 mm

length

Each circle has a diameter of 10 cm.

She leaves 5 mm between each circle and 5 mm from each
circle to the edge of the fabric.

a What is the length of the rectangle?
   Give your answer in centimetres.

b Dionne draws this regular pattern onto each circular mat.

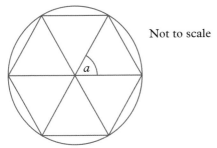

Not to scale

a

   i Without measuring, explain fully why angle *a* is 60°.

  ii The diameter of a mat is 10 cm.
   Calculate the **total** length of the lines that Dionne
   draws on one mat.

c It costs Dionne £1·60 to make each mat.
   She adds 50% of the cost for her profit.
   Calculate the price at which Dionne sells each mat.

(OCR, 2013)

16 The diagram shows a coastline, CL.
A and B are two rocks in the sea.

Scale: 1 cm represents 500 m
Rosie is sailing her boat.
She sails on a course towards the coast so that she is an equal
distance from the rocks, A and B.

When she is less than 1 km from the coast she turns
and sails due West.
She now sails so that she is between 500 m and 1 km
from the coast.

Copy the diagram and construct a route that Rosie could take.
You must leave in all your construction lines.

(OCR, 2013)

17 Here is a regular polygon.
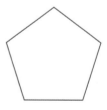

  a What is the mathematical name of this polygon?

  b How many lines of symmetry does this polygon have?

  c What is the order of rotation symmetry of this polygon?

  d In this polygon each side has length x cm.

    Write down an algebraic expression, in cm, for the perimeter
    of this polygon.

(OCR, 2013)

18 Here is a quadrilateral.

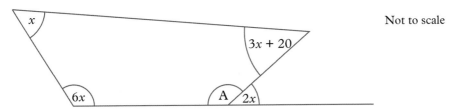

Not to scale

a Explain why angle A in the quadrilateral is $180 - 2x$.
b Work out the size of the angle marked x.
Show all your working.

(OCR, 2013)

19 The diagram shows a triangle ABC.
AB = 14·7 cm, BC = 11·5 cm and AC = 19·4 cm

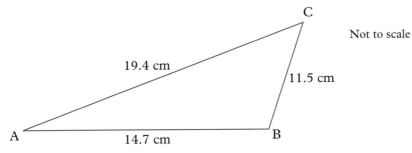

Not to scale

Show that triangle ABC is **not** a right-angled triangle.

(OCR, 2013)

20 *XYZ* is a right-angled triangle.

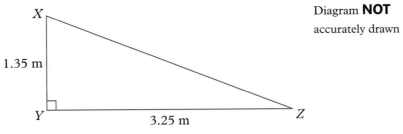

Diagram **NOT**
accurately drawn

Calculate the length of *XZ*.
Give your answer correct to 3 significant figures.

(Edexcel, 2013)

8 Statistics 2

Collecting, displaying and interpreting data accurately is essential for modern science.

8.1 The data handling cycle

1 Specifying the problem and planning	Start with a problem or a question you wish to answer, for example • Do most children go to the secondary school of their choice? • Do adults watch more television than children?
2 Collecting data	• You have to decide what data (information) you need to collect to answer the question. • You may decide to collect the data yourself. When you collect data by a survey or by experiment the data are called **primary data**. • If you look up published data that someone else has collected in a book or on the Internet the data are called **secondary data**.
3 Processing and presenting the data	• You usually have to **process** the data by, for example, calculating averages or frequencies so that you can **represent** the data in a pictorial form such as a pie chart or a frequency diagram.
4 Interpreting and discussing results	• Finally you need to **interpret** the results and draw conclusions so that you can answer your original question. • You need to look for patterns in the data and for any possible exceptions. • Sometimes the results you find suggest that you need to modify the way the survey was conducted.

8.1.1 Surveys

Statisticians frequently carry out surveys to investigate a characteristic of a population. A population is the set of all possible items to be observed, not necessarily people.

Surveys are done for many reasons.

- Newspapers seek to know voting intentions before an election.
- Advertisers try to find out which features of a product appeal most to the public.
- Research workers testing new drugs need to know their effectiveness and their possible side effects.
- Government agencies conduct tests for the public's benefit such as the percentage of first class letters arriving the next day or the percentage of '999' calls answered within 5 minutes.

Census

A census is a survey in which data are collected from every member of the population of a country. A census is done in Great Britain every ten years; 1971, 1981, 1991, 2001, … . Data about age, educational qualifications, race, distance travelled to work, etc., is collected from every person in the country on the given 'Census Day'. Census forms take several years to prepare and many years for a thorough analysis of the data collected.

> Because no person must be omitted, a census is a huge and very expensive task.

Samples

If a full census is not possible or practical, then it is necessary to take a sample from the population. Two reasons are:

- It will be much cheaper and quicker than a full census
- The object being tested may be destroyed in the test, like testing the average lifetime of a new light bulb or the new design of a car tyre.

The choice of a sample prompts two questions.

1   Does the sample truly represent the whole population?
2   How large should the sample be?

After taking a sample, it is assumed that the result for the sample reflects the whole population. For example, if 20% of a sample of 1000 people say they will vote in a particular way, then it is assumed that 20% of the whole electorate would do so likewise, subject of course to some error.

Bias

> ▶   A sample of data is **biased** when it does not give a true representation of the main population.

Bias can come from a variety of sources including not sampling the population randomly. For example, for data collected by a street survey, the time of day and location may well mean that there is a sector or sectors of the population who are not questioned.

A common cause of bias occurs when the questions asked in the survey are not clear or are leading questions.

Exercise 1 ❶

In questions **1** to **4**, decide whether the method of choosing people to answer questions is satisfactory or not. Consider whether or not the sample suggested might be **biased** in some way. Where necessary, suggest a better way of finding a sample.

1 A teacher, with responsibility for school meals, wants to hear students' opinions on the meals currently provided. She waits next to the dinner queue and questions the first 50 students as they pass.

2 An opinion pollster wants to canvass opinion about our European neighbours. He questions drivers as they are waiting to board their ferry at Dover.

3 A journalist wants to know the views of local people about a new one-way traffic system in the town centre. She takes the electoral roll for the town and selects a random sample of 200 people.

4 A pollster working for the BBC wants to know how many people are watching a new series. She questions 200 people as they are leaving a shop between 10:00 and 12:00 one Thursday.

8.1.2 Collecting data

Questionnaires

A questionnaire is a set of questions on a form that is asked of a number of people to gather information which is then analysed.

● Surveys are made using questionnaires to find the popularity of various TV programmes. Advertisers are prepared to pay a large sum for a 30-second advertisement in a programme with an audience of 10 million people.

● Supermarkets use questionnaires to discover what things are most important to their customers. They might want to find out how people feel about ease of car parking, price of food, quality of food, length of time queueing to pay, etc.

● Most surveys are conducted using questionnaires. It is very important to design the questionnaire well so that:
  a people will cooperate and will answer the questions honestly.
  b the questions are not biased.
  c the answers to the questions can be analysed and presented for ease of understanding.

> You need to think about how many people to survey. Too many will take a long time!

Checklist

A Provide an introduction to the sheet so that the person you are asking knows the purpose of the questionnaire.

> **'Proposed new traffic lights'**

B Make the questions easy to understand and specific to answer.
Do **not** ask vague questions like this.
The answers could be:

> 1 Did you see much of the Olympics on TV?

    'Yes, a lot'
    'Not much'
    'Only the best bits'
    'Once or twice a day'

You will find it hard to analyse this sort of data.

A **better** question is:

1 'How much of the Olympic coverage did you watch?' Tick one box			
Not at all	Up to 1 hour per day	1 to 2 hours per day	More than 2 hours per day
☐	☐	☐	☐

C Make sure that the questions are not **leading** questions.
Remember that the survey is to find out opinions of other people, not to support your own opinions.
Do **not** ask:

    'Do you agree that BBC has the best sports coverage?'

A better question is:

'Which of these channels has the best sports coverages?'			
BBC	ITV	Channel 4	Satellite TV
☐	☐	☐	☐

You might ask for one tick or possibly numbers 1, 2, 3, 4 to show an order of preference.

D If you are going to ask sensitive questions (about age or income, for example), design the question with care so as not to offend or embarrass.
Do **not** ask:

    'How old are you?'
or 'Give your date of birth'

A better question is:

Tick one box for your age group.

15–17	18–20	21–30	31–50
☐	☐	☐	☐

E Do not ask more questions than necessary and put the easy questions first.

Exercise 2 ①

In questions **1** to **7**, explain why the question is not suitable for a questionnaire. Write an improved question in each case.

● Write some questions with 'yes/no' answers and some questions which involve multiple responses.

● Remember to word your questions simply.

1 Which sort of holiday do you like best?

2 What do you think of the new head teacher?

3 For how long do you watch television each day?

2–3 hours	3–4 hours	5–6 hours

4 How much would you pay to use the new car park?

☐ less than £1   ☐ more than £2·50

5 Do you agree that English and Maths are the most important subjects at school?

6 Do you or your parents often buy DVDs from a shop?

7 Do you agree that you get too much homework?

8 Some students designed a questionnaire to find out peoples' views about television. Comment on the questionnaire below and write an improved version.

Name _____   Sex   M/F

Age   _____

1 How much television do you watch?

☐   ☐   ☐

not much   quite a lot   a lot

2 What is your favourite programme on TV?

3 Do you agree that there should be more stations like BBC 1?

☐ agree   ☐ disagree

4 Do you like nature programmes?

☐   ☐   ☐   ☐

No   Not really   Sometimes   I love them

---

9 Write a suitable question to find out what **type** of TV programme people of your age watch most. For example: comedy, romance, sport etc.

10 Here is another style of question which can be useful.

> Which of the following statements best describes your attitude to using a computer?
> **A**   I like using them for all sorts of things.
> **B**   I use them when I have to.
> **C**   I hate them
> Please circle: **A**   **B**   **C**

Write a similar style question about peoples' attitude to any topic of your own choice.

Tally charts

A **tally chart** is often used as part of a data collection sheet. Here is a tally chart for the responses to a question about the amount of homework set to students in a school.

Response	Tally	Frequency
Not enough	JHI IIII	9
About right	JHI JHI JHI JHI II	22
Too much	JHI JHI III	13
Don't know	III	3

You could represent these data in a bar chart or pie chart.

● Here is another question which you can answer using statistical methods. 'Do different newspapers use words of different lengths or sentences of different lengths?'

You could conduct an experiment by choosing a similar page from different newspapers.

Number of words in a sentence	1–5	6–10	11–15	16–20	21+
Guardian					
Mirror					
Mail					

8.2   Frequency diagrams for discrete and continuous data

Discrete and continuous data

Data can be either **discrete** or **continuous**.
Discrete data can take only certain values, for example
● the number of peas in a pod
● the number of children in a class
● shoe size.

Continuous data comes from measuring and can take any value, for example
● the height of a child
● the weight of an apple
● time taken to boil a kettle.

EXAMPLE

The marks obtained by 36 students in a test were

1	3	2	3	4	2	1	3	0
5	3	0	1	4	0	4	4	3
3	4	3	1	3	4	3	1	2
1	3	4	0	4	3	2	5	3

Show the marks

a   on a tally chart   b   on a frequency diagram.

> This is discrete data as the marks are counted.

a

Mark	Tally	Freq.
0	IIII	4
1	IIII I	6
2	IIII	4
3	IIII IIII II	12
4	IIII III	8
5	II	2

b

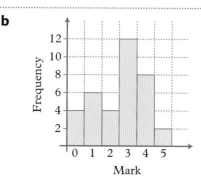

Exercise 3 ①

1 In a survey, the number of occupants in the cars passing a school was recorded. The bar chart shows the results.

a   How many cars had 3 occupants?
b   How many cars had fewer than 4 occupants?
c   How many cars were in the survey?
d   What was the total number of occupants in all the cars in the survey?
e   What fraction of the cars had only one occupant?

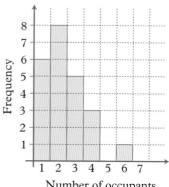

2 A dice was thrown 30 times. These were the scores.

1	2	5	5	3	6	1	5	2	4
3	6	1	5	6	5	6	5	1	4
1	3	5	5	1	4	6	3	2	2

Draw a tally chart, like the one in the example box, and use it to draw a frequency diagram.

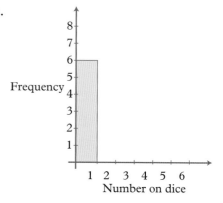

3 The number of letters in each word on one page of a book
was recorded and the results were

2	6	3	5	2	4	1	6	3	7	5	4	3	8
6	1	5	1	4	3	6	4	7	4	1	7	5	4
3	5	4	6	1	4	2	6	1	5	6	2	8	4
3	2	6	4	3	7	3	9	2	4	9	4	10	2

a Draw a tally chart for these results.

b What was the most frequent number of letters in a word?

4 Jake threw two dice sixty times and recorded the total score.

2	3	5	4	8	6	4	7	5	10
7	8	7	6	12	11	8	11	7	6
6	5	7	7	8	6	7	3	6	7
12	3	10	4	3	7	2	11	8	5
7	10	7	5	7	5	10	11	7	10
4	8	6	4	6	11	6	12	11	5

a Draw a tally chart to show the results
of the experiment.
The tally chart is started here.

b Draw a frequency diagram to illustrate
the results. Plot the frequency on
the vertical axis.

Score	Tally marks	Frequency
2	II	2
3	IIII	4
4		
.		
.		

5 The bar chart shows the profit or loss
made by a toy shop from September
2014 to April 2015.

a Estimate the total profit in this period.

b Describe what is happening to
the shop's profits in this period.
Try to think of an explanation
for the shape of the bar chart.

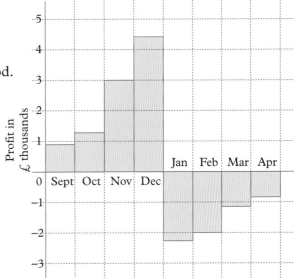

Grouped data

Sometimes the data have a wide range of values. In such cases, it is useful to put the data into groups before drawing a tally chart and frequency diagram.

Grouping data is particularly essential when the data are continuous as measuring gives so many different answers.

EXAMPLE

A group of 21 students measured their hand spans in centimetres.

14·8	20·0	16·9	20·7	18·1	17·5	18·7
19·0	19·8	17·8	14·3	19·2	21·7	17·4
16·0	15·9	18·5	19·3	16·6	21·2	18·4

This is continuous data because it is measured.

Show the data grouped in

a a tally chart

b a frequency diagram.

a

Class intervals	Tally
$14 \leq s < 16$	|||
$16 \leq s < 18$ |卌	
$18 \leq s < 20$	卌 |||
$20 \leq s < 22$	||||

b

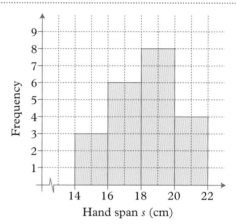

Hand span s (cm)

Note that '20·0' goes into the last group $20 < s < 22$.

$\leq$ means 'less than or equal to' and $<$ means 'less than'

Note: the horizontal axis does not start at zero. The zig-zag in the axis reminds you of this.

The data in this example is continuous so there are no gaps between the bars.

The data must be grouped into equal class intervals so that the height of each bar represents the frequency.

Exercise 4 ①

1 The graph shows the heights of students in a class.

    a How many students were over 150 cm tall?

    b How many students had a height between 135 cm and 155 cm?

    c How many students were in the class?

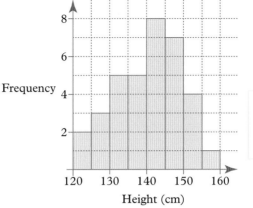

> The horizontal axis starts at 120 rather than 0.

2 In a survey, the heights of children aged 15 were measured in four countries, A, B, C and D, around the world. A random sample of children was chosen by computer, not necessarily the same number from each country.

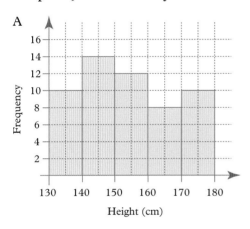

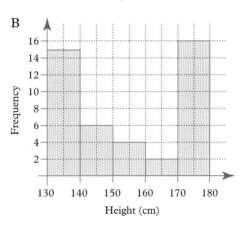

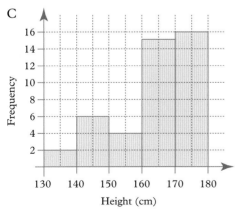

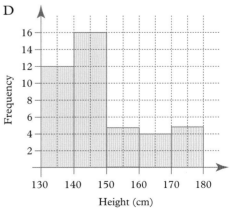

Use the graphs to identify the country in each of these statements.

    a Country _____ is poor and the diet of children is not good.
Two-thirds of the children were less than 150 cm tall.

    b There were 54 children in the sample from Country _____.

   c In Country _____, the heights were spread fairly evenly across the range 130 cm to 180 cm.

   d The smallest sample of children came from Country _____.

   e In Country _____, there were similar numbers of very short and very tall children.

3 Scientists have developed a new fertiliser which is supposed to increase the size of carrots. A farmer grew carrots in two adjacent fields, A and B, and treated one of the fields with the new fertiliser. A random sample of 50 carrots was taken from each field and weighed. Here are the results, in grams, for Field A.

118	91	82	105	72	92	103	95	73	109
63	111	102	116	101	104	107	119	111	108
112	97	100	75	85	94	76	67	93	112
70	116	118	103	65	107	87	98	105	117
114	106	82	90	77	88	66	99	95	103

Make a tally chart for this data using the groups given here.

Weight, g	Tally	Frequency
$60 \leq w < 70$		
$70 \leq w < 80$		
$80 \leq w < 90$		
$90 \leq w < 100$		
$100 \leq w < 110$		
$110 \leq w < 120$		

This is the frequency graph for Field B.

Draw a similar frequency graph for Field A.
Which field do you think was treated with the new fertiliser? Give your reasons.

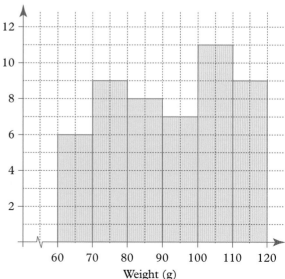

Frequency

Weight (g)

4 Karine and Jackie intend to go skiing in February. They have information about the expected snowfall for two possible places.

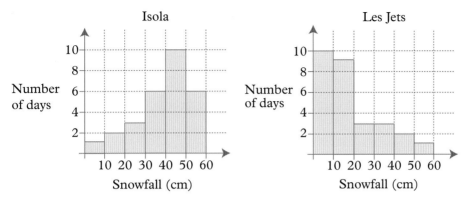

Decide where you think they should go. It doesn't matter where you decide, but you **must** say why, using the charts above to help you explain.

5 Some people think that children's IQs increase when they eat extra vitamins.
In an experiment, 52 children took an IQ test before and then after a course of vitamin pills. Here are the results.

> **Vitamins benefit IQ**
> Vitamins take Johnny to the top of the class!

Before

81	107	93	104	103	96	101	102	93	105	82	106	97
108	94	111	92	86	109	95	116	92	94	101	117	102
95	108	112	107	106	124	125	103	127	118	113	91	113
113	114	109	128	115	86	106	91	85	119	129	99	98

After

93	110	92	125	99	127	114	98	107	128	103	91	104
103	83	125	91	104	99	102	116	98	115	92	117	97
126	100	112	113	85	108	97	101	125	93	102	107	116
94	117	95	108	117	96	102	87	107	94	103	95	96

a Put the scores into appropriate groups between 80 and 130.
b Draw two frequency graphs to display the results.
c Write a conclusion. Did the vitamin pills make a significant difference?

> You could use 80–89, 90–99, . . . or 80–84, 85–89, 90–94, . . . It is up to you!

8.3 Correlation and scatter diagrams

Sometimes it is interesting to discover if there is a relationship
(or **correlation**) between two sets of data.

Examples

- Do tall people weigh more than short people?
- If you spend longer revising for a test, will you get a higher mark?
- Do people who have credit cards have more debts than other
  people?
- Do tall parents have tall children?
- If there is more rain, will there be less sunshine?
- Does the number of Olympic gold medals won by
  British athletes affect the rate of inflation?

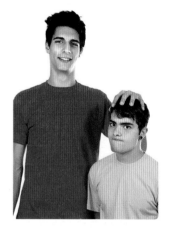

If there is a relationship, it will be easy to spot if your data are plotted
on a scatter diagram.

▶ A **scatter diagram** is a graph in which one set of data is plotted on the
horizontal axis and the other set on the vertical axis.

EXAMPLE

Each month the average outdoor temperature was recorded
together with the amount of gas (in therms) used to heat a house.
The results are plotted on this scatter diagram.

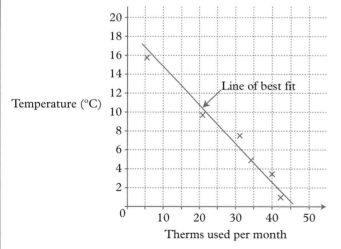

There is a high degree of **correlation** between these figures.
Gas companies do use weather forecasts to predict future gas
consumption over the whole country.

A **line of best fit** has been drawn 'by eye'. You can estimate that, if
the outdoor temperature was 12 °C, then about 17 therms of gas
would be used.

Note: You can only
predict within the range
of values given. If you
extend the line below
0 °C, the line of best fit
predicts that about 60
therms would be used
when the temperature
is −4 °C. But −4 °C is
well outside the range
of the values plotted,
so the prediction is not
valid. Perhaps at −4 °C
many **people** might
stay in bed and the gas
consumption would not
increase by much. The
point is you don't know!

▶ The line in the example has a negative gradient, and so there is **negative correlation**.

▶ If the line of best fit has a positive gradient there is **positive correlation**.

▶ Some data when plotted on a scatter diagram does not appear to fit any line at all. In this case, there is no correlation.

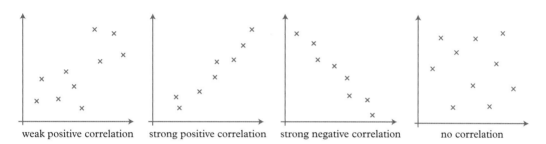

weak positive correlation   strong positive correlation   strong negative correlation   no correlation

A correlation does not **prove** that changes in one variable are **causing** changes in the other. However, it does suggest that further investigation might be useful.

Exercise 5 ① ②

1   For this question, you need to make some measurements of people in your class.

   a Measure everyone's height and armspan to the nearest cm.

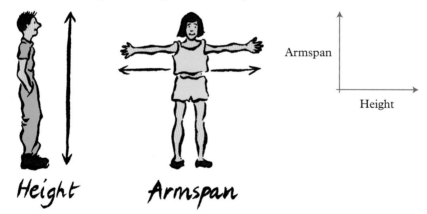

      Plot the measurements on a scatter graph.
      Is there any correlation?

   b Now measure everyone's 'head circumference' just above the eyes.
      Plot head circumference and height on a scatter graph.
      Is there any correlation?

   c Decide as a class which other measurements (for example, pulse rate) you can (fairly easily) take and plot these to see if any correlation exists.

   d Which pair of measurements gave the best correlation?
      What conclusions can you make?

2 Plot the points in each table on a scatter graph on axes as shown. Draw axes with values from 0 to 20. Describe the correlation, if any, between the values of t and z (for example, 'strong positive', 'weak negative' etc.).

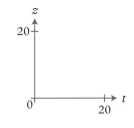

a

t	8	17	5	13	19	7	20	5	11	14
z	9	16	7	13	18	10	19	8	11	15

b

t	4	9	13	16	17	6	7	18	10
z	5	3	11	18	6	11	18	12	16

c

t	12	2	17	8	3	20	9	5	14	19
z	6	13	8	15	18	2	12	9	12	6

3 Describe the correlation, if any, in these scatter graphs.

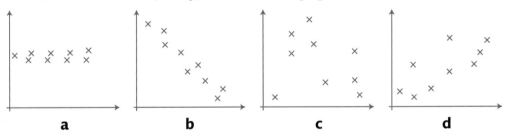

    a   **b**   **c**   **d**

4 Plot the points in the table on a scatter graph, on axes with values from 0 to 20.

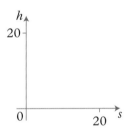

s	3	13	20	1	9	15	10	17
h	6	13	20	6	12	16	12	17

a Draw a line of best fit.
b What value would you expect for h when s is 6?

5 The marks of seven students in the two papers of a physics examination are shown in the table.

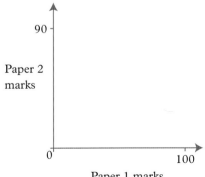

Paper 1	20	32	40	60	71	80	91
Paper 2	15	25	40	50	64	75	84

a Plot the marks on a scatter diagram, using a scale of 1 cm to 10 marks, and draw a line of best fit.
b A student scored a mark of 50 on Paper 1. What would you expect her to get on Paper 2?

6 The table shows
  i the engine size in litres of various cars
  ii the distance travelled in km on one litre of petrol.

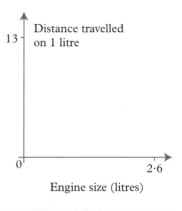

Engine	0·8	1·6	2·6	1·0	2·1	1·3	1·8
Distance	13	10·2	5·4	12	7·8	11·2	8·5

  a Plot the figures on a scatter graph using a scale of
  5 cm to 1 litre across the page and 1 cm to 1 km
  up the page. Draw a line of best fit.
  b A car has a 2·3 litre engine. How far would you
  expect it to go on one litre of petrol?

7 The data show the latitude of 10 cities in the northern
hemisphere and the average high temperatures.

City	Latitude (degrees)	Mean high temperature (°F)
Bogota	5	66
Casablanca	34	72
Dublin	53	56
Hong Kong	22	77
Istanbul	41	64
St Petersburg	60	46
Manila	15	89
Mumbai	19	87
Oslo	60	50
Paris	49	59

  a Draw a scatter diagram and draw a line of best fit. Plot
  latitude across the page with a scale of 2 cm to 10°.
  Plot temperature up the page from 40 °F to 90 °F with a
  scale of 2 cm to 10 °F.
  b Which city lies well off the line of best fit?
  Do you know what factor might cause this apparent discrepancy?
  c The latitude of Shanghai is 31°N. What do you think its
  mean high temperature is?

8 What sort of pattern would you expect if you took readings
of these variables and drew a scatter diagram?
  a cars on roads – accident rate
  b sales of perfume – advertising costs
  c birth rate – rate of inflation
  d petrol consumption of car – price of petrol
  e outside temperature – sales of ice cream.

Test yourself

1 A company sells ice cream.
The average midday temperature and the sales for each
month in 2011 are shown.

	Jan	Feb	Mar	Apr	May	Jun	Jul	Aug	Sep	Oct	Nov	Dec
Average midday temperature (°C)	8	6	11	14	17	21	22	29	20	14	10	4
Sales (tonnes)	23	24	23	30	33	37	39	47	36	28	22	23

a Copy and complete the scatter
diagram by plotting the values
for July to December. The values
for January to June have been done
for you.

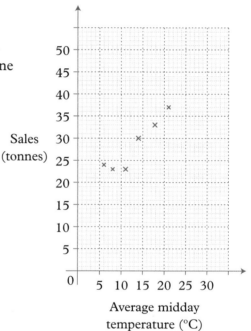

b In July 2012, the average midday temperature is predicted to be 25 °C.
Use the graph to estimate the sales of ice cream in July 2012.
Show clearly how you obtain your answer.

c In December 2012, the average midday temperature is predicted
to be 5°C higher than in December 2011.
Should the company increase its production of ice cream for
December 2012?

Choose a box.

Yes ☐   No ☐

Give a reason for your answer.

(AQA, 2012)

2 The scatter diagram shows the height, in cm, and the weight, in kg, for each of 20 members of a sports club.

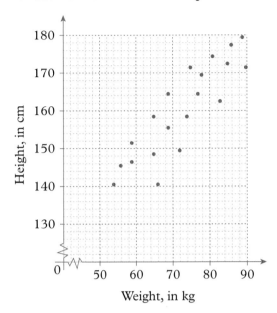

a Write down the height and weight of the **heaviest** of the 20 members of the sports club.

b Write down the type of correlation shown by the scatter diagram.

c Estimate the weight of a person of height 155 cm.

d Is it possible to estimate the weight of a person with a height of 210 cm from the scatter diagram?

You must give a reason for your answer.

(WJEC, 2011)

3 Mason is doing a survey to find out how many magazines people buy. He uses this question on his questionnaire.

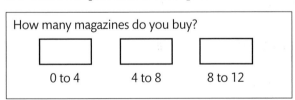

How many magazines do you buy?

0 to 4   4 to 8   8 to 12

a Write down **two** things wrong with this question.

b Write a better question for Mason to use on his questionnaire to find out how many magazines people buy.

Mason asks his friends at school to do his questionnaire.

This may **not** be a good sample to use.

c Give **one** reason why.

(Edexcel, 2013)

4 Jamie is doing a survey on how people travel to work during the week.
  a Here is one of his questions.

> Which form of transport do you use to travel to work?
> Tick one box only.
>
> Car   Bus   Train   Walk
> ☐   ☐   ☐   ☐

This is not a good question and set of response boxes. Explain why.
  b Here is another of his questions.

> How long, in minutes, does it take you to travel to work?
> Tick one box only.
>
> | 0 to 10 | 10 to 20 | 20 to 30 | 30 to 50 |
> |---------|----------|----------|----------|
> | | | | |

Give two reasons why some people may find it difficult to decide which box to tick.
  c Jamie does his survey at the train station on a Tuesday morning. Explain why this is not sensible

(OCR, 2013)

5 Kurt has to undertake a survey to find out the most popular drink in Wales. He carries out his survey by asking students at a school disco to answer this question.

> Which is your favourite drink?
> Tick the appropriate box
>
> Tea ☐   Coffee ☐   Cola ☐

  a State **one** reason why the design of this question is unsuitable for his survey.
  b State **two** reasons why his survey is likely to be biased.

(WJEC)

6 Joe and Pam planted crocus bulbs in their gardens.
They shared a bag of 250 crocus bulbs.
The table shows the colour of the flower from each bulb.

	Yellow	Purple	White	Totals
Joe	64	40		125
Pam	56		32	125
Totals	120			250

  a Copy and complete the table.
  b Write the ratio 64 : 56 as simply as possible.

(OCR, 2012)

7 a The numbers of people who went paintballing each day for four weeks are recorded below.

128 57 67 98 120 48 46 122 38 47 108 94 78 86

68 53 90 84 49 127 82 105 64 117 111 67 54 104

Copy and complete the frequency chart.

Number of people	Tally	Frequency
30 – 49		
50 – 69		
70 – 89		
90 – 109		
110 – 129		

b On how many days did 90 or more people go paintballing?

(OCR, 2013)

8 Geta did a survey of the type and weight of tea bought by 100 people. She displayed her results in a table.

Copy and complete the table.

Tea type / Weight	Regular tea bags	Decaffeinated tea bags	Loose leaf tea	Total
50 g	2	0	5	
100 g	35	18		60
200 g	16			
Total		25		100

(OCR, 2012)

9 Number 3

In this unit you will:

- revise how to convert between fractions, decimals and percentages
- learn how to calculate with compound measures
- revise metric and imperial units
- revise rounding and estimation.

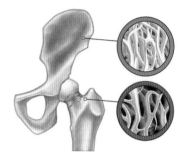

Doctors can measure the density of bones to see if they are weak or strong.

9.1 Fractions, decimals and percentages

Percentages are a way of expressing fractions or decimals.
You should be able to convert readily from one form to another.

> ▶ To change a fraction to a decimal, do a long or short division.
> ▶ To change a decimal to a fraction, write as tenths or hundredths and cancel.
> ▶ To change a fraction or a decimal to a percentage, multiply by 100%.

EXAMPLE

Change

a $\frac{7}{8}$ to a decimal **b** $0\cdot35$ to a fraction **c** $\frac{3}{8}$ to a percentage

d $0\cdot85$ to a percentage **e** $\frac{5}{6}$ to a decimal.

------

a Divide 8 into 7 $8\overline{)7\cdot000}$ $0\cdot875$

b $0\cdot35 = \frac{35}{100}$

$ = \frac{7}{20}$

c $\frac{3}{8} = \frac{3}{8} \times 100\%$

$\phantom{\frac{3}{8}} = \frac{300}{8} = 37\frac{1}{2}\%$

d $0\cdot85 = 0\cdot85 \times 100\%$

$ = 85\%$

e $6\overline{)5\cdot000}$ $0\cdot8333...$

This is a **recurring** decimal.
You write $\frac{5}{6} = 0\cdot\dot{8}\dot{3}$. Similarly,
$0\cdot525252... = 0\cdot\dot{5}\dot{2}$

Note that **all** recurring decimals are exact fractions but that not all exact fractions are recurring decimals.

Exercise 1 ① ②

1 Change the fractions to decimals.

  a $\dfrac{1}{4}$ **b** $\dfrac{2}{5}$ **c** $\dfrac{3}{8}$ **d** $\dfrac{5}{12}$ **e** $\dfrac{1}{6}$ **f** $\dfrac{2}{7}$

2 Change the decimals to fractions and simplify.

  a $0 \cdot 2$ **b** $0 \cdot 45$ **c** $0 \cdot 36$ **d** $0 \cdot 125$ **e** $1 \cdot 05$ **f** $0 \cdot 007$

3 Change to percentages.

  a $\dfrac{1}{4}$ **b** $\dfrac{1}{10}$ **c** $0 \cdot 72$ **d** $0 \cdot 075$ **e** $0 \cdot 02$ **f** $\dfrac{1}{3}$

4 In July, 360 000 people visited Bali for their holiday.

  a One-eighth of the people were American.
Find the number of American visitors.

  b 11% of the people were French. How many people was that?

  c There were 12 000 people from Japan.
What fraction of the total were from Japan?

5 Copy and complete the table.

	Fraction	Decimal	Percentage
a	$\dfrac{1}{4}$		
b		$0 \cdot 2$	
c			80%
d	$\dfrac{1}{100}$		
e			30%
f	$\dfrac{1}{3}$		

6 Here are some fractions.

  $\dfrac{4}{10}$ $\dfrac{11}{33}$ $\dfrac{1}{5}$ $\dfrac{7}{12}$

  Which of these fractions is

  a equal to $0 \cdot 2$ **b** equal to 40%

  c equal to $\dfrac{1}{3}$ **d** greater than $\dfrac{1}{2}$.

(?) **7** Max wants 3 bottles of Coke, which normally costs 90p per bottle.
Which of the three offers is the cheapest for 3 bottles?

A
3 for the
price of **2** !!!

B
30% off
marked price

C
BUY ONE get
the 2nd HALF PRICE !

(?) **8** Two shops had sale offers on an article which previously
cost £69. One shop had '$\frac{1}{3}$ off' and the other had '70%
of old price'. Which shop had the lower price?

(?) **9** Shareholders in a company can opt for either '$\frac{1}{6}$ of £5000'
or '15% of £5000'. Which is the greater amount?

10 A photocopier increases the sides of a square
in the ratio 4:5. By what percentage are the
sides increased?

*11 In an alloy, the ratio of copper to iron to lead
is 5:7:3. What percentage of the alloy is lead?

9.2 Compound measures

9.2.1 Speed

When a train moves at a constant speed of 20 metres per second,
it moves a distance of 20 metres in 1 second. In 2 seconds the
train moves 40 metres, and so on.

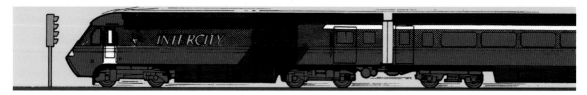

▶   The distance moved is equal to the speed multiplied by the time taken.

$$\text{Distance} = \text{Speed} \times \text{Time} \qquad \text{Speed} = \frac{\text{Distance}}{\text{Time}} \qquad \text{Time} = \frac{\text{Distance}}{\text{Speed}}$$

EXAMPLE

Distance = Speed × Time

Speed = $\dfrac{\text{Distance}}{\text{Time}}$

Time = $\dfrac{\text{Distance}}{\text{Speed}}$

D

S × T

1   A bird takes 20 s to fly a distance of 100 m. Calculate the average speed of the bird.

$S = \dfrac{D}{T}$ Average speed $= \dfrac{100}{20}$

$= 5 \,\text{m/s}$

2   A car travels a distance of 200 m at a speed of 25 m/s. How long does it take?

$T = \dfrac{D}{S}$

Time taken $= \dfrac{200}{25} = 8 \,\text{seconds}$

3   A boat sails at a speed of 12 km/h for 2 days. How far does it travel?

2 days = 48 hours

D = S × T

Distance travelled = 12 × 48

$= 576 \,\text{km}$

Exercise 2 ① ②

1 Find the distance travelled.
  a 55 mph for 2 hours
  b 17 m/s for 20 seconds
  c 63 km/h for 5 hours
  d 5 cm/day for 12 days.

2 Find the speed.
  a 98 miles in 7 hours
  b 364 km in 8 hours
  c 250 m in 10 seconds
  d 63 cm in 6 minutes.

3 A car travels at a constant speed of 40 mph for three hours. How far does it go?

4 An athlete runs at a steady speed of 5 m/s for 100 s. How far does he run?

5 How far will a train travel in 15 s if it is travelling at a steady speed of 20 m/s?

6 A ball travels for 30 s at a speed of 12 m/s. Find the distance it covers.

7 An aircraft flies at a speed of 800 km/h for $2\frac{1}{2}$ h. How far does it fly?

8 How far will a ship sail in half an hour if it is going at a steady speed of 24 km/h?

9 A car travelling at a steady speed takes 4 hours to travel 244 km. What is the speed of the car?

10 Joseph runs 750 m in a time of 100 s. At what speed does he run?

11 A train takes 6 hours to travel 498 km. What is the speed of the train?

12 After a meal an earthworm moves a distance of 45 cm in 90 s.
At what speed does the worm move?

13 A plane flies 720 miles at a speed of 240 mph. How long does it take?

14 An octopus swims 18 km at a speed of 3 km/h. How long does it take?

15 A rocket is flying at a speed of 1000 km/h. How far does it go in 15 minutes?

16 Find the time taken.

    a 360 km at 20 km/h   **b** 56 miles at 8 mph

    c 200 m at 40 m/s   **d** 60 km at 120 km/h.

17 A car takes 15 minutes to travel 22 miles.
Find the speed in mph.

18 An athlete runs at 9 km/h for 30 minutes.
How far does he run?

9.2.2 Density, mass and volume

▶ Density in $g/cm^3 = \dfrac{\text{Mass of object (in grams)}}{\text{Volume of object (in } cm^3)}$

EXAMPLE

The cuboid shown has a mass of 480 g.
Find the density of the material.

Volume of cuboid $= 4 \times 5 \times 3 = 60\ cm^3$

Density $= \dfrac{\text{Mass}}{\text{Volume}} = \dfrac{480}{60} = 8\ g/cm^3$

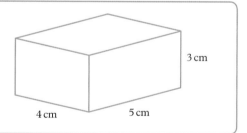

3 cm

4 cm   5 cm

Exercise 3 ②

Where necessary give answers correct to 1 d.p.

1 Find the density of each object.

    a   **b**   **c**

      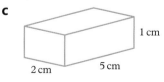

1 cm

2 cm   5 cm

    a Sphere mass = 320 g   **b** Mass = 2500 g   **c** Mass = 56 g

           Volume = 40 cm^3 Volume = 1000 cm^3

2 A steel ball has volume $1000\,\text{cm}^3$. The density of steel is $8\,\text{g/cm}^3$.
Find the mass of the ball.

3 A cylindrical bar has a cross-sectional area of $12\,\text{cm}^2$ and a length of $2\,\text{m}$.
  a Calculate the volume of the bar.

  b The bar is made of metal with density $9\,\text{g/cm}^3$.
    Find the mass of the bar.

2 m

area
$12\,\text{cm}^2$

4 A circular piece of flat metal has a hole of diameter $1\cdot6\,\text{cm}$
in the middle. The metal is $2\,\text{mm}$ thick.
  a Find the area of the flat surface (shown shaded).

  b Calculate the volume of the object.

  c The metal has density $8\,\text{g/cm}^3$. Find the mass of the object.

1.6 cm

2 cm

5 Copy and complete the table

Object	Volume	Density	Mass
Sphere	$20\,\text{cm}^3$	$4\,\text{g/cm}^3$	*
Cone	$650\,\text{cm}^3$	*	$1300\,\text{g}$
Cylinder	*	$8\,\text{g/cm}^3$	$80\,\text{g}$
Box	$480\,\text{cm}^3$	*	$1440\,\text{g}$
Jar	$300\,\text{cm}^3$	$7\cdot5\,\text{g/cm}^3$	*

6 A dog has a volume of $10,800\,\text{cm}^3$ and a mass of $64\cdot8\,\text{kg}$.
Calculate the density of the dog in g/cm^3.

7 The density of silk is $1\cdot5\,\text{g/cm}^3$. What is the mass of $250\,\text{cm}^3$ of silk?

8 A solid bar is made of steel with a density of $8\cdot2\,\text{g/cm}^3$.
Work out the mass of a steel bar of volume $520\,\text{cm}^3$.

9 A farmer makes £$2\cdot45$ per m^2 from the crops he grows
in this field. How much will he make in total?

40 m

60 m

40 m

100 m

10 Material for curtains costs £$15\cdot50$ per m^2. Work out the
cost of a rectangular curtain with dimensions $200\,\text{cm} \times 55\,\text{cm}$.

11 Which has the greater mass and by how much:
  a $365\,\text{cm}^3$ of metal with density $8\cdot2\,\text{g/cm}^3$ or $610\,\text{cm}^3$ of
    rubber with density $3\cdot7\,\text{g/cm}^3$?

  b A solid metal cube of density $7\,\text{g/cm}^3$ with sides of $8\,\text{cm}$
    or a sphere of volume $365\,\text{cm}^3$ and density $9\cdot5\,\text{g/cm}^3$.

12 A ballet dancer has a mass of $58\,\text{kg}$ and density $8\cdot1\,\text{g/cm}^3$.
Calculate the volume of this dancer.

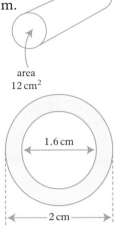

9.2.3   Other compound measures

Exercise 4 ②

1 Silver plating costs £9 per cm². How much will it cost to plate this metal rectangle?

2 cm

5 cm

2 You can hire a satellite at a cost of £250 per second. How much will it cost to hire the satellite for one minute?

3 Telephone cable costs £3·50 per metre. Find the cost of 200 m of this cable.

4 A multi-millionaire earns £120 million in one year. On average, how much did he earn per day? Give your answer to the nearest thousand pounds.

5 Rooney Farm has a rectangular field measuring 200 m by 500 m. The land can be sold at £3000 per hectare (1 hectare = 10 000 m²). Find the price of the field.

*6 The town of Vernon has a population of 30 500 and an area of 33 square miles.

  a Work out the population density in people/square mile.

  b The area of the town is planned to increase to 47 square miles. Calculate the increased population of the town if the population density remains the same. Give your answer to a sensible degree of accuracy.

9.3   Metric and imperial units

Years ago, measurements were made using parts of the body. The inch was measured using the thumb and the foot by using the foot. Even today many people, when asked their height or weight, will say '5 feet 3 inches' or '9 stone' rather than '1 metre 60 cm' or '58 kg'.

These are called imperial units.

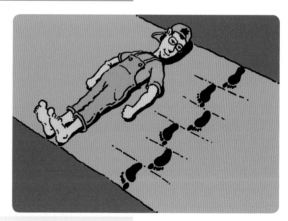

9.3.1   Metric units

Length	Mass	Volume
10 mm = 1 cm	1000 g = 1 kg	1000 ml = 1 litre
100 cm = 1 m	1000 kg = 1 t	1 000 000 cm³ = 1 m³
1000 m = 1 km	(t for tonne)	
Area		
10 000 cm² = 1 m		

Exercise 5 ①

Copy and complete.

1 4 m = __ cm

2 2·4 km = __ m

3 0·63 m = __ cm

4 25 cm = __ m

5 70 mm = __ cm

6 2 cm = __ mm

7 1·2 km = __ m

8 815 cm = __ m

9 650 m = __ km

10 25 mm = __ cm

11 5 kg = __ g

12 4·2 kg = __ g

13 6·4 kg = __ g

14 3 kg = __ g

15 0·8 kg = __ g

16 400 g = __ kg

17 2 t = __ kg

18 250 g = __ kg

19 0·5 t = __ kg

20 0·62 t = __ kg

21 7 kg = __ g

22 1500 g = __ kg

23 8 litres = __ ml

24 2 litres = __ ml

25 1000 ml = __ litres

26 4·5 litres = __ ml

27 3·2 m = __ cm

28 55 mm = __ cm

29 1·4 kg = __ g

30 11 cm = __ mm

31 $20\,000\,\text{cm}^2 = \text{__ m}^2$

32 $12·5\,\text{m}^2 = \text{__ cm}^2$

33 $0·5\,\text{m}^3 = \text{__ cm}^3$

34 $1\,000\,000\,\text{mm}^3 = \text{__ m}^3$

35 Write the most appropriate metric unit for measuring
  a the distance between Glasgow and Leeds
  b the capacity of a wine bottle
  c the mass of raisins needed for a cake
  d the diameter of a small drill
  e the mass of a car
  f the area of a football pitch.

9.3.2 Imperial units

Length		Mass		Volume
12 inches	= 1 foot	16 ounces	= 1 pound	8 pints = 1 gallon
3 feet	= 1 yard	14 pounds	= 1 stone	
1760 yards	= 1 mile	2240 pounds	= 1 ton	

Exercise 6 ① ②

1 How many inches are there in two feet?

2 How many ounces are there in a pound?

3 How many feet are there in ten yards?

4 How many pounds are there in two tons?

5 How many pints are there in six gallons?

6 How many yards are there in ten miles?

7 How many inches are there in one yard?

8 How many pounds are there in five stones?

9 How many pints are there in half a gallon?

10 How many yards are there in half a mile?

In questions **11** to **20**, copy each statement and fill in the missing numbers.

11 9 feet =__yards

12 16 pints = __gallons

13 2 miles = __yards

14 5 pounds = __ounces

15 10 stones = __pounds

16 4 yards = __feet

17 4 feet = __inches

18 10 tons = __pounds

19 1 mile = __feet

20 6 feet = __yards

9.3.3 Changing units

Although the metric system is replacing the imperial system, you still need to be able to convert from one set of units to the other.

Try to remember these approximate conversions.

1 inch ≈ 2·5 cm	1 gallon ≈ 5 litres
1 kg ≈ 2 pounds	1 km ≈ $\frac{5}{8}$ mile
1 ounce ≈ 30 g	(or 8 km ≈ 5 miles)

Remember that ≈ means 'is approximately equal to.'

EXAMPLE

a Change 16 km into miles.

$$1 \text{ km} \approx \frac{5}{8} \text{ mile}$$

$$\text{so } 16 \text{ km} \approx \frac{5}{8} \times 16$$

$$16 \text{ km} \approx 10 \text{ miles}$$

b Change 2 feet into cm.

$$2 \text{ feet} \approx 24 \text{ inches}$$

$$1 \text{ inch} \approx 2\cdot5 \text{ cm}$$

$$\text{so } 2 \text{ feet} \approx 2\cdot5 \times 24 \text{ cm}$$

$$\approx 60 \text{ cm}$$

Exercise 7 ① ②

Copy each statement and fill in the missing numbers.

1 10 inches = __ cm

2 10 gallons = __ litres

3 3 kg = __ pounds

4 4 kg = __ pounds

5 8 km = __ miles

6 2 pounds = __ kg

7 32 km = __ miles

8 4 inches = __ cm

9 8 pounds = __ kg

10 1 gallon = __ litres

11 40 km = __ miles

12 100 litres = __ gallons

13 3 kg = __ pounds

14 5 miles = __ km

15 400 litres = __ gallons

16 60 g = __ ounces

17 10 oz = __ g

18 5 litres = __ gallons

19 20 kg = __ pounds

20 25 cm = __ inches

21 1 foot = __ cm

22 A car handbook calls for the oil to be changed every 8000 km. How many miles is that?

23 On an Italian road, the speed limit is 80 km/h. Convert this into a speed in mph.

24 Tomatoes are sold in Supermarket A at 85p per kilo and in Supermarket B at 30p per pound. Which supermarket has the better buy?

25 Here is the recipe for a pie in imperial units. Write the recipe with the correct metric quantities.

In questions **26** to **30**, copy each sentence and choose the number which is the best estimate.

26 A one pound coin has a mass of about [1 g, 10 g, 1 kg].

27 The width of the classroom is about [100 inches, 7 m, 50 m].

28 A can of cola contains about [500 ml, $\frac{1}{2}$ gallon].

29 The distance from London to Birmingham is about [20 miles, 20 km, 100 miles].

30 The thickness of a one pound coin is about [3 mm, 6 mm, $\frac{1}{4}$ inch].

31 Here are scales for changing: **A** kilograms and pounds, **B** litres and gallons. In this question give your answers to the **nearest whole number**.

  a About how many kilograms are there in 6 pounds?

  b About how many litres are there in 3·3 gallons?

  c About how many pounds are there in 1·4 kilograms?

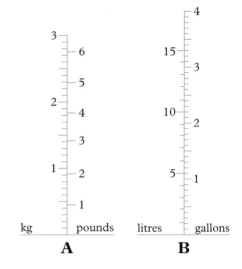

A   **B**

*32 A fitter is doing a job that requires a 3 mm drill. He has no metric drills but he does have these imperial sizes (in inches): $\frac{1}{16}, \frac{1}{8}, \frac{3}{16}$.

Which of these drills is the nearest in size to 3 mm?

9.4 Measurement is approximate

● Suppose you measure a line as 14 mm, correct to the nearest mm. The actual length could be any value from 13·5 mm to 14·5 m.

  We write   13·5 cm ≤ length < 14·5 cm

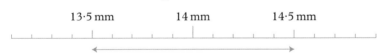

| 13·5 mm | 14 mm | 14·5 mm |

> In these cases, the measurement expressed to a given unit is in **possible error** of **half a unit**.

● If you measure the length of some fabric for a dress, you might say the length is 145 cm **to the nearest** cm. The actual length could be anything from 144·5 cm to 145·49999…cm. This is effectively 145·5.

  We write   144·5 cm ≤ length <145·5 cm

> The 'unit' is 1 so half a 'unit' is 0·5.

● Similarly, if you say you weigh 57 kg to the nearest kg, you could actually weigh anything from 56·5 kg to 57·5 kg. If your brother was weighed on more sensitive scales and the result was 57·2 kg, his actual weight could be any value from 57·15 kg to 57·25 kg.

  We write   57·15 kg ≤ weight < 57·25 kg.

> The 'unit' is 0·1 so half a 'unit' is 0·05.

● Here are some further examples.

1 The diameter of a CD is 12 cm to the nearest cm.

2 The mass of a coin is 6·2 g to the nearest 0·1 g.

Lower limit	Upper limit	Limits of accuracy
11·5 cm	12·5 cm	11·5 cm ≤ diameter <12·5 cm
6·15 g	6·25 g	6·15 g ≤ mass < 6·25 g
325 m	335 m	325 m ≤ length < 335 m

3 The length of a fence is 330 m to the nearest 10 m.

Exercise 8 ②

1 In a DIY store, the height of a door is given as 195 cm to the nearest cm. Write the greatest possible height of the door.

2 A vet weighs a sick goat at 37 kg to the nearest kg. What is the least possible weight of the goat?

3 A surveyor using a laser beam device can measure distances to the nearest 0·1 m. What is the least possible length of a warehouse which he measures at 95·6m?

4 A cook's weighing scales weigh to the nearest 0·1 kg. What is the greatest possible weight of a chicken which she weighs at 3·2 kg?

5 In the county sports, Jill was timed at 28·6 s for the 200 m race. What is the greatest time she could have taken?

6 Copy and complete the table using this information.

   a Temperature in a fridge = 2 °C, to nearest degree

Lower limit	Upper limit	Limits of accuracy
		——— ≤ ——— < ———

   b Mass of an acorn = 2·3 g, to 1dp

   c Length of telephone cable = 64 m, to nearest m

   d Time taken to run 100 m = 13·6 s, to nearest 0·1 s.

7 The length of a telephone is measured as 193 mm, to the nearest mm. Which is correct, A, B or C?
   A 192 mm ≤ length < 194 mm
   B 192·5 mm ≤ length < 193·5 mm
   C 188 mm ≤ length < 198 mm

8 The weight of a labrador is 35 kg, to the nearest kg. Which is correct, A, B or C?
   A 30 kg ≤ weight < 40 kg
   B 34 kg ≤ weight < 36 kg
   C 34·5 kg ≤ weight < 35·5 kg

9 Liz and Julie each measure a different worm and they both say that their worm is 11 cm long to the nearest cm.
   a Does this mean that both worms are the same length?
   b If not, what is the maximum possible difference in the length of the two worms?

10 A card measuring 11·2 cm (to the nearest 0·1 cm) is to be posted in an envelope which is 12 cm (to the nearest cm).

11·2 cm   12 cm

Can you guarantee that the card will fit inside the envelope?
Explain your answer.

In questions **11** to **24**, you are given a measurement. Write the upper and lower limits of the number. For example if you are given a length as 13 cm, you can write 'length is between 12·5 cm and 13·5 cm' or
$12.5 \text{ cm} \leq \text{length} < 13.5 \text{ cm}$.

11 mass = 17 kg

12 $d = 256$ km

13 length = 2·4 m

14 $m = 0.34$ grams

15 $v = 2.04$ m/s

16 $x = 8$ cm

17 $T = 81.4$ °C

18 $M = 0.3$ kg

19 $d = 4.80$ cm

20 $y = 0.07$ m

21 mass = 0·7 tonne

22 $t = 615$ seconds

23 $d = 7.13$ m

24 $n = 52\,000$ (nearest thousand)

9.5 Solving numerical problems 3

EXAMPLE

A shopkeeper buys potatoes at a wholesale price of £180 per tonne and sells them at a retail price of 22p per kg.
How much profit does he make on one kilogram of potatoes?

He pays £180 for 1000 kg of potatoes.

So he pays £180 ÷ 1000 for 1 kg of potatoes.
So he pays 18p for 1 kg.

He sells at 22p per kg.
His profit is 4p per kg.

Exercise 9

 Find the profit in each case.

	Commodity	Retail price	Wholesale price	Profit
1	cans of drink	15p each	£11 per 100	profit per can?
2	rulers	24p each	£130 per 1000	profit per ruler?
3	birthday cards	22p each	£13 per 100	profit per card?
4	soup	27p per can	£8·50 for 50 cans	profit per can?
5	newspapers	22p each	£36 for 200	profit per paper?
6	box of matches	37p each	£15·20 for 80	profit per box?
7	potatoes	22p per kg	£160 per tonne	profit per kg?
8	carrots	38p per kg	£250 per tonne	profit per kg?
9	T-shirts	£4·95 each	£38·40 per dozen	profit per T-shirt?
10	eggs	96p per dozen	£50 per 1000	profit per dozen?
11	oranges	5 for 30p	£14 for 400	profit per orange?
12	car tyres	£19·50 each	£2450 for 200	profit per tyre?
13	wine	55p for 100 ml	£40 for 10 litres	profit per 100 ml?
14	sand	16p per kg	£110 per tonne	profit per kg?
15	wire	23p per m	£700 for 10 km	profit per m?
16	cheese	£2·64 per kg	£87·50 for 50 kg	profit per kg?
17	copper tube	46p per m	£160 for 500 m	profit per m?
18	apples	9p each	£10·08 per gross	profit per apple?
19	carpet	£6·80 per m²	£1600 for 500 m²	profit per m²?
20	tin of soup	33p per tin	£72 for 400 tins	profit per tin?

Exercise 10 ① ②

 1 Write each calculation and find the missing digits.

a
```
    5  7  □  2
 +  □  6  9  □
 ─────────────
    8  □  2  8
```

b
```
    8  □  5
 -  2  6  □
 ──────────
    □  7  3
```

c □ □ □ ÷ 7 = 35

2 A hotel manager was able to buy loaves of bread at £4·44
per dozen, whereas the shop price was 43p per loaf.
How much did he save on each loaf ?

3 A high performance car uses one litre of petrol every 2·5 miles. How much petrol does it use on a journey of 37·5 miles?

4 John Lowe made darts history in 1984 with the first ever perfect game played in a tournament. He scored 501 in just nine darts. He won a special prize of £100 000 from the sponsors of the tournament. His first eight darts were six treble 20s, treble 17 and treble 18.

   a What did he score with the ninth dart?

   b How much did he win per dart thrown, to the nearest pound?

5 An engineering firm offers all of its workers a choice of two pay rises. Workers can choose either an 8% increase on their salaries or they can accept a rise of £1600.

   a A fitter earns £10 400 a year. Which pay rise should he choose?

   b The personnel manager earns £23 000 a year. Which pay rise should he choose?

6 A map is 280 mm wide and 440 mm long. When reduced on a photocopier, the copy is 110 mm long. What is the width of the copy?

7 I have 213 mugs and one tray takes 9 mugs. How many trays do I need?

8 How many prime numbers are there between 120 and 130?

9 Write these answers, without using a calculator.

   a $0·03 \times 10$ **b** $0·03 \div 10$

   c $115 \div 1000$ **d** $0·07 \times 1$ million

10 Work out $\frac{2}{5} + 0·14 + \frac{3}{4}$ and write the answer as a decimal.

Exercise 11 ① ②

1 A maths teacher bought 40 calculators at £8·20 each and a number of other calculators costing £2·95 each. In all she spent £387. How many of the cheaper calculators did she buy?

2 The total mass of a jar one-quarter full of jam is 250 g. The total mass of the same jar three-quarters full of jam is 350 g. What is the mass of the empty jar?

$\frac{1}{4}$ full
250 g

$\frac{3}{4}$ full
350 g

3 I have many 1p, 2p, 3p and 4p stamps. How many different combinations of stamps can I make which total 5p?

4 8% of 2500 + 37% of P = 348. Find the value of P.

5 Eggs are packed twelve to a box. A farmer has enough eggs to fill 316 boxes with unbroken eggs and he has 62 cracked eggs left over. How many eggs had he to start with?

6 Booklets have a mass of 19 g each, and they are posted in an envelope of mass 38 g. Postage charges are shown in the table.

Mass (in grams) not more than	60	100	150	200	250	300	350	600
Postage (in pence)	24	30	37	44	51	59	67	110

a A package consists of 15 booklets in an envelope. What is the total mass of the package?

b The mass of a second package is 475 g. How many booklets does it contain?

c What is the postage charge on a package of mass 320 g?

d The postage on a third package was £1·10. What is the largest number of booklets it could contain?

7 A cylinder has a volume of 200 cm^3 and a height of 10 cm. Calculate the area, A, of its base.

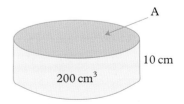

A

10 cm

200 cm^3

8 You are told that 41 × 271 = 11 111. Use this to work out **in your head** the answer to 82 × 271.

9 Here is a subtraction using the digits 2, 3, 4, 5, 6.
Which subtraction using all the digits 2, 3, 4, 5, 6 has the smallest positive answer?

$$362 - 45$$

10 One litre of petrol costs 131·5 p and one litre of oil costs £4·52.
   a Find the cost of 100 litres of petrol.
   b Find the cost of 10 litres of oil.

Exercise 12 ① ②

1 Copy and complete the additions.

a
```
    4  3  ☐
    2  ☐  1
 +  ☐  3  5
 ─────────────
 ☐  3  9  8
```

b
```
    4  3  ☐
    5  ☐  1
 +  ☐  1  5
 ─────────────
 ☐  0  7  2
```

2 Use the clues to find the mystery number:
   ● the sum of the digits is 6
   ● the number reads the same forwards as backwards
   ● the number is less than 2000
   ● the number has four digits.

3 The diagrams show magic squares in which the sum of
the numbers in any row, column or diagonal is the same.
Find the value of x in each square.

a

	x	6
3		7
		2

b

4		5	16
x		10	
	7	11	2
1			13

4 A 20p coin is 2 mm thick. Graham has a pile of 20p coins
which is 18 cm tall. What is the value of Graham's pile?

5 The students in a school were given a general knowledge
quiz. A mark of 20 or more was a 'pass'. Some of the
results are given in the table. Copy and complete the table
with the missing entries.

	Passed	Failed	Total
Boys		245	595
Girls	416		
Total	766		1191

6 a Find two consecutive numbers with a product of 342.

  b Find four consecutive numbers with a total of 102.

  c Find **any** two numbers with a product of 407.

  d Find a pair of numbers with a sum of 19 and a product of 48.

> Reminder:
> The **product** of 3 and 5 is 3 × 5 = 15.

7 In a supermarket, five boxes of chocolates can be bought for £21·25. How many boxes can be bought for £34?

8 I think of a number. If I double the number and then take away 10, the answer is 230. What number was I thinking of?

9 What is the fourth angle in a quadrilateral with angles 70°, 106° and 59°?

10 The test results of 100 students are shown in the table.

Result	5	6	7	8	9	10
Number of students	8	14	21	20	21	16

The pass mark in the test was 8. What percentage of the students passed?

Test yourself

You may use a calculator in this exercise.

1 Copy and complete this table.

Fraction	Decimal	Percentage
$\frac{1}{2}$	0·5	–
–	0·75	75%
$\frac{1}{4}$	–	–
–	–	9%

(OCR, 2003)

2 Write the metric unit that you would use to measure
  a the amount of butter in a pack
  b the length of a needle
  c the weight of a lorry.

(MEI, Spec)

3 Write down the metric unit **best** used to measure
  a the distance from Milan to Venice,
  b the weight of a person,
  c the volume of a bucket,
  d the length of a classroom.

(WJEC, 2011)

4 Pat works out the answer to $23{\cdot}6 \times 36{\cdot}2$ on a calculator.
Her answer is shown on the calculator.
  a Round her answer to the nearest 10.
  b Round her answer to the nearest 100.
  c Round her answer to one decimal place.

(AQA, 2007)

5

Millimetres	**Grams**	**Kilometres**	**Centimetres**
Kilograms	**Metres**	**Litres**	**Millilitres**

Which of these metric units is best to use for measuring
  a the length of a pen,
  b the weight of an egg,
  c the capacity of a small drinking glass,
  d the distance from Chesterfield to Dover,
  e the weight of a large bag of potatoes?

(OCR, 2009)

6 a Some mobile phones show a number of world clocks.

London   New York   Sydney

Monday   Monday   Monday
11:30   06:30   22:30

When it is 3 p.m. on Wednesday in New York, what day and
time will it be in Sydney?
Explain your reasoning.
  b Some mobile phones can convert measurements.
Covert 600 metres per minute to kilometres per hour.
You must show all your working.

(WJEC, 2011)

7 a Write down a sensible **metric** unit for measuring
   i the distance from London to Paris,
   ii the amount of water in a swimming pool.
 b i Change 5 centimetres to millimetres.
   ii Change 4000 grams to kilograms.

(Edexcel, 2008)

8 Zala has bought a sack of potatoes and a bottle of cooking oil.
The sack holds 20 kg of potatoes measured correct to the nearest kilogram.
The bottle contains 750 ml of cooking oil measured correct to the
nearest 10 ml.
Copy and complete the table below to show the least and greatest
weight of potatoes and the least and greatest amount of cooking
oil that Zala could have bought.

	Least Value	**Greatest Value**
Potatoes	........................kg	........................kg
Cooking oil	........................ml	........................ml

(WJEC, 2011)

9 As part of her training, an athlete runs for 5 minutes and then
walks for 1 minute. She repeats this without stopping for a
period of one hour.

Her average running speed is 18 km per hour.
Her average walking speed in 6 km per hour.

Calculate how many kilometres she will complete during the hour.

(WJEC, 2011)

10 The video tape of *Jurassic Park* is 72 m long and can be
rewound in 4 minutes.
 a What is the speed of rewinding in cm per second?

The film *Gone with the Wind* is rewound at the same speed.
It takes 5 minutes to rewind.
 b How long in metres is the video tape of *Gone with the Wind?*

(CCEA)

11 a Find the value of $\sqrt{2 \cdot 25}$.

 b Change $\dfrac{17}{100}$ into a decimal.

 c Work out $\dfrac{3}{5}$ of 145.

 d Work out 29% of £4·35.

(OCR, 2009)

12 a i This diagram is part of a thermometer marked in °C.
What temperature does the arrow point to?

 ii The diagram below is part of a measuring tape marked
in centimetres.
What measurement does the arrow point to?

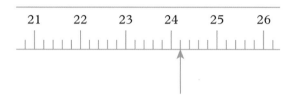

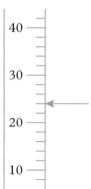

 b Vernon needs $2\dfrac{1}{2}$ litres of water to make some tomato fertiliser.
His measuring jug holds 500 ml of water when full.
How many full jugs of water will he need?

(OCR, 2013)

13 The width of a wardrobe was given as 115 centimetres,
measured to the nearest centimetre.

 a Write the lower and upper limits of the width.

 The wardrobe was bought to fit into a space between
two walls. The width of the space had been measured
as 1·2 metres, to the nearest tenth of a metre.

 b Explain why the wardrobe might not fit into the space.

(CCEA)

14 An airline allows a maximum weight limit of 20 kg on each
passenger's luggage.

 a A passenger has luggage weighing 18 kg.
He wishes to add an item weighing 1200 g to his luggage.
Find out how much under or over the airline's weight limit
this will make his luggage.

 b Anton's luggage weighs 50 lb. (1 kg is approximately 2.2 lb.)
Showing all your calculations, say whether his luggage is
under or over the airline's weight limit.

(WJEC, 2011)

10 Probability

Predicting the weather at sea uses complex probability models.

In this unit you will:
- learn about sets, Venn diagrams and possibility spaces
- revise ideas of probability and chance
- use the probability scale
- use probability to list outcomes of two or more events
- learn how to find the probabilities of mutually exclusive, independent and conditional events
- calculate probabilities.

10.1   Sets

10.1.1   Venn diagrams

A **set** is the mathematical name for a collection of distinct objects. The **universal set** contains all the objects in the collection and has the symbol, ξ.

You can define a set in words or symbols and you write it with a capital letter.

Here are three ways of writing the set N.

N is the set of even numbers between 1 and 15.

N = {2, 4, 6, 8, 10, 12, 14}

N = {x : even numbers, $1 < x < 15$}

The symbols ∈ and ∉ mean 'is a member of' and 'is not a member of'. So 4 ∈ N and 5 ∉ N.

Venn diagrams were first used by John Venn (1834 – 1923) to illustrate sets. The members of the sets are written inside circles (or a rectangle for the universal set, ξ). Two sets need two circles which may overlap or may be separate or may have one entirely inside the other.

When two sets A and B have members in common, then the set of common members is called the **intersection** (or overlap) of A and B, which is written A ∩ B.

The set that includes all the members of A and B combined together is the **union** of A and B, which is written A ∪ B.

EXAMPLE

Given that ξ = {integers from 0 to 10}, A = {odd numbers}, B = {1, 2, 3, 4} and
C = {6, 8}, draw a Venn diagram and find the sets A ∩ B, A ∪ B, B ∩ C and B ∪ C.

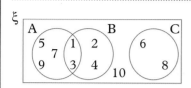

The set with no members
is called the **empty set**,
written as { } or Ø.

A ∩ B = {1, 3}   A ∪ B = {1, 2, 3, 4, 5, 7, 9}

B ∩ C = { }   B ∪ C = {1, 2, 3, 4, 6, 8}

Exercise 1 ❶

1   M = {all multiples of 3} and N = {odd numbers between 0 and 10}
    a   List the members of these two sets.
    b   Which set has an infinite number of members?
    c   Which of these three statements is false?
        6 ∈ M   6 ∈ N   9 ∉ M

2   Describe these sets in words.
    a   {Jack, John, James, Julian, ...}   b   {5, 10, 15, 20, 25, 30, ...}
    c   {1, 4, 9, 25, 36, 49, 64, 81}   d   {R, O, Y, G, B, I, V}

3   List all the members of these sets.
    a   {even numbers between 5 and 15}   b   {the first four letters of the alphabet}
    c   {cube numbers less than 100}   d   {the letters used in the word
                                                 MATHEMATICS}

4   Say whether these statements are *true* or *false*.
    a   June ∈ {girls names}   b   June ∈ {months of the year}
    c   2 ∉ {even numbers}   d   2 ∈ {square numbers}
    e   13 ∉ {prime numbers]   f   (4, 5) ∈ {points on the line $y = x + 1$}

5   For each pair of sets, choose the most appropriate Venn diagram and write
    in the members of the sets. For each pair of sets, find the members of A ∩ B.

   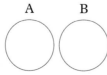

    a   A = {1, 2, 3, 4}, B = {3, 4, 5, 6, 7}   b   A = {2, 4, 6, 8}, B = {10, 15, 20}
    c   A = {3, 6, 9, 12, 15}, B = {3, 9}   d   A = {5, 10, 50, 100}, B = {1, 10, 100, 1000}

6 Copy this diagram and write in the members of these three sets.

X = {1, 2, 3, 4, 8}   Y = {3, 4, 5, 6, 7}   Z = {3, 5, 6, 8, 9}

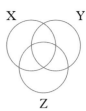

List the members of these sets.

a X∩Y   **b** Y∪Z   **c** X∪Y∪Z

d X∩Y∩Z   **e** X∪(Y∩Z)   **f** X∩(Y∪Z)

10.1.2 Possibility spaces

The **possibility space** is the list of all the possible outcomes of an event. Possibility spaces can be found by listing all the possibilities systematically and by using a **grid** or a **tree diagram** to help.

For example, a family has four children, Amy, Ben, Cathy and Dan. Two of the children are chosen, the first to do the washing-up after a meal, the second to go and buy something at the shop. In how many different ways can these choices be made?

A grid uses two axes labelled A, B, C, D and the squares of the grid give the possible choices. A tree diagram starts with the first choice and then branches out to give the second choice.

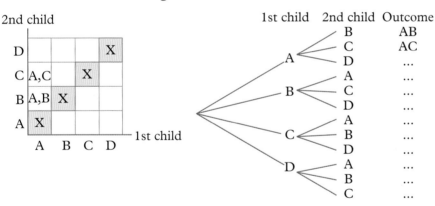

Both diagrams show that the same child should not be chosen twice.

The possibility space can also be written as:

AB, AC, AD, BA, BC, BD, CA, CB, CD, DA, DB, DC,

where the first child washes up and the second child goes shopping. There are 12 different ways.

Exercise 2 ① ②

1 In the previous example, the two tasks of washing-up and shopping could be done by the same child. How many different ways are there now of making the choice?

2 You toss a coin and throw a dice. Using the letters H and T for heads and tails, copy this diagram and complete it to find all the possible outcomes.

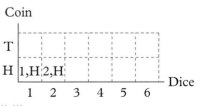

a How many different outcomes are possible?

b How many outcomes have a head with an even score?

c Draw your own tree diagram to generate the same possibility space.

3 **a** Olivia tosses two coins. Construct a grid with two axes and show the possible outcomes on the grid. Use H for heads and T for tails. List the possibility space.

b Draw a tree diagram to show the same possibility space.

4 A family decides to buy two pets. The children of the family cannot decide between a dog, a cat and a rabbit and they cannot decide whether they would like two of the same kind.

The letters D, C, R stand for dog, cat, rabbit.

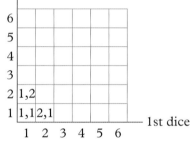

a Use this grid to list the possibilities.

b What is the size of the possibility space?

c How many possibilities include one or two dogs?

d What fraction of all the possibilities include a least one cat?

5 You roll two dice, one after the other. Copy and complete this diagram to find the possibility space. How many possibilities

a have a score of 1 on either or both dice

b have a score of 1 on the first dice

c have a score of 1 only on the first dice

d do not have a score of 1 on either dice?

What fraction of the possibilities

e has just one 6 on either dice

f gives a double 6?

6 Some families have three children in them.

a Copy and complete this tree diagram to show all the possible ways of listing the boys (B) and girls (G) in a three-child family.

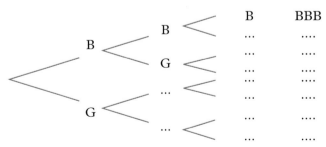

b Copy and complete this table to show the different possibilities.

	3 boys	2 boys and 1 girl	1 boy and 2 girls	3 girls
Number of possibilities				

c Why is a grid diagram (as used in the example at the start of this exercise) not a good method to use in this problem?

7 Sam tosses three coins, one after the other. Construct a tree diagram similar to that in **6** to show all the possible outcomes. Use H for heads and T for tails. How many outcomes have

   a   three heads   **b**   two heads and one tail?

What fraction of the possibility space has

   c   one head and two tails   **d**   three tails?

The product rule for counting states that if there are m ways for the first outcome and n ways for the second outcome, then the total number of possible outcomes is $m \times n$.

> **EXAMPLE**
>
> A café menu has 5 types of hot drink and 3 types of scone to choose from. How many possible combinations are there?
>
> ....
>
> There are $5 \times 3 = 15$ different combinations.

8 A bag contains equal numbers of 1p and 2p coins. A second bag contains equal numbers of 5p, 10p and 20p coins. Julie selects one coin at random from each bag.
Draw a tree diagram and list the possible combinations of copper coins and silver coins that she may choose.

9 a A young toddler has 4 shirts (A, B, C, D) and 2 pairs of trousers (Y, Z). List all the possible ways that his parent can dress him in shirt and trousers. How many different ways are there?

   b Ellie's dad takes her to a rescue home for pets. She can choose one kitten and one puppy. There are 3 kittens (K, L, M) and 3 puppies (P, Q, R) to choose from. How many different ways can she choose her kitten and puppy?

10 How many possible outcomes are there when you choose

   a   one man and one woman from four men and five women

   b   one toffee and one mint from six toffees and five mints

   c   one consonant and two different vowels (a, e, i, o, u) from the English alphabet

   d   one consonant and two vowels (which can be the same or different) from the alphabet?

> Welsh has more than five vowels.

10.2 Probability scale and experiments

10.2.1 How likely is it?

> ▶ The probability of something happening is the **likelihood** or **chance** that it might happen. It does **not** tell you what is definitely going to happen.

When thinking about probability, you ask questions like:

'What are the chances of Chelsea winning the F.A. cup?'
'How likely is it that I will pass my driving test first time?'

You do not know the answer to either of these questions yet.

Some events are **certain**. Some events are **impossible**.
Some events are in between certain and impossible.

> ▶ The probability of an event occurring is measured on a scale like this.
>
>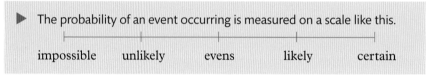

Where on this scale you would put these events?
1   The sun will rise tomorrow.
2   My teacher will win the National Lottery next week.
3   You spin a coin and get a head.

Exercise 3 ①

Draw a probability scale like this.

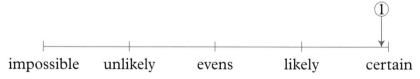

Draw an arrow to show the chance of these events happening.
(The arrow for question **1** has been done for you.)

1 You will do some writing in this lesson.

2 You will watch television this evening.

3 You toss a coin and get a tail.

4 You draw a card from a pack and get a king.

5 You draw a card from a pack and get a heart.

6 It will rain tomorrow.

7 You will have a new pair of shoes in the next six months.

8 You roll a dice and get a 1.

9 You have the same birthday as the Prime Minister.

10 It will snow next week.

11 The next car to pass the school will be red.

12 The letter 'e' appears somewhere on the next page of this book.

10.2.2   Probability as a number

People in different countries have different words for 'likely' and 'unlikely'. All over the world, people use numbers or percentages on a scale from 0 to 1 (or 0% to 100%) to measure probability, rather than using words.

impossible					evens				certain
0	0·1	0·2	0·3	0·4	0·5	0·6	0·7	0·8	0·9	1
0	10%	20%	30%	40%	50%	60%	70%	80%	90%	100%

10.2.3   Experimental probability

The chance of certain events occurring can easily be predicted. For example, the chance of tossing a head with an ordinary coin. Many events cannot be so easily predicted, and you need to do an experiment to work out an experimental probability of the event occurring.

Experiment   To find the experimental probability of a drawing pin landing point up when dropped onto a hard surface.

Step 1   Do 50 **trials**.
Step 2   If the pin lands point up on a table, this is a **success**.
Step 3   Make a tally chart like this.

> Remember to put a mark in the 'trials' box every time you drop the pin.

Number of trials	Number of successes
IIII IIII II	IIII I

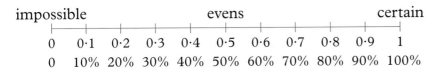

▶ **Experimental probabillity** $= \dfrac{\text{Number of trials in which a success occurs}}{\text{Total number of trials made}}$

> The experimental probability of an event is called its **relative frequency**.

You get a better estimate of the experimental probability of an event if you do a **large number** of trials.

Exercise 4 ❶ ❷

Carry out experiments to work out the experimental probability of some of the events in Exercise 1. Give your answers as a fraction, a decimal or a percentage. Use a tally chart to record your results. Don't forget to record how many times you do the experiment (the number of trials).

1 Roll a dice. What is the probability of rolling a six?
Do 60 trials. Copy and complete this table and fill in the second row.

Number of trials (rolls of the dice)	12	24	36	48	60
Number of 6s you expect					10
Number of 6s you actually got					

   a What was the experimental probability of rolling a 6?
   b Would you expect to get the same result if you did the experiment again?
   c How could you get a more accurate value for the experimental probability of rolling a 6?

2 **a** Toss two coins one hundred times. Write your experimental probability of tossing two tails, giving your answer as a percentage.
   b How could you get a more accurate answer?

3 Pick a counter from a bag containing counters of different colours. What is the experimental probability of picking a red counter? Do 100 trials.

4 Roll a pair of dice. What is the chance of rolling a double? Do 100 trials.

5 This frequency tree shows the number of times you would expect to roll a 'six' from 300 rolls of a fair dice.

The numbers on top of the box show the number of successes of that outcome.

Sam rolls a fair dice 400 times.

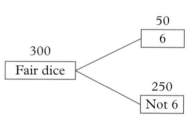

Draw a frequency tree to show how many times you would expect him to roll
   a a 'four'
   b an even number.

10.3 Working out probabilities

10.3.1 Theoretical probability

For simple events, like throwing a dice or tossing a coin, you can work out the expected probability of an event occurring without doing an experiment. For a fair dice the **theoretical probability** of throwing a '3' is $\frac{1}{6}$. For a normal coin, the theoretical probability of tossing a 'head' is $\frac{1}{2}$ or 50%.

▶ Theoretical probability $= \dfrac{\text{The number of ways the event can happen}}{\text{The number of possible outcomes}}$

Random choice

If a card is chosen at random from a pack, it means that every card has an equal chance of being chosen.

If you do an experiment like tossing a coin, the experimental probability (or **relative frequency**) of getting a 'head' will get nearer and nearer to the theoretical probability (0·5) as you do more and more trials.

For example, if you tossed a coin 10 times, you might get 8 'heads'. If you tossed the coin 1000 times, you would be likely to get between 480 and 520 'heads', which is about the expected number as $\frac{1}{2} \times 1000 = 500$.

EXAMPLE

Seven identical balls numbered 1, 2, 3, 4, 5, 6 and 7 are put into a bag. One ball is selected at random.
What is the probability of selecting
a a '3'
b an even number?

There are 7 possible equally likely outcomes.

a The probability of selecting a '3' $= \frac{1}{7}$

This may be written p (selecting a '3') $= \frac{1}{7}$

b p (selecting an even number) $= \frac{3}{7}$

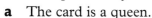

EXAMPLE

A single card is drawn from a pack of 52 playing cards.
Find the probability of these results.

a The card is a queen.
b The card is a club.
c The card is the jack of hearts.

There are 52 equally likely outcomes of the trial.

13 Spades

a $p(\text{queen}) = \dfrac{4}{52} = \dfrac{1}{13}$ or 0·77

13 Hearts

13 Diamonds

b $p(\text{club}) = \dfrac{13}{52} = \dfrac{1}{4}$ or 0·25

13 Clubs

c $p(\text{jack of hearts}) = \dfrac{1}{52}$ or 0·02

Total = 52

Exercise 5 ①

1 A bag contains a red ball, a white ball and a black ball.
One ball is chosen at random.
Copy and complete these sentences.

a The probability that the red ball is chosen is $\dfrac{\square}{3}$.

b The probability that the white ball is chosen is $\dfrac{\square}{\square}$.

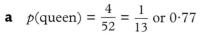

2 One ball is chosen at random from a bag which contains a
red ball, a black ball, a grey ball and a white ball. Write the
probability that the chosen ball will be

a red   **b** grey   **c** black.

3 A bag contains 2 red balls and 1 white ball. One ball is
chosen at random. Find the probability that it is

a red   **b** white.

4 Izzy rolls a fair dice. Write the chance of her getting

a a 6   **b** a 2   **c** a 7.

5 Here is a spinner with 8 equal sectors. Write the probability
of spinning

a a 7   **b** a 3   **c** an even number.

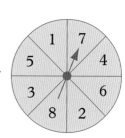

6 Write the probability of the arrow pointing at a shaded section when each of these spinners stops spinning.

a
b
c
d

e
f
g
h

7 Here are two spinners.
Say whether the following statements are true or false.
Explain why in each case.

a 'Dan is more likely to spin a 4 than Nick.'
b 'Dan and Nick are equally likely to spin an odd number.'
c 'If Nick spins his spinner six times he is bound to get at least one 6.'

Dan's spinner   Nick's spinner

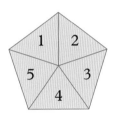 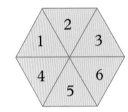

8 If one card is picked at random from a pack of 52 playing cards, what is the probability that it is
a a king   **b** the ace of clubs   **c** a heart?

9 Nine counters numbered 1, 2, 3, 4, 5, 6, 7, 8, 9 are placed in a bag. One is taken out at random. What is the probability that it is
a a 5   **b** divisible by 3
c less than 5   **d** divisible by 4?

10 A bag contains 5 grey balls, 2 red balls and 4 white balls. One ball is taken out at random. What is the probability that it is
a grey   **b** red   **c** white?

11 A cash bag contains two 20p coins, four 10p coins, five 5p coins, three 2p coins and three 1p coins. Find the probability that one coin selected at random is
a a 10p coin   **b** a 2p coin   **c** a silver coin.

12 A bag contains 8 orange balls, 5 green balls and 4 silver balls.
Find the probability that a ball picked out at random is
   a silver   **b** orange   **c** green.

13 One card is selected at random from this hand.

> You will not be tested on your knowledge of playing cards in your examination.

Find the probability of selecting
   a a heart   **b** an ace
   c the 10 of clubs   **d** a spade
   e a heart or a diamond.

14 A pack of playing cards is well shuffled and a card is drawn.
Find the probability that the card is
   a a jack
   b a queen or a jack
   c the ten of hearts
   d a club higher than the 9 (count the ace as high).

EXAMPLE

Cards with numbers 1, 2, 3, 4, 5, 6, 7, 8 are shuffled and then placed face down in a line. The cards are then turned over one at a time from the left. In this example, the first card is a 4.

Find the probability that the next card turned over will be
   a the 5
   b a number less than 4
   c the 4.

....

   a The next card can be either 1, 2, 3, 5, 6, 7 or 8.
      p(turning over a 5) $= \dfrac{1}{7}$
   b You can turn over a 1, a 2 or a 3.
      p(turning over a number less than 4) $= \dfrac{3}{7}$
   c p(turning over a 4) $= 0$

Exercise 6 ①

Questions **1**, **2** and **3** are about cards with numbers 1, 2, 3, 4, 5, 6, 7 and 8. The cards are shuffled and then placed face down in a line. The cards are then turned over one at a time.

1 In this case, the first card is a 3.

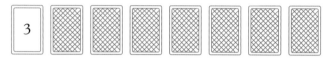

Find the probability that the next card will be

  a the 6   **b** an even number   **c** higher than 1.

2 Suppose the second card is a 7.

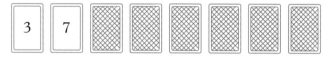

Find the probability that the next card will be

  a the 5   **b** an odd number   **c** higher than 6.

3 Suppose the first three cards are

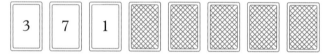

Find the probability that the next card will be

  a the 8   **b** less than 4   **c** the 5 or the 6.

4 The numbers of matches in ten boxes are
48, 46, 45, 49, 44, 46, 47, 48, 45, 46
One box is selected at random. Find the probability of the box containing

  a 49 matches   **b** 46 matches   **c** more than 47 matches.

5 One ball is selected at random from those shown here.

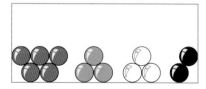

Find the probability of selecting

  a a white ball   **b** a grey or a black ball

  c a ball which is not red.

6 a A bag contains 5 red balls, 6 green balls and 2 black balls.
Find the probability of selecting
    i   a red ball   **ii**   a green ball.

  b One black ball is removed from the bag. Find the new
probability of selecting
    i   a red ball   **ii**   a black ball.

7 A pack of playing cards is split so that all the picture cards
(kings, queens, jacks) are in Pile A and all the other cards
are in Pile B.
Find the probability
of selecting

  a the queen of clubs
from pile A

  b the seven of spades
from pile B

  c any heart from pile B.

Pile A

Pile B

8 A bag contains 12 white balls, 12 green balls and 12 purple balls.
After 3 white balls, 4 green balls and 9 purple balls have been
removed, what is the probability that the next ball to be selected
will be white?

9 Jade has 3 queens and 1 king. She shuffles
the cards and takes one without looking.
Jade asks two of her friends about the
probability of getting a king.

Kim says:

'It is $\frac{1}{3}$ because there
are 3 queens and 1 king.'

Megan says:

'It is $\frac{1}{4}$ because
there are 4 cards
and only 1 king.'

Which of her friends is right?

10 A bag contains 9 balls, all of which are black or white. Jane selects
a ball and then replaces it. She repeats this several times.
Here are her results (B = black, W = white):

B   W   B   W   B   B   B   W   B   B   W   B   B   W   B

B   B   W   W   B   B   B   B   W   B   W   B   B   W   B

How many balls of each colour do you think there were in
the bag?

11 A large firm employs 3750 people. One person is chosen at random.
What is the probability that that person's birthday is on a Monday in the year 2008?

12 There are eight balls in a bag. Asif takes a ball from the bag, notes its colour and then returns it to the bag.
He does this 25 times. The results are in the table.

Red	4
White	11
Black	10

  a What is the smallest number of red balls there could be in the bag?

  b Asif says 'There cannot be any blue balls in the bag because there are no blues in my table.' Explain why Asif is wrong.

13 The numbering on a set of 28 dominoes is like this.

| 6 | 6 | 6 | 6 | 6 | 6 | 6 | | 5 | 5 | 5 |
| 6 | 5 | 4 | 3 | 2 | 1 | 0 | | 5 | 4 | 3 |

| 5 | 5 | 5 | | 4 | 4 | 4 | 4 | 4 | | 3 | 3 |
| 2 | 1 | 0 | | 4 | 3 | 2 | 1 | 0 | | 3 | 2 |

| 3 | 3 | | 2 | 2 | 2 | | 1 | 1 | | 0 |
| 1 | 0 | | 2 | 1 | 0 | | 1 | 0 | | 0 |

  a What is the probability of drawing a domino from a full set with
    i at least one six on it?   **ii** at least one four on it?
    iii at least one two on it?

  b What is the probability of drawing a 'double' from a full set?

  c If I draw a double five, which I do not return to the set, what is the probability of drawing another domino with a five on it?

14 A bag contains a number of green, red and blue discs. When a disc is selected at random, the probability of it being blue is 0·3.
There are 50 discs in the bags. How many of the discs are blue?

10.4 Listing possible outcomes of two or more events

When a 10p coin and a 50p coin are tossed together, two events are occurring.

  1 Tossing the 10p coin

  2 Tossing the 50p coin

10p	50p
head	head
head	tail
tail	head
tail	tail

The result of tossing the 10p coin does not affect the result of the 50p coin.
The two events are **independent**.
All the possible outcomes are shown in the table. This is the possibility space.

EXAMPLE

A red dice and a black dice are thrown together. Show all
the possible outcomes. Find the probability of throwing a double.

You could list them in pairs with the red dice first.

(1, 1), (1, 2), (1, 3), ... (1, 6)
(2, 1), (2, 2), ...
and so on.

In this example, it is easier to see all the possible
outcomes when they are shown on a grid.

There are 36 possible outcomes in the possibility space.

The × shows 6 on the red dice and 2 on the black.

The ○ shows 3 on the red dice and 5 on the black.

There are six double scores: (1, 1), (2, 2) up to (6, 6).

So the probability of throwing a double = $\frac{6}{36} = \frac{1}{6}$.

Exercise 7 ①

1 Three coins (10p, 20p, 50p) are tossed together.
   a Use a tree diagram to list all the possible ways in
      which they could land.
   b What is the probability of getting three heads?

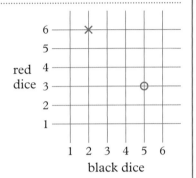

2 Use a tree diagram to list all the possible outcomes when four
   coins are tossed together. How many outcomes are there
   altogether? What is the probability of getting four heads?

3 A black dice and a white dice are thrown together.
   a Draw a grid to show all the possible outcomes (see the example
      above) and write how many possible outcomes there are.
   b How many ways can you get a total of nine on the two dice?
   c What is the probability that you get a total of nine?

4 A red spinner and a white spinner are spun together.

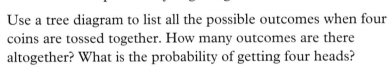

Red   White

   a List all the possible outcomes in a copy of the table.
   b In how many ways can you get a total of 4?
   c What is the probability that you get a total of 4?

5 Four friends, Wayne, Xavier, Yves and Zara, each write their
name on a card and the four cards are placed in a hat. Two cards
are chosen to decide who does the maths homework that night.
List all the possible combinations. A tree diagram or grid
may help you.

6 The spinner is spun and the dice is thrown at the same time.

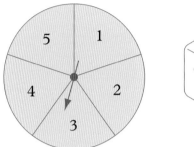

    a Draw a grid to show all the possible outcomes.

    b A 'win' occurs when the number on the spinner
is greater than or equal to the number on the dice.
Find the probability of a win.

7 The menu in a restaurant has two choices of starter,
three choices of main course and two choices of dessert.

    a List all the different combinations I could choose from
this menu.

Gourmet Greg always chooses his three courses at
random.

    b What is the probability that he chooses Melon,
Chicken pie and Gateau?

Starters	
Melon	*(A)*
or Pineapple	*(B)*
Main course	
Steak	*(C)*
or Cod fillet	*(D)*
or Chicken pie	*(E)*
Dessert	
Ice cream	*(F)*
or Gateau	*(G)*

8 By a strange coincidence, the Branson family, the Green family
and the Webb family all have the same first names for the five
members of their families: James, Don, Samantha, Laura and Kate.
One year, Father Christmas decides to give each person
a embroidered handkerchief with two initials.

    a How many different handkerchiefs does he need
for these three families?

    b How many different combinations are there if
any first name and **any** surname is possible
[for example 'Zak Quilfeldt']?

9 Keith, Len, Mike and Neil enter a cycling race.

    a List all the possible orders in which they could finish.
State the number of different finishing orders.

    b In how many of these ways does Mike finish in front of Len?

10 Shirin and Dipika are playing a game in which three coins are tossed. Shirin wins if there are no heads or one head. Dipika wins if there are either two or three heads. Is the game fair to both players?

*11 Students Xavier, Yolanda and Zachary play a game in which four coins are tossed.

X wins if there is 0 or 1 head.
Y wins if there are 2 heads.
Z wins if there are 3 or 4 heads.

Is the game fair to all three players?

10.5 Exclusive events

▶ Events are **mutually exclusive** if they cannot occur at the same time.

Example 1 Selecting an ace and selecting a ten from a pack of cards when choosing just one card.

Example 2 Tossing a head and tossing a tail on one throw

Example 3 Getting a total of 5 on two dice and getting a total of 7 on two dice when tossing the dice together just once.

▶ The probability of something not happening is 1 minus the probability of it happening.
▶ The sum of the probabilities of all mutually exclusive events is 1.

EXAMPLE

A bag contains 3 red balls, 4 black balls and 1 white ball. Find the probability of selecting

a a white ball
b a ball which is not white
c a red ball
d a ball which is not red.

There are 8 balls in the bag.

a p (white ball) $= \dfrac{1}{8}$

b p (not white ball) $= 1 - \dfrac{1}{8} = \dfrac{7}{8}$

c p (red ball) $= \dfrac{3}{8}$

d p (not red ball) $= 1 - \dfrac{3}{8} = \dfrac{5}{8}$

Exercise 8

1   A bag contains a large number of balls including some red balls.

   The probability of selecting a red ball is $\dfrac{1}{5}$. What is the probability of selecting a ball which is not red?

2   A card is selected from a pack of 52. Find the probability of selecting
   a a king
   b a card which is not a king
   c any picture card (king, queen or jack)
   d a card which is not a picture card.

3   In a game, the probability of scoring 21 is $\dfrac{1}{36}$.

   What is the probability of not scoring 21?

4   A motorist does a survey at some traffic lights on his way to work every day. He finds that the probability that the lights are red when he arrives is $0\cdot24$. What is the probability that the lights are not red?

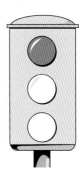

5   U.K. birth statistics show that the probability of a woman giving birth to a boy is $0\cdot506$.
   What is the probability of having a girl?

6 The spinner has 8 equal sectors.
Find the probability of
  a spinning a 5
  b not spinning a 5
  c spinning a 2
  d not spinning a 2
  e spinning a 7
  f not spinning a 7.

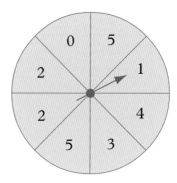

7 The king of clubs is removed from a normal pack of cards. One card is selected from the remaining cards. Find the probability of
  a selecting a king **b** not selecting a king **c** selecting a club.

8 A bag contains a large number of balls coloured red, white, black or green. The probabilities of selecting each colour are shown in the table.

Colour	red	white	black	green
Probability	0·3	0·1		0·3

Find the probability of selecting a ball
  a which is black   **b** which is not white.

9 In a survey, the number of people in cars is recorded. When a car passes the school gates, the probability of it having 1, 2, 3, … occupants is shown in the table.

Number of people	1	2	3	4	more than 4
Probability	0·42	0·23		0·09	0·02

  a Find the probability that the next car past the school gates contains
    i three people   **ii** fewer than 4 people.
  *b One day 2500 cars passed the gates. How many of the cars would you expect to have 2 people inside?

10.6 Probabilities of independent events

10.6.1 Independent events

> ▶ Two events are **independent** if the occurrence of one event is unaffected by the occurrence of the other.

So, obtaining a head on one coin and a tail on another coin when the coins are tossed at the same time are independent. Similarly 'tossing a coin and getting a head' and 'rolling a dice and getting a 2' are independent events.

10.6.2   The 'AND' rule: the multiplication rule

▶ For independent events A and B
   $p\,(A \text{ and } B) = p\,(A) \times p\,(B)$

The multiplication rule only works for **independent** events.
Two coins are tossed and the possible outcomes are listed.
HH   HT   TH   TT

The probability of tossing two heads is $\dfrac{1}{4}$.

By the multiplication rule for independent events
$p(\text{two heads}) = p(\text{head on first coin}) \times p(\text{head on second coin})$

So $p\,(\text{two heads}) = \dfrac{1}{2} \times \dfrac{1}{2} = \dfrac{1}{4}$ as before.

EXAMPLE

A fair coin is tossed
and a fair dice is rolled.
Find the probability
of obtaining a 'head'
and a 'six'.

The two events are independent.

$p\,(\text{head } \textbf{and } \text{six}) = p\,(\text{head}) \times p\,(\text{six}) = \dfrac{1}{2} \times \dfrac{1}{6} = \dfrac{1}{12}$

EXAMPLE

When the two spinners are spun, what is the probability of
getting a B on the first and a 3 on the second?

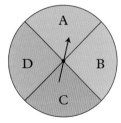 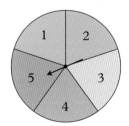

The events 'B on the first spinner' and '3 on the second spinner'
are independent.

$\therefore\; p\,(\text{spinning B and 3}) = p\,(B) \times p\,(3) = \dfrac{1}{4} \times \dfrac{1}{5} = \dfrac{1}{20}$

Exercise 9 ②

1 A card is drawn from a pack of playing cards and a dice is thrown. Events A and B are
A: 'a jack is drawn from the pack'
B: 'a one is thrown on the dice'.
  a Write the values of p (A), p (B).
  b Write the value of p (A and B).

2 A coin is tossed and a dice is thrown. Write the probability of obtaining
  a a 'head' on the coin
  b an odd number on the dice
  c a 'head' on the coin and an odd number on the dice.

3 Box A contains 3 blue balls and 3 white balls.
Box B contains 1 blue ball and 4 white balls.

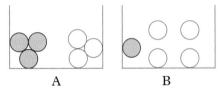

One ball is randomly selected from Box A and one from Box B.
What is the probability that both balls are blue?

4 In an experiment, Ann draws a card from a pack of playing cards and throws a dice.
Find the probability of Ann obtaining
  a a card which is an ace and a six on the dice
  b the king of clubs and an even number on the dice
  c a heart and a 'one' on the dice.

5 Joe takes a card at random from a pack of playing cards and replaces it. After shuffling, he takes a second card. Find the probability of Joe obtaining
  a two cards which are clubs
  b two kings
  c two picture cards.

6 Maria takes a ball at random from a bag containing 3 red balls, 4 black balls and 5 green balls. She replaces the first ball and takes a second.
Find the probability of Maria obtaining
  a two red balls
  b two green balls.

> p (two red balls)
> = p (red ball 1st draw and red ball 2nd draw)

7 The letters of the word 'INDEPENDENT' are written on individual cards and the cards are put into a box. A card is selected and then replaced and then a second card is selected. Find the probability of obtaining

a the letter 'P' twice

b the letter 'E' twice.

8 A game at a fairground uses 3 buckets. Each bucket contains 15 pieces of fruit. You win if you pull out an apple from each bucket. The first bucket has 3 apples, the second has 4 apples and the third has 2 apples.

Find the probability of winning.

9 Three coins are tossed and two dice are thrown at the same time. Find the probability of obtaining

a three heads and a total of 12 on the dice

b three tails and a total of 9 on the dice.

10 A coin is biased so that it shows 'heads' with a probability of $\frac{2}{3}$. The same coin is thrown three times. Find the probability of obtaining

a two tails on the first two throws

b a head, a tail and a head (in that order).

11 A fair dice and a biased dice are thrown together. The probabilities of throwing the numbers 1 to 6 are shown for the biased dice.

Fair

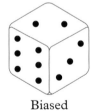

Biased

$$p(6) = \frac{1}{4}; \; p(1) = \frac{1}{12}$$

$$p(2) = p(3) = p(4) = p(5) = \frac{1}{6}$$

Find the probability of obtaining a total of 12 on the two dice.

12 Philip and his sister toss a coin to decide who does the washing up. If the result is heads Philip does it. If the result is tails his sister does it. What is the probability that Philip does the washing up every day for a week (7 days)?

13 Here is the answer sheet for five questions in a multiple choice test.
What is the probability of getting all five correct by guessing?

Answer sheet
1. (A) (B) (C)
2. (A) (B)
3. (A) (B) (C)
4. (A) (B) (C) (D)
5. (A) (B)

14 This spinner has six equal sectors.
Find the probability of getting
 a 20 Qs in 20 trials
 b no Qs in n trials
 c at least one Q in n trials.

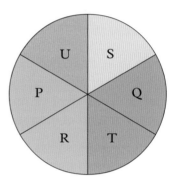

10.6.3 Tree diagrams

Tree diagrams are used to list the outcomes of a series of events and so find their probabilities. Some events may have equally likely outcomes; others may not.

EXAMPLE

A dice is rolled three times. Find the probability of getting two odd scores and one even score, in any order.

Rolling an odd score and rolling an even score are equally likely events.
The branches of the tree give 8 outcomes that are equally likely to occur.
The branches marked ∗ give two odd and one even score.

So p (two odd and one even score) $= \dfrac{3}{8}$

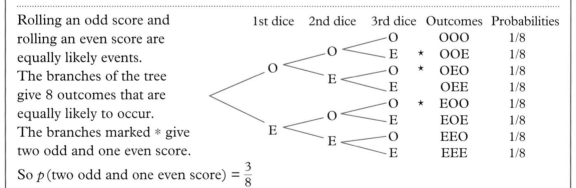

When the branches of the tree diagram are *not* equally likely, the probability of each branch is written on the branch.

EXAMPLE

A bag contains 5 red balls and 3 green balls.
A ball is drawn at random and then replaced.
Another ball is drawn.

What is the probability that both
balls are green?

....

The branch marked ★ involves the selection
of a green ball twice.
You find the probability of this event
by simply multiplying the fractions
on the two branches.

$\therefore p$ (two green balls) $= \dfrac{3}{8} \times \dfrac{3}{8} = \dfrac{9}{64}$

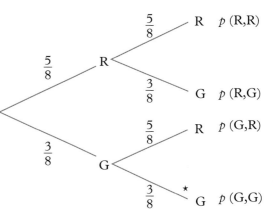

EXAMPLE

A bag contains 5 red balls and 3 green balls. A ball is
selected at random and **not** replaced.
A second ball is then selected.
Find the probability of selecting

a two green balls
b one red ball and one green ball.

....

Notice that since the first ball is not
replaced there are only 7 balls in the bag
for the second selection.

a p(two green balls) $= \dfrac{3}{8} \times \dfrac{2}{7}$

$= \dfrac{3}{28}$

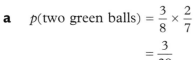

b p(one red, one green) $= \left(\dfrac{5}{8} \times \dfrac{3}{7}\right) + \left(\dfrac{3}{8} \times \dfrac{5}{7}\right)$

$= \dfrac{15}{28}$

Notice that you can add here because the events 'red then green'
and 'green then red' are mutually exclusive.

As a check, all the fractions at the ends of the branches should add up to 1.

So $\left(\dfrac{5}{8} \times \dfrac{4}{7}\right) + \left(\dfrac{5}{8} \times \dfrac{3}{7}\right) + \left(\dfrac{3}{8} \times \dfrac{5}{7}\right) + \left(\dfrac{3}{8} \times \dfrac{2}{7}\right) = \dfrac{20}{56} + \dfrac{15}{56} + \dfrac{15}{56} + \dfrac{6}{56} = 1$

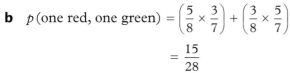

Exercise 10 ②

1 Sam buys three sweets. She chooses a toffee or a mint
by tossing a coin. If she gets a head, she buys a toffee.
If she gets a tail, she buys a mint.

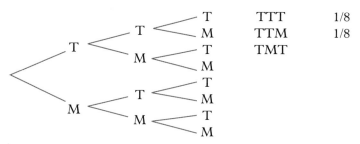

1st choice 2nd choice 3rd choice Outcomes Probabilities

Copy and complete this tree diagram to show her choices.
Find the probability of her choosing
a a mint followed by two toffees
b a mint and two toffees in any order.

2 A bag contains 10 discs; 5 are black and 5 are white.
A disc is selected and then replaced. A second disc
is selected.
Copy and complete the tree diagram showing all
the outcomes and probabilities.
Find the probability that
a both discs are black
b both discs are white.

3 A bag contains 10 discs; 7 are black
and 3 white. A disc is selected, and then
replaced. A second disc is selected. Copy
and complete the tree diagram showing
all the probabilities and outcomes.
Find the probability that
a both discs are black
b both discs are white.

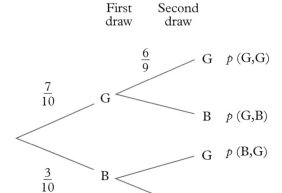

4 A bag contains 5 red balls and 3 green balls. A ball is drawn and
then replaced before a ball is drawn again.
Draw a tree diagram to show all the possible outcomes.
Find the probability that
a two green balls are drawn
b the first ball is red and the second is green.

5 A bag contains 5 black balls, 3 blue balls and 2 white balls.
A ball is drawn and then replaced. A second ball is drawn.
Find the probability of drawing
   a two black balls
   b one blue ball and one white ball (in any order)
   c two white balls.

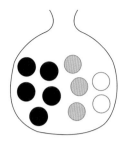

6 A bag contains 7 green discs and 3 blue discs.
A disc is drawn and **not** replaced. A second
disc is drawn. Copy and complete the
tree diagram.
Find the probability that
   a both discs are green
   b both discs are blue.

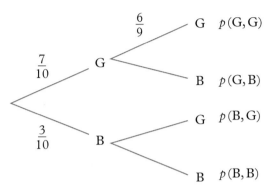

7 A bag contains 4 red balls, 2 green balls
and 3 blue balls. A ball is drawn and
not replaced. A second ball is drawn.
Find the probability of drawing
   a two blue balls
   b two red balls
   c one red ball and one blue ball (in any order)
   d one green ball and one red ball (in any order).

8 A six-sided dice is thrown three times.
Complete the tree diagram, showing at
each branch the two events:
'three' and 'not three'.
Find the probability of throwing a total of
   a three threes
   b no threes
   c one three
   d at least one three (use part **b**)

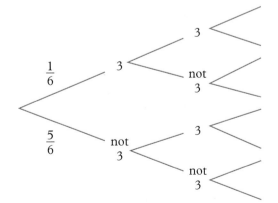

9 A card is drawn at random from a pack of 52 playing cards.
The card is replaced and a second card is drawn. This card is
replaced and a third card is drawn. What is the probability of drawing
   a three hearts   **b** at least two hearts   **c** exactly one heart?

10 A bag contains 6 red marbles and 4 yellow marbles. A marble is drawn at random and **not** replaced. Two further draws are made, again without replacement.
Find the probability of drawing
   a   three red marbles
   b   three yellow marbles
   c   no red marbles
   d   at least one red marble (use part **c**).

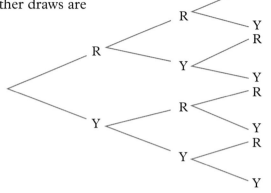

11 When a cutting is taken from a geranium the probability that it grows is $\frac{3}{4}$. Two cuttings are taken.
What is the probability that
   a   both cuttings grow   **b**   neither of them grows?

12 A dice has its six faces marked 0, 1, 1, 1, 6, 6.
Two of these dice are thrown together and the total score is recorded.
Draw a tree diagram.
   a   How many different totals are possible?
   b   What is the probability of obtaining a total of 7?

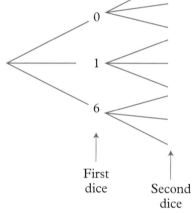

First dice   Second dice

13 Louise loves the phone! When the phone rings, $\frac{2}{3}$ of the calls are for Louise, $\frac{1}{4}$ are for her brother, Herbert, and the rest are for me.
On Christmas Day, I answered the phone twice.
Find the probability that
   a   both calls were for Louise   **b**   one call was for me and one was for Herbert.

14 Bag A contains 3 red balls and 1 white ball.
Bag B contains 2 red balls and 3 white balls.
A ball is chosen at random from each bag in turn.
Find the probability of taking
   a   a white ball from each bag
   b   two balls of the same colour.

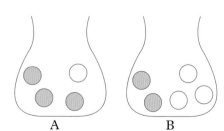

A   B

15 A teacher decides to award exam grades A, B or C by a new,
  fairer method. Out of 20 students, three are to receive As, five Bs
  and the rest Cs. She writes the grades on 20 pieces of paper and
  invites the students to draw their exam result, going through the
  class in alphabetical order.
  Find the probability that
  a the first three students all get grade A
  b the first three students all get grade B
  c the first four students all get grade B.
  (Do not cancel down the fractions.)

EXAMPLE

There are 10 boys and 12 girls in a class.
Two students are chosen at random.
What is the probability that one boy and one girl are chosen in
either order?

Here two students are chosen. It is like
choosing one student
without replacement and then
choosing a second.
Here is the tree diagram.

$$p(B, G) + p(G, B) = \left(\frac{10}{22} \times \frac{12}{21} \right) + \left(\frac{12}{22} \times \frac{10}{21} \right)$$

$$= \frac{40}{77}$$

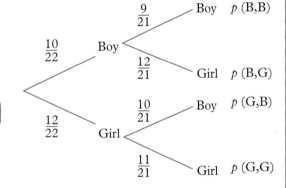

Exercise 11 ②

1 There are 1000 components in a box of which 10 are known
  to be defective. Two components are selected at random.
  What is the probability that
  a both are defective
  b neither is defective
  c just one is defective?
    (Do **not** simplify your answers.)

> **Remember:**
> When a question says
> '2 balls are drawn' or '3
> people are chosen', you
> draw a tree diagram
> **without** replacement

2 There are 10 boys and 15 girls in a class. Two students are chosen
  at random. What is the probability that
  a both are boys
  b both are girls
  c one is a boy and one is a girl?

3 Two similar spinners are spun together and the scores are added.

   a What is the most likely total score on the two spinners?

   b What is the probability of obtaining this score on three successive spins of the two spinners?

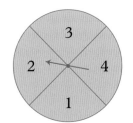

4 A bag contains 3 red, 4 white and 5 green balls.
Three balls are selected without replacement.
Find the probability that the three balls chosen are

   a all red   **b** all green   **c** one of each colour.

   d If the selection of the three balls was carried out 1100 times, how often would you expect to choose three red balls?

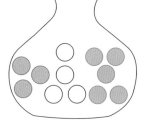

5 The diagram represents 15 students in a class and the diagram shows the sets G, S and F.
G represents those who are girls
S represents those who are swimmers
F represents those who believe in Father Christmas.
A student is chosen at random. Find the probability that the student

   a can swim   **b** is a girl swimmer

   c is a boy swimmer who believes in Father Christmas.

Two students are chosen at random. Find the probability that

   d both are boys   **e** neither can swim.

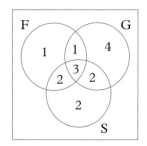

6 There are 500 ball bearings in a box of which 100 are known to be undersize. Three ball bearings are selected at random.
What is the probability that

   a all three are undersize   **b** none is undersize?

Give your answers as decimals correct to three significant figures.

7 There are 9 boys and 15 girls in a class. Three students are chosen at random. What is the probability that

   a all three are boys   **b** all three are girls

   c one is a boy and two are girls?

Give your answers as fractions.

8 A box contains x milk chocolates and y plain chocolates.
Two chocolates are selected at random. Find, in terms of x and y, the probability of choosing

   a a milk chocolate on the first choice

   b two milk chocolates

   c one of each sort

   d two plain chocolates.

9 A pack of z cards contains x 'winning cards'. Two cards are selected at random. Find, in terms of x and z, the probability of choosing
   a a 'winning' card on the first choice
   b two 'winning cards' in the two selections
   c exactly one 'winning' card in the pair.

10 When a golfer plays any hole, he will take 3, 4, 5, 6, or 7 strokes with probabilities of $\frac{1}{10}, \frac{1}{5}, \frac{2}{5}, \frac{1}{5}$, and $\frac{1}{10}$ respectively.
   He never takes more than 7 strokes.
   Find the probability of these events:
   a scoring 4 on each of the first three holes
   b scoring 3, 4 and 5 (in that order) on the first three holes
   c scoring a total of 28 for the first four holes
   d scoring a total of 10 for the first three holes.

11 Assume that births are equally likely on each of the seven days of the week. Two people are selected at random. Find the probability that
   a both were born on a Sunday
   b both were born on the same day of the week.

12 a A playing card is drawn from a standard pack of 52 and then replaced. A second card is drawn. What is the probability that the second card is the same as the first?
   b A card is drawn and **not** replaced. A second card is drawn. What is the probability that the second card is **not** the same as the first?

10.7 Conditional probability

Two events are **dependent events** if the outcome of the first event affects the outcome of the second event.

> ▶ The **conditional probability** of an event measures the probability of event B given that event A has already happened. The probability of 'event B given A' is written as $p(B|A)$.

There are three types of diagram that can be used to show the possibility spaces of such events: two-way tables (grids), Venn diagrams and tree diagrams.

EXAMPLE

A group of 41 students can study French or Spanish, but not both. Of the 21 who study French, there are 13 boys and 11 girls. Of the 17 who study Spanish, there are 8 boys and 9 girls.
Their teacher chooses one student at random. Write the probability of choosing

a a boy
b a boy, given that it must be a French student.

	French	Spanish	
Boys	13	8	21
Girls	11	9	20
	24	17	41

....

a The possibility space includes all 41 students.
$$p(\text{a boy}) = \frac{22}{41}$$

b The possibility space is reduced to only the 21 French students. Of these 21, there are 13 boys.
$$p(\text{a boy} \mid \text{a French student}) = \frac{13}{21}$$

The same group of 41 students who study French and Spanish can be represented on this Venn diagram where B = {boys} and F = {French students}.

The 13 boys who study French are written in the intersection B ∩ F. The 11 French girl students complete set F.
The 8 Spanish boy students complete set B.
The 9 Spanish girl students are outside both B and F.

As before, the probability of choosing a boy at random considers the universal set of all 42 students. But the probability of choosing a boy, given that it must be a French student, considers only the 13 + 8 = 21 students within set F.

Exercise 12 ②

1 There are 50 members of a youth club. This table shows the numbers of boys and girls who go overseas or stay in the UK for their summer holiday.
Write the probability of choosing at random

a a girl
b a girl who went overseas for her holiday
c a member of the club who stayed in the UK for their holiday
d a boy, given that they were one of those who had an overseas holiday.

	UK	overseas	
Boys	14	13	27
Girls	12	11	23
	26	24	50

2 A company makes large and small computer monitors.
A sample of 45 monitors is tested at random for faults
before they leave the factory. This table shows the results.
Write the probability of choosing from this sample

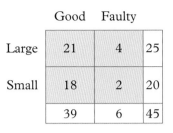

a a monitor that is faulty
b a large monitor that is faulty
c a faulty monitor, given that it is a small one
d a large monitor, given that it is faulty.

3 A bag of 24 chocolate buttons has passed its 'best by' date.
The buttons are made from brown chocolate or white
chocolate and some of them have gone cloudy. This table
shows the contents of the bag.
Write the probability of choosing a button that is

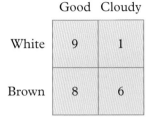

a both white and cloudy
b either white or cloudy (or both)
c cloudy, given that it is white
d brown, given that it is cloudy.

4 This Venn diagram shows how many of the 40
members of a sports club enjoy soccer and tennis.
ξ = {all members of the club}
S = {those who enjoy soccer}
T = {those who enjoy tennis}
The social secretary of the club is selected at random.
Find the probability that the secretary will

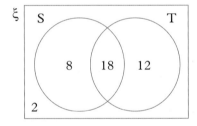

a enjoy soccer but not tennis
b enjoy soccer
c enjoy tennis, given that they enjoy soccer
d do not enjoy soccer, given that they enjoy tennis.

5 An advertisement for a job receives 25 applications
of which 20 are from qualified people, both male
and female. This Venn diagram shows the detailed
breakdown where
ξ = {all applicants}, Q = {qualified applicants} and
F = {female applicants}.
The applicants are interviewed in random order.
Find the probability that the first applicant to be
interviewed is

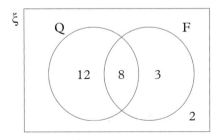

a a female applicant
b a qualified female applicant
c a qualified applicant, given that the first interviewee is female
d an unqualified applicant, given that the first interviewee is male.

6 A farmer rears both sheep and cows. There are 34 new-born animals in spring. This Venn diagram shows the numbers of animals where
ξ = {all animals}, S = {sheep} and N = {new-born animals}.
One animal escapes from the farm through a hedge.
Find the probability that it is

a a sheep

b a new-born sheep

c a new-born animal, given that the farmer can see that it is a sheep

d not a new-born animal, given that the farmer can see that it is a cow.

Venn diagram: ξ, with circles S and N. S region: 42, overlap: 30, N region: 4, outside: 24.

Test yourself

1 Copy and complete the sentences.
Use words from this list.

impossible	certain	fifty-fifty
	likely	unlikely

a It is _____ that when you spin an ordinary coin it lands on heads.

b It is _____ that when you roll an ordinary 6-sided dice, you will get a number between 1 and 5.

c It is _____ that, one day, your maths teacher will become Prime Minister.

d It is _____ that if you drop a ball it will move downwards not upwards.

(OCR, 2009)

2 a The diagram shows five probabilities marked on a probability line.

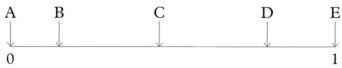

Copy these terms and match each one with the correct letter from the diagram.

Certain   Even   Impossible   Likely   Unlikely

b Jade picks a ball at random from this box.

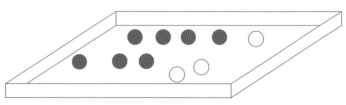

Which of these words describes the probability that she picks a red ball?

Certain   Even   Impossible   Likely   Unlikely

(MEI, Spec)

3 Here is a fair 6-sided spinner.

Jake is going to spin the spinner once.

a Write down the probability that the spinner will land

   i on 4

   ii on a number greater than 10.

Liz is going to spin the spinner 120 times.

b Work out an estimate for the number of times the spinner will land on 7.

(Edexcel, 2013)

4 In a raffle 520 tickets are sold. They are numbered from 1 to 520.
The winning number is chosen at random.
Work out the probability that it is

a 99

b greater than 240.

(MEI, Spec)

5 A fair ordinary dice is rolled and then a fair spinner is spun.
The spinner is split into 2 equal sections, blue (b) and
yellow (y).

a List all the possible outcomes of rolling the dice and
spinning the spinner in a table like this one.
The first one has been done for you.

Dice	Spinner
1	b

b **i** What is the probability of getting an even number and blue?
Give your answer in its simplest form.

   ii Which of these is more likely?

     ● Getting a number less than 6 and yellow

        or

     ● Getting an odd number and blue.

   Explain how you decide.

(OCR, 2013)

6 Sophie has a spinner.

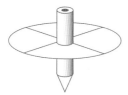

The spinner is coloured so that

● **Red** is opposite **White**, and
● **Yellow** is opposite **Purple**.

The disc of the spinner is as shown here, with two straight lines passing through the centre of the spinner.
A table to show the probabilities of Sophie obtaining **Red**, **White**, **Yellow** and **Purple** has been started.
Copy and complete the table and indicate how the disc should be coloured by labelling each sector.

Color	Red	White	Yellow	Purple
Probability	0·2			

(WJEC, 2011)

7 The diagram shows a fair octagonal spinner.
   a What is the probability that the spinner lands on 2?
   b Dave spins the spinner 20 times.
     The results are shown in this table.

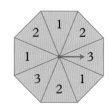

Spin	1	2	3	4	5	6	7	8	9	10	11	12	13	14	15	16	17	18	19	20
Result	1	3	1	2	3	2	1	2	1	2	3	3	2	1	2	2	1	2	3	1

     i What is the relative frequency of the spinner landing on 1?
     ii Steve also spins the spinner 20 times.
       Explain why Steve may not get the same results as Dave.
   c How many times would you expect a result of 3 if you spin the spinner 1000 times?

(AQA, 2007)

8 Janet and Phil go to a summer fair.

Three cakes are left for sale on the cake stall.

Fruit cake (F)	Chocolate cake (C)	Sponge cake (S)

Janet buys one of these cakes at random and then Phil buys one of the remaining cakes at random.

a List all the possible ways they could do this in a table like this one.
   The first one has been done for you.

Janet	Phil
F	C

b What is the probability that either Janet or Phil buys the chocolate cake?

(OCR, 2013)

9 a The probability that Rovers will win their next match is 0·6.
The probability that Rovers will lose their next match is 0·15.
Work out the probability that Rovers will draw their next match.

b A weather man says 'There is a 40% chance of rain today.'
Tracy says 'That means there is a 60% chance of it being sunny today.'
Explain why Tracy is **not** correct.

(OCR, 2008)

10 Two square shaped spinners A and B have numbers written on them.

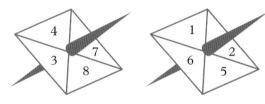

Spinner A   Spinner B

In a game, a player spins both spinners and adds double the number showing on spinner A to the number showing on spinner B to get the score for the game. For example, if the number on spinner A is 4 and the number on spinner B is 5, the player works out $2 \times 4 + 5 = 13$ and the player scores 13.

a Copy and complete the following table to show all the possible scores.

	6	......	......	......	......
Spinner	5	......	......	......	......
B	2	8	10	16	18
	1	7	9	15	17
		3	4	7	8

Spinner A

b What is the probability that a player scores less than 10?
c A player wins a prize by getting a score of less than 10.
Eighty people play the game once.
How many will be expected to win a prize?

(WJEC, 2011)

11 Tomas has one spin of the circular spinner shown here. Two of the lines shown on the diagram are diameters of the circle.

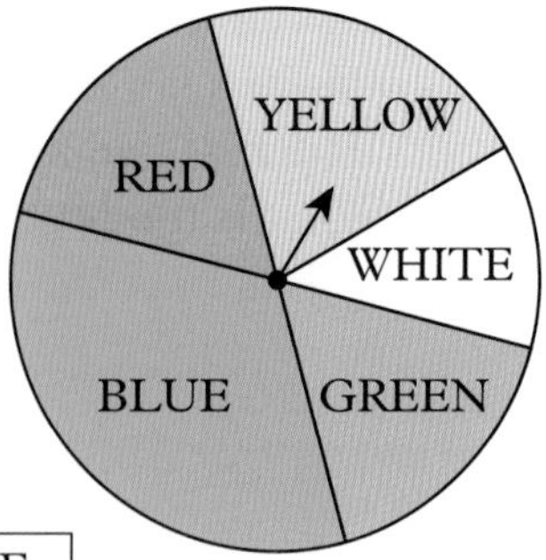

a The table below shows the probabilities of Tomas obtaining YELLOW, WHITE and GREEN with one spin of the spinner.

Copy and complete the table.

Colour	YELLOW	WHITE	GREEN	RED	BLUE
Probability	0·2	0·12	0·18		

b Find the probability of obtaining either WHITE or GREEN on the spinner.

(WJEC, 2011)

12 The table shows the probabilities of how a randomly selected adult in a particular town would vote in an election.

Party	Conservative	Liberal Democrat	Labour	Other
Probability	0·4	0·17	0·35	

a Copy and complete the table.
b Work out the probability that a randomly selected adult would vote Conservative or Liberal Democrat.
c There are 2500 adults in the town.
   How many of these adults might you expect to vote Labour?

(OCR, 2013)

13 In a game at a fete a spinner is spun. The score obtained may be any one of 2, 4, 6, 8, 10, 12, 14, the probabilities of which are given in the table.

Number	2	4	6	8	10	12	14
Probability	0·11	0·09	0·14	0·31	0·23	0·07	0·05

a Calculate the probability that the score obtained will be greater than 8.
b The spinner is spun 300 times. About how many times will the score be 6?

(WJEC)

11 Problems, proofs and puzzles

11.1 Problem-solving tasks

Here are a few guidelines for investigating a problem.

- If the set problem is too complicated try an easier case.
- Draw your own diagrams.
- Make tables of your results and be systematic.
- Look for patterns.
- Is there a rule or formula to describe the results?
- Can you **predict** further results?
- Can you **explain** any rules which you may find?
- Where possible extend the task further by asking questions like 'what happens if...'.

> **Systematic** means work through things in order.

11.1.1 Opposite corners

Here the numbers are arranged in 9 columns.

1	2	3	4	5	6	7	8	9
10	11	12	13	14	15	16	17	18
19	20	21	22	23	24	25	26	27
28	29	30	31	32	33	34	35	36
37	38	39	40	41	42	43	44	45
46	47	48	49	50	51	52	53	54
55	56	57	58	59	60	61	62	63
64	65	66	67	68	69	70	71	72
73	74	75	76	77	78	79	80	81
82	83	84	85	86	87	88	89	90

You can start with simpler cases:
In the 2×2 square...

6	7
15	16

$6 \times 16 = 96$
$7 \times 15 = 105$
...the difference between them is 9.

In the 3×3 square...

22	23	24
31	32	33
40	41	42

$22 \times 42 = 924$
$24 \times 40 = 960$
...the difference between them is 36.

Investigate to see if you can find any rules or patterns connecting the size of square chosen and the difference. If you find a rule, use it to **predict** the difference for larger squares.
Test your rule by looking at squares like 8×8 or 9×9.

x	?
?	?

Can you **generalise** the rule?
(What is the difference for a square of size $n \times n$?)

Can you **prove** the rule?

> In a 3×3 square...

What happens if the numbers are arranged in six columns or seven columns?

1	2	3	4	5	6
7	8	9	10	11	12
13	14	15	16	17	18
19	_	_	_	_	_
_	_	_	_	_	_

1	2	3	4	5	6	7
8	9	10	11	12	13	14
15	16	17	18	19	20	21
22	_	_	_	_	_	_
_	_	_	_	_	_	_

11.1.2 Hiring a car

You are going to hire a car for one week (7 days). Which of the firms below should you choose?

Gibson car hire	Snowdon rent-a-car	Hav-a-car
£270 per week unlimited mileage	£20 per day 10p per mile	£100 per week 500 miles without charge 30p per mile over 500 miles

Work out as detailed an answer as possible.

11.1.3 Half-time score

The final score in a football match was 3−2. How many different scores were possible at half-time?
Investigate for other final scores where the difference between the teams is always one goal (1−0, 5−4, etc.). Is there a pattern or rule which would tell you the number of possible half-time scores in a game which finished 8−7?
Suppose the game ends in a draw. Find a rule which would tell you the number of possible half-time scores if the final score was 10–10.

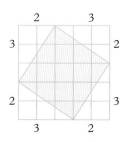

Investigate for other final scores (3−0, 5−1, 4−2, etc.).
Find a rule which gives the number of different half-time scores for **any** final score (say a–b).

11.1.4 Squares inside squares

Here is a square drawn inside a square. The inside square is set at an angle to the outside square. Is there a connection between the **area** of the inside square and the numbers $\binom{2}{3}$ that describe the angle through which the square is rotated?

Method

Calculate the area of the inside square by subtracting the areas of triangles A, B, C and D from the area of the outside square.

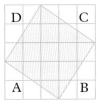

Area of outside square $= 5 \times 5 = 25$ squares

Area of each triangle $= \dfrac{1}{2} \times 2 \times 3 = 3$ squares

So, area of inside square $= 25 - (4 \times 3)$

$\qquad\qquad\qquad\qquad = 13$ squares

Your task is to investigate the areas of different squares drawn inside other squares in a similar way.

A Start with simple cases.

 2 3 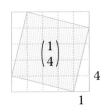 4

Find the areas of the inside squares where the top number is 1. Can you find a connection or rule which you can use to find the area of the inside square **without** all the working with subtracting areas of triangles?

B Look at more difficult cases.

Again try to find a rule which you can use to find the area of the inside square directly.

 2

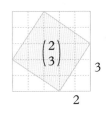 3

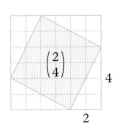 4

C Look at **any** size square.

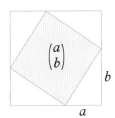 b

Try to find a rule and, if possible, use algebra to show why it **always** works.

11.1.5 Maximum box

A You have a square sheet of card 24 cm by 24 cm.
You can make a box (without a lid) by cutting squares
from the corners and folding up the sides.
What size corners should you cut out so that the volume
of the box is as large as possible?
Try different sizes for the corners and record the results
in a table.

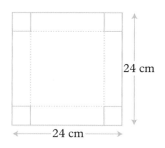

24 cm

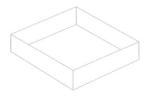

Length of the side of the corner square (cm)	Dimensions of the open box (cm)	Volume of the box (cm³)
1	22 × 22 × 1	484
2		
–		
–		

Now consider boxes made from different sized cards
15 cm × 15 cm and 20 cm × 20 cm.

What size corners should you cut out this time so that the
volume of the box is as large as possible?
Is there a connection between the size of the corners cut out
and the size of the square card?

B Investigate the situation when the card is not square.
Take rectangular cards where the length is twice the width
(20 × 10, 12 × 6, 18 × 9, etc.).

Again, for the maximum volume, is there a connection between
the size of the corners cut out and the size of the original card?

11.1.6 Diagonals

In a 4 × 7 rectangle, the diagonal passes through 10 squares.

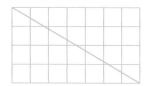

Draw rectangles of your own choice and count the number
of squares through which the diagonal passes.
A rectangle is 640 × 250. How many squares will the diagonal
pass through?

11.1.7 Painting cubes

The large cube on the right consists of 27 unit cubes.

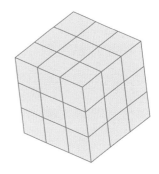

All six faces of the large cube are painted red.
The large cube is dismantled into its unit cubes.

1 How many unit cubes have 3 red faces?
2 How many unit cubes have 2 red faces?
3 How many unit cubes have 1 red face?
4 How many unit cubes have 0 red faces?

Answer the four questions for the cube which is $n \times n \times n$.

11.1.8 Conjectures
Exercise 1

In questions **1** to **8**, consider the conjecture given. Some are
true and some are false. Write a counter-example where the
conjecture is false. If you cannot find a counter-example, state
that the conjecture is 'not proven' as you are **not** asked to prove it.

> Remember ≠ means
> 'is not equal to'.

1 $(n + 1)^2 = n^2 + 1$ for all values of n.
2 $\dfrac{1}{x}$ is always less than x.
3 $\sqrt{n+1}$ is always smaller than $\sqrt{n} + 1$.
4 For all values of n, $2n$ is greater than $n - 2$.
5 For any set of numbers, the median is always smaller than the mean.
6 $\sqrt{x}$ is always less than x.
7 The diagonals of a parallelogram never intersect at right angles.
8 If n is even, $5n^2 + n + 1$ is odd.

11.2 Puzzles and games

11.2.1 Crossnumbers

Draw four copies of this crossnumber pattern and work
out the answers using the clues. You can check your
working by doing **all** the across and **all** the down clues.
There are four sets of clues for the same grid.

Part A

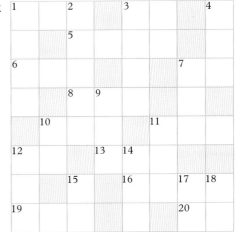

	Across			*Down*
1	327 + 198		**1**	3280 + 1938
3	245 ÷ 7		**2**	65 720 − 13 510
5	3146 − 729		**3**	3·1 × 1000
6	248 − 76		**4**	1284 ÷ 6
7	2^6		**7**	811 − 127

8	$850 \div 5$	**9**	65×11
10	$102 + 12$	**10**	$(122 - 8) \div 8$
11	$3843 \div 7$	**11**	$(7^2 + 1^2) \times 11$
12	$1000 - 913$	**12**	$7 + 29 + 234 + 607$
13	$37 \times 5 \times 3$	**14**	$800 - 265$
16	$152\,300 \div 50$	**15**	$1 + 2 + 3 + 4 + 5 + 6 + 7 + 8 + 13$
19	3^6	**17**	$(69 \times 6) \div 9$
20	$100 - \left(\dfrac{17 \times 10}{5} \right)$	**18**	$3^2 + 4^2 + 5^2 + 2^4$

Part B

Draw decimal points on the lines between squares where necessary.

Across

1 $4 \cdot 2 + 1 \cdot 64$
3 $7 \times 0 \cdot 5$
5 $20 \cdot 562 \div 6$
6 $(2^3 \times 5) \times 10 - 1$
7 $0 \cdot 034 \times 1000$
8 $61 \times 0 \cdot 3$
10 $8 - 0 \cdot 36$
11 19×50
12 $95 \cdot 7 \div 11$
13 $8 \cdot 1 \times 0 \cdot 7$
16 $(11 \times 5) \div 8$
19 $(44 - 2 \cdot 8) \div 5$
20 Number of inches in a yard

Down

1 $62 \cdot 6 - 4 \cdot 24$
2 $48 \cdot 73 - 4 \cdot 814$
3 $25 + 7 \cdot 2 + 0 \cdot 63$
4 $2548 \div 7$
7 $0 \cdot 315 \times 100$
9 $169 \times 0 \cdot 05$
10 $770 \div 100$
11 $14 \cdot 2 + 0 \cdot 7 - 5 \cdot 12$
12 $11 \cdot 4 - 2 \cdot 64 - 0 \cdot 18$
14 $0 \cdot 0667 \times 10^3$
15 $0 \cdot 6 + 0 \cdot 7 + 0 \cdot 8 + 7 \cdot 3$
17 $0 \cdot 73\,\text{m}$ written in cm
18 $0 \cdot 028 \times 200$

Part C

Across

1 Eleven squared take away six

3 Next in the sequence
21, 24, 28, 33

5 Number of minutes in a day

6 $2 \times 13 \times 5 \times 5$

7 Next in the sequence
92, 83, 74

8 5% of $11\,400$

10 $98 + 11^2$

Down

1 Write $18 \cdot 6\,\text{m}$ in cm

2 Fifty-one thousand and fifty-one

3 Write $3 \cdot 47\,\text{km}$ in m

4 $1\frac{1}{4}$ as a decimal

7 $7\,\text{m} - 54\,\text{cm}$ (in cm)

9 $0 \cdot 0793 \times 1000$

10 2% of 1200

11 $\frac{1}{5}$ of 3050

12 $127 \div 100$

11 $(120 - 9) \times 6$

12 $1\frac{2}{5}$ as a decimal

13 $2387 \div 7$

16 $9{\cdot}05 \times 1000$

19 $8\,m - 95\,cm$ (in cm)

20 3^4

14 Number of minutes between
12:00 and 20:10

15 4% of 1125

17 $7^2 + 3^2$

18 Last two digits of (67×3)

Part D

Across

1 $1\frac{3}{4}$ as a decimal

3 Two dozen

5 Forty less than ten thousand

6 Emergency

7 5% of 740

8 Nine pounds and five pence

10 1·6m written in cm

11 $5649 \div 7$

12 One-third of 108

13 $6 - 0{\cdot}28$

16 A quarter to midnight
on the 24 h clock

19 $5^3 \times 2^2 + 1^5$

20 $3300 \div 150$

Down

1 Twelve pounds 95 pence

2 Four less than sixty thousand

3 245×11

4 James Bond

7 Number of minutes between
09:10 and 15:30

9 $\frac{1}{20}$ as a decimal

10 Ounces in a pound

11 8·227 to two decimal places

12 $4\,m - 95\,cm$ (in cm)

14 Three to the power 6

15 20·64 to the nearest whole number

17 $\left(6\frac{1}{2}\right)^2$ to the nearest whole number

18 Number of minutes between
14:22 and 15:14

11.2.2 Designing square patterns

The aim is to design square patterns of different sizes. The patterns are all to be made from smaller tiles, all of which are themselves square.

Designs for a 4 × 4 square:

This design consists of four tiles each 2 × 2.
The pattern is rather dull.

Suppose we say that the design must contain at least one 1 × 1 square.

This design is more interesting and consists
of seven tiles.

1 Try a 5 × 5 square. Design a pattern which divides the 5 × 5 square into eight smaller squares.

2 Try a 6 × 6 square. Here you must include at least one 1 × 1 square. Design a pattern which divides the 6 × 6 square into nine smaller squares. Colour in the final design to make it look interesting.

3 The 7 × 7 square is more difficult. With no restrictions, design a pattern which divides the 7 × 7 square into nine smaller squares.

4 Design a pattern which divides an 8 × 8 square into ten smaller squares. You must not use a 4 × 4 square.

5 Design a pattern which divides a 9 × 9 square into ten smaller squares. You can use only one 3 × 3 square.

6 Design a pattern which divides a 10 × 10 square into eleven smaller squares. You must include a 3 × 3 square.

7 Design a pattern which divides an 11 × 11 square into eleven smaller squares. You must include a 6 × 6 square.

12 Revision

12.1   Revision exercises

Exercise 1

1 Copy this bill and complete it by filling in the four blank spaces. 8 rolls of wallpaper at
£3·20 each   = £☐
3 tins of paint at £☐
each   = £ 20·10
☐ brushes at £2·40
each   = £ 9·60
                          Total = £☐

2 Write each sequence and find the next two numbers.
  a   2, 9, 16, 23, ___, ___
  b   20, 18, 16, 14, ___, ___
  c   −5, −2, 1, 4, ___, ___
  d   128, 64, 32, 16, ___, ___
  e   8, 11, 15, 20, ___, ___

3 Joshua buys 450 pencils at 3 pence each. What change does he receive from £20?

4 Every day at school Stephen buys a roll for 34p, crisps for 21p and a drink for 31p.
How much does he spend in pounds in the whole school year of 200 days?

5 An athlete runs 25 laps of a track in 30 minutes 10 seconds.
  a   How many seconds does he take to run 25 laps?
  b   How long does he take to run one lap, if he runs the 25 laps at a constant speed?

6 A pile of 250 tiles is 2 m thick. What is the thickness of one tile in cm?

7 Work out
  a   20% of £65
  b   37% of £400
  c   8·5% of £2000.

8 In a test, the marks of nine students were 7, 5, 2, 7, 4, 9, 7, 6, 6. Find
  a   the mean mark
  b   the median mark
  c   the modal mark.

9 Work out
  a   −6 − 5   **b**   −7 + 30
  c   −13 + 3   **d**   −4 × 5
  e   −3 × (−2)   **f**   −4 + (−10)

10 Given $a = 3$, $b = -2$ and $c = 5$, work out
  a $b + c$ **b** $a - b$
  c ab **d** $a + bc$

11 Solve the equations.
  a $x - 6 = 3$ **b** $x + 9 = 20$
  c $x - 5 = -2$ **d** $3x + 1 = 22$

12 Which of these nets can be used to make a cube?
  a   **b**

  c

> Cut out the nets on square grid paper.

13 a Draw the next diagram in this sequence.

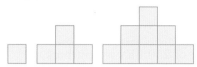

b Write the number of squares in each diagram.

c Describe in words the sequence you obtain in part **b**.

d How many squares will there be in the diagram which has 13 squares on the bottom row?

Exercise 2

1 Solve the equations.

a $3x - 1 = 20$ **b** $4x + 3 = 4$

c $5x - 7 = -3$

2 Copy the diagrams and then calculate x, correct to 3 sf.

a

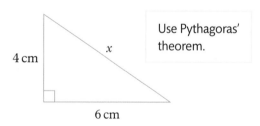

4 cm

x

6 cm

Use Pythagoras' theorem.

b

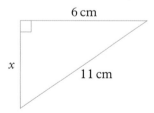

6 cm

x

11 cm

c

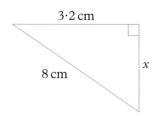

3·2 cm

8 cm

x

3 A bag contains 3 red balls and 5 white balls. Find the probability of selecting

a a red ball

b a white ball.

4 A box contains 2 yellow discs, 4 blue discs and 5 green discs. Find the probability of selecting

a a yellow disc

b a green disc

c a blue or a green disc.

5 Work out on a calculator, correct to 4 sf.

a $3·61 - (1·6 \times 0·951)$

b $\dfrac{4.65 + 1·09}{3·6 - 1·714}$

6 Look at this diagram.

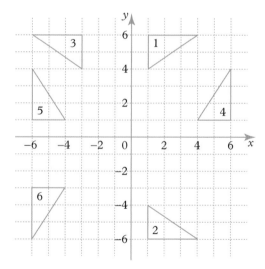

Describe fully these transformations.

a $\triangle 1 \to \triangle 2$

b $\triangle 1 \to \triangle 3$

c $\triangle 1 \to \triangle 4$

d $\triangle 5 \to \triangle 1$

e $\triangle 5 \to \triangle 6$

f $\triangle 4 \to \triangle 6$

7 Plot and label these triangles on an appropriate grid.

Δ1: (−3, −6), (−3, −2), (−5, −2)
Δ2: (−5, −1), (−5, −7), (−8, −1)
Δ3: (−2, −1), (2, −1), (2, 1)
Δ4: (6, 3), (2, 3), (2, 5)
Δ5: (8, 4), (8, 8), (6, 8)
Δ6: (−3, 1), (−3, 3), (−4, 3)

Describe fully these transformations.

a   Δ1 → Δ2
b   Δ1 → Δ3
c   Δ1 → Δ4
d   Δ1 → Δ5
e   Δ1 → Δ6
f   Δ3 → Δ5
g   Δ6 → Δ2

Exercise 3

1 The tables show the rail fares for adults and part of a rail timetable for trains between Cambridge and Bury St. Edmunds.

Fares for one adult

Cambridge				
£2·00	Dullingham			
£2·40	80p	Newmarket		
£2·60	£2·00	£1·20	Kennett	
£4·00	£2·60	£2·40	£1·60	Bury St. Edmunds

Train times

Cambridge	11:20
Dullingham	11:37
Newmarket	11:43
Kennett	11:52
Bury St. Edmunds	12:06

a   How much would it cost for four adults to travel from Dullingham to Bury St. Edmunds?
b   How long does this journey take?

2 This is a sketch of a clock tower.

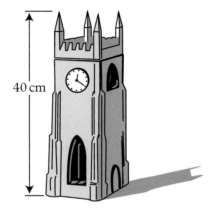

40 cm

Sarah makes a model of the tower using a scale of 1 to 20.

a   The minute hand on the tower clock is 40 cm long. What is the length of the minute hand on the model?
b   The height of the model is 40 cm. What is the height, in metres, of the actual clock tower?

3 Look at this number pattern.

$(2 \times 1) - 1 = 2 - 1$
$(3 \times 3) - 2 = 8 - 1$
$(4 \times 5) - 3 = 18 - 1$
$(5 \times 7) - 4 = 32 - 1$
$(6 \times a) - 5 = b - 1$

i   What number does the letter a stand for?
ii   What number does the letter b stand for?
iii   Write the next line in the pattern.

4 **a** Plot and label these triangles.
Δ1: $(-3, 4)$, $(-3, 8)$, $(-1, 8)$
Δ5: $(-8, -2)$, $(-8, -6)$,
$(-6, -2)$

b Draw Δ2, Δ3, Δ4, Δ6 and Δ7 using these transformations.

i Δ1 → Δ2:
translation $\begin{pmatrix} 9 \\ -4 \end{pmatrix}$

ii Δ2 → Δ3:
translation $\begin{pmatrix} -4 \\ -8 \end{pmatrix}$

iii Δ3 → Δ4: reflection in the line $y = x$

iv Δ5 → Δ6: rotation 90° anticlockwise, centre $(-4, -1)$

v Δ6 → Δ7: rotation 180°, centre $(0, -1)$

c Write the coordinates of Δ2, Δ3, Δ4, Δ6, and Δ7.

5 The faces of a round clock and square clock are exactly the same area. If the round clock has a radius of 10 cm, how wide is the square clock?

6 A metal ingot is in the form of a solid cylinder of length 7 cm and radius 3 cm.

a Calculate the volume, in cm^3, of the ingot. Give your answer to 1 dp.

The ingot is to be melted down and used to make cylindrical coins of thickness 0·3 cm and radius 1·2 cm.

b Calculate the volume, in cm^3, of each coin. Give your answer to 1 dp.

c Calculate the number of coins which can be made from the ingot, assuming that there is no wastage of metal.

Exercise 4

1 In December 2014, a factory employed 220 men, each man being paid £300 per week.

a Calculate the total weekly wage bill for the factory.

b In January 2015, the work force of 220 was reduced by 10 per cent. Find the number of men employed at the factory after the reduction.

c Also in January 2015, the weekly wage of £300 was increased by 10 per cent. Find the new weekly wage.

d Calculate the total weekly wage bill in January 2015.

2 $1 + 3 = 2^2$ $1 + 3 + 5 = 3^2$

a $1 + 3 + 5 + 7 = x^2$
Calculate x.

b $1 + 3 + 5 + \ldots + n = 100$
Calculate n.

3 A motorist travelled 800 miles during May, when the cost of petrol was 90 pence per litre. In June the cost of petrol increased by 10% and he reduced his mileage for the month by 5%.
   a What was the cost, in pence per litre, of petrol in June?
   b How many miles did he travel in June?

4 The distance–time graphs for several moving objects are shown. Decide which line represents each of these.
   ● Hovercraft from Dover
   ● Car ferry from Dover
   ● Cross-channel swimmer
   ● Marker buoy outside harbour
   ● Train from Dover
   ● Car ferry from Calais

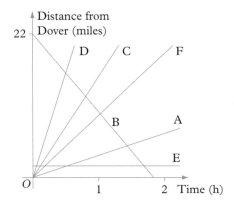

5 The mean mass of 10 boys in a class is 56 kg.
   a Calculate the total mass of these 10 boys.
   b Another boy, whose mass is 67 kg, joins the group. Calculate the mean mass of the 11 boys.

6 This electricity bill is not complete.

> METER READING on
> 07-11-14   26819 units
> METER READING on
> 04-02-15   ☐A ☐ units
>   ELECTRICITY USED   1455 units
>   1455 units at 5·44 pence per unit   £ ☐B☐
>   Quarterly charge   £ 6·27
>   TOTAL (now due)   £ ☐C☐

   a Find the correct amounts, A, B and C, to be placed in each box.
   b In 2014, in what month was the meter read?

Exercise 5

1 $a = \dfrac{1}{2}$, $b = \dfrac{1}{4}$. Which one of these expressions has the greatest value?
   a ab **b** $a + b$ **c** $\dfrac{a}{b}$
   d $\dfrac{b}{a}$ **e** $(ab)^2$

2 **a** Calculate the speed (in metres per second) of a slug which moves a distance of 300 cm in 1 minute.
   b Calculate the time taken for a bullet to travel 8 km at a speed of 5000 m/s.
   c Calculate the distance flown, in a time of four hours, by a pigeon which flies at a speed of 12 m/s.

3 Given $a = 3$, $b = 4$ and $c = -2$, evaluate

a $2a^2 - b$ **b** $a(b - c)$

c $2b^2 - c^2$

4 In the diagram, the equations of the lines are $y = 3x$, $y = 6$, $y = 10 - x$ and $y = \dfrac{1}{2}x - 3$.

Find the equation corresponding to each line.

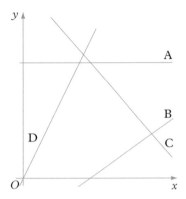

5 Given that $s - 3t = rt$, express

a s in terms of r and t

b r in terms of s and t.

6 Find the area of this shape.

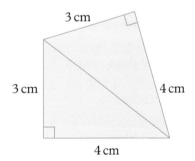

3 cm

3 cm   4 cm

4 cm

7 A cylinder of radius 8 cm has a volume of 2 litres. Calculate the height of the cylinder.
(1 litre = 1000 cm^3)
Give your answer to 1 dp.

Exercise 6

1 The pump shows the price of petrol in a garage.

Total £	19·28
Litres	20·00
Price p per litre	96·40

One day I buy £20 worth of petrol. How many litres do I buy?

2 Given that $x = 4$, $y = 3$, $z = -2$, evaluate

a $2x(y + z)$ **b** $(xy)^2 - z^2$

c $x^2 + y^2 + z^2$ **d** $(x + y)(x - z)$

3 Twenty-seven small wooden cubes fit exactly inside a cubical box without a lid. How many of the cubes are touching the sides or the bottom of the box?

4 Each diagram in the sequence below consists of a number of dots.

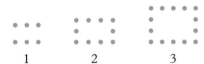

1   2   3

Diagram number

a Draw diagram number 4, diagram number 5 and diagram number 6.

b Copy and complete this table.

Diagram number	Number of dots
1	6
2	10
3	
4	
5	
6	

c Without drawing the diagrams, state the number of dots in
  i diagram number 10
  ii diagram number 15.
d Write x for the diagram number and n for the number of dots and write a formula involving x and n.

5 Write each incorrect statement and make corrections where necessary.
  a $t + t + t = t^3$
  b $a^2 \times a^2 = 2a^2$
  c $2n \times n = 2n^2$

6 The table shows the number of students in a class who scored marks 3 to 8 in a test.

Marks	3	4	5	6	7	8
Number of students	2	3	6	4	3	2

Find
a the mean mark
b the modal mark
c the median mark.

7 The diagram shows a regular octagon with centre O. Find angles a and b.

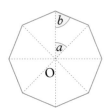

8 a Given that $x - z = 5y$, express z in terms of x and y.
  b Given that $mk + 3m = 11$, express k in terms of m.
  c For the formula $T = \dfrac{C}{z}$, write z in terms of T and C.

Exercise 7

1 This graph shows a car journey from Gateshead to Middlesbrough and back again.

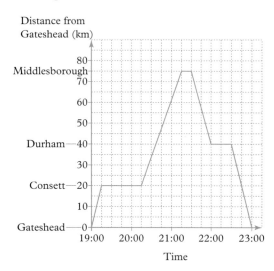

a Where is the car
  i at 19:15
  ii at 22:15
  iii at 22:45?
b How far is it
  i from Consett to Middlesbrough
  ii from Durham to Gateshead?
c At what speed does the car travel
  i from Gateshead to Consett
  ii from Consett to Middlesbrough
  iii from Middlesbrough to Durham
  iv from Durham to Gateshead?
d For how long, in total, is the car stationary during the journey?

2 Work out the difference between one ton and one tonne.
Give your answer to the nearest kg.

> 1 tonne = 1000 kg
> 1 ton = 2240 Ib
> 1 Ib = 454 g

3 A motorist travelled 200 miles in five hours. Her average speed for the first 100 miles was 50 mph. How long did she take for the second 100 miles?

4 Evaluate these and give the answers to 3 significant figures.

a $\sqrt{(9 \cdot 61 \times 0 \cdot 0041)}$

b $\dfrac{1}{9 \cdot 5} - \dfrac{1}{11 \cdot 2}$

c $\dfrac{15 \cdot 6 \times 0 \cdot 714}{0 \cdot 0143 \times 12}$

d $\sqrt[3]{\dfrac{1}{5 \times 10^3}}$

5 If $m = 3$ and $p = 2$, work out the value of

a mp

b $m + 2p$

c $m^2 - p^2$

6 Estimate the answer correct to one significant figure. Do not use a calculator.

a $(612 \times 52) \div 49 \cdot 2$

b $(11 \cdot 7 + 997 \cdot 1) \times 9 \cdot 2$

c $\sqrt{\dfrac{91 \cdot 3}{10 \cdot 1}}$

> For part **a** write
> $(600 \times 50) \div 50$

d $\pi \sqrt{5 \cdot 2^2 + 18 \cdot 2^2}$

7 In the quadrilateral PQRS, PQ = QS = QR, PS is parallel to QR and $\angle QRS = 70°$. Calculate

a $\angle RQS$

b $\angle PQS$

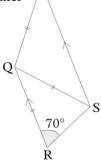

8 A bag contains x green discs and 5 blue discs. A disc is selected. Find, in terms of x, the probability of selecting a green disc.

9 In the diagram, the equations of the lines are $2y = x - 8$, $2y + x = 8$, $4y = 3x - 16$ and $4y + 3x = 16$, but not in that order. Find the equation corresponding to each line.

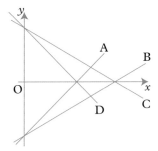

Exercise 8

1 A supermarket sells their 'own-label' raspberry jam in two sizes.

Which jar represents the better value for money? (1 kg = 2·2 lb)

2 A photo 21 cm by 12 cm is enlarged as shown.

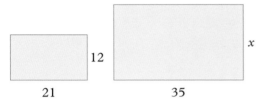

12

21

x

35

  a What is the scale factor of the enlargement?

  b Work out the length *x*.

3 Nadia said: 'I thought of a number, multiplied it by 6, then added 15. My answer was less than 200.'

  a Write Nadia's statement in symbols, using *x* as the starting number.

  b Nadia actually thought of a prime number. What was the largest prime number she could have thought of?

4 Find the area of the red part of the flag. All lengths are in cm.

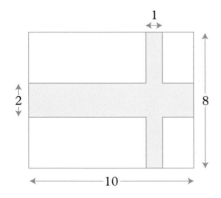

5 Point B is on a bearing 120° from point A. The distance from A to B is 110 km.

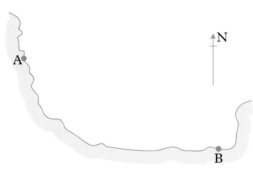

  a Draw a diagram showing the positions of A and B. Use a scale of 1 cm to 10 km.

  b Ship S is on a bearing 072° from A. Ship S is on a bearing 325° from B. Show S on your diagram and state the distance from S to B.

6 Evaluate these using a calculator (answers to 4 sf).

  a $\dfrac{0 \cdot 74}{0 \cdot 81 \times 1 \cdot 631}$
  b $\sqrt{\dfrac{9 \cdot 61}{8 \cdot 34 - 7 \cdot 41}}$

  c $\left(\dfrac{0 \cdot 741}{0 \cdot 8364}\right)^{3}$
  d $\dfrac{8.4 - 7.642}{3.333 - 1.735}$

7 The mean of four numbers is 21.

  a Calculate the sum of the four numbers.

  Six other numbers have a mean of 18.

  b Calculate the mean of all the ten numbers.

8 Tins of peaches are packed 24 to a box. How many boxes are needed for 1285 tins?

12.2 Multiple choice tests

Test 1

1 How many mm are there in 1 m 1 cm?

A 1001
B 1110
C 1010
D 1100

2 If $x = 3$, then $x^2 - 2x =$

A 3
B 4
C 0
D 5

3 The gradient of the line $y = 2x - 1$ is

A 2
B -1
C $\dfrac{1}{2}$
D -2

4 The mean weight of a group of 11 men is 70 kg. What is the mean weight of the remaining group when a man of weight 90 kg leaves?

A 80 kg
B 72 kg
C 68 kg
D 62 kg

5 A, B, C and D are points on the sides of a rectangle. Find the area in cm^2 of quadrilateral ABCD.

A $27\dfrac{1}{2}$
B 28
C $28\dfrac{1}{2}$
D It cannot be found

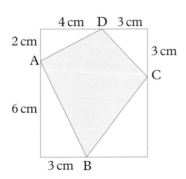

6 The area of the triangle is

A 6
B 10
C 12
D dependent on x

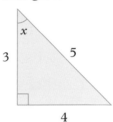

7 The formula $\dfrac{x}{a} + b = c$ is rearranged to make x the subject. What is x?

A $a(c - b)$
B $ac - b$
C $\dfrac{c - b}{a}$
D $ac + ab$

8 If $m = 2$ then $3m - m^2 =$

A -1
B 10
C 1
D 2

9 The sum of the lengths of the edges of a cube is 36 cm. The volume, in cm^3, of the cube is

A 36
B 27
C 64
D 48

10 In the triangle, the size of angle x is

A 35°
B 70°
C 110°
D 40°

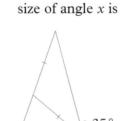

11 A man paid tax on £9000 at 30%. He paid the tax in 12 equal payments. Each payment was

A £2·25
B £22·50
C £225
D £250

12 The approximate value of $\dfrac{3 \cdot 96 \times (0 \cdot 5)^2}{97 \cdot 1}$ is

A 0·01
B 0·02
C 0·04
D 0·1

13 Given that $\dfrac{3}{n} = 5$, then $n =$

A 2
B -2
C $1\dfrac{2}{3}$
D 0·6

14 Cube A has side 2 cm.
Cube B has side 4 cm.
$\dfrac{\text{Volume of B}}{\text{Volume of A}} =$

A 2
B 4
C 8
D 16

15 How many tiles of side 50 cm will be needed to cover this floor?

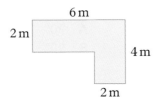

6 m
2 m
4 m
2 m

A 16
B 32
C 64
D 84

16 The equation $ax + 6 = 0$ has a solution $x = 3$.
What is a?

A 1
B -2
C $\sqrt{2}$
D 2

17 Which of these is /are correct?

1 $\sqrt{0 \cdot 16} = 0 \cdot 4$

2 $0 \cdot 2 \div 0 \cdot 1 = 0 \cdot 2$

3 $\dfrac{4}{7} > \dfrac{3}{5}$

A **1** only
B **2** only
C **3** only
D **1** and **2**

18 How many prime numbers are there between 30 and 40?

A 0
B 1
C 2
D 3

19 Kevin is paid £180 per week after a pay rise of 20%.
What was he paid before the rise?

A £144
B £150
C £160
D £164

20 A car travels for 20 minutes at 45 mph. How far does the car travel?

A 900 miles
B $2\dfrac{1}{4}$ miles
C 9 miles
D 15 miles

21 The point $(3, -1)$ is reflected in the line $y = 2$. The new coordinates are

A $(3, 5)$
B $(1, -1)$
C $(3, 4)$
D $(0, -1)$

22 Given the equation $5^x = 120$, the best approximate solution is $x =$

A 2
B 3
C 4
D 22

23 Here are four statements about the diagonals of a rectangle. The statement which is not **always** true is

A They are equal in length.
B They divide the rectangle into four triangles of equal area.
C They cross at right angles.
D They bisect each other.

24 The shaded area in cm^2 is

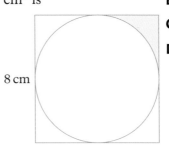

8 cm

8 cm

A $16 - 2\pi$
B $16 - 4\pi$
C $\dfrac{4}{\pi}$
D $64 - 8\pi$

25 An estimate of the value of
$$\frac{204 \cdot 7 \times 97 \cdot 5}{1064 \cdot 2},$$
to one significant figure, is

  A 2
  B 20
  C 200
  D 2000

Test 2

1 What is the value of the expression $(x - 2)(x + 4)$ when $x = -1$?

  A 9
  B −9
  C 5
  D −5

2 The perimeter of a square is 36 cm. What is its area?

  A $36\,\text{cm}^2$
  B $324\,\text{cm}^2$
  C $81\,\text{cm}^2$
  D $9\,\text{cm}^2$

3 The gradient of the line $2x + y = 3$ is

  A 3
  B −2
  C $\dfrac{1}{2}$
  D $-\dfrac{1}{2}$

4 The shape consists of four semicircles placed round a square of side 2 m. The area of the shape, in m², is

  A $2\pi + 4$
  B $2\pi + 2$
  C $4\pi + 4$
  D $\pi + 4$

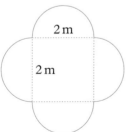

2 m

2 m

5 A firm employs 1200 people, of whom 240 are men. The percentage of employees who are men is

  A 40%
  B 10%
  C 15%
  D 20%

6 A car is travelling at a constant speed of 30 mph. How far will the car travel in 10 minutes?

  A $\dfrac{1}{3}$ mile
  B 3 miles
  C 5 miles
  D 6 miles

7 What are the coordinates of the point $(1, -1)$ after reflection in the line $y = x$?

  A $(-1, 1)$
  B $(1, 1)$
  C $(-1, -1)$
  D $(1, -1)$

8 $\dfrac{1}{3} + \dfrac{2}{5} =$
[no calculators!]

  A $\dfrac{2}{8}$
  B $\dfrac{3}{8}$
  C $\dfrac{3}{15}$
  D $\dfrac{11}{15}$

9 In the triangle, the size of the largest angle is

  A 30°
  B 90°
  C 120°
  D 80°

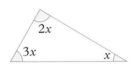

2x

3x   x

10 800 decreased by 5% is

  A 795
  B 640
  C 760
  D 400

11 What is the area, in m², of a square with each side 0·02 m long?

  A 0·0004
  B 0·004
  C 0·04
  D 0·4

12 Given $a = \dfrac{3}{5}$, $b = \dfrac{1}{3}, c = \dfrac{1}{2}$, then

  A abc
  B acb
  C abc
  D acb

13 The **larger** angle between south-west and due east is

A
B
C
D

14 In a triangle PQR, $\angle$PQR = 50° and point X lies on PQ such that QX = XR. Calculate $\angle$QXR.

A
B
C
D

15 What is the value of $1 - 0{\cdot}05$ as a fraction?

A $\dfrac{1}{20}$

B $\dfrac{9}{10}$

C $\dfrac{19}{20}$

D $\dfrac{5}{100}$

16 I start with x, then square it, multiply by 3 and finally subtract 4. The final result is

A $(3x)^2 - 4$
B $(3x - 4)^2$
C $3x^2 - 4$
D $(3x - 4)^2$

17 Given that $m = 2$ and $n = -3$, what is mn^2?

A -18
B 18
C -36
D 36

18 How many prime numbers are there between 50 and 60?

A 1
B 2
C 3
D 4

19 £240 is shared in the ratio $2:3:7$. The largest share is

A £130
B £140
C £150
D £160

20 Adjacent angles in a parallelogram are x and $3x$. The smallest angles in the parallelogram are each

A 30°
B 45°
C 60°
D 120°

21 A square has sides of 10 cm. When the sides of the square are increased by 10%, the area is increased by

A 10%
B 20%
C 21%
D 15%

22 The volume, in cm^3, of the cylinder is

A 9π
B 12π
C 600π
D 900π

6 cm ⟷ 1 m

23 $3a + 2a - 7 - a = 21$
The value of a is

A $3\dfrac{1}{2}$

B 7

C $4\dfrac{2}{3}$

D 8

24 Four people each toss a coin. What is the probability that the fourth person will toss a tail?

A $\dfrac{1}{2}$

B $\dfrac{1}{4}$

C $\dfrac{1}{8}$

D $\dfrac{1}{16}$

25 What is the next term in the sequence 1, 3, 7, 15?

A 23
B 21
C 31
D 24

Answers

1 Number 1

page 1 **Exercise 1** ①

1 a 3 **b** 4 **c** 539
 d 2000 + 400 + 10 + 6
2 a 80 **b** 3000
3 a Five hundred and twenty three
 b Six thousand, four hundred and ten
 c Twenty five thousand
4 a 217 **b** 4250 **c** 5 000 000 **d** 6020
5 Roydon, Penton, Quarkby
6 a 29, 85, 290, 314 **b** 564, 645, 1666, 2000, 5010
 c 7510, 8888, ten thousand, 60 000
7 a 258, 285, 528, 582 **b** 852
8 a 1348 **b** 8431 **c** 483
9 2 × 20p, 2 × 2p
10 25 000 **11** 4110 **12** 510 212
13 70 **14 a** 7320 **b** 2037
15 a 75 423 **b** 23 574
16 a 257 **b** 3221 **c** 704
17 a 1392 **b** 26 611 **c** 257 900
18 a 0 **b** 52 000
19 a 2058, 2136, 2142, 2290
 b 5029, 5299, 5329, 5330
 c 25 000, 25 117, 25 171, 25 200, 25 500
20 a 96 + 84 + 73 or 94 + 86 + 73 or 96 + 83 + 74
 or 94 + 83 + 76 or 93 + 86 + 74 or 93 + 84 + 76
 b 974 + 863 or 964 + 873 or 973 + 864
 or 963 + 874
21 100 **22** 10

page 4 **Exercise 2** ①

1 a 1, 3, 9 **b** 1, 2, 5, 10 **c** 1, 2, 3, 4, 6, 12
2 a 1, 2, 7, 14 **b** 1, 2, 4, 5, 10, 20
 c 1, 2, 3, 5, 6, 10, 15, 30
3 1 and 5
4 a

Number	Factors
12	1, 2, 3, 4, 6, 12
20	1, 2, 4, 5, 10, 20

 b 1, 2, 4
5 b i 3, 6, 9, 12 **ii** 4, 8, 12, 16
 iii 10, 20, 30, 40
6 16 **7** 76 **8** 22 **9** 34
10 2 **11 a** even **b** odd

page 5 **Exercise 3** ①

1 a 2, 4, 6, 8, 10, 12 **b** 5, 10, 15, 20, 25, 30
 c 10
2 a 4, 8, 12, 16 **b** 12, 24, 36, 48 **c** 12
3 a 18 **b** 24 **c** 70
 d 12 **e** 10 **f** 63
4 12 **5** 6
6 a 6 **b** 11 **c** 9
 d 6 **e** 12 **f** 10
7 a 6 **b** 40
 c Many possible answers, e.g. 22 and 33
 d 2 and 5 or 1 and 10
8 15 **9** 21

page 7 **Exercise 4** ①

1

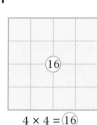

4 × 4 = (16)   5 × 5 = (25)

2 a 25 **b** 100 **c** 36
 d 13 **e** 50 **f** 96
3 a 4 **b** 9 **c** 1
4 a 3 **b** 5 **c** 10 **d** 7
5 a 8 **b** 7 **c** 36
6 a True **b** True **c** False **d** True
7 a 104 **b** 28 **c** 18
 d 900 **e** 6 **f** 10 001
8 a 4 9 (48) 64 **b** 16 25 36 (55)
 c 1 (8) 16 100 **d** 4 (10) 49 81
 e 25 64 (120) 144 **f** 16 36 (108) 121
9 a 9 **b** 25 **c** 36 **d** 16
 e 4 **f** 100 **g** 49 **h** 81
10 a 25 **b** 27 **c** 125
11 13

page 9 **Exercise 5** ①

1 2, 7, 17 **2** 2, 3, 5, 7, 11, 13, 17, 19, 23, 29
3 2 **4** 31, 37

5 a 11 **b** 17 **c** 29 **d** 23 **e** 37
  f 47 **g** 59 **h** 41

6 a 7 ⑨ 13 17   **b** 2 13 ㉑ 23

  c 11 13 19 ㉗   **d** ⑮ 19 29 31

  e 31 37 ㊴ 41   **f** 23 43 47 ㊾

7 a

  b
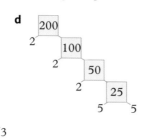

  c
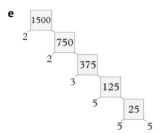

  d

  e
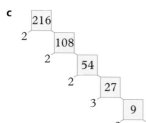

8 $2 \times 2 \times 2 \times 2 \times 3 \times 5 \times 5$

9 a $2 \times 2 \times 7$ **b** $2 \times 2 \times 2 \times 2 \times 2$
  c 2×17 **d** $3 \times 3 \times 3 \times 3$
  e $2 \times 3 \times 7 \times 7$

10

	Even number	Square number	Prime number
Factor of 14	2	1	7
Multiple of 3	6	9	3
Between 3&9	8	4	5

11

	Factor of 10	More than 3	Factor of 18
Square number	1	4	9
Even number	2	8	6
Prime number	5	7	3

12 Multiple solutions

13 29, 31 41, 43 59, 61 71, 73

14 a $= 9 \ = 3^2$
    $= 16 = 4^2$
  b $1 + 3 + 5 + 7 + 9 = 25 = 5^2$
    $1 + 3 + 5 + 7 + 9 + 11 = 36 = 6^2$
    $1 + 3 + 5 + 7 + 9 + 11 + 13 = 49 = 7^2$
    $1 + 3 + 5 + 7 + 9 + 11 + 13 + 15 = 64 = 8^2$
    $1 + 3 + 5 + 7 + 9 + 11 + 13 + 15 + 17 = 81 = 9^2$

15 $13 + 15 + 17 + 19 = 64 = 4^3$
   $21 + 23 + 25 + 27 + 29 = 125 = 5^3$
   $31 + 33 + 35 + 37 + 39 + 41 = 216 = 6^3$

16 No, e.g. $2 + 3 = 5$, an odd number.

17 For example, $7 = 3 + 2^2$, $9 = 5 + 2^2$, $11 = 7 + 2^2$,
   $13 = 5 + 2^3$, $15 = 7 + 2^3$, $17 = 13 + 2^2$,
   $9 = 11 + 2^3$, $21 = 17 + 2^2$

page 12 Exercise 6 ❶

1 67	**2** 96	**3** 13	**4** 121	**5** 144
6 83	**7** 225	**8** 23	**9** 231	**10** 333

page 13 Exercise 7 ❶

1 58	**2** 67	**3** 251	**4** 520	**5** 961
6 337	**7** 496	**8** 511	**9** 320	**10** 992
11 647	**12** 1071	**13** 328	**14** 940	**15** 197
16 2384	**17** 3312	**18** 5335	**19** 7008	**20** 8193
21 1031	**22** 3121	**23** 3541	**24** 827	**25** 6890
26 1021	**27** 13011	**28** 21844	**29** 115387	
30 19885				

page 13 Exercise 8 ❶

1 34	**2** 28	**3** 23	**4** 82	**5** 111
6 204	**7** 57	**8** 15	**9** 56	**10** 23
11 137	**12** 461	**13** 381	**14** 542	**15** 301
16 113	**17** 533	**18** 123	**19** 522	**20** 81
21 265	**22** 5646	**23** 4819	**24** 6388	**25** 7832
26 384	**27** 399	**28** 5804	**29** 1361	**30** 548
31 50				

page 14 Exercise 9 ❶

1

3	8	1
2	4	6
7	0	5

2

4	11	6
9	7	5
8	3	10

3

8	1	6
3	5	7
4	9	2

4

7	2	9
8	6	4
3	10	5

5

12	7	14
13	11	9
8	15	10

6

17	10	15
12	14	16
13	18	11

7

1	12	7	14
8	13	2	11
10	3	16	5
15	6	9	4

8

15	6	9	4
10	3	16	5
8	13	2	11
1	12	7	14

9

3	10	12	17
14	15	5	8
9	4	18	11
16	13	7	6

10

18	9	12	7
13	6	19	8
11	16	5	14
4	15	10	17

11

11	24	7	20	3
4	12	25	8	16
17	5	13	21	9
10	18	1	14	22
23	6	19	2	15

12

16	23	10	17	4
3	15	22	9	21
20	2	14	26	8
7	19	6	13	25
24	11	18	5	12

page 16 Exercise 10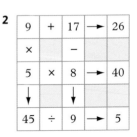

1 63	**2** 96	**3** 252	**4** 140	**5** 639
6 230	**7** 1230	**8** 168	**9** 1477	**10** 2114
11 1065	**12** 1923	**13** 168	**14** 1884	**15** 1179
16 1712	**17** 4920	**18** 3684	**19** 12846	**20** 15125
21 2592	**22** 4501	**23** 2655	**24** 6410	**25** 8460
26 2200	**27** 4417	**28** 7965	**29** 3976	**30** 12918

page 16 Exercise 11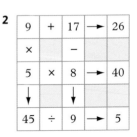

1 23	**2** 143	**3** 211	**4** 115	**5** 178
6 232	**7** 527	**8** 528	**9** 83	**10** 497
11 273	**12** 6024	**13** 604	**14** 271	**15** 415
16 383	**17** 824	**18** 936	**19** 321	**20** 2142
21 9486	**22** 2314	**23** 241	**24** 7005	**25** 837
26 6145	**27** 2638	**28** 415	**29** 15 r 1	**30** 21 r 3
31 28 r 1	**32** 62 r 1	**33** 41 r 1	**34** 24 r 4	**35** 56 r 1
36 130 r 3	**37** 535 r 3	**38** 1283 r 3	**39** 1506 r 3	
40 689 r 1				

page 16 Speed tests

page 17 Test 1

1 22	**2** 45	**3** 8	**4** 58	**5** 77
6 48	**7** 36	**8** 9	**9** 110	**10** 42
11 48	**12** 7	**13** 116	**14** 21	**15** 900

page 17 Test 2

1 22	**2** 27	**3** 54	**4** 45	**5** 143
6 9	**7** 5	**8** 1300	**9** 198	**10** 50
11 57	**12** 21	**13** 49	**14** 37	**15** 12

page 17 Test 3

1 40	**2** 40	**3** 10	**4** 81	**5** 98
6 90	**7** 6	**8** 35	**9** 52	**10** 190
11 5	**12** 8	**13** 110	**14** 195	**15** 32

page 17 Test 4

1 35	**2** 18	**3** 83	**4** 8	**5** 32
6 89	**7** 29	**8** 12	**9** 100	**10** 154
11 55	**12** 11	**13** 5000	**14** 225	**15** 63

page 17 Exercise 12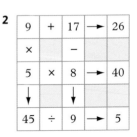

1 135 cm **2** 20p, 10p, 5p, 2p **3** £27

4 57 **5 a** £8·37 **b** £1·63 **6** 893

7 £6·44 **8** 900 g **9** 42

10 a 145 **b** 135 **c** 145 + 135 = 280

11 *e.g. a* = 30, *b* = 25 *a* = 225, *b* = 1

There are other possible answers.

page 18 Exercise 13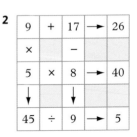

1

11	+	4	→	15
×		÷		
6	÷	2	→	3
↓		↓		
66	×	2	→	132

2

9	+	17	→	26
×		−		
5	×	8	→	40
↓		↓		
45	÷	9	→	5

3

14	+	17	→ 31
×		+	
4	×	23	→ 92
↓		↓	
56	−	40	→ 16

4

15	÷	3	→ 5
+		×	
22	×	5	→ 110
↓		↓	
37	−	15	→ 22

5

9	×	10	→ 90
+		÷	
11	÷	2	→ $5\frac{1}{2}$
↓		↓	
20	×	5	→ 100

6

26	×	2	→ 52
−		×	
18	×	4	→ 72
↓		↓	
8	÷	8	→ 1

7

5	×	12	→ 60
×		÷	
20	+	24	→ 44
↓		↓	
100	×	$\frac{1}{2}$	→ 50

8

7	×	6	→ 42
÷		÷	
14	−	3	→ 11
↓		↓	
$\frac{1}{2}$	×	2	→ 1

9

19	×	2	→ 38
−		÷	
12	×	4	→ 48
↓		↓	
7	−	$\frac{1}{2}$	→ $6\frac{1}{2}$

page 19 **Exercise 14** ①

1 a
```
   2 8 5
+ [5] 1 4
---------
  7 [9] [9]
```
b
```
   6 3 [7]
+ [2] 5 2
---------
  8 [8] 9
```
c
```
  [6] 3 5
+  3 4 [4]
---------
  9 [7] 9
```

2 a
```
   3 5 6
+  5 [2] 6
---------
  8 8 [2]
```
b
```
   2 [2] 4
+  5 3 7
---------
  [7] 6 1
```
c
```
   3 8 8
+ [4] 2 [5]
---------
  8 [1] 3
```

3 a
```
    4 [8]
  ×    3
  -------
  1 4 4
```
b
```
    3 [3]
  ×    7
  -------
  2 3 1
```
c
```
   [3] [2] 1
  ×        5
  ---------
  1 6 0 5
```

4 a $[1][5][0] \div 3 = 50$ **b** $[1][5] \times 4 = 60$

c $9 \times [9] = 81$ **d** $[1][1][5][2] \div 6 = 192$

5 a
```
    4 [4] 5
  + 2 8 [5]
  ---------
  [7] 3 0
```
b
```
    4 [2] 7
  + [1] 7 [7]
  ---------
  6 0 4
```
c
```
   [5] 3 [5]
  + 2 [6] 4
  ---------
  7 9 9
```

6 a $[3][5] \times 7 = 245$ **b** $[5][8] \times 10 = 580$

c $32 \div [4] = 8$ **d** $[9][5][0] \div 5 = 190$

7 a $[7][2] + 29 = 101$ **b** $[1][0][8] - 17 = 91$

c
```
   [8] 8 9
  −3 [4] 6
  ---------
   5 4 [3]
```
d
```
    3 3 5
  −2 1 [8]
  ---------
  [1][1] 7
```

8 Many possible answers

9 a $[4] \times [4] - [4] = 60$ **b** $[8] \div [8] + [8] = 9$

c $[8] \times [8] + [8] = 72$

10 a 5 → ×6 → +9 → 39

b 6 → +2 → ×5 → 40

c 2 → +7 → ×4 → 36

d 7 → ×3 → −11 → 10

11 a $7 \times 4 \,(-)\, 3 = 25$ **b** $8 \times 5 \,(\div)\, 2 = 20$

c $7 \,(\times)\, 3 - 9 = 12$ **d** $12 \,(\div)\, 2 + 4 = 10$

e $75 \div 5 \,(+)\, 5 = 20$

12 a $5 \times 4 \times 3 \,(+)\, 3 = 63$ **b** $5 + 4 \,(-)\, 3 \,(-)\, 2 = 4$

c $5 \times 2 \times 3 \,(+)\, 1 = 31$

page 22 **Exercise 15** ①

1 a $\frac{3}{10}, 0 \cdot 3$ **b** $\frac{1}{10}, 0 \cdot 1$ **c** $\frac{1}{5}, 0 \cdot 2$ **d** $\frac{7}{10}, 0 \cdot 7$

e $\frac{4}{5}, 0 \cdot 8$ **f** $\frac{3}{5}, 0 \cdot 6$ **g** $\frac{2}{5}, 0 \cdot 4$ **h** $\frac{9}{10}, 0 \cdot 9$

2 a $\frac{1}{2}$ **b** 20   **c** 7   **d** $\frac{9}{10}$

e $\frac{1}{100}$ **f** $\frac{7}{10}$ **g** 30   **h** $\frac{8}{100} = \frac{2}{25}$

3 a 0·6, 0·7   **b** 0·8, 1·0   **c** 2·5, 3·0   **d** 0·9, 1·1
4 a 0·2   **b** 0·5   **c** 0·9
   d 1·1   **e** 1·3   **f** 1·7
5 a 0·2   **b** 0·13   **c** 0·02
   d 0·15   **e** 0·155   **f** 0·227

page 23 Exercise 16 ❶

1 1·5, 3·7, 4, 12   **2** 1·3, 3·1, 13, 31
3 0·2, 5·2, 6, 11   **4** 0·11, 0·4, 1, 1·7
5 2, 2·2, 20, 22   **6** 0·12, 0·21, 0·31
7 0·04, 0·35, 0·4   **8** 0·67, 0·672, 0·7
9 0·045, 0·05, 0·07   **10** 0·089, 0·09, 0·1
11 0·57, 0·705, 0·75   **12** 0·041, 0·14, 0·41
13 0·8, 0·809, 0·81   **14** 0·006, 0·059, 0·6
15 0·143, 0·15, 0·2   **16** 0·04, 0·14, 0·2, 0·53
17 0·12, 0·21, 1·12, 1·2   **18** 0·08, 0·75, 2·03, 2·3
19 0·26, 0·3, 0·602, 0·62   **20** 0·5, 1·003, 1·03, 1·3
21 0·709, 0·79, 0·792, 0·97   **22** 0·312, 0·321, 1·04, 1·23
23 0·007, 0·008, 0·09, 0·091   **24** 2, 2·046, 2·05, 2·5
25 CARWASH
26 a 32·51   **b** 0·853   **c** 1·16
27 a £3·50   **b** £0·15   **c** £0·03
   d £0·10   **e** £12·60   **f** £0·08
28 a True   **b** False   **c** True   **d** True

page 23 Exercise 17 ❶

1 c   **2** a   **3** c   **4** b   **5** b
6 a   **7** False   **8** True   **9** True   **10** True
11 True   **12** False   **13** True   **14** True   **15** True
16 False   **17** True   **18** False   **19** True   **20** False
21 False   **22** True   **23** True   **24** True   **25** True
26 False   **27** False   **28** True   **29** False   **30** False
31 True   **32** True   **33** True   **34** True

35 number line 0 to 1 with 0·3, 0·5, 0·7, 0·9

36 number line 0 to 2 with 0·4, 0·7, 1·4, 1·7

37 number line 1·0 to 1·2 with 1·05, 1·1, 1·15, 1·02, 1·19

38 number line 0·9 to 1·1 with 0·94, 1·0, 1·05, 0·91, 0·97, 1·08

page 24 Exercise 18 ❶

1 4·3   **2** 0·7   **3** 9·4   **4** 1·2   **5** 16
6 10·7   **7** 17·4   **8** 128   **9** 375   **10** 0·24 m
11 1·92 m   **12** 5·2 mg   **13** 0·06 cm   **14** 1·76 mm
15 3·16 kg   **16** 105 km   **17** 50 g   **18** 125 kg

page 25 Exercise 19 ❶

1 6·8   **2** 8·8   **3** 11·7   **4** 14·9   **5** 6·6
6 12·81   **7** 8·77   **8** 10·19   **9** 14·54   **10** 9·07
11 10·14   **12** 20·94   **13** 26·71   **14** 216·95   **15** 9·6
16 23·1   **17** 12·25   **18** 17·4   **19** 0·062   **20** 85·47
21 28   **22** 27·02   **23** 2·44   **24** 275·9   **25** 96·8
26 12·078   **27** 5·83   **28** 22·61   **29** 560·357   **30** 55·2
31 58·4   **32** 20·055   **33** 35·24   **34** 1008·09   **35** 22·03
36 257·4   **37** 35·5   **38** 29·66   **39** 43·55   **40** 583·6

page 26 Exercise 20 ❶

1 7·2   **2** 5·4   **3** 7·1   **4** 5·5   **5** 6·8
6 2·7   **7** 4·5   **8** 2·7   **9** 9·11   **10** 7·08
11 1·11   **12** 4·36   **13** 2·41   **14** 10·8   **15** 1·36
16 6·23   **17** 2·46   **18** 12·24   **19** 8·4   **20** 15·96
21 2·8   **22** 2·2   **23** 10·3   **24** 21·8   **25** 0·137
26 0·054   **27** 6·65   **28** 4·72   **29** 0·57   **30** 3·6
31 7·23   **32** 6·53   **33** 0·43   **34** 48·01   **35** 1·3
36 7·96   **37** 9·34   **38** 2·43   **39** 16·65   **40** 7·53
41 £8·96

page 27 Exercise 21 ❶

1 6·34   **2** 8·38   **3** 81·5   **4** 7·4   **5** 7245
6 32   **7** 6·3   **8** 142   **9** 4·1   **10** 30
11 710   **12** 39·5   **13** 0·624   **14** 0·897   **15** 0·175
16 0·236   **17** 0·127   **18** 0·705   **19** 1·3   **20** 0·08
21 0·007   **22** 21·8   **23** 0·035   **24** 0·0086   **25** 95
26 111·1   **27** 32   **28** 70   **29** 5·76   **30** 9·99
31 660   **32** 1   **33** 0·42   **34** 6200   **35** 0·009
36 0·0555   **37 a** 0   **b i** 5 and 2   **ii** 5, 2 and 0   **iii** 0, 0, 5 and 2

page 28 Exercise 22 ❶

1 3·9   **2** 4·8   **3** 20·4   **4** 30·5   **5** 28·8
6 7·8   **7** 4·9   **8** 9·9   **9** 3·36   **10** 8·15
11 21·7   **12** 0·84   **13** 81·9   **14** 17·12   **15** 8·65
16 8·16   **17** 200·4   **18** 494·9   **19** 23·8   **20** 8·36
21 £16·25   **22** £31   **23** £66·15   **24** 12 cm^2

page 29 Exercise 23 ❶

1 0·06   **2** 0·15   **3** 0·12   **4** 0·006   **5** 1·8
6 3·5   **7** 1·8   **8** 0·8   **9** 0·36   **10** 0·014
11 1·26   **12** 2·35   **13** 8·52   **14** 3·12   **15** 0·126
16 127·2   **17** 0·170   **18** 0·327   **19** 0·126   **20** 0·34
21 0·055   **22** 0·52   **23** 1·30   **24** 0·001

page 30 **Exercise 24** ①

1 2·1	**2** 2·3	**3** 2·5	**4** 1·5	**5** 45·7
6 3·45	**7** 3·8	**8** 2·7	**9** 2·15	**10** 1·2
11 1·84	**12** 1·3	**13** 2·4	**14** 4·2	**15** 2·6
16 6·2	**17** 6·8	**18** 6·4	**19** 4·6	**20** 3·6
21 0·18	**22** 0·14	**23** 0·35	**24** 0·15	

page 30 **Exercise 25** ①

1 2·1	**2** 3·1	**3** 4·36	**4** 4	**5** 4
6 2·5	**7** 16	**8** 200	**9** 70	**10** 9·2
11 30·5	**12** 6·2	**13** 12·5	**14** 122	**15** 212
16 56	**17** 60	**18** 1500	**19** 0·3	**20** 0·7
21 0·5	**22** 3·04	**23** 5·62	**24** 0·78	**25** 0·14
26 3·75	**27** 0·075	**28** 0·15	**29** 1·22	**30** 163·8
31 1·75	**32** 18·8	**33** 8	**34** 88	**35** 580

page 31 **Exercise 26** ①

1 4·32	**2** 5·75	**3** 9·16	**4** 1·008	**5** 0·748
6 20·24	**7** 10·2	**8** 2·95	**9** 4·926	**10** 34
11 0·621	**12** 8·24	**13** 0·1224	**14** 12·15	**15** 2·658
16 66·462	**17** 34100	**18** 0·0041	**19** 2·104	**20** 0·285
21 0·25884	**22** 3·27	**23** 2·247	**24** 0·54	**25** 0·027
26 6·6077	**27** 6·56	**28** 7·84		

page 31 **Exercise 27** ①

1 A

[1]9	1	[2]2	1	[3]4	0
4		[4]1	7		5
	[5]6	4	[6]8		[7]9
[8]7	0	[9]3	1	[10]8	
4		8		[11]3	6
[12]4	3	[13]7		[14]8	1
0		[15]2	3	[16]6	3

B

[1]9	6	[2]1	5	[3]7	3
3		[4]5	6		1
	[5]8	3	[6]3		[7]8
[8]6	6	[9]5	9	[10]4	
3		6		[11]7	0
[12]1	7	[13]4		[14]8	9
8		[15]7	0	[16]9	3

C

[1]2	6	[2]6	3	[3]1	4
0		[4]3	7		8
	[5]2	5	[6]2		[7]6
[8]8	0	[9]3	1	[10]5	
4		2		[11]7	7
[12]6	0	[13]2		[14]1	0
3		[15]5	7	[16]6	4

page 32 **Exercise 28** ①

1 a 608	**b** 6080	**c** 6·08
2 a 203·04	**b** 203·04	**c** 20304
3 a 1892·9	**b** 189·29	**c** 18·929
4 a 36·173	**b** 3·6173	**c** 0·36173
5 a 485080	**b** 48·508	**c** 36·2
6 a 17544	**b** 175·44	**c** 215

page 33 **Exercise 29** ①

1 $+5°$	**2** $-1°$	**3** $-4°$	**4** $-7°$
5 $-4°$	**6** $-4°$	**7** $+4°$	**8** $+12°$
9 $-5°$	**10** $0°$	**11** $-11°$	**12** $-15°$

13 a $-7°C, -2°C, 5°C$　　**b** $-1°C, 0°C, 2°C$
　　c $-11°C, -8°C, 3°C$　　**d** $-7°C, -4°C, -2°C, -1°C$
　　e $-5°C, -4°C, -2°C, 4°C$

14 a $-5, -3, 0, 2, 6$　　**b** $-8, -6, -1, 3, 11$
　　c $-20, -15, -10, -2, 5$　　**d** $-15, -10, -5, 18, 23$

15 a $2, 0, -2$　　**b** $0, -3, -6$　　**c** $-2, -3, -4$
　　d $-4, -6, -8$　　**e** $-6, -12, -18$　　**f** $0, 1, 2$
　　g $-2, 0, 2$　　**h** $-2, -6, -10$

page 34 **Exercise 30** ①

1 -4	**2** -12	**3** -11	**4** -3	**5** -5
6 4	**7** -5	**8** -8	**9** 19	**10** -17
11 -4	**12** -5	**13** -11	**14** 6	**15** -4
16 6	**17** 0	**18** -18	**19** -3	**20** -11
21 -12	**22** 4	**23** 4	**24** 0	**25** -8
26 -3	**27** 3	**28** -12	**29** 18	**30** -5
31 -66	**32** 98			

page 35 **Exercise 31** ①

1 -8	**2** -7	**3** 1	**4** 1	**5** 9
6 11	**7** -8	**8** 42	**9** 4	**10** 15
11 -7	**12** -9	**13** -1	**14** -7	**15** 0
16 11	**17** -14	**18** 0	**19** 17	**20** 3
21 -1	**22** -3	**23** 12	**24** -9	**25** 3
26 0	**27** 8	**28** 2		

29 a

+	−5	1	6	−2
3	−2	4	9	1
−2	−7	−1	4	−4
6	1	7	12	4
−10	−15	−9	−4	−12

b

+	−3	2	−4	7
5	2	7	1	12
−2	−5	0	−6	5
10	7	12	6	17
0	−3	2	−4	7

page 36 **Exercise 32** ①

1 -6	**2** -4	**3** -15	**4** 9	**5** -8
6 -15	**7** -24	**8** 6	**9** 12	**10** -18
11 -21	**12** 25	**13** -60	**14** 21	**15** 48
16 -16	**17** -42	**18** 20	**19** -42	**20** -66

21 −4 **22** −3 **23** 3 **24** −5 **25** 4
26 −4 **27** −4 **28** −1 **29** −2 **30** 4
31 −16 **32** −2 **33** −4 **34** 5 **35** −10
36 11 **37** 16 **38** −10 **39** 1 **40** −5
41 64 **42** −27 **43** −600 **44** 40 **45** 2
46 36 **47** −2 **48** −8

49a

×	4	−3	0	−2
−5	−20	15	0	10
2	8	−6	0	−4
10	40	−30	0	−20
−1	−4	3	0	2

b

×	−2	5	−1	−6
3	−6	15	−3	−18
−3	6	−15	3	18
7	−14	35	−7	−42
2	−4	10	−2	−12

page 37 **Test 1**

1 −16 **2** 64 **3** −15 **4** −2 **5** 15
6 18 **7** 3 **8** −6 **9** 11 **10** −48
11 −7 **12** 9 **13** 6 **14** −18 **15** −10
16 8 **17** −6 **18** −30 **19** 4 **20** −1

page 37 **Test 2**

1 −16 **2** 6 **3** −13 **4** 42 **5** −4
6 −4 **7** −12 **8** −20 **9** 6 **10** 0
11 36 **12** −10 **13** −7 **14** 10 **15** 6
16 −18 **17** −9 **18** 15 **19** 1 **20** 0

page 37 **Test 3**

1 100 **2** −20 **3** −8 **4** −7 **5** −4
6 10 **7** 9 **8** −10 **9** 7 **10** 35
11 −20 **12** −24 **13** −10 **14** −7 **15** −19
16 −1 **17** −5 **18** −13 **19** 0 **20** 8

page 38 **Exercise 33** 1

1 11 **2** 1 **3** −5 **4** 12 **5** 21
6 2 **7** 17 **8** 24 **9** 9 **10** 30
11 30 **12** 25 **13** 8 **14** 5 **15** 6
16 8 **17** 6 **18** 3 **19** 7 **20** −2
21 −4 **22** 14 **23** 13 **24** 0 **25** 52
26 11 **27** 10 **28** 20 **29** 5 **30** 5

page 39 **Exercise 34** 1

1 15 **2** 10 **3** 5 **4** 9 **5** 11
6 1 **7** 7 **8** 8 **9** 8 **10** 4
11 0 **12** 1 **13** 18 **14** 18 **15** 12
16 27 **17** 8 **18** 6 **19** 1 **20** 22
21 9 **22** 0 **23** 5 **24** 0 **25** 20
26 10 **27** 16 **28** 52 **29** 40 **30** 111
31 51 **32** 30 **33** 11 **34** 9 **35** 28
36 106 **37** 54 **38** 4 **39** 4 **40** 153
41 59 **42** 165 **43** 85 **44** 12 **45** 33
46 64 **47** 67 **48** 1172 **49** 52 **50** 5
51 4 **52** 16 **53** 8 **54** 2

page 40 **Exercise 35** 1

1 3^4 **2** 5^2 **3** 6^3 **4** 10^5
5 1^7 **6** 8^4 **7** 7^6 **8** $2^3 \times 5^2$
9 $3^2 \times 7^4$ **10** $3^2 \times 10^3$ **11** $5^4 \times 11^2$ **12** $2^2 \times 3^3$
13 $3^2 \times 5^3$ **14** $2^2 \times 3^3 \times 11^2$
15 a 16 **b** 36 **c** 100 **d** 27 **e** 1000
16 a 81 **b** 441 **c** 1·44 **d** 0·04 **e** 9·61
 f 10000 **g** 625 **h** 75·69 **i** 0·81 **j** 6625·96
17 a $169\,\text{cm}^2$ **b** $6·25\,\text{cm}^2$ **c** $129·96\,\text{cm}^2$
18 a a^3 **b** n^4 **c** s^5 **d** p^2q^3 **e** b^7
19 a 216 **b** 256 **c** 243 **d** 100 000
 e 64 **f** 0·001 **g** 8·3521 **h** 567 **i** 1250
20 10^{10} **21** 2^7
22 a $2^1, 2^2, 2^3, 2^4$ **b** $2^{25}p = £335\,544·32$
23 Sean is right.

page 42 **Exercise 36** 2

1 a 4 **b** 6 **c** 1 **d** 10
2 a 9 cm **b** 7 cm **c** 12 cm
3 a 3·2 **b** 5·4 **c** 10·3 **d** 4·4
 e 49·1 **f** 7·7 **g** 0·4 **h** 0·9
4 12·2 cm **5** 447 m **6** 7·8 cm
7 a 4 **b** 5 **c** 10
8 5·8 cm **9** 10

page 43 **Exercise 37** 2

1 5^6 **2** 6^5 **3** 10^9 **4** 7^8 **5** 3^{10}
6 8^6 **7** 2^{13} **8** 3^8 **9** 5^5 **10** 7^{10}
11 5^8 **12** 3^4 **13** 6^{11} **14** 5^{10} **15** 7^{12}
16 7^2 **17** 6^5 **18** 8 **19** 5^8 **20** 10^2
21 9^2 **22** 3^2 **23** 2^4 **24** 3^2 **25** 7^3
26 3^3 **27** 5^7 **28** 8^6 **29** 5^3 **30** 6^3
31 1 **32** 1 **33** 3^7 **34** 2^7 **35** 7^2 **36** 5^{12}

page 44 **Exercise 38** 2

1 ± 3 **2** 1 **3** 3 **4** 0 **5** 3 **6** 1
7 2 **8** 3 **9** 5 **10** 3 **11** 3 **12** 2

page 45 **Exercise 39** 1

1 a £6·10 **b** £6·40 **c** £4·70 **d** £116
 e £129·30 **f** £0·04
2 a 2·5 h **b** 4·25 h **c** 3·75 h **d** 0·1 h
 e 0·2 h **f** 0·25 h
3 a 24·75 h **b** 22·75 h **c** 2·9 h **d** 2·75 h
 e 2·5 h **f** 1·75 h

page 45 **Exercise 40** 1

1 4·2 **2** 15·9 **3** 0·6 **4** 5·3 **5** 4·0
6 12·7 **7** 0·5 **8** 5·6 **9** 14·0 **10** 2·1
11 14·1 **12** 1·2 **13** 9·9 **14** 9·1 **15** 9·5

16 0·6 **17** 23·0 **18** 11·4 **19** 7·4 **20** 5·5
21 11·5 **22** 11·7 **23** 10·9 **24** 1·9 **25** 13·0
26 4·9 **27** 18·8 **28** 3·4 **29** 2·4 **30** 2·9

page 47 Exercise 41 ①

1 9·1 **2** 11·4 **3** 4·9 **4** 12·4 **5** 1·5
6 4·7 **7** 2·2 **8** 2·6 **9** 0·7 **10** 1·4
11 3·7 **12** 15·1 **13** 9·3 **14** 10·0 **15** 6·0
16 6·9 **17** 1·0 **18** 0·2 **19** 5·4 **20** 80·6
21 7·8 **22** 16·6 **23** 7·3 **24** 12·5 **25** 64·1
26 1·6 **27** 14·1 **28** 2·5 **29** 0·6 **30** 2·7
31 86·6 **32** 44·9 **33** 1·038 **34** 1·0 **35** 6·3
36 0·8 **37** 2·3 **38** 9·9 **39** 13·4 **40** 1·5

page 48 Exercise 42 ①

1 6×10^2 **2** $8·2 \times 10^3$ **3** 5×10^4
4 2×10^5 **5** 2×10^6 **6** $5·5 \times 10^2$
7 $6·18 \times 10^5$ **8** 3×10^9 **9** $2·22 \times 10^6$
10 $6·18 \times 10^3$ **11** 1×10^8 **12** 7×10^5
13 7×10^{-3} **14** $5·5 \times 10^{-2}$ **15** $8·1 \times 10^{-4}$
16 7×10^{-5} **17** 4×10^{-6} **18** 7×10^{-1}
19 $1·11 \times 10^{-2}$ **20** 2×10^{-2} **21** 7000
22 820 **23** 0·06 **24** 0·8
25 480 000 **26** 0·013 **27** 826 000
28 0·00382 **29** 301 m
30 a $3·844 \times 10^5$ km **b** $3·844 \times 10^7$ m
31 230 000 ants
32 0·001, 0·06, 22 000, 30 000, 100 000

page 49 Exercise 43 ① ②

1 1×10^{-3} **2** $1·8 \times 10^6$ cm³ **3** *e, d, f, c*
4 a $5·62 \times 10^4$ **b** $3·2 \times 10^{-3}$ **c** $1·2 \times 10^8$
d $3·7 \times 10^{-5}$
5 $2·8 \times 10^{11}$ cm³ **6** £$5·623 \times 10^6$
7 ii (9×10^8) **8** 0·04 m/s

page 50 Exercise 44 ① ②

1 a $2·8 \times 10^9$ **b** $6·4 \times 10^{11}$ **c** $4·5 \times 10^{12}$
d $3·3 \times 10^{12}$ **e** $1·6 \times 10^5$ **f** $6·2 \times 10^{-5}$
g 3×10^3 **h** 4×10^3 **i** 2×10^3
j $2·1 \times 10^6$ **k** 1×10^4 **l** 3×10^8
2 a 150 **b** 600 **c** 300·5
3 $2·75 \times 10^{13}$ pence
4 5×10^{-3}, 4×10^2, 2×10^3, 4×10^4
5 4×10^9 hours **6** *r, q, p*
7 a 730 000 **b** 1·03
c 7 130 000 000 000 **d** 400 000 000
e 0·065 **f** 8 000 000 000 000 000
8 $1·8 \times 10^{10}$ **9** $2·5 \times 10^6$ minutes = 5·7 years
10 *a, b, c*

page 51 Exercise 45 ①

1 0·7, 2·44, 3·5 g **2** 23 **3** 130
4 £28·50 **5** 14·55 **6** 15 h 5 mins
7 £10·35
8 a

6	13	8
11	9	7
10	5	12

b

11	8	5	10
2	13	16	3
14	1	4	15
7	12	9	6

9 128 cm² **10** £34·75

page 53 Exercise 46 ①

1 a
$$\begin{array}{r} 5\ \boxed{1}\ 3 \\ +\ 3\ 4\ \boxed{6} \\ \hline \boxed{8}\ 5\ 9 \end{array}$$
b
$$\begin{array}{r} 3\ 3\ 4 \\ +\ \boxed{3}\ 4\ \boxed{6} \\ \hline 6\ \boxed{8}\ 0 \end{array}$$

c $\boxed{1}\ \boxed{6}\ \boxed{0} \div 5 = 32$
2 a 16 384 **b** 4096 **3** £4050
4 a 3·32 **b** 1·61 **c** 1·46
d 4·4 **e** 6·2 **f** 2·74
5 18
6 a

−1	−2	3
4	0	−4
−3	2	1

b

4	3	−1
−3	2	7
5	1	0

c

0	1	−4
−5	−1	3
2	−3	−2

7 0·05, 0·2, 0·201, 0·21, 0·5
8 13°C, −2°C, 22°C, −21°C
9 9 h 15 mins **10** 96 523

page 54 Exercise 47 ① ②

1 £2·60; £3·60; 7 jars; Total = £16·45
2 a 3^4 **b** 1^7 **c** 7^5 **d** 2^4 **e** 10^6 **f** 10^4
3 a 9 **b** −5 **c** −6 **d** −10 **e** 7 **f** −10
4 20
5 a 50p, 20p, 5p, 2p **b** 50p, 20p, 10p, 5p, 1p
c £1, 50p, 5p, 1p, 1p or 50p, 50p, 50p, 5p, 2p
6 *m, 9, z* **7** 24
8

	Prime Number	Multiple of 3	Factor of 16
Number >5	7	9	8
Odd number	5	3	1
Even number	2	6	4

9 1·50 m
10 a What time do we finish **b** Spurs are rubbish
c We are under attack

page 56 **Exercise 48**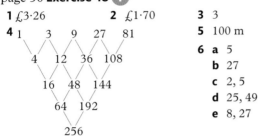

1 £3·26   **2** £1·70   **3** 3
4
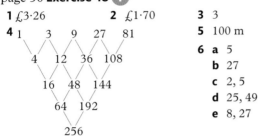

7 10 h 30 mins
8 a 0·54 **b** 40 **c** 0·004 **d** 2·2 **e** £9 **f** £40
9 a 782 or 827 or 872   **b** 278 or 287
  c 287 or 827
10 a 3 and 4 **b** 6 and 7 **c** 4 and 8 **d** 2 and 24

page 57 **Exercise 49**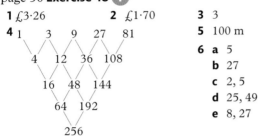

1 a 69   **b** 65   **c** Many possibilities e.g. SAT
2 8·155 kg   **3** 360 000 kg **4** 16
5

1	2	3	4
2	3	4	1
3	4	1	2
4	1	2	3

6 64 mph

  Other solutions are possible
7 a Yes   **b** No   **c** Yes   **d** Yes
  e Yes   **f** Yes   **g** No   **h** No
8 £5·12

page 59 **Exercise 50**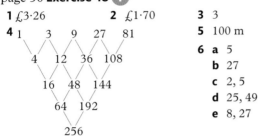

1 a 45 mins   **b** 30 mins   **c** 30 mins   **d** 20 mins
2 2 h 10 mins   **3** 20·05   **4** Mr Turner **5** 15 mins
6 3 h 45 mins   **7** 21·10   **8** 5
9 9 h 45 mins   **10** 18·00   **11** 15 mins

page 60 **Test yourself**

1 1, 2, 4, 5, 10 and 20
2 a 16, 4   **b** 27, 3
3 a 17   **b** 55   **c** 9 **d**   180
4 a 6, 17, 24, 168   **b** 0·5, 1·8, 3·71, 12·4
5 a 8°C   **b** −12°C
6 a 2085   **b** Five thousand one hundred and eight
  c Nine tenths   **d** 160
7 2 × 2 × 3 × 3 × 7
8 a 3000 **b** 4681   **c** Five thousand and sixty
9 4 bags of crisps
10 a i 6, 20 or 26   **ii** 9
    iii 26 and 11   **iv** 51 and 26
  b i 20   **ii** 5 × 4   **c i** 11   **ii** 3 × 11

11 £2400
12 a

−1	4	−3
−2	0	2
3	−4	1

b 12

13 a 15:02   **b** 4 hrs 47 min   **c** Yes
14 25 people
15 a 50 litres   **b** 12 litres
16 Large pot costs 0·107 pence per g and small pot costs
  0·104 pence per g
  ∴ Small pot is a better value.
17 a 10   **b** 29   **c** 125   **d** 4
18 11·94117647
19 54·32 s
20 a i 27   **ii** 10
  b i 15 − (6 − 4) = 13   **ii** 2 + 2 × (3 + 8) = 24

2   Algebra 1

page 64 **Exercise 1**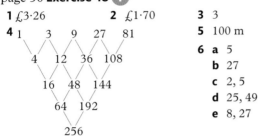

1 $2n$ **2** $y + 4$ **3** $x - 7$ **4** $x + 100$
5 $5y$ **6** $100s$ **7** $3t$ **8** $z + 11$
9 $p - 9$ **10** $n + x$ **11** $4n$ **12** $2x + 3$
13 $2n - 12$ **14** $3m + 2$ **15** $20y$ **16** $3x + 3$
17 $2y - 7$ **18** $3k + 10$

page 65 **Exercise 2**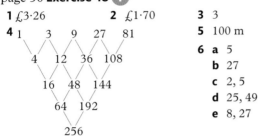

1 $3(x + 4)$ **2** $5(x + 3)$ **3** $6(y + 11)$
4 $\dfrac{x+3}{4}$ **5** $\dfrac{x-7}{3}$ **6** $\dfrac{y-8}{5}$
7 $\dfrac{2(4a+3)}{4}$ or $\dfrac{4a+3}{2}$ **8** $\dfrac{3(m-6)}{4}$ **9** $x^2 - 6$
10 $\dfrac{x^2+3}{4}$ **11** $(n + 2)^2$ **12** $(w - x)^2$
13 $\dfrac{x^2-7}{3}$ **14** $(x - 9)^2 + 10$ **15** $\dfrac{(y+7)^2}{x}$
16 $\dfrac{(a-x)^3}{y}$ **17** $l - 3$ cm   **18** $15 - x$ cm
19 $l + 200 - m$ kg   **20** $4(n + 2)$ **21** $6w$ kg
22 xl kg   **23** $\dfrac{n}{6}$ pence **24** £$\dfrac{p}{5}$ **25** $\dfrac{12}{n}$ kg

page 67 **Exercise 3**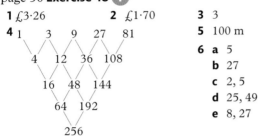

1 $5a$ **2** $11a$ **3** $9a$ **4** $6n$
5 $5n$ **6** $9n$ **7** $4x$ **8** $16x$
9 $3x$ **10** $6a + 4b$ **11** $9a + 6b$ **12** $5x + 5y$
13 $9x + 7y$ **14** $5m + 3n$ **15** $16m + 2n$ **16** $5x + 11$
17 $13x + 8$ **18** $8x + 10$ **19** $16x + 4y$ **20** $5x + 8$
21 $9x + 5$ **22** $7x + 4$ **23** $7x + 4$ **24** $7x + 7$
25 $8x + 12$ **26** $12x - 6$ **27** $2x + 5$ **28** $2x - 5$

29 $2x - 5$ **30** $13a + 3b - 1$ **31** $10m + 3n + 8$

32 $3p - 2q - 8$ **33** $2s - 7t + 14$ **34** $2a + 1$

35 $x + y + 7z$ **36** $5x - 4y + 4z$ **37** $5k - 4m$

38 $4a - 9 + 5b$ **39** $a - 4x - 5e$ **40** $3n + 3$

41 $2x + 8$ **42** $2x + 7$ **43** $2x + 14$

44 $3a + b + 3$ **45** $6x + 2y + 12$ **46** $10x + 8$

page 69 **Exercise 4** ①

1 a ab **b** nm **c** xy **d** ht

   e adn **f** $3ab$ **g** $4nm$ **h** $3abc$

2 a $6ab$ **b** $15ab$ **c** $12cd$ **d** $10nm$

   e $14pq$ **f** $30an$ **g** $20st$ **h** $36ab$

   i $32uv$

3 a $3y^2$ **b** y^2 **c** $6x^2$ **d** $24t^2$

   e $6a^2$ **f** $10y^2$ **g** $2x^2$ **h** $3y^2$

   i $100x^2$

page 69 **Exercise 5** ①

1 $2x + 6$ **2** $3x + 15$ **3** $4x + 24$ **4** $4x + 2$

5 $10x + 15$ **6** $12x - 4$ **7** $12x - 12$ **8** $15x - 6$

9 $15x - 20$ **10** $14x - 21$ **11** $4x + 6$ **12** $6x + 3$

13 $5x + 20$ **14** $12x + 12$ **15** $8x - 2$ **16** $2a + 6b$

17 $6a + 15b$ **18** $10m + 15n$ **19** $14a - 21b$

20 $11a + 22b$ **21** $24a + 16b$ **22** $x^2 + 5x$

23 $x^2 - 2x$ **24** $x^2 - 3x$ **25** $2x^2 + x$ **26** $3x^2 - 2x$

27 $3x^2 + 5x$ **28** $2x^2 - 2x$ **29** $2x^2 + 4x$ **30** $6x^2 + 9x$

31 $7x + 10$ **32** $8x + 2$ **33** $5a - 3$ **34** $11a + 17$

35 $8a - 10$ **36** $8t + 4$ **37** $5x + 8$ **38** $7x + 18$

39 $8x - 16$ **40** $8a - 6$ **41** $2x^2 + 4x + 6$

42 $2x^2 + 2x + 5$ **43** $3a^2 + 6a - 4$

44 $5y^2 + 4y - 3$ **45** $5x^2 + 2x$ **46** $3a^2 + 4a$

page 70 **Exercise 6** ①

1 a $-m - n$ **b** $-a - b$ **c** $-2a - b$

   d $-m + n$ **e** $-a + b$ **f** $-3a + b$

   g $-a - b + c$ **h** $-2a - b + 2c$ **i** $-3x + y + 2$

2 $a + 4b$ **3** $4a + b$ **4** $4a + 7b$

5 $a + 4b$ **6** $a + 3b$ **7** $5a + b$

8 $m + 2n$ **9** $4m + 11n$ **10** $11m + 4n$

11 $6m + 8n$ **12** $x - 2y$ **13** $4x - 3y$

14 $9a + 35b$ **15** $3x + 11y$

16 a $4a + 8$ **b** $10x + 4$ **c** $8a - 2$

17 a $6x + 2y + 4$ **b** $6x + 2y$

18 a $l = 2x - 2$ **b** $l = 2x + 2$ **c** $l = 1$

   d $l = x - 1$ **e** $l = x + 2$ **f** $l = x - 3$

   g $l = x + 3$ **h** $l = x + 3$

page 72 **Exercise 7** ①

1 True   **2** False $\left(\text{unless } n = 1\frac{1}{2}\right)$

3 True   **4** True

5 False(unless $n = \sqrt[4]{3}$ or 0)   **6** True

7 False (unless $m = n$)   **8** True

9 False (unless $n = 0$ or $\sqrt[4]{3}$)   **10** False (unless $m = 0$)

11 False$\left(\text{unless } c = 1\frac{1}{2}\right)$ **12** True

13 True   **14** False(unless $n = \pm 2$)

15 True   **16** False (unless $n = 0$)

17 True   **18** False (unless $n = 0$ or 1)

19 a $n + n, 4n - 2n$ **b** $n \times n^2, n \times n \times n$

   c $3n \div 3, n^2 \div n$ **d** $4 \div n$

   e Many possible answers e.g. $\boxed{n \times n + n \times n}$

20 a $n \rightarrow \boxed{\times 6} \rightarrow \boxed{-1} \rightarrow 6n - 1$

   b $n \rightarrow \boxed{\times 8} \rightarrow \boxed{+10} \rightarrow 8n + 10$

   c $n \rightarrow \boxed{\div 2} \rightarrow \boxed{+3} \rightarrow \frac{n}{2} + 3$

   d $n \rightarrow \boxed{\times 2} \rightarrow \boxed{+5} \rightarrow \boxed{\times 3} \rightarrow 3(2n + 5)$

   e $n \rightarrow \boxed{\times 2} \rightarrow \boxed{-4} \rightarrow \boxed{\times 5} \rightarrow 5(2n - 4)$

   f $n \rightarrow \boxed{+4} \rightarrow \boxed{\div 7} \rightarrow \frac{(n + 4)}{7}$

21 3   **22** 1   **23** n **24** n

25 $2a + b + c$ **26** $2n^2$ **27** $2mn$ **28** n^2

29 3   **30** a^3 **31** n

32 $3t - 3p + 3$ **33** 1   **34** $4n + 2$ **35** $2n + 8$

page 73 **Exercise 8** ①

1 4 kg   **2** 3 kg   **3** 3 kg   **4** 2 kg   **5** 3 kg

6 4 kg   **7** 4 kg   **8** 3 kg   **9** 2 kg   **10** 2 kg

page 74 **Exercise 9** ①

1 3   **2** 17   **3** 14   **4** 16   **5** 7

6 7   **7** 3   **8** 13   **9** 4   **10** 0

11 31   **12** 8   **13** 10   **14** 8   **15** 8

16 3   **17** 5   **18** 3   **19** 2   **20** 4

21 1   **22** 0   **23** 2   **24** 1

page 75 **Exercise 10** ①

1 $\frac{4}{5}$ **2** $2\frac{1}{3}$ **3** $7\frac{1}{2}$ **4** $1\frac{5}{6}$ **5** 0

6 $\frac{5}{9}$ **7** 1   **8** $\frac{1}{5}$ **9** $\frac{2}{7}$ **10** $\frac{3}{4}$

11 $\frac{2}{3}$ **12** $1\frac{1}{4}$ **13** $1\frac{1}{5}$ **14** $1\frac{5}{9}$ **15** $\frac{1}{3}$

16 $\frac{1}{2}$ **17** $\frac{1}{10}$ **18** $-\frac{3}{8}$ **19** $\frac{9}{50}$ **20** $\frac{1}{2}$

21 $\frac{3}{5}$ **22** $-\frac{4}{9}$ **23** 0   **24** $4\frac{5}{8}$ **25** $-1\frac{3}{7}$

26 $2\frac{1}{3}$ **27** $\frac{3}{4}$ **28** 1   **29** $3\frac{3}{5}$ **30** $\frac{1}{3}$

31 $2\frac{1}{14}$ **32** -1 **33** $-\frac{5}{6}$ **34** $8\frac{1}{4}$ **35** -55

page 75 Exercise 11 ②

1 $2\frac{3}{4}$ **2** $1\frac{2}{3}$ **3** 2 **4** $\frac{1}{5}$ **5** $\frac{1}{2}$

6 2 **7** $5\frac{1}{3}$ **8** $1\frac{1}{5}$ **9** 0 **10** $\frac{2}{9}$

11 $1\frac{1}{2}$ **12** $\frac{1}{6}$ **13** $1\frac{1}{3}$ **14** $\frac{6}{7}$ **15** $\frac{4}{7}$

16 7 **17** $\frac{5}{8}$ **18** 5 **19** $\frac{2}{5}$ **20** $\frac{1}{3}$

21 4 **22** -1 **23** 1 **24** $\frac{6}{7}$ **25** $1\frac{1}{4}$

26 1 **27** $\frac{7}{9}$ **28** $-1\frac{1}{2}$ **29** $\frac{2}{9}$ **30** $-1\frac{1}{2}$

page 76 Exercise 12 ②

1 3 **2** 5 **3** $10\frac{1}{2}$ **4** -8 **5** $\frac{1}{3}$

6 $-4\frac{1}{2}$ **7** $3\frac{1}{3}$ **8** $3\frac{1}{2}$ **9** $3\frac{2}{3}$ **10** -2

11 $-5\frac{1}{2}$ **12** $4\frac{1}{5}$ **13** $\frac{3}{7}$ **14** $\frac{7}{11}$ **15** $4\frac{4}{5}$

16 5 **17** 9 **18** $-2\frac{1}{3}$ **19** $\frac{2}{5}$ **20** $\frac{3}{5}$

21 -1 **22** 13 **23** 9 **24** $4\frac{1}{2}$ **25** $3\frac{1}{3}$

page 77 Exercise 13 ②

1 $\frac{3}{5}$ **2** $\frac{4}{7}$ **3** $\frac{11}{12}$ **4** $\frac{6}{11}$ **5** $\frac{2}{3}$

6 $\frac{5}{9}$ **7** $\frac{7}{9}$ **8** $1\frac{1}{3}$ **9** $\frac{1}{2}$ **10** $\frac{2}{3}$

11 3 **12** $1\frac{1}{2}$ **13** 24 **14** 15 **15** -10

16 21 **17** 21 **18** $2\frac{2}{3}$ **19** $4\frac{3}{8}$ **20** $1\frac{1}{2}$

21 $3\frac{3}{4}$ **22** 11 **23** 13 **24** 10 **25** $\frac{1}{3}$

26 1 **27** $3\frac{2}{3}$ **28** 28 **29** $4\frac{1}{2}$ **30** 1

31 220 **32** -500 **33** $-\frac{98}{99}$ **34** 6 **35** 30

36 $1\frac{1}{2}$ **37** 84 **38** 6 **39** $\frac{5}{7}$ **40** $\frac{3}{5}$

page 77 Exercise 14 ①

1 3 **2** $\frac{3}{4}$ **3** $4\frac{1}{2}$ **4** $-\frac{3}{10}$

5 $-\frac{1}{2}$ **6** $17\frac{2}{3}$ **7** $\frac{1}{6}$ **8** 5

9 12 **10** $3\frac{1}{3}$ **11** $4\frac{2}{3}$ **12** -9

page 79 Exercise 15 ① ②

1 $\frac{3}{4}$ **2** $1\frac{1}{2}$ **3** $1\frac{3}{8}$ cm   **4** $1\frac{1}{4}$ cm   **5** 7 cm

6 a $3\frac{3}{5}$ **b** 0 **c** 7

7 a 41 **b** 31 **c** 65

8 29 **9** $55, 56, 57$ **10** $41, 42, 43, 44$

11 a i $x - 3$ cm   **ii** $2x - 6$ cm   **b** $12 \cdot 5$

12 $x = 8$; perimeter $= 60$

13 a -6 **b** 2 **c** A and C

14 11 **15** £6   **16** 3

page 81 Exercise 16 ②

1 a £240   **b** £440   **2** 60 cm

3 a 15 cm   **b** 33 cm

4 a 30 cm²   **b** 16 cm²   **c** 35 cm²

5 a 4000 grams   **b** 6100 grams   **c** 7750 grams

6 a £20   **b** £60

page 82 Exercise 17 ②

1 a 14 cm   **b** 16 cm   **2 a** 10 cm² **b** 21 cm²

3 a 13   **b** 53   **c** 33

4 a 60   **b** 140   **c** 88

5 a 56   **b** 17   **c** 120   **d** 74

6 a 600   **b** 300

7 a 1 hr 30 min   **b** $T = 8n + 50$

page 84 Exercise 18 ②

1 36 **2** 29 **3** 8 **4** 18

5 84 **6** 52 **7** 165 **8** 181

9 $1 \cdot 62$ (2 dp)   **10** 650

page 85 Exercise 19 ①

1 11 **2** 29 **3** 33 **4** 3 **5** 45

6 6 **7** 9 **8** 46 **9** 304 **10** 10

11 a 11 **b** 18 **c** 27

12 a 3 **b** $3 \cdot 5$ **c** 3

13 a 2 **b** 6 **c** 12

14 a 9 **b** 0 **c** 1 **d** 5 **e** 1 **f** 10

page 85 Exercise 20 ①

1 2 **2** 14 **3** 0 **4** 4 **5** 6

6 -6 **7** 17 **8** 2 **9** 4 **10** 1

11 24 **12** 13 **13** 3 **14** -1 **15** 17

16 -9 **17** 30 **18** 10 **19** 3 **20** -3

21 45 **22** 23 **23** 41 **24** 9 **25** 4

26 8 **27** 21 **28** -12 **29** 14 **30** 27

31 -7 **32** 8 **33** 9 **34** 25

page 86 Exercise 21 ①

1 4 **2** 4 **3** 9 **4** 16 **5** 8

6 -8 **7** -27 **8** 64 **9** 8 **10** 16

11 8 **12** 16 **13** 18 **14** 36 **15** 48

16 16 **17** 20 **18** 54 **19** 144 **20** 24

21 13 **22** 10 **23** 1 **24** 18 **25** 13

26 19 **27** 10 **28** 32 **29** 16 **30** 144

31 36 **32** 36 **33** 4 **34** 1 **35** 2

36 -14 **37** -5 **38** -5 **39** -10 **40** 10

41 0 **42** 4 **43** 50 **44** 4 **45** -10

46 -4 **47** -6 **48** -16 **49** 28 **50** 44

page 87 **Exercise 22** ②

1 2	**2** $\frac{1}{2}$	**3** 0	**4** 18
5 −3	**6** 8	**7** 26	**8** −24
9 2	**10** 0	**11** $\frac{1}{4}$	**12** $-\frac{2}{3}$
13 17	**14** $\frac{1}{3}$	**15** −3	**16** $\frac{3}{4}$

page 87 **Exercise 23** ①

1 a 16	**b** 14	**c** 3	**d** 17	
2 13	**3** 20	**4** 14	**5** 5	
6 25	**7** 7	**8** 26	**9** 45	
10 11	**11** 32	**12** 81	**13** 20 000	
14 5	**15** 12·5	**16** 121	**17** 27	
18 18	**19** 6	**20** 16	**21** 9	
22 25	**23** 3, 81	**24** 0, 32		

25 0·1, 1000

page 88 **Exercise 24** ①

1 a 2, 5, 8, 11, 14   **b** 30, 25, 20, 15, 10
   c 1, 2, 4, 8, 16   **d** 1, 10, 100, 1000, 10 000
   e 35, 28, 21, 14, 7   **f** 64, 32, 16, 8, 4
   g −10, −8, −6, −4, −2
2 11, 16, 21, 26, 31
3 a 96, 98, 100, 102, 104   **b** 100, 99, 88, 77, 66
   c 10, 20, 40, 80, 160   **d** −6, −3, 0, 3, 6
4 a Add 3   **b** Subtract 5   **c** Multiply by 2
   d Add 7   **e** Subtract 4
5 a 16   **b** 10   **c** 3000   **d** 2
   e 16   **f** 0   **g** 14   **h** 8
6 a Add 4   **b** Subtract 10   **c** Multiply by 4
   d Divide by 2   **e** Subtract 5   **f** Add 1
   g Divide by 2   **h** Divide by 10

page 89 **Exercise 25** ① ②

1 a 15   **b** 30   **c** 19   **d** 16
   e 40   **f** 13   **g** $\frac{1}{2}$ **h** 14
2 13, 21, 34, 55
3 a 11, 18, 29, 47, 76, 123   **b** 12, 19, 31, 50
4 a $6 \times 7 = 6 + 6^2$ **b** $10 + 10^2$
     $7 \times 8 = 7 + 7^2$ $30 + 30^2$
5 $5 + 9 \times 1234 = 11\,111$
    $6 + 9 \times 12\,345 = 111\,111$
    $7 + 9 \times 123\,456 = 1\,111\,111$
6 63, 3968   **7** 3, 5, 5
8 a 16   **b** 15   **c** 26   **d** 25
9 a $1^3 + 2^3 + 3^3 + 4^3 = (1 + 2 + 3 + 4)^2 = 100$
     $1^3 + 2^3 + 3^3 + 4^3 + 5^3 = (1 + 2 + 3 + 4 + 5)^2 = 225$
     $1^3 + 2^3 + 3^3 + 4^3 + 5^3 + 6^3$
     $= (1 + 2 + 3 + 4 + 5 + 6)^2 = 441$
   b 3025
10 a Yes   **b i** 5   **ii** 10   **iii** 1331

page 92 **Exercise 26** ① ②

1 a $2n$ **b** $10n$ **c** $3n$ **d** $11n$
   e $100n$ **f** $6n$ **g** $22n$ **h** $30n$
2 a $11 \to 55$ **b** $20 \to 180$ **c** $12 \to 1200$
    $n \to 5n$ $n \to 9n$ $n \to 100n$
3 a $10 \to 20 \to 23$ **b** $20 \to 60 \to 61$
    $n \to 2n \to 2n + 3$ $n \to 3n \to 3n + 1$
4 a $3 \to 18 \to 20$ **b** $3 \to 15 \to 13$
    $4 \to 24 \to 26$ $4 \to 20 \to 18$

5 a

Term number (n)		$5n$		Term
1	→	5	→	6
2	→	10	→	11
3	→	15	→	16
4	→	20	→	21
5	→	25	→	26
⋮		⋮		⋮
n	→	$5n$	→	$5n + 1$

   b i 51   **iii** $5n + 1$

6 a A

Term number (n)		$3n$		Term
1	→	3	→	1
2	→	6	→	4
3	→	9	→	7
⋮				
n	→	$3n$	→	$3n - 2$

B

Term number (n)		$4n$		Term
1	→	4	→	6
2	→	8	→	10
3	→	12	→	14
⋮				
n	→	$4n$	→	$4n + 2$

C

Term number (n)		$7n$		Term
1	→	7	→	5
2	→	14	→	12
3	→	21	→	19
⋮				
n	→	$7n$	→	$7n - 2$

b **A** Multiply by 3 and subtract 2
 B Multiply by 4 and add 2
 C Multiply by 7 and subtract 2
 c (A) $3n - 2$ (B) $4n + 2$ (C) $7n - 2$
7 a A10 = 80 **b** B10 = 76 **c** $8n$ **d** $8n - 4$
8 a 79 **b** 81 **c** $4n - 1$ **d** $4n + 1$

page 95 **Exercise 27** ❶

1 $w = r + 4$ **2** $w = 2r + 6$

3

r	w
8	4
9	6
10	8
11	10

$w = 2r - 12$

4

t	m
1	3
2	5
3	2
4	9
5	11
6	13
7	15
8	17
9	19
10	21

$m = 2t + 1$

5

t	m
1	5
2	8
3	11
4	14

$m = 3t + 2$

6 $t = s - 2$ or $s = t + 2$
7 a $p = 5n - 2$ **b** $k = 7n + 3$ **c** $w = 2n + 11$
8 $m = 8c + 4$

page 97 **Test yourself**

1 a $3x - 18$ **b** $5(y - 2)$ **c** $13 - 3w$
2 $y + y \rightarrow 2y$ $5y - y \rightarrow 4y$ $y + 2y \rightarrow 2y + y$
 $y \div 2 \rightarrow \dfrac{y}{2}$
3 a i $x = 7$ **ii** $x = 3$ **iii** $x = 2\cdot5$
 b i $y = 41$ **ii** $M = 4$
4 a $2p$ **b i** $x = 75$ **ii** $y = 2$
 c 2 **d** $M = 9$
5 a $x = 10$ **b** $y = 8\cdot5$ **c** $w = 32$ **d** $6 + 3t$
6 a $5d$ **b** $2y^2$ **c** $12a - 28$
 d t^3 **e** m^2
7 a i 25 **ii** Add 6
 b i 48 **ii** Multiply by 2
8 a $3a + 2b$ **b i** x^2y **ii** $2x^2 + 4xy$
9 a $y = 3k - 1$ **b** 5
10 a $3a + 2b$ **b** $6\dfrac{1}{2}$ **c i** -7 **ii** $4\dfrac{1}{2}$
11 a 14 **b** 53
 c $k = 2 \times$ any even number will demonstrate
 this, e.g. $k = 20 \Rightarrow \dfrac{1}{2}k + 1 = 11$ which is odd.
12 a $1 + 3 + 5 + 7 + 9 = 25$,
 $1 + 3 + 5 + 7 + 9 + 11 = 36$
 b Square numbers

13 a Bryani, because $4x^2$ means square x,
 then multiply by 4.
 b 64
14 Angela **15** $3x - 2 = 13$ **16** 37
17 a 75 **b** 3
18 a i 98p **ii** £4·12 **b** 6

3 Geometry 1

page 102 **Exercise 1** ❶

1 A(5, 1) B(1, 4) C(4, 4) D(1, 2) E(7, 3)
 F(3, 0) G(2, 1) H(0, 3) I(6, 5) J(6, 0)
 K(3, 5)
2 2: (1, 6), (2, 7), (4, 7), (5, 6), (5, 4), (1, 0), (5, 0)
 S: (10, 2), (11, 1), (12, 1), (13, 2), (13, 3), (12, 4),
 (11, 4), (10, 5), (10, 6), (11, 7), (12, 7), (13, 6)
3 A Parrot

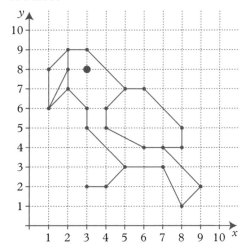

4 A Mouse

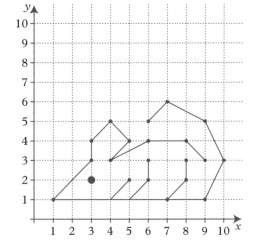

5 A hollowed out cuboid

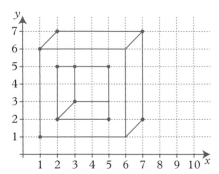

6 A car

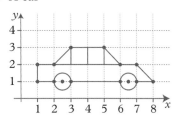

7 What do you call a man with a spade? Doug.

page 103 **Exercise 2** ➊

1 A(3, − 2) B(4, 3) C(−2, 3) D(−4, − 2)
E(5, − 3) F(− 4, 2) G(1, −4) H(−2, −4)
I(−2, 1)

2 A face

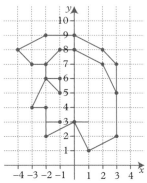

3

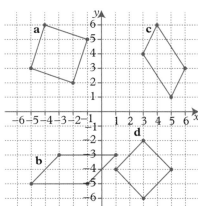

4 a (−4, 0) **b** (1, 0) **c** (3, 1)
d (2, 1). Another possible answer is (0, 7).
5 (4, 6)
6 a (5, 2) **b** $(3, 1\frac{1}{2})$ **c** (6, 5)
d (2, 5) **e** (4, 1) **f** (2, 2)

page 105 **Exercise 3** ➊

1 True **2** True **3** False, 70° **4** False, 80°
5 True **6** True **7** False, 135° **8** False, 75°
9 True **10** True **11** True **12** True
13 True **14** True **15** False, 110° **16** True
17 False, 60° **18** False, 330° **19** True **20** False, 120°

page 106 **Exercise 4** ➊

1 Acute **2** Right angle **3** Acute
4 Acute **5** Obtuse **6** Reflex
7 Obtuse **8** Acute **9** Obtuse
10 Reflex **11** Obtuse **12** Reflex
13 Right angle **14** Reflex **15** Obtuse
16 Acute **17** Acute **18** Reflex
19 Acute **20** Obtuse

page 107 **Exercise 5** ➊

1 70° **2** 100° **3** 70° **4** 100°
5 55° **6** 70° **7** 70° **8** $33\frac{1}{3}$°
9 30° **10** 35° **11** 155° **12** 125°
13 44° **14** 80° **15** 40° **16** 48°
17 40° **18** 35° **19** $a = 40°, b = 140°$
20 $y = 72°, × = 108°$

page 109 **Exercise 6** ➊

1 50° **2** 70° **3** 40° **4** 26°
5 130° **6** 73° **7** 18° **8** 75°
9 29° **10** 30° **11** 70° **12** 42°
13 120° **14** 100° **15** 45° **16** 72°
17 40° **18** $s = 55°, t = 70°$
19 $u = 72°, v = 36°$ **20** $w = z = 55°$
21 140° **22** 75° **23** 60°
24 $d = 122°, y = 116°$ **25** 135° **26** 75°
27 30°, 60°, 90° **28** 28°

page 111 **Exercise 7** ➊

1 72° **2** 98° **3** 80° **4** 74°
5 86° **6** 88° **7** $x = 95°, y = 50°$
8 $a = 87°, b = 74°$ **9** $a = 65°, c = 103°$
10 $a = 68°, b = 42°$ **11** $y = 65°, z = 50°$
12 $a = 55°, b = 75°, c = 50°$

page 111 **Exercise 8** ➊

1 42° **2** 68° **3** 100° **4** 73° **5** 120°
6 64° **7** 20° **8** $a = 70°, b = 60°$

9 $x = 58°, y = 109°$
10 $66°$ **11** $65°$ **12** $e = 70°, f = 75°$
13 $x = 72°, y = 36°$
14 $a = 68°, b = 72°, c = 68°$
15 $4°$ **16** $28·5°$ **17** $x = 60°, y = 48°$
18 $a = 65°, b = 40°$ **19** $x = 49°, y = 61°$
20 $a = 60°, b = 40°$ **21** $136°$ **22** $80°$

page 113 Exercise 9 ① ②

1, 2, 3 – Check students' work.

page 115 Exercise 10 ①

1 $40°$ **2** $65°$ **3** $110°$ **4** $72°$
5 A = $60°$, B = $66°$, c = $54°$ AB = $4·8$ cm,
  BC = $5·1$ cm AC = $5·4$ cm
6 D = $38°$, E = $112°$, F = $30°$
  DE = $5·4$ cm, EF = $6·6$ cm, DF = 10 cm
7 G = $48°$, I = $54°$, H = $78°$
  GH = $8·1$ cm, HI = $7·5$ cm, IG = $9·8$cm
8 J = $108°$, K = $35°$, L = $37°$
  JK = $5·4$ cm, KL = $8·5$ cm, LJ = $5·1$ cm
9 Check students' work.

page 117 Exercise 11 ①

1 $7·4$ cm   **2** $7·9$ cm   **3** $8·0$ cm
4 $10·3$ cm   **5** $6·4$ cm   **6** $6·8$ cm
7 $9·0$cm   **8** $9·6$ cm   **9** 7cm
10 $7·4$ cm   **11** $7·6$cm   **12** $9·1$ cm

page 117 Exercise 12 ②

2 Check students' work.   **3** $61°$
4 $85°$ **5** $72°$ **6** $121°$

page 118 Exercise 13 ②

1 Check students' work.
2 Two triangles are possible with
  the angle opposite the 8 cm
  side being $59°$ and $121°$ in
  each case.

3 Two triangles are possible
  with the angle opposite the
  7 cm side being $63°$ and
  $117°$ in each case.

4 AAS, SSA or RHS.

page 119 Exercise 14 ①

1 a, b, d
2

3

4 The net can be completed by any
  one of squares ① – ④

5 a A 6 vertices   5 faces
    B 5 vertices   5 faces
  b A is a prism and B a pyramid.
6 Check students' work.

page 121 Exercise 15 ②

1–8   Check student's work.

page 123 Exercise 16 ① ②

1 A and G, B and O, C and F, H and P, I and N, J and K
2

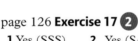

Shapes labelled with same numbers are congruent

3

Many other answers are possible.
4 A and G, C and J, K and M, L and F
5 Any enlargement of I
6 The ratio of the sides for the first figure is $3 : 2$
  whereas for the second figure it is $4 : 1$.

page 126 Exercise 17 ②

1 Yes (SSS)   **2** Yes (SAS)   **3** No   **4** Yes (AAS)
5 No   **6** Yes (AAS)   **7** CDA, DEB, EAC, ABD
8

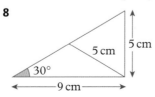

9 AB = BC
AD = DC
BD is common
Therefore, ΔABD ≡ ΔCBD (SSS)
∠A corresponds to ∠C & therefore under the congruency, A = C.

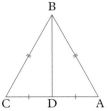

10 ∠OAT = 90° = ∠OBT; OA = OB (radii);
OT is common.
Therefore, ΔOAT ≡ ΔOBT (RHS). Therefore
AT = BT as required.

11 Since LY = LX, LM = LN and angle L is common, the two triangles are congruent (SAS).

12 ∠AXD = ∠BCD (corresponding angles)
∴ ∠AXD = ∠ADX
∴ ΔADX is isosceles (base angles equal)

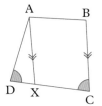

13 ∠NYM = ∠MYZ and
∠MZN = ∠NZY.
But triangle XZY is
isosceles, so ∠Z = ∠Y.
Thus ∠NYM = ∠MYZ
= ∠MZN = ∠NZY.
Therefore, ΔMYZ ≡ ΔNZY (AAS i.e.
∠NZY = ∠MYZ, ∠NYZ = ∠MZY and ZY
is common)
Therefore YM = ZN.

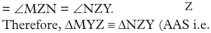

14a DX = XC, so ΔDXC is isosceles, i.e.
angle C = angle D. But angle C = ∠XAB and
angle D = ∠XBA. Therefore,
∠XAB = ∠XBA so ΔABX is isosceles. Therefore
AX = BX.
b AX = BX and XC = XD,
so AX + XC = BX + XD, i.e. AC = BD
c DZ = DV + VZ = ZC + VZ = VC
AC = BD (proven in **b**) and angle
D = angle C as DXC is isosceles.
Therefore, ΔDBZ ≡ ΔCAV (SAS).

15 ∠ABX = ∠XDC   (alternate angles)
∠BAX = ∠XCD   (alternate angles)
AB = CD   (parallelogram)
∴ΔABX ≡ ΔCDX (AAS)
∴ BX = XD and AX = XC

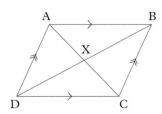

16 PQ = QA
QR = QB
Angle PQR = 90°
Angle AQR = 90° − 60° = 30°
Therefore, ∠AQB = 30° + 60° = 90° = ∠PQR
∴ ΔPQR ≡ ΔAQB (SAS)
∴ PR = AB as required.

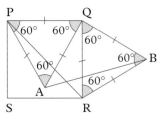

page 127 Exercise 18 ①

1 Students' own drawing
2 b i True   **ii** True
3 You cannot fit pentagons together around a point
with no gaps. You can fit hexagons together at a point.
4a Yes   **b** Yes

page 128 Exercise 19 ①

1 a 1	**b** 1		**2 a** 1	**b** 1
3 a 4	**b** 4		**4 a** 2	**b** 2
5 a 0	**b** 6		**6 a** 0	**b** 2
7 a 0	**b** 2		**8 a** 4	**b** 4
9 a 0	**b** 4		**10 a** 4	**b** 4
11 a 6	**b** 6		**12 a** infinite	**b** infinite

13 a

  or   or

b

14 a 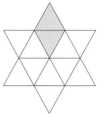   **b**

Other answers are possible.

15 a   **b**

Other answers are possible.

page 130 **Exercise 20** ❶

1   **2**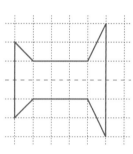

3   **4**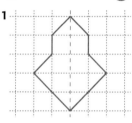

5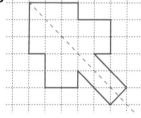

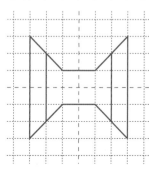

6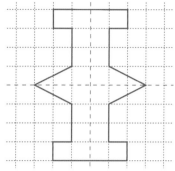

7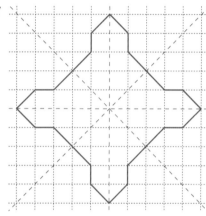

8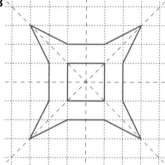

9 Check students' work.

page 132 **Exercise 21** ❶

1 A Isosceles trapezium   **B** Rhombus
  C Rectangle   **D** Trapezium
  E Parallelogram

2

	Diagonals always equal	Diagonals always perpendicular	Diagonals always bisect the angles	Diagonals always bisect each other
Square	✓	✓	✓	✓
Rectangle	✓			✓
Parallelogram				✓
Rhombus	✓	✓	✓	✓
Trapezium				

3 a 30° **b** 90° **c** 115°

4 a 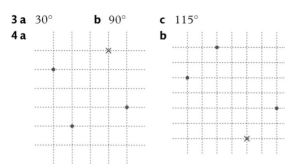 **b**

c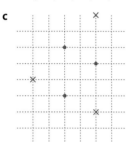

5 Quadrilaterals
6 Square
7 Rectangle
8 Trapezium
9 Equilateral triangle
10 Isosceles triangle
11 Rhombus
12 True
13 False

14 a Rectangle **b** Square **c** Trapezium
d Parallelogram
15 a Trapezium **b** Rectangle **c** Square
d Parallelogram **e** Isosceles triangle
16 Any trapezium with 2 right angles
17 (5, 1) and (0, 0)

page 134 **Exercise 22** ❶
1 a 34° **b** 56°
2 a 35° **b** 35°
3 a 72° **b** 108° **c** 80°
4 a 40° **b** 30° **c** 110°
5 a 116° **b** 32° **c** 58°
6 a 55° **b** 55°
7 a 26° **b** 26° **c** 77°

page 135 **Exercise 23** ❶ ❷
1–3 Check students' work.
4 Multiple answers – check students' work.
5 Multiple answers – check students' work.
6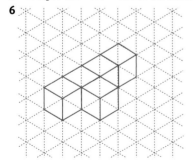

7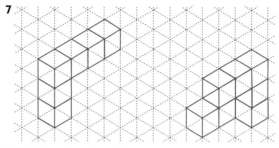

8 Check students' work.

9 B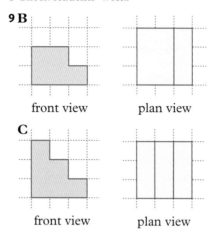

front view   plan view

C

front view   plan view

10 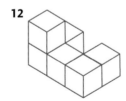 **11**

12 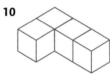 **13**

page 138 **Exercise 24** ❶
1 8 cm² **2** 14 cm² **3** 10 cm² **4** 6 cm²
5 8 cm² **6** 7 cm² **7** $9\frac{1}{2}$ cm² **8** 12 cm²
9 34 cm²
10 a 19 cm² **b** 15 cm² **c** 16 cm² **d** 21 cm²

page 140 **Exercise 25** ❶
1 24 cm² **2** 24 cm² **3** 20 cm² **4** 8 cm²
5 10 cm² **6** 25 cm² **7** 35 cm² **8** 6000 cm²
9 40 000 m² **10** 12 cm² **11** 28 cm²

page 141 **Exercise 26** ❶
1 17 cm² **2** 34 cm² **3** 33 cm² **4** 54 cm²
5 18 cm² **6** 20 cm² **7** 25 cm² **8** 23 cm²
9 31 cm² **10** 35 cm² **11** 54 cm² **12** 51 cm²

13a B **b** **A** = 6 cm^2, **C** = 4 cm^2 **D** = 6 cm^2
  c Multiple answers – check students' work.
14 36 cm^2
15a 4 cm **b** 5 cm **c** 5 cm
16 £340·10 **17** 25 m^2 **18** 248 cm^2

page 143 **Exercise 27** ❶

1 5 cm^2 **2** 14 cm^2 **3** 20 cm^2
4 9 cm^2 **5** 12 cm^2 **6** 25 cm^2
7 14 cm^2 **8** 33 cm^2 **9** 39 cm^2
10 39 cm^2 **11** 15 cm^2 **12** $23\frac{1}{2}$ cm^2
13 33 cm^2
14a 6 cm **b** 5 cm **c** 7 cm
15a b Multiple answers – check students' work.
16 Mark

page 145 **Exercise 28** ❶

1 b A = 10, B = 6, C = 3 square units
  c 36 square units **d** 17 square units
2 b A = 5, B = 14, C = 6 square units
  c 42 square units **d** 17 square units
3 $13\frac{1}{2}$ square units **4** $14\frac{1}{2}$ square units

5 24 square units **6** 21 square units

7 21 square units

page 146 **Exercise 29** ❶

1 42 cm^2 **2** 56 cm^2 **3** 103 cm^2
4 55 cm^2 **5** 143 cm^2 **6** 18 cm^2
7 47 cm^2 **8** 75·75 cm^2

page 147 **Exercise 30** ❶

1 a 20 cm **b** 30 cm **c** 38 cm **d** 28 cm
2 a 12 cm **b** 12 cm **c** 13 cm
3 a 16 cm (A) 14 cm (B) 18 cm (C)
  b Check students' work.
4 a 12 cm **b** 12 cm **c** 12 cm
5 50 cm **6** 8 m **7** 26 cm
8 34 cm **9** 30 cm
10a 5 **b** 4·12 (to 2 dp)
11 40 cm **12** 32 cm **13** 28 cm
14 28 cm

page 149 **Test yourself**

1 a i 40 cm^2 **ii** 28 cm
  b 7 cm, 4 cm **c** 7 cm^2
2 a 8 cm^2 **b** 34 cm
3 x = 115°
4 a i 25° **ii** 130° **b** 65°

5 110 km
6 a 36 **b** 27 cm^3
7 a Shape F; square **b** 9 cm^2
8 a (1, 3) **b** **D** at (5, 3) **c** 14 cm

9 a Scale drawing: 9 cm; 7·5 cm; 7·5 cm
  Yes, he can stand up.
10 a A **b** 2 **c** Tessellating hexagons
11 x = 1·5 cm
12 18°
13a Cylinder **b** Rectangle, 2 cm by 4 cm

4 Statistics 1

page 154 **Exercise 1** ❶

1 Athlete **a** 40 m **b** 80 m **c** 8 m/s
  Cyclist **a** 50 m **b** 100 m **c** 10 m/s
  Car **a** 120 m **b** 240 m **c** 24 m/s
2 a i 10·00 **ii** 08·30 **iii** 10·45
  b i Farnham **ii** Dorking **iii** Reigate
  c 40 mph
3 a i 15·00 **ii** 13·30 **iii** 15·15
  b i Westbury **ii** Southampton
  iii Bristol **iv** Portsmouth
  c 40 km **d** 1 hour **e** 40 km/hour
4 a i 13·30 **ii** 11·00
  b i Stump Cross **ii** Mountfitchet
  c 20 km **d** 1 hour **e** 20 km/hour
5 a i 16·30 **ii** 19·00 **iii** 16·07
  b i Sevenoaks
  ii Chiddingstone Hoath **iii** Maidstone
  c 80 km **d** 80 km/hour **e** 40 km
  f 20 km/hour
6 a i 15·30 **ii** 16·30 **iii** 14·30
  b i Chipping Norton **ii** Furzy Leaze
  iii Long Compton **c** 30 km
  d 20 km/hour **e** 20 km **f** 10 km/hour
7 a i 08·45 **ii** 09·15 **iii** 11·00
  b i Great Bricett **ii** Bury St. Edmunds
  iii Shimpling **c** 30 km
  d 20 km/hour **e** 20 km **f** 10 km/hour
  g 14·3 km/hour (1 dp)
8 a 30 minutes
  b i 30 km/hour
  ii 40 km/hour **iii** 70 km/hour
  c i 12·00 **ii** 10·30 **iii** 09·30
9 a 45 mins **b** 15·15 **c** 14·15
  d i 40 km/hour **ii** 50 km/hour
  iii 70 km/hour **e** 16·15

10 a 40 km **b** 60 km
  c York and Scarborough **d** 15 minutes
  e i 11·00 **ii** 13·45
  f i 40 km/hour **ii** 60 km/hour
    iii 100 km/hour
11 a i 14·00 **ii** 13·45
  b i 15·45 **ii** Towards Aston
  c i 15 km/hour **ii** 40 km/hour
    iii 40 km/hour **iv** 20 km/hour
  d 16·8 km/hr (1 dp)

page 160 Exercise 2 ①

1 a 1 m/s^2 **b** 2 m/s^2 **c** 5 m/s^2 Straight line
2 a $P: 1 \text{ m/s}^2$ $Q: 2 \text{ m/s}^2$ $R: -2 \text{ m/s}^2$
  b R is slowing down
3 a 0 s, 12 s **b** 16 s **c** 6 s
d

	Acceleration, m/s² given by		
	OA	**AB**	**BC**
P	3	0	$-1\frac{1}{2}$
Q	$\frac{1}{3}$	0	$-1\frac{1}{2}$

4

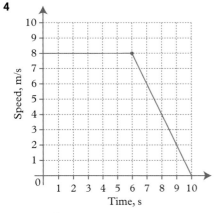

-2 m/s^2

5

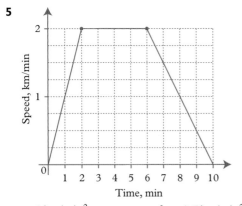

a 1 km/min^2 **b** -0.5 km/min^2

6

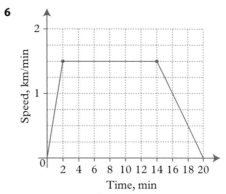

a 0.75 km/min^2 **b** -0.25 km/min^2

page 162 Exercise 3 ① ②

1 a i £2·80 **ii** £2·30
  b i €4·25 **ii** €2 **c** £35 000
2 a i £25 **ii** £44 **iii** £30
    iv £40 **v** £10 **vi** £15
  b i $32 **ii** $80 **iii** $10
    iv $26 **v** $58 **vi** $54
  c £45
3 a i £3·60 **ii** £8·50 **iii** £5·40
    iv £4·90 **v** £7·70 **vi** £0·50
  b i 620 rupees **ii** 580 rupees
    iii 360 rupees **iv** 160 rupees
    v 780 rupees **vi** 515 rupees
  c 700 rupees
4 a 11·2 cm **b** May and October
  c November, could have been heavy rainfall, or snow followed by a thaw.
5 a 15·2°C **b** October **c** April and November
  d September and October **e** 21°C
6 a 30 litres
  b i 8 miles per litre
    ii n 6 miles per litre
7 a

X	0	50	100	150	200	250	300
C	35	45	55	65	75	85	95

b

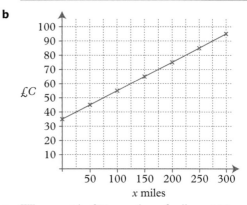

c When cost is £71, number of miles = 180.

8 a

h	0	1	2	3
C	60	105	150	195

b

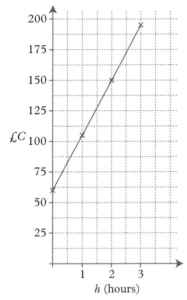

c When $C = £127·50$, Jeff worked $1\frac{1}{2}$ hours.

page 165 Exercise 4 ①

1 B

2 a D   **b** C **c**   F
   d E   **e** A **f**   B

3 D

4 a C   **b** A   **c** D   **d** B

5 a i B   **ii** A
   b from $7\frac{1}{2}$ secs to $18\frac{1}{2}$ secs **c** B **d**   A

6 X → B   Y → C   Z → A

7 P

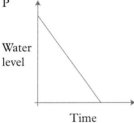

Q

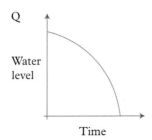

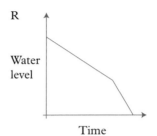

R

Water level / Time

8

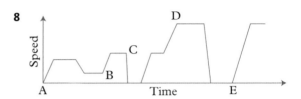

page 170 Exercise 5 ①

1 a 7   **b** 5   **c** 15   **d** 6·5
   e − 1   **f** 7   **g** 3

2 a 16 years **b** 170 cm **c** 50 kg

3 a 5   **b** 15

4 3·5   Sally does with 20p.

5 Multiple answers possible. **6** 17 kg

page 172 Exercise 6 ①

1 a 1   **b** 6   **c** 4   **d** 4
   e red   **f** Z

2 6

3 a 6   **b** 16°C   **c** 31

4 a 3   **b** 3

5 Multiple answers possible.

6

Number of chicks	Tally	Frequency
0	‖	2
1	‖‖	4
2	‖‖‖	6
3	‖‖‖‖	7
4	‖‖‖‖ ‖‖‖‖	9
5	‖‖‖	3
6	‖	1

Modal number of chicks per nest = 4

page 173 Exercise 7 ①

1 a 7   **b** 10 **c**   4
   d 7·8   **e** 1·3   **f** 4

2 Girls = 7   Boys = $6\frac{2}{3}$ (or 6·7 to 1 dp)

3 44·2 kg

4 Multiple answers possible.

5 a 10   **b** 13

6 a 5   **b** 35

7 5·9 grams   **8** 70 kg   **9** 161 cm

page 174 **Exercise 8** ①

1 $10 - 2 = 8$

2 a 26   **b** 49   **c** 41

3 7°C

4 Multiple answers possible.

5 a Impossible   **b** Possible   **c** True

6 a Impossible   **b** Possible   **c** Possible

7 a 17   **b** 10 to 19

   c We do not actually know the highest or lowest scores obtained.

8 a 15; 13; 5; exclude   **b** 3; 6; 2; exclude

   c 67; 38; 13; exclude   **d** 2; 12; 5; exclude

   e 9; 9; 3; exclude

page 176 **Exercise 9** ①

1

	Mean	Median	Mode
a	6	5	4
b	9	7	7
c	6·5	8	9
d	3·5	3·5	4

   e Median, as it is central to the data and not affected by the one very high value of 12.

2 a 2°C   **b** 5°C

3 Mean = 17   Median = 3

   Median best describes the set.

4 a 11   **b** 4

5 2 or 45   **6** 4

7 6

8 a 1·60 m   **b** 1·634 m

9 a 7·2   **b** 5   **c** 6

10 £2·10

11 Multiple answers possible. e.g. 4 4 5 7 10

12 a False   **b** Possible   **c** False

   d Possible

13 a Mean = £47,920   Median = £22,500

   b The mean does not give a fair average, as 4 of the 5 employees earn less than half of this amount.

14 a Mean = 157·1 kg (1 dp) Median = 91 kg

   b The farmer has used the mean weight. Not fair, as 25 of the 32 cattle are less than this weight.

page 179 **Exercise 10** ①

1 96·25 g   **2** 51·9 p   **3** 4·82 cm

4 a Mean = 3·025   Median = 3   Mode = 3

   b Mean = 17·75   Median = 17   Mode = 17

5 a Mean = 6·6 (1 dp)   Median = 8   Mode = 3

   b The mode

6 a 9   **b** 15

page 181 **Exercise 11** ①

1 a

Number of words	Frequency f	Midpoint x	fx
1–5	6	3	18
6–10	5	8	40
11–15	4	13	52
16–20	2	18	36
21–25	3	23	69
Totals	20	–	215

   b 10·75 words

2 68·25   **3** 3·8 letters (1 dp)

page 182 **Exercise 12** ①

1 a Population   **b** Sample and Population

   c Sample   **d** Population   **e** Sample

2 Check students' own answers

3 a categorical   **b** continuous   **c** discrete

   d discrete   **e** continuous   **f** continuous

   g categorical   **h** discrete

4 a primary   **b** secondary   **c** primary

   d primary   **e** secondary

page 183 **Exercise 13** ①

1 a Sharon   **b** £11   **c** Half of a £ symbol

2 a 2

   b

Make	Number of cars	
Ford	4	🚗🚗
Renault	6	🚗🚗🚗
Toyota	6	🚗🚗🚗
Audi	3	🚗🚗

3 Any suitable pictogram – check students' work.

page 184 **Exercise 14** ①

1 $\frac{5}{6}$ **2** $\frac{3}{4}$ **3** $\frac{1}{2}$ **4** $\frac{4}{5}$

5 $\frac{3}{5}$ **6** $\frac{5}{6}$ **7** $\frac{4}{5}$ **8** $\frac{1}{3}$

9 $\frac{1}{4}$ **10** $\frac{5}{12}$ **11** $\frac{1}{2}$ **12** $\frac{2}{3}$

13 $\frac{5}{6}$ **14** $\frac{1}{6}$ **15** $\frac{1}{8}$ **16** $\frac{1}{9}$

17 $\frac{2}{9}$ **18** $\frac{7}{12}$ **19** $\frac{7}{72}$ **20** $\frac{3}{20}$

21 12   **22** 45   **23** 40   **24** 20

25 11   **26** 36   **27** 240°   **28** 300°

29 135°   **30** 30°   **31** 252°   **32** 150°

page 185 **Exercise 15** ①

1 a i $\frac{1}{6}$ **ii** $\frac{1}{3}$ **iii** $\frac{1}{8}$ **iv** $\frac{3}{8}$

   b i $\frac{1}{6}$ **ii** $\frac{1}{3}$ **iii** $\frac{1}{8}$ **iv** $\frac{3}{8}$

   c i 20   **ii** 40   **iii** 15

   d i 20   **ii** 40   **iii** 15   **iv** 45

2 a i $\frac{1}{4}$ **ii** $\frac{1}{6}$ **iii** $\frac{1}{3}$

b i $\frac{1}{4}$ **ii** $\frac{1}{6}$ **iii** $\frac{1}{4}$ **iv** $\frac{1}{3}$

c i 9 **ii** 6 **iii** 9 **iv** 12

3 a i $\frac{1}{8}$ **ii** $\frac{1}{6}$ **iii** $\frac{5}{12}$

iv $\frac{1}{12}$ **v** $\frac{1}{8}$ **vi** $\frac{1}{12}$

b i £45 **ii** £60 **iii** £150 **iv** £30

4 a £425 **b** £150 **c** £250 **d** £75

5 i £13,333 (nearest pound) **ii** £15,000

iii £6,667 (nearest pound) **iv** £10,000

v £12,000 **b** Rent **c** Other items

6 a i 8 mins **ii** 34 mins **iii** 10 mins **b** 18°

page 187 Exercise 16 ①

1 a i 45° **ii** 200° **iii** 110° **iv** 5°

b Check pie chart has a title and that angles are
Spurs 45°, Chelsea 200° Man Utd 110°, York 5°.

2 a $\frac{3}{10}, \frac{2}{5}, \frac{1}{5}, \frac{1}{10}$

b Check pie chart has a title and that angles are
sixth form 108°, employment 144°, FE colleges
72°, unemployed 36°.

3 $x = 60°$, $y = 210°$

4 Check pie chart has a title and that angles are Oats
90°, Barley = 60°, Wheat = 45°, Rye = 165°.

5 a Number of boys choosing red = 20
Number of girls choosing red = 12
∴ Tony is wrong.

b Number of boys choosing blue = 25
Number of girls choosing blue = 15
∴ Mel is right.

page 189 Exercise 17 ①

1 a Beef **b** Rabbit **c** 35 **d** $\frac{1}{7}$

e

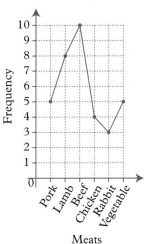

2 a, d

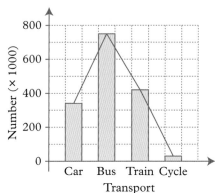

b 1000 **c** 63%

3 a Year 7 **b** Approx 400(397)

c Year 9 **d** 1320

4 a

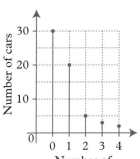

b Bus **c** 50%

5 a

Hair colour	Blonde	Auburn	Brown	Black	Grey
Frequency	3	4	6	5	2

b Brown **c** 20 **d** 15%

6 a

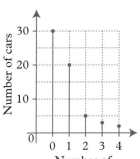

b 60 cars **c** 0 **d** 50% **e** Yes

7 a

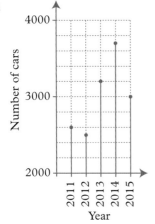

b 2014   **c** 15 000 cars   **d** 20%

page 192 **Exercise 18** ❶

1 squares: shaded 33·3%, unshaded 66·6%
triangles: shaded 50%, unshaded 50%

2 a

	Football	**Hockey**	**Swimming**	**Total**
Boys	5	2	1	8
Girls	2	3	2	7
Total	7	5	3	15

b 8   **c** 25%

3 a

	Girls	**Boys**	**Total**
Can Cycle	95	120	215
Cannot Cycle	179	82	216
Total	274	202	476

b 35%   **c** 59%

4 a

	Men	**Women**	**Total**
Had accident	75	88	163
Had no accident	507	820	1327
Total	582	908	1490

b 13%   **c** 10%
d Men are more likely to have accidents.

page 193 **Exercise 19** ❷

1 a July   **b** Upwards
   c July, December; Spending on his summer holiday and Christmas
2 a 104·5°F   **b** 27 hours
   c 6·6 degrees   **d** downward
3 a 18°C   **b** 19°C, 19·4°C
   d The trend is neither up nor down

page 196 **Exercise 20** ❶

1

```
1 | 5
2 | 3  7  8  9
3 | 2  5  6  8  9
4 | 0  1  2  5  6  7  8
5 | 1  2  3  4  9
6 | 5  6
```

2 a

```
2 | 0  4  5  8
3 | 1  7  9
4 | 0  4  6
5 | 2  5  8  9
6 | 1  5  7  8
7 | 3  5
```

b

```
2 | 2  8  9
3 | 0  5  8
4 | 1  4  6  7  7
5 | 3  4  9
6 | 7
7 | 2
```

3 a 50 kg   **b** 15 **c**   50 kg

4   **a** 4·5   **b** 5·3

```
1 | 4  8
2 | 4  4  8
3 | 1  3  3  7  8
4 | 0  5  5  6  6  7  9
5 | 1  2  5  8
6 | 2  3  7
```

5 a 13   **b** 78
   c The pulse rate of women is on average higher than the pulse rate of men.

page 197 **Exercise 21** ❶

1 a Pearce family: Median 54 kg, Range 20 kg; Taylor family: Median 63 kg, Range 40 kg.
   b The median weight for the Pearce family is less than that for the Taylor family and the range for the Taylor family is much greater. The weights of the Taylor family are more spread out.
2 a Class 10 M: Median 73, Range 31;
Class 10 S: Median 58, Range 47
   b The median for Class 10 M is much higher than that for Class 10 S and the range is much less so the marks in Class 10 M are higher and less spread out.

page 198 **Test yourself**

1 a   **b** 10 boys

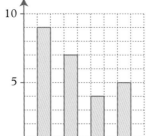

2 a 2   **b** 4

   c The median, as there are some extreme values.

3 a 6   **b** 6·2

4 a

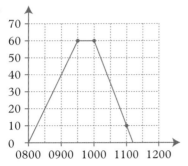

   b Yes

5 a 15   **b** 10

6

Coin	Tally	Frequency
£1	|||	3
50p		0
20p	|||||		6
10p	|||	3
5p	||	2
2p		0
1p	||	2

7 a 26 min   **b** 19 min   **c** No, higher median.

8 B; C; A

9 a 38·6 min

   b i December   **ii** November

5   Number 2

page 202 Exercise 1 ①

1 805	**2** 459	**3** 650	**4** 1333
5 2745	**6** 1248	**7** 4522	**8** 30 368
9 28 224	**10** 8568	**11** 46 800	**12** 66 281
13 57 602	**14** 89 516	**15** 97 525	

page 203 Exercise 2 ①

1 32	**2** 25	**3** 18	**4** 13
5 35	**6** 22 r 2	**7** 23 r 24	**8** 18 r 10
9 27 r 18	**10** 13 r 31	**11** 35 r 6	**12** 23 r 24
13 64 r 37	**14** 151 r 17	**15** 2961 r 15	

page 203 Exercise 3 ①

1 £47·04   **2** 46   **3** 7592

4 15 stamps, 20p change   **5** 8

6 £85   **7** £14 million   **8** £80·64

page 205 Exercise 4 ①

1 a a quarter, $\frac{1}{4}$ **b** a half, $\frac{1}{2}$

   c four fifths, $\frac{4}{5}$ **d** three quarters, $\frac{3}{4}$

   e one tenth, $\frac{1}{10}$ **f** one third, $\frac{1}{3}$

   g two thirds, $\frac{2}{3}$ **h** three tenths, $\frac{3}{10}$

2 a A $= \frac{1}{4}$, B $= \frac{3}{4}$ **b** C $= \frac{4}{10}$, D $= \frac{9}{10}$

   c E $= \frac{1}{3}$, F $= \frac{2}{3}$

3 Multiple answers possible – check students' work.

page 206 Exercise 5 ①

1 2	**2** 2	**3** 2	**4** 6
5 5	**6** 4	**7** 6	**8** 8

9 a $\frac{1}{2} = \frac{2}{4} = \frac{4}{8} = \frac{8}{16}$ **b** $\frac{1}{2} = \frac{2}{4} = \frac{3}{6} = \frac{4}{8} = \frac{5}{10}$

10 $\frac{3}{4}$	**11** $\frac{1}{4}$	**12** $\frac{4}{5}$	**13** $\frac{2}{5}$	**14** $\frac{1}{4}$
15 $\frac{2}{3}$	**16** $\frac{3}{5}$	**17** $\frac{1}{3}$	**18** $\frac{1}{3}$	**19** $\frac{8}{9}$
20 $\frac{4}{5}$	**21** $\frac{1}{3}$	**22** $\frac{2}{9}$	**23** $\frac{2}{3}$	**24** $\frac{1}{3}$
25 $\frac{2}{5}$	**26** $\frac{3}{5}$	**27** $\frac{7}{8}$	**28** $\frac{3}{5}$	**29** $\frac{7}{9}$
30 $\frac{2}{7}$	**31** $\frac{4}{5}$	**32** $\frac{2}{5}$	**33** $\frac{1}{4}$	**34** $\frac{3}{4}$
35 $\frac{1}{100}$	**36** $\frac{1}{2}$	**37** $\frac{1}{4}$	**38** $\frac{1}{6}$	**39** $\frac{3}{10}$
40 $\frac{5}{12}$	**41** $\frac{7}{30}$	**42** $\frac{1}{10}$	**43** $\frac{1}{2}$	

44 Sam $\frac{5}{8}$ Sue $\frac{3}{8}$

45 a $\frac{4}{15}$ **b** $\frac{2}{5}$ **c** $\frac{1}{2}$ **d** $\frac{3}{4}$ **e** $1\frac{1}{2}$ **f** $1\frac{1}{4}$

46 $\frac{1}{4}$ **47** $\frac{3}{7}$ **48** $\frac{2}{5}; \frac{3}{5}$ **49** $\frac{2}{5}$

50 $1\frac{1}{2}$ **51** $1\frac{1}{2}$ **52** $1\frac{1}{4}$ **53 a** $1\frac{1}{3}$ **b** 3:4

54 a $2\frac{1}{2}$ **b** 5:2   **55 a** $\frac{2}{4}$ **b** $\frac{2}{97}$

56 a $\frac{8}{12}, \frac{6}{12}, \frac{9}{12}$ **b** $\frac{1}{2}, \frac{2}{3}, \frac{3}{4}$

57 a $\frac{10}{12}, \frac{8}{12}, \frac{3}{12}$ **b** $\frac{1}{4}, \frac{2}{3}, \frac{5}{6}$

58 $\frac{1}{2}, \frac{3}{5}, \frac{7}{10}$ **59** $\frac{7}{12}, \frac{5}{6}, \frac{5}{4}$ **60** $\frac{3}{8}, \frac{1}{2}, \frac{3}{4}$

61 $\frac{2}{15}, \frac{1}{3}, \frac{7}{15}$ **62** $\frac{3}{8}, \frac{2}{5}, \frac{3}{7}, \frac{4}{9}$ **63** $\frac{7}{12}, \frac{8}{13}, \frac{11}{17}, \frac{2}{3}$

page 210 **Exercise 6** ①

1 $3\frac{1}{2}$ 2 $1\frac{2}{3}$ 3 $2\frac{1}{3}$ 4 $1\frac{1}{4}$ 5 $2\frac{2}{3}$

6 $1\frac{1}{3}$ 7 3   8 $4\frac{1}{2}$ 9 $2\frac{1}{4}$ 10 5

11 $1\frac{2}{3}$ 12 $1\frac{3}{7}$ 13 $1\frac{5}{8}$ 14 $2\frac{1}{3}$ 15 2

16 12   17 $3\frac{1}{7}$ 18 $1\frac{2}{3}$ 19 $2\frac{2}{5}$ 20 $1\frac{1}{2}$

21 $\frac{5}{4}$ 22 $\frac{4}{3}$ 23 $\frac{9}{4}$ 24 $\frac{8}{3}$ 25 $\frac{15}{8}$

26 $\frac{5}{3}$ 27 $\frac{22}{7}$ 28 $\frac{13}{6}$ 29 $\frac{19}{4}$ 30 $\frac{15}{2}$

31 $\frac{29}{8}$ 32 $\frac{22}{5}$ 33 $\frac{17}{5}$ 34 $\frac{33}{4}$ 35 $\frac{13}{10}$

page 210 **Exercise 7** ①

1

	Fraction of quantity required	Divide the quantity by
a	$\frac{1}{2}$	2
b	$\frac{1}{3}$	3
c	one-quarter	4
d	$\frac{1}{10}$	10
e	$\frac{1}{5}$	5

2 £6   3 £10   4 5 litres   5 5 kg

6 24 cm   7 £4   8 75 miles   9 £10

10 70 kg   11 6 litres   12 3 kg   13 £20

14 6 kg   15 70 eggs   16 110 hens

17 111 miles   18 £12   19 40 pages

20 2, 3, 5, 6, 9, APPLE

page 212 **Exercise 8** ①

1 £46   2 £48   3 9 kg   4 £52

5 48 miles   6 60 cm   7 £144   8 15 kg

9 35 hens   10 300 cm   11 48 pence

12 400 miles   13 28p   14 £2500   15 27p

16 £80   17 16 kg   18 603 km   19 54

20 180   21 £4   22 72

23 12, 15, 16, 18, 30, 32, 40 BECKHAM

page 214 **Exercise 9** ①

1 a $\frac{6}{7}$ b $\frac{5}{9}$ c $\frac{3}{5}$ d 1

e $\frac{5}{8}$ f $\frac{5}{11}$ g $\frac{1}{3}$ h $\frac{4}{15}$

2 a $\frac{2}{8}+\frac{1}{8}=\frac{3}{8}$ b $\frac{5}{10}+\frac{4}{10}=\frac{9}{10}$

c $\frac{6}{15}+\frac{5}{15}=\frac{11}{15}$ d $\frac{5}{10}+\frac{2}{10}=\frac{7}{10}$

e $\frac{6}{8}+\frac{1}{8}=\frac{7}{8}$ f $\frac{3}{6}+\frac{1}{6}=\frac{4}{6}=\frac{2}{3}$

3 a $\frac{5}{6}$ b $\frac{5}{12}$ c $\frac{7}{12}$ d $\frac{9}{20}$

e $1\frac{1}{15}$ f $\frac{13}{20}$ g $\frac{11}{14}$ h $3\frac{3}{4}$

4 a $\frac{1}{6}$ b $\frac{5}{12}$ c $\frac{3}{10}$ d $\frac{5}{12}$

e $\frac{1}{10}$ f $\frac{2}{15}$

5 $\frac{9}{10}$ m   6 $\frac{3}{4}$ 7 $\frac{1}{4}$ 8 5

9 Length $=\frac{7}{8}$ inch, Width $=\frac{1}{2}$ inch

page 216 **Exercise 10** ①

1 a $\frac{2}{15}$ b $\frac{3}{28}$ c $\frac{4}{15}$ d $\frac{5}{42}$

e $\frac{5}{27}$ f $\frac{2}{33}$ g $\frac{7}{32}$ h $\frac{2}{9}$

2 a $\frac{2}{5}$ b $\frac{2}{7}$ c $\frac{1}{6}$ d $\frac{7}{9}$

e $1\frac{1}{5}$ f $\frac{5}{14}$ g $\frac{9}{13}$ h 1

3 a $\frac{3}{4}$ b $\frac{7}{8}$ c $\frac{3}{4}$ d $\frac{7}{10}$

e $1\frac{1}{2}$ f $1\frac{1}{5}$ g $3\frac{1}{2}$ h 6

4 a $1\frac{1}{2}$ b $1\frac{4}{5}$ c $3\frac{1}{3}$ d $2\frac{2}{3}$

e $\frac{5}{18}$ f $\frac{5}{21}$ g $\frac{3}{10}$ h $\frac{4}{35}$

5

×	$\frac{2}{3}$	$\frac{3}{4}$	$\frac{1}{5}$
$\frac{1}{2}$	$\frac{1}{3}$	$\frac{3}{8}$	$\frac{1}{10}$
$\frac{1}{4}$	$\frac{1}{6}$	$\frac{3}{16}$	$\frac{1}{20}$
$\frac{2}{5}$	$\frac{4}{15}$	$\frac{3}{10}$	$\frac{2}{25}$

6 £53·33   7 192 cm   8 $\frac{1}{5}$

9

$\boxed{\frac{1}{16}+\frac{5}{16}}$ → $\boxed{\frac{1}{2}-\frac{1}{8}}$

$=\frac{3}{8}$

$\boxed{-\frac{1}{8}+\frac{1}{2}}$ $\boxed{3\div 8}$

10 a 2   b $\frac{1}{3}$ c $1\frac{5}{7}$ d $\frac{3}{10}$

11 $\frac{1}{36}$

page 217 **Exercise 11** ①

1 i $\dfrac{45}{100} = \dfrac{9}{20}$ **ii** $\dfrac{22}{100} = \dfrac{11}{50}$ **iii** $\dfrac{18}{100} = \dfrac{9}{50}$

2 60% **3** 27% **4** 33%

5 a 15% **b** 5% **c** 45%
  d 60% **e** 75%

6 a $\dfrac{30}{100} = \dfrac{3}{10}$ **b** $\dfrac{75}{100} = 75\%$ **c** $33\dfrac{1}{3}\%$

  d $\dfrac{1}{100}$ **e** $\dfrac{80}{100}$ or $\dfrac{8}{10}$ or $\dfrac{4}{5}$ **f** 10%

7 A a $\dfrac{1}{2}$ **b** 50%

  B a $\dfrac{4}{5}$ **b** 80%

  C a $\dfrac{1}{4}$ **b** 25%

  D a $\dfrac{1}{5}$ **b** 20%

  E a $\dfrac{3}{10}$ **b** 30%

  F a $\dfrac{1}{4}$ **b** 25%

  G a $\dfrac{1}{3}$ **b** $33\dfrac{1}{3}\%$

  H a $\dfrac{2}{5}$ **b** 40%

8 a 25% **b** 75% **c** 60% **d** 20%
9 a 75% **b** 60% **c** 75%
  d $33\dfrac{1}{3}\%$ **e** 50% **f** 25%

10 Multiple answers possible – check students' work.

page 220 **Exercise 12** ①

1 75% **2** 40% **3** $37\dfrac{1}{2}\%$

4 90% **5** 85% **6** 25%
7 68% **8** 15% **9** 98%
10 7% **11** 25% **12** 32%

13 $67\dfrac{1}{2}\%$ **14** $12\dfrac{1}{2}\%$ **15** 23.5%

16 64% **17** 90% **18** 40%

19 $22\dfrac{1}{2}\%$ **20** 34% **21** 93%
22 a 44% **b** 65% **23** 21%
24 a 50% **b** 40% **c** 10%
25 a 25 **b** 44% **c** 56%
26 Susan 70%, Jane 54%, Jackie 52%

page 221 **Exercise 13** ①

1 £12 **2** £8 **3** £10 **4** £3
5 £2.40 **6** £24 **7** £45 **8** £72
9 £244 **10** £9.60 **11** $42 **12** $88

13 8 kg **14** 12 kg **15** 272 g **16** 45 m
17 40 km **18** $710 **19** 4.94 kg **20** 60 g
21 €204 **22** £5 **23** 48 **24** 156
25 £12 **26** £70

page 222 **Exercise 14** ①

1 £63 **2** £864 **3** £87.45 **4** £104

5 £1960 **6** £792 **7** £132 **8** £45.75

9 £110.30 **10** £42 **11** £12.41 **12** £266.40

13 £6.90 **14** £8160 **15** £12.88 **16** £79.20

17 £31.87 **18** £8.89 **19** £8.93 **20** £14.21

page 223 **Exercise 15** ①

1 £35.20 **2** £5724 **3** £171.50

4 £88.35 **5** £58.50 **6** 24

7 59,400 **8** £20.72 **9** 3.348 kg

10 13.054 kg **11** £2762.50

12 a 480 cm² **b** 384 cm²

page 224 **Exercise 16** ①

1 £17.88 **2** £41.52 **3** £79.62 **4** £189.29

page 225 **Exercise 17** ①

1 8% **2** 10% **3** 25% **4** 2%
5 4% **6** 2.5% **7** 20% **8** 50%
9 15% **10** 80% **11** 25% **12** 20%
13 12.5% **14** $33\dfrac{1}{3}\%$ **15** 80% **16** 5%
17 6% **18** 20% **19** 5% **20** 2.5%

page 225 **Exercise 18** ①

1 36.4% **2** 19.0% **3** 19.4% **4** 22.0%
5 12.2% **6** 9.4% **7** 14.0% **8** 17.4%
9 32.7% **10** 10.2% **11** 7.7% **12** 35.3%
13 30.8% **14** 5.2% **15** 14.1% **16** 14.5%
17 19.1% **18** 3.6% **19** 31.1% **20** 6.5%

page 226 **Exercise 19** ①

1 12% **2** 29% **3** 30% **4** 0.25%
5 15% **6** 61.1% **7** 15%

page 227 **Exercise 20** ①

1 a £285.60 **b** £367.50 **c** £378
  d £414.10 **e** £426.40
2 a £206 **b** £210.12 **c** £214.32
3 £9751.20, £10 043.74

page 228 **Exercise 21** ①

1 £420   **2** £17 500
3 a £50   **b** £200   **c** £54
   d £150   **e** £2400
4 210 000 people   **5** 85 kg
6 a £9·79   **b** £32
   c £59·80   **d** £38·34
7 Width: 8 cm Height 10 cm
8 350 g   **9** £1550   **10** 198 000

page 229 **Exercise 22** ①

1 462 g   **2** £36·85   **3** 35 500
4 320 kg   **5** 411·8 cm^3 **6** £145 000
7 a £10·50   **b** £15   **c** £22
   d £159·62   **e** £48   **f** £10·50
8 £3·22

page 230 **Exercise 23** ②

1 £2420
2 a £5250   **b** £5788·13
3 £23152·50
4 a £3200   **b** £2048
5 a £4775·44   **b** £297·14   **c** £159 181·20
6 12 years
7 a £424   **b** £449·44   **c** £716·34
8 £159·63   **9** £120·00
10 a £14025   **b** £7321·14

page 231 **Exercise 24** ①

1 a 3 : 2   **b** 3 : 5   **c** 1 : 4   **d** 5 : 2
   e 12 : 11   **f** 3 : 4   **g** 8 : 5   **h** 3 : 7
   i 10 : 7   **j** 8 : 11   **k** 3 : 2 : 4   **l** 8 : 1 : 3
   m 6 : 5 : 4   **n** 3 : 2 : 3
2 3 : 4   **3** 4 : 1
4 a 2 : 5   **b** 1 : 5   **c** 1 : 5
   d 1 : 3   **e** 4 : 1   **f** 1 : 5
   g 1 : 10   **h** 1 : 6   **i** 1 : 3
5 $\frac{3}{8}$ **6** $\frac{1}{5}$ **7** $\frac{3}{5}$
8 a $\frac{1}{4}$ **b** $\frac{1}{3}$ **c** $\frac{5}{12}$ **9** $\frac{2}{3}$

page 232 **Exercise 25** ①

1 £10 : £20
2 £45 : £15
3 Cat 330 g; Dog 550 g
4 Sam $480; Chris $600
5 Steven 36 litres; Dave 90 litres.
6 £10 : £20 : £30
7 £70

8 £50
9 3250
10 a 2 : 7   **b** 3 : 5   **c** 2 : 6
11 a 2 : 9   **b** 3 : 8   **c** 2 : 8
12 a 2 : 3   **b** $\frac{2}{3}$ **13 a** 1 : 4   **b** $\frac{1}{4}$
14 $\frac{1}{3}$, 1 : 3   **15** $\frac{1}{3}$, 1 : 3

page 234 **Exercise 26** ①

1 8 girls   **2** 5 women   **3** 9 screws
4 30 g zinc and 40 g of tin   **5** 24
6 2·4 kg   **7** 6 eggs   **8** 18 cm
9 22·5 cm   **10** 12 cm   **11** 300 g
12 5 : 3   **13** £200   **14** 42 p
15 £175 000   **16** 0·25 m^3

page 235 **Exercise 27** ①

1 a 5 SF   **b** 8 SF   **c** 12 SF   **d** 6·2 SF
   e £4   **f** £3·40   **g** £2·60   **h** £4·60
2 a 2·2 SF   **b** $k = 2·5$ 8 500 SF
3 a £1072·50   **b** £72·73
   c The bank may charge a fee
4 a 1·24 euros   **b** $k = 1·24$ 9·92 euros
5 a $\frac{15.5}{7} = \frac{31}{14}$ **b** 26·6 lbs
6 18333, 6417, £16·37
7 Hardware shop
8 3·5 litres
9 Pack of 6 litres
10 a Bank B   **b** Bank A
11 8 litres red, 12 litres white
12 20 litres red, 8 litres black
13 16·7 litres
14 9 ripe 6 unripe
15 a $\frac{5}{6}$ **b** 15 litres   **c** 4 litres
16 a 7:3   **b** 3 litres
17 a 0·5 litres   **b** $\frac{1}{11}$

page 238 **Exercise 28** ①

1 200 m   **2** 500 m
3

	Map scale	Length on map	Actual length on land
a	1 : 10 000	10 cm	1 km
b	1 : 2000	10 cm	200 m
c	1 : 25 000	4 cm	1 km
d	1 : 10 000	6 cm	0·6 km

4 63 m   **5** 24 km   **6** 120 m
7 a 3·5 km   **b** 4·35 km   **c** 3·7 km

page 239 Exercise 29 ①

1 1·5m   **2** 1·25m   **3** 28cm   **4** 5·9cm
5 a 60cm   **b** 84cm   **c** 56cm
 d 140cm   **e** 100cm
6 2·5cm

page 239 Exercise 30 ①

1, 2 Students' own drawing
2 $x = 72$ cm
3 a i 14m **ii** 6m   **iii** 4m
 b 8m   **c** 3m   **d** 2cm
 e 12m   **f** 42m^2

page 242 Exercise 31 ① ②

1 a 4, 6, 8, 10, 20   **b** 8, 16, 32, 40, 80
 c 1·20, 1·80, 3·00, 6·00, 7·20
2 £24   **3** £1·08   **4** 315p
5 a £1·26   **b** £4·20
6 a £2·20   **b** £22   **7** £97·50
8 2750g   **9** 1400   **10** 4·5 litres
11 £3·45   **12** £1·61   **13** £3·99
14 125 seconds or 2 mins 5 secs   **15** $1\frac{1}{2}$ hours

page 243 Exercise 32 ① ②

1 a 2, 4, 6, 10, 20   **b** 3, 6, 9, 15, 30
 c 4, 8, 12, 20, 40
2 10   **3** 10   **4** 12m^2
5 9   **6** 100   **7** 160
8 450   **9 a** £267   **b** 11
10 a £127·50   **b** 2
11 a £2·24   **b** £4·20
12 a £1·60   **b** £3·60
13 35 litres   **14** 200 litres
15 70 gallons

page 245 Exercise 33 ① ②

1 a 30, 20, 15, 12, 6   **b** 60, 40, 24, 12, 10
 c 12 000, 6 000, 4 800, 4 000, 3 000
 d Same total
2 3 days   **3** 16 minutes   **4** 15 days
5 4 hours   **6** 3 cm   **7** 4 metres
8 15 days   **9** 27 rows   **10** 7·5 hours
11 £9·72
12 a 6, 4, 2, 1 $y = \dfrac{12}{x}$ **b** 12, 8, 6, 4 $y = \dfrac{48}{x}$
13 $y \propto x$ **a, b, e, g** $y \propto \dfrac{1}{x}$ **c, d, f, h**

14 a   **b**   **c**   **d**

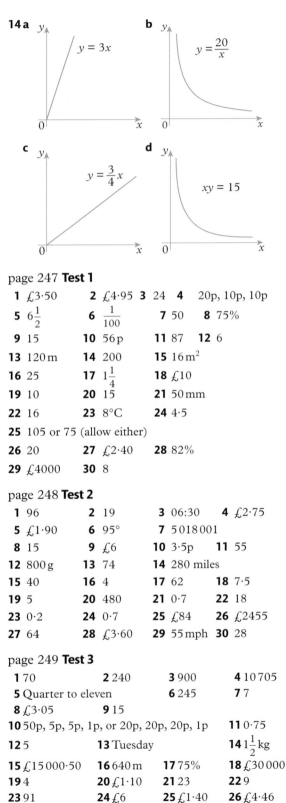

page 247 Test 1

1 £3·50   **2** £4·95 **3** 24   **4** 20p, 10p, 10p
5 $6\frac{1}{2}$ **6** $\frac{1}{100}$ **7** 50   **8** 75%
9 15   **10** 56p   **11** 87   **12** 6
13 120m   **14** 200   **15** 16m^2
16 25   **17** $1\frac{1}{4}$ **18** £10
19 10   **20** 15   **21** 50mm
22 16   **23** 8°C   **24** 4·5
25 105 or 75 (allow either)
26 20   **27** £2·40   **28** 82%
29 £4000   **30** 8

page 248 Test 2

1 96   **2** 19   **3** 06:30   **4** £2·75
5 £1·90   **6** 95°   **7** 5 018 001
8 15   **9** £6   **10** 3·5p   **11** 55
12 800g   **13** 74   **14** 280 miles
15 40   **16** 4   **17** 62   **18** 7·5
19 5   **20** 480   **21** 0·7   **22** 18
23 0·2   **24** 0·7   **25** £84   **26** £2455
27 64   **28** £3·60   **29** 55 mph **30** 28

page 249 Test 3

1 70   **2** 240   **3** 900   **4** 10 705
5 Quarter to eleven   **6** 245   **7** 7
8 £3·05   **9** 15
10 50p, 5p, 5p, 1p, or 20p, 20p, 20p, 1p   **11** 0·75
12 5   **13** Tuesday   **14** $1\frac{1}{2}$ kg
15 £15 000·50   **16** 640 m   **17** 75%   **18** £30 000
19 4   **20** £1·10   **21** 23   **22** 9
23 91   **24** £6   **25** £1·40   **26** £4·46
27 £3·30   **28** £42   **29** 34   **30** 64

page 249 **Test 4**

1 £8·05	**2** 75	**3** 25	**4** 0·1 cm
5 24p	**6** 104	**7** 40 pence	
8 £18	**9** Ten to six		**10** 270°
11 North-east	**12** 92%	**13** £4·25	**14** 998
15 20 miles	**16** 200	**17** 22·5 cm	
18 75 pence	**19** 10	**20** 16	**21** 20
22 £9·75	**23** 25 minutes		**24** 15:40
25 15	**26** 70p	**27** 200	**28** 35%
29 100 minutes	**30** £2500		

page 250 **Test 5**

1 160	**2** 106	**3** 6011	**4** £2·10
5 92	**6** 12	**7** £1·01	**8** 1·55 m
9 £30·11	**10** 4 cm	**11** 1500 m	**12** £25
13 7	**14** $2\frac{1}{2}$	**15** 64	**16** 12
17 12 litres	**18** 23%	**19** 31	**20** 3
21 64	**22** 84 square yards	**23** 100 000	
24 27 hours	**25** 6	**26** £10	**27** 500
28 180	**29** 9 or 18 or 27 etc.	**30** Saturday	

page 251 **Test 6**

1 60°	**2** 15	**3** 75%	**4** 8000
5 £26	**6** 11	**7** Half past 7	
8 20%	**9** 8·15 pm	**10** 0·11	**11** 27
12 £2·50	**13** 1·8 litres (allow 1·795)		
14 2104	**15** £15	**16** 150	**17** 36°
18 60	**19** 270°	**20** £9	**21** 50 mph
22 250	**23** 25	**24** 37	**25** 96p
26 £20	**27** 12	**28** 50p, 20p, 5p, 1p	
29 35°	**30** 20		

page 252 **Test 7**

1 82°	**2** 66p	**3** 2107	**4** 1000
5 75%	**6** 0·23	**7** 15	**8** 8
9 24	**10** 0·9	**11** Trapezium	
12 89 990	**13** False	**14** $\frac{5}{6}$	**15** 21
16 63 m²	**17** 6	**18** 1000	**19** 4 hours
20 9	**21** £8·70	**22** 5·5	
23 2550 grams	**24** £40 000		
25 120°	**26** 1100 or 1000 (Accept either)		
27 8	**28** 5	**29** $\frac{1}{4}$	**30** 2·1

page 253 **Test 8**

1 61	**2** 39	**3** 10	**4** 0·82
5 120	**6** 55	**7** £3·80	**8** 63
9 154	**10** 1100	**11** 15	**12** 165
13 355	**14** 4·1	**15** 40	**16** 200
17 £2·10	**18** 19	**19** 24	**20** 24

page 253 **Test 9**

1 130	**2** 77	**3** 1	**4** 2300
5 49	**6** £9·50	**7** 342	**8** 0·8
9 300	**10** 27	**11** 30	**12** 10000
13 7	**14** 25	**15** 40	**16** 5·5
17 19·2	**18** 8	**19** 45	**20** 99

page 253 **Test 10**

1 84	**2** 35	**3** 36	**4** 100
5 210	**6** 5000	**7** 84	**8** 20
9 90	**10** 22	**11** 415	**12** 1
13 1000	**14** 227	**15** 1600	**16** 60
17 23	**18** 8·8	**19** 5·2	**20** 26

page 253 **Test 11**

1 220	**2** 20	**3** 56	**4** 3·85
5 200	**6** 199	**7** 121	**8** 315
9 500	**10** 60	**11** 400	**12** 6
13 1800	**14** 69	**15** 32	**16** 101
17 10·7	**18** 1	**19** 3600	**20** 120

page 254 **Exercise 34** 1

1 18	**2** 23	**3** 42	**4** 3
5 225	**6** 36	**7** 8	**8** 57
9 4	**10** 2300	**11** 200	**12** 100
13 2500	**14** 2000	**15** 700	**16** 100
17 5900	**18** 5700	**19** 200	**20** 65 500
21 14	**22** 129	**23** 153	**24** 10
25 4	**26** 33	**27** 2	**28** 4
29 44	**30** 6	**31** 24	**32** 57
33 34	**34** 28	**35** 331	**36** 37
37 18	**38** 12	**39** 23	**40** 8

page 255 **Exercise 35** 1

1 200	**2** 400	**3** 5000	**4** 7000
5 8000	**6** 5000	**7** 400	**8** 30
9 40	**10** 200	**11** 400	**12** 4000
13 700	**14** 700	**15** 7000	**16** 30
17 4000	**18** 20 000	**19** 9000	**20** 700
21 500	**22** 2000	**23** 7000	**24** 90
25 200	**26** 60	**27** 20	**28** 60
29 5000	**30** 2000	**31** 6000	**32** 1000

33 a 1493·2 → 1500 m³   23·41→ 23°C
2108 → 2000   5173 → 5000
b 100, 58·23, 2012

page 256 **Exercise 36** 1

1 2·35	**2** 0·814	**3** 26·2	**4** 35·6
5 113	**6** 211	**7** 0·825	**8** 0·0312
9 5·9	**10** 1·2	**11** 0·55	**12** 0·72

13 0·14   **14** 1·8   **15** 25   **16** 31

17 487   **18** 500   **19** 2·89   **20** 3·11

21 0·0715   **22** 3·04   **23** 2460   **24** 489 000

25 0·513   **26** 5·8   **27** 66   **28** 588

29 0·6   **30** 0·07   **31** 5·84   **32** 88

33 2500   **34** 52 700   **35** 0·006   **36** 7000

page 257 **Exercise 37** ❶

1 5·38   **2** 11·05   **3** 0·41   **4** 0·37

5 8·02   **6** 87·04   **7** 9·01   **8** 0·07

9 8·4   **10** 0·7   **11** 0·4   **12** 0·1

13 6·1   **14** 19·5   **15** 8·1   **16** 7·1

17 8·16   **18** 3·0   **19** 0·545   **20** 0·0056

21 0·71   **22** 6·83   **23** 0·8   **24** 19·65

25 0·0714   **26** 60·1   **27** 7·3   **28** 5·42

29 a i length 6 cm, width 3·2 cm

     ii length 5 cm, width 3 cm

   b i 19·2 cm²   **ii** 15 cm²

page 258 **Exercise 38** ❶

1 0·57   **2** 3·45   **3** 431   **4** 19·3

5 0·22   **6** 3942·7   **7** 53   **8** 18·4

9 0·059   **10** 1·1   **11** 6140   **12** 127·89

13 20·3   **14** 47·6   **15** 71·1   **16** 0·16

page 259 **Exercise 39** ❶

1 B   **2** A   **3** C   **4** B   **5** C

6 A   **7** B   **8** B   **9** A   **10** C

11 B   **12** A   **13** A   **14** C   **15** C

16 B   **17** C   **18** A   **19** B   **20** B

21 C   **22** B   **23** B   **24** A   **25** B

26 B   **27** B   **28** C   **29** A   **30** C

31 B   **32** C   **33** C

page 260 **Exercise 40** ❶

1 £8000   **2** £6   **3** £440   **4** £5200

5 a 89·89   **b** 4·2   **c** 358·4   **d** 58·8

   e 0·3   **f** 2·62

6 a 4·5   **b** 462   **c** 946·4   **d** 77·8

   e 0·2   **f** 21

7 B   **8** C   **9** B   **10** B

11 C   **12** B   **13** A   **14** A

15 Area ≈ 30 × 40 = 1200 m²

     Cost ≈ 1200 × 7 = £8400

     Estimate job at £8500

16 £3900 (accept £4000)

page 261 **Exercise 41** ❶

1 a 1670   **b** 90·8   **c** 32·6

   d 5·29 (2dp)   **e** 44·7

2 a smaller   **b** larger   **c** smaller

3 a OK   **b** OK   **c** OK

   d Highly unlikely   **e** Impossible

   f Highly unlikely

page 262 **Exercise 42** ❶

1 42 kg   **2** £120

3 a False, unless $n = 1\frac{1}{2}$ **b** True

   c True   **d** False, unless $n = 1\frac{1}{2}$

   e False, unless $a = b$

   f False, unless $n = 0$ or $\frac{1 \pm \sqrt{5}}{2}$

4 24 tonnes   **5** £345

6 83 200   **7** £1·80   **8** £204

9 a   **b**   **c**

page 263 **Exercise 43** ❶

1 a 15   **b i** 16·7% (1 dp)   **ii** 30%

2 a Multiply by 3 **b** 177 147   **c** 1 594 323

3 a 36   **b** 24   **c** 240

4 a 38,62   **b** 64,125   **c** 81,64   **d** $\frac{6}{7}, \frac{7}{8}$

5 120°

6 (1, 2) (2, 1·1) (3, 1·1) (4, 2) (4, 3) (3, 4) (4, 5) (4, 6)

   (3, 6·9) (2, 6·9) (1, 6)

7 2   **8** 6; 14p   **9** 100

10 a 90   **b** 1   **c** 23   **d** 44

    e 77   **f** 111

page 264 **Exercise 44** ❶

1 410 calories

2 a 64   **b** 1   **c** 100   **d** 3000

   e 32   **f** 81

3 a 13, 14   **b** 7, 8, 9   **c** 10, 11, 12, 13

4 10 cm²

5 a 273   **b** 7457   **c** 84·5   **d** 305

6 a 30   **b** 32   **c** 5

7 474·4 Euros

8 5 hours 34 minutes   **9** 50 m   **10** 60

page 266 **Exercise 45** ❶

1 a 80g   **b** 5·2 calories   **c** 416 calories

2 a 7   **b** 50   **c** 1   **d** 5

3 a 12   **b** 8   **c** 48

4 Put 2 coins on the scales. If they don't balance, the heavier one is the fake. If they do balance, the 3rd coin is the fake.

5 200 litres   **6** £21 600   **7** 5·4 km

8

×	6	3	4	7
5	30	15	20	35
9	$\sqrt{4}$	27	36	63
2	12	6	8	14
8	48	24	32	56

9 16   **10** A solution is :

Finish

Start

page 267 **Exercise 46** ①

1a 5 m   **b** 50 m   **c** 6 km

2 9   **3** 51·4°

4a 40 acres   **b** 15 acres   **c** 10%   **d** 22·5%

5 £8176   **6** £369   **7** 78   **8** £5·85

9a $99 + \frac{9}{9}$ **b** $6 + \frac{6}{6}$ **c** $55 + 5$

d $55 + 5 + \frac{5}{5}$ **e** $77 \div 77$ **f** $88 \div 8$

10a

2	3
+ 5	4
7	7

b

1	7
+ 4	6
6	3

c

5	8	2
+ 1	3	6
7	1	8

d

4	7	4
+ 3	5	0
8	2	4

e

8	6
− 3	4
5	2

f

8	8	2
− 6	5	0
2	3	2

page 269 **Test yourself**

1a 40%   **b** 11 girls

c No. No total given for class A

2 2·5 times quantity. Not enough mincemeat, needs 700 g

3a 75%   **b** £125 : £175   **c** £422·10

4 $\frac{7}{25}$ **5** $5\frac{5}{12}$

6 Test 1 : 75%; Test 2: 70%

7a 295   **b** 11676   **c** 60

8 £6

9 25% off total offer is better (£42·73) compared to £42·99

10 £780   **11** 6·21

12a 720 euros   **b** £10 (£450 in France)

13 £9720

14 London: £140000 Paris £139 130 equivalent

15a 2·8 Bars   **b** 7 Bars

16 Becks £7200, Clare £9000

17a i 150 g   **ii** 3 packs

b No later than 1240   **c** 70 mugs

18 24%

19a 10 cars   **b** £94

20 £500 is better, 2% is £480

21a $\frac{4}{15}$ **b** 0·575757...

22a i 5 trains   **ii** 1233   **iii** 53 minutes

b 3 hours 2 mins   **c** 50 p

6 Algebra 2

page 274 **Exercise 1** ①

1 $2(3x + 2y)$ **2** $3(3x + 4y)$

3 $2(5a + 2b)$ **4** $4(x + 3y)$

5 $5(2a + 3b)$ **6** $6(3x - 4y)$

7 $4(2u - 7v)$ **8** $5(3s + 5t)$

9 $8(3m + 5n)$ **10** $9(3c - 8d)$

11 $4(5a + 2b)$ **12** $6(5x - 4y)$

13 $3(9c - 11d)$ **14** $7(5u + 7v)$

15 $4(3s - 8t)$ **16** $8(5x - 2t)$

17 $12(2x + 7y)$ **18** $4(3x + 2y + 4z)$

19 $3(4a - 2b + 3c)$ **20** $5(2x - 4y + 5z)$

21 $4(5a - 3b - 7c)$ **22** $8(6m + n - 3x)$

23 $7(6x + 7y - 3z)$ **24** $3(2x^2 + 5y^2)$

25 $5(4x^2 - 3y^2)$ **26** $7(a^2 + 4b^2)$

27 $9(3a + 7b - 4c)$ **28** $6(2x^2 + 4xy + 3y^2)$

29 $8(8p - 9q - 5r)$ **30** $12(3x - 5y + 8z)$

31 $3x(3 + 2y - x)$ **32** $x(x - 5)$

33 $x(2x - 3)$ **34** $x(7x + 1)$

35 $y(y + 4)$ **36** $2x(x + 4)$

37 $4y(y - 1)$ **38** $p(p - 2)$

39 $2a(3a + 1)$ **40** $a(2b - 1)$

41 $x(3y + 2)$ **42** $3t(1 + 3t)$

43 $4(1 - 2x^2)$ **44** $5x(1 - 2x^2)$

45 $\pi r(4r + h)$ **46** $\pi r(r + 2)$

page 276 **Exercise 2** ①

1 $x = e - b$ **2** $x = m + t$

3 $x = a + b + f$ **4** $x = A + B - h$

5 $x = y$ **6** $x = b - a$

7 $x = m - k$

8 $x = w + y - v$

9 $x = \dfrac{b}{a}$

10 $x = \dfrac{m}{h}$

11 $x = \dfrac{(a + b)}{m}$

12 $x = \dfrac{(c - d)}{k}$

13 $x = \dfrac{(e + n)}{v}$

14 $x = \dfrac{(y + z)}{3}$

15 $x = \dfrac{r}{p}$

16 $x = \dfrac{(h - m)}{m}$

17 $x = \dfrac{(a - t)}{a}$

18 $x = \dfrac{(k + e)}{m}$

19 $x = \dfrac{(m + h)}{u}$

20 $x = \dfrac{(t - q)}{e}$

21 $x = \dfrac{(v^2 + u^2)}{k}$

22 $x = \dfrac{(s^2 - t^2)}{g}$

23 $x = \dfrac{(m^2 - k)}{a}$

24 $x = \dfrac{(m + v)}{m}$

25 $x = \dfrac{(c - a)}{b}$

26 $x = \dfrac{(y - t)}{s}$

27 $x = \dfrac{(z - y)}{c}$

28 $x = \dfrac{a}{h}$

29 $x = \dfrac{2b}{m}$

30 $x = \dfrac{(cd - ab)}{k}$

31 $x = \dfrac{c}{a} + b$

32 $x = \dfrac{e}{c} + d$

33 $x = \dfrac{n^2}{m} - m$

34 $x = \dfrac{t}{k} + a$

35 $x = \dfrac{k}{n} + h$

36 $x = \dfrac{n}{m} - b$

37 $x = 2a$

38 $x = \dfrac{d}{c} - a$

39 $x = \dfrac{e}{m} - b$

page 276 **Exercise 3** ➊

1 $x = tm$

2 $x = en$

3 $x = pa$

4 $x = tam$

5 $x = abc$

6 $x = ey^2$

7 $x = a(b + c)$

8 $x = t(c - d)$

9 $x = m(s + t)$

10 $x = k(h + i)$

11 $x = \dfrac{ab}{c}$

12 $x = \dfrac{mz}{y}$

13 $x = \dfrac{hc}{d}$

14 $x = \dfrac{em}{n}$

15 $x = \dfrac{hb}{e}$

16 $x = c(a + b)$

17 $x = m(h + k)$

18 $x = \dfrac{mu}{y}$

19 $x = t(h - k)$

20 $x = (a + b)(z + t)$

21 $x = \dfrac{e}{t}$

22 $x = \dfrac{e}{a}$

23 $x = \dfrac{h}{m}$

24 $x = \dfrac{cb}{a}$

page 277 **Exercise 4** ➊

1 $x = a - y$

2 $x = h - m$

3 $x = z - q$

4 $x = b - v$

5 $x = k - m$

6 $x = \dfrac{h - d}{c}$

7 $x = \dfrac{y - c}{m}$

8 $x = \dfrac{k - h}{e}$

9 $x = \dfrac{a^2 - d}{b}$

10 $x = \dfrac{m^2 - n^2}{t}$

11 $x = \dfrac{v^2 - w}{a}$

12 $x = y - y^2$

13 $x = \dfrac{k - m}{t^2}$

14 $x = \dfrac{b - e}{c}$

15 $x = \dfrac{h - z}{g}$

16 $x = \dfrac{c - a - b}{d}$

17 $x = \dfrac{v^2 - y^2}{k}$

18 $x = \dfrac{d - h}{f}$

19 $x = b - \dfrac{c}{a}$

20 $x = m - \dfrac{n}{h}$

page 277 **Exercise 5** ➊

1 a $a = \dfrac{v - u}{t}$

b 2

2 a $k = \dfrac{py}{m}$

b $y = \dfrac{mk}{p}$

3 a $n = pR + d$

b $n = 1255$

page 278 **Exercise 6** ➊

1 $x = \dfrac{h + d}{a}$

2 $y = \dfrac{m - k}{z}$

3 $y = \dfrac{f}{d} - e$

4 $k = \dfrac{d}{m} - a$

5 $m = \dfrac{c - a}{b}$

6 $e = \dfrac{b}{a}$

7 $t = \dfrac{z + a}{y}$

8 $x = \dfrac{e + c}{a}$

9 $y = \dfrac{b + n}{m}$

10 $z = \dfrac{b}{a} - a$

11 $x = \dfrac{a}{d}$

12 $k = mt$

13 $u = mn$

14 $x = \dfrac{y}{d}$

15 $m = \dfrac{a}{t}$

16 $g = \dfrac{d}{n}$

17 $t = k(a + b)$ **18** $e = \dfrac{v}{y}$

19 $y = \dfrac{m}{c}$ **20** $a = mb$

21 $m = \dfrac{b}{g} - a$ **22** $g = \dfrac{x^2}{h} - h$

23 $t = y - z$ **24** $e = \dfrac{c}{m}$

25 $x = \dfrac{t}{a} - y$ **26** $v = \dfrac{y^2 + t^2}{u}$

27 $k = \dfrac{c - t}{p}$ **28** $w = k - m$

29 $n = \dfrac{b - c}{a}$ **30** $y = \dfrac{c}{m} - a$

31 $x = pq - ab$ **32** $k = \dfrac{a^2 - t}{b}$

33 $z = \dfrac{w}{v^2}$ **34** $u = t - c$

35 $c = \dfrac{t}{x}$ **36** $w = \dfrac{k}{m} - n$

37 $m = \dfrac{v - t}{x}$ **38** $y = \dfrac{c}{a} - b$

39 $c = a - \dfrac{e}{m}$ **40** $a = \dfrac{ce}{b}$

41 $p = \dfrac{a}{q}$ **42** $n = \dfrac{a}{e}$

page 279 Exercise 7 ②

1 a $<$ **b** $>$ **c** $>$
 d $>$ **e** $>$ **f** $>$
2 a $x > 2$ **b** $x \le 5$ **c** $x < 100$
 d $-2 \le x \le 2$ **e** $x \ge -6$ **f** $3 < x \le 8$

3 a

b

c

d

e

4 a $A \ge 16$ **b** $3 < A \le 70$
 c $150 < T < 175$ **d** $h \ge 1 \cdot 75$
5 a True   **b** True   **c** True

page 280 Exercise 8 ②

1 $x > 13$ **2** $x < -1$ **3** $x < 12$
4 $x \le 2\frac{1}{2}$ **5** $x > 3$ **6** $x \ge 8$

7 $x < \dfrac{1}{4}$ **8** $x \ge -3$ **9** $x < -8$
10 $x < 4$ **11** $x > -9$ **12** $x < 8$
13 $x > 3$ **14** $x \ge 1$ **15** $x < 1$
16 $x > 2\frac{1}{3}$ **17** $x < -3$ **18** $x > 7\frac{1}{2}$
19 $x > 0$ **20** $x < 0$ **21** $x > 5$

22 $5 \le x \le 9$ **23** $-1 < x < 4$ **24** $5\frac{1}{2} \le x \le 6$

25 $1\frac{1}{3} < x < 8$ **26** $-8 < x < 2$ **27** $\frac{1}{4} < x < 1\frac{1}{5}$

page 281 Exercise 9 ②

1 $x > 8$ **2** 1, 2, 3, 4, 5, 6
3 7, 11, 13, 17, 19   **4** 4, 9, 16, 25, 36, 49
5 $-4, -3, -2, -1$
6 2, 3, 4, 5, 6, 7, 8, 9, 10, 11, 12
7 2, 3, 5, 7, 11
8 2, 4, 6, 8, 10, 12, 14, 16, 18
9 1, 2, 3, 4
10 5
11 a 16   **b** -16 **c** 20   **d** -5
12 $>$
13 $\frac{1}{2}$ (other answers are possible)

14 19   **15** 17   **16** $x > 3\frac{2}{3}$ **17** 7
18 5   **19** 6   **20** 3, 4, 5

page 282 Exercise 10 ①

1 $A: y = 7$, $B: y = 3$, $C: y = 1$
2 $P: x = 3$, $Q: x = 5$, $R: x = -3$
3

(b) $(2, 3)$
(c) $(5, 1)$
(d) $(8, 6)$

4 a $x = 1$ **b** $y = 7$ **c** $x = 2$ **d** $x = 7$
 e $x = 3$ **f** $y = 3$ **g** $y = 5$ **h** $y = 0$

page 284 **Exercise 11** ①

1

x	0	1	2	3	4	5	6
y	3	5	7	9	11	13	15
coordinates	(0, 3)	(1, 5)	(2, 7)	(3, 9)	(4, 11)	(5, 13)	(6, 15)

$y = 2x + 3$

2

x	0	1	2	3	4	5	6	7
y	3	4	5	6	7	8	9	10
coordinates	(0, 3)	(1, 4)	(2, 5)	(3, 6)	(4, 7)	(5, 8)	(6, 9)	(7, 10)

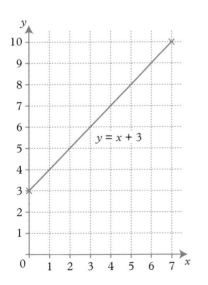

$y = x + 3$

3

x	0	1	2	3	4	5
y	0	2	4	6	8	10
coordinates	(0, 0)	(1, 2)	(2, 4)	(3, 6)	(4, 8)	(5, 10)

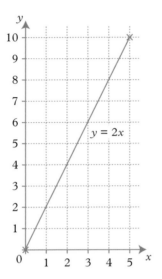

$y = 2x$

4

x	0	1	2	3	4	5
y	4	6	8	10	12	14

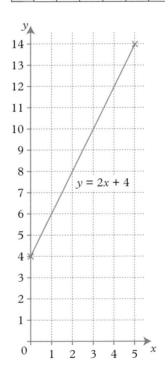

$y = 2x + 4$

5

x	0	1	2	3	4	5
y	−1	1	3	5	7	9

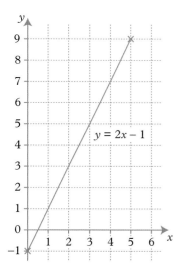

7

x	0	1	2	3	4	5
y	2	4	6	8	10	12

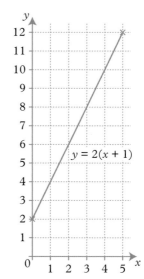

6

x	0	1	2	3
y	1	4	7	10

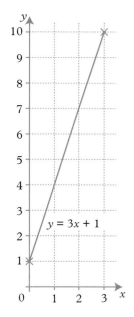

8

x	0	1	2	3	4
y	6	9	12	15	18

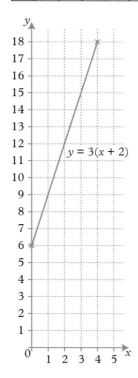

9

x	0	1	2	3	4	5	6	7
y	0	$\frac{1}{2}$	1	$1\frac{1}{2}$	2	$2\frac{1}{2}$	3	$3\frac{1}{2}$

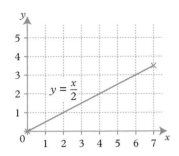

10

x	0	1	2	3	4	5
y	5	6	7	8	9	10

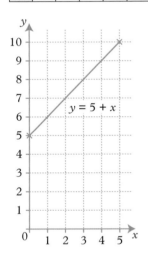

11

x	0	1	2	3	4	5	6
y	6	5	4	3	2	1	0

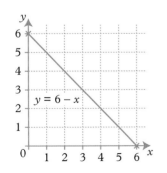

12

x	0	1	2	3	4	5
y	-3	-1	1	3	5	7

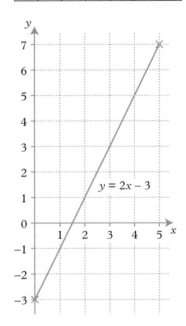

page 285 **Exercise 12** ❶

1

x	-1	0	1	2	3	4
y	-1	1	3	5	7	9

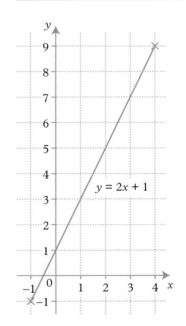

2

x	-2	-1	0	1	2	3
y	-4	-1	2	5	8	11

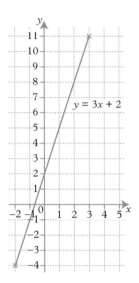

$y = 3x + 2$

3

x	-2	-1	0	1	2	3	4	5
y	-5	-4	-3	-2	-1	0	1	2

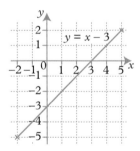

$y = x - 3$

4

x	-3	-2	-1	0	1	2	3
y	3	4	5	6	7	8	9

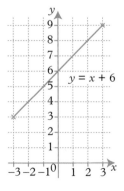

$y = x + 6$

5

x	-3	-2	-1	0	1	2	3
y	-7	-5	-3	-1	1	3	5

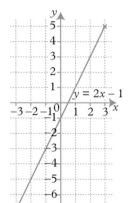

$y = 2x - 1$

6

x	-2	-1	0	1	2
y	-7	-3	1	5	9

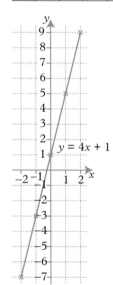

$y = 4x + 1$

7

x	0	1	2	3	4	5
y	5	4	3	2	1	0

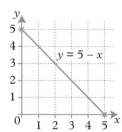

$y = 5 - x$

8

x	0	1	2	3	4	5	6
y	10	8	6	4	2	0	-2

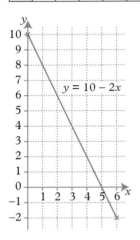

$y = 10 - 2x$

9

x	0	1	2	3
y	4	2	0	-2

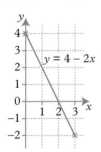

$y = 4 - 2x$

10

x	0	1	2	3	4	5	6
y	12	10	8	6	4	2	0

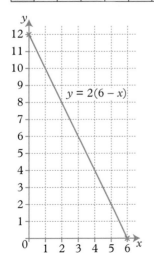

$y = 2(6 - x)$

11

x	-2	-1	0	1	2	3	4
y	-2	0	2	4	6	8	10

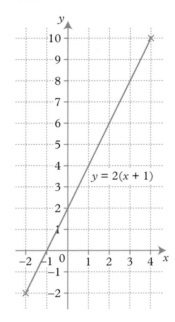

$y = 2(x + 1)$

12

x	-2	-1	0	1	2
y	-9	-6	-3	0	3

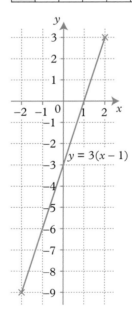

$y = 3(x - 1)$

13

x	0	1	2	3	4
y	1	3	5	7	9

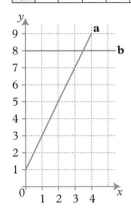

c $\left(3\frac{1}{2}, 8\right)$

14

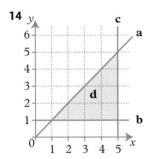

15

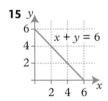

16

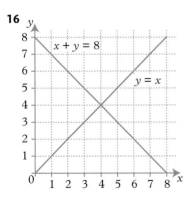

b $(4, 4)$

page 287 **Exercise 13** ❶

1 AB $\frac{1}{5}$ BC $\frac{5}{2}$ AC $-\frac{4}{3}$

2 PQ $\frac{4}{5}$ PR $-\frac{1}{6}$ QR -5

3 a 3 **b** $\frac{3}{2}$ **c** $\frac{1}{2}$ **d** $-\frac{5}{2}$

4 1

5 a $\frac{n-4}{3}$ **b** 4

page 288 **Exercise 14** ❷

1 a $y = 2x + 1$ **b** $y = \frac{1}{2}x$

c $y = 4x - 2$ **d** $y = -2x + 5$
e $y = 7x$ **f** $y = -x + 1$
g $y = 2x + 3$ **h** $y = x - 4$
i $y = 3x - 10$

2 $y = 3x - 7$ and $y = 3x$

3 a $y = 4x - 3$

b $y = -3x + 5$

c $y = \frac{1}{3}x - 2$

4 Any line of the form $y = 5x + c$

page 290 **Exercise 15** ❷

1 a 1 **b** 3 **2 a** 1 **b** –2

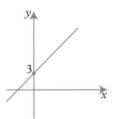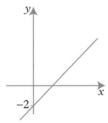

3 a 2 **b** 1 **4 a** 2 **b** –5

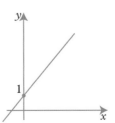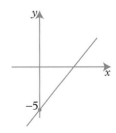

5 a 3   **b** 4   **6 a** $\frac{1}{2}$ **b** 6

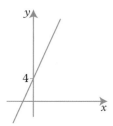   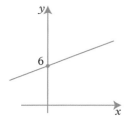

7 a 3   **b** −2   **8 a** 2   **b** 0

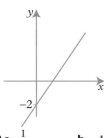   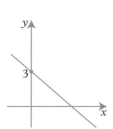

9 a $\frac{1}{4}$ **b** −4   **10 a** −1   **b** 3

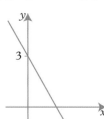   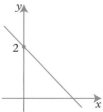

11 a −2   **b** 6   **12 a** −1   **b** 2

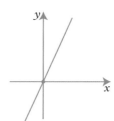

13 a −2   **b** 3   **14 a** −3   **b** −4

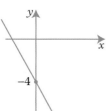

15 a $\frac{1}{2}$ **b** 3   **16 a** $-\frac{1}{3}$ **c** 3

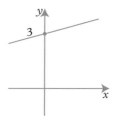   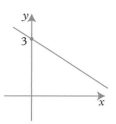

17 a 4   **c** −5   **18 a** $\frac{3}{2}$ **b** −4

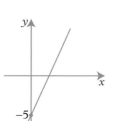   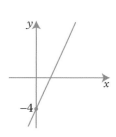

19 a 10   **b** 0   **20 a** 0   **b** 4

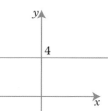

21 A: $y = 3x - 4$ **22** C: $y = \frac{2}{3}x - 2$
B: $y = x + 2$ D: $y = -2x + 4$

page 291 Exercise 16 ②

1 10°C/min   **2** 5 litres/min   **3** 200 m/min

4

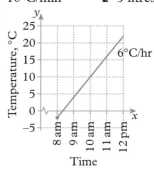

5 a 1, 3, 5, 7, 9

b

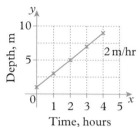

6 a 4, 1, –2, –5, –8   **b** –3 m/min

page 294 **Exercise 17** 2

1 a $x = 3, y = 7$ **b** $x = 1, y = 3$
  c $x = 11, y = –1$
2 $x = 2, y = 4$
3 $x = 2, y = 3$ **4** $x = 3, y = 1$ **5** $x = 1, y = 5$
6 $a = 5, b = 3$
7 a $x = 4, y = 0$ **b** $x = 1, y = 6$ **c** $x = –2, y = –3$
  d $x = 8, y = –1$ **e** $x = –0.6, y = 1.2$

page 295 **Exercise 18** 2

1 $x^2 + 4x + 3$ **2** $x^2 + 5x + 6$
3 $y^2 + 9y + 20$ **4** $x^2 + 7x + 12$
5 $x^2 + 7x + 10$ **6** $x^2 + 10x + 16$
7 $a^2 + 12a + 35$ **8** $z^2 + 11z + 18$
9 $x^2 + 6x + 9$ **10** $k^2 + 22k + 121$
11 $2x^2 + 7x + 3$ **12** $3x^2 + 10x + 8$
13 $2y^2 + 5y + 3$ **14** $49y^2 + 14y + 1$
15 $9x^2 + 12x + 4$ **16** $20 + 9x + x^2$
17 $x^2 + 8x + 16$ **18** $x^2 + 4x + 4$
19 $x^2 + 2x + 1$ **20** $4x^2 + 4x + 1$
21 $y^2 + 10y + 25$ **22** $9y^2 + 6y + 1$
23 $3x^2 + 12x + 12$
24 a $2n$ = even $2n – 1$ = odd
  b $2n + 1$; $4n$ is a multiple by 4

page 296 **Exercise 19** 2

1 $(x + 2)(x + 5)$ **2** $(x + 3)(x + 4)$
3 $(x + 3)(x + 5)$ **4** $(x + 3)(x + 7)$
5 $(x + 2)(x + 6)$ **6** $(y + 5)(y + 7)$
7 $(y + 3)(y + 8)$ **8** $(y + 5)^2$
9 $(y + 3)(y + 12)$ **10** $(a – 5)(a + 2)$
11 $(a – 4)(a + 3)$ **12** $(z + 3)(z – 2)$
13 $(x – 7)(x + 5)$ **14** $(x – 8)(x + 3)$
15 $(x – 4)(x – 2)$ **16** $(y – 3)(y – 2)$
17 $(x – 3)(x – 5)$ **18** $(a – 3)(a + 2)$
19 $(a + 9)(a + 5)$ **20** $(b – 7)(b + 3)$
21 $(x – 4)^2$ **22** $(y + 1)^2$
23 $(y – 7)(y + 4)$ **24** $(x – 5)(x + 4)$

page 297 **Exercise 20** 2

1 $(y + a)(y – a)$ **2** $(m + n)(m – n)$
3 $(x + t)(x – t)$ **4** $(y – 1)(y + 1)$
5 $(x – 3)(x + 3)$ **6** $(a – 5)(a + 5)$
7 $\left(x + \frac{1}{2}\right)\left(x – \frac{1}{2}\right)$ **8** $\left(x + \frac{1}{3}\right)\left(x – \frac{3}{17}\right)$
9 $(2x – y)(2x + y)$ **10** $(a – 2b)(a + 2b)$
11 $(5x – 2y)(5x + 2y)$ **12** $(3x – 4y)(3x + 4y)$
13 $(2x – z)(2x + z)$ **14** $x(x + 1)(x – 1)$
15 $a(a – b)(a + b)$ **16** $x(2x + 1)(2x – 1)$
17 $2x(2x + y)(2x – y)$ **18** $y(y + 3)(y – 3)$
19 1 200 000   **20** 12 000 000
21 $10000 – 9 = (100 + 3)(100 – 3) = 103 × 97$

page 298 **Exercise 21** 2

1 $x = –3$ or $x = –4$ **2** $x = –2$ or $x = –5$
3 $x = 3$ or $x = –5$ **4** $x = 2$ or $x = –3$
5 $x = 2$ or $x = 6$ **6** $x = –3$ or $x = –7$
7 $x = 2$ or $x = 3$ **8** $x = 5$ or $x = –1$
9 $x = 2$ or $x = –7$ **10** $x = 8$ or $x = 7$
11 $x = \frac{1}{2}$ or $x = \frac{5}{6}$ **12** $x = 7$ or $x = –9$
13 $x = –1$ **14** $x = 3$
15 $x = –5$ **16** $x = 7$
17 $x = 13$ or $x = –5$ **18** $x = 1$

page 299 **Exercise 22** 2

1 $x = 0$ or $x = 3$ **2** $x = 0$ or $x = –7$
3 $x = 0$ or $x = 1$ **4** $x = \frac{1}{3}$ or $x = 0$
5 $x = 4$ or $x = –4$ **6** $x = 7$ or $x = –7$
7 $x = \frac{1}{2}$ or $x = –\frac{1}{2}$ **8** $x = \frac{2}{3}$ or $x = –\frac{2}{3}$
9 $y = 0$ or $y = –1\frac{1}{2}$ **10** $a = 0$ or $a = 1\frac{1}{2}$
11 $x = 0$ or $x = 5\frac{1}{2}$ **12** $x = \frac{1}{4}$ or $x = –\frac{1}{4}$
13 $x = \frac{1}{2}$ or $x = –\frac{1}{2}$ **14** $x = 0$ or $x = \frac{5}{8}$
15 $x = 0$ or $x = \frac{1}{12}$ **16** $x = 6$ or $x = 0$
17 $x = 11$ or $x = 0$ **18** $x = 0$ or $x = 1\frac{1}{2}$
19 $x = 0$ or $x = 1$ **20** $x = 0$ or $x = 4$
21 $x + 5$ **22** $x = 11$

page 300 **Exercise 23** 2

1 8, 11   **2** 11, 13   **3** 12 cm   **4** 6 cm
5 $x = 3$ **6** 8, 9, 10   **7** $x = 4$

page 302 **Exercise 24** 2

1 a quadratic, x positive **b** cubic, x positive
  c quadratic, x negative **d** reciprocal, x positive
  e quadratic, x positive **f** cubic, x negative

2 a

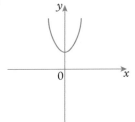

b

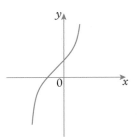

c

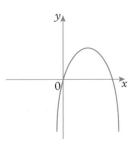

d

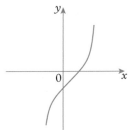

e

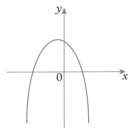

f

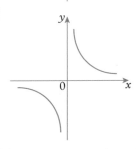

3 a

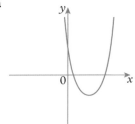

4 A $y = x^3$ **B** $y = x(x - 2)$

  C $y = \dfrac{8}{x}$ **D** $y = 2 + x - x^2$

  E $y = (x - 2)(x - 4)$ **F** $y = x^3 - 4x$

5

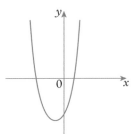

6 a

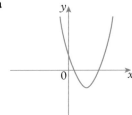

b

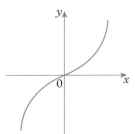

c

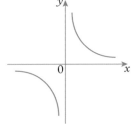

d

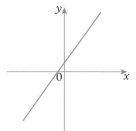

e

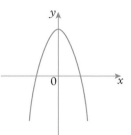

f

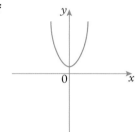

page 304 **Exercise 25** 2

1 $y = x^2 + 4$

x	-3	-2	-1	0	1	2	3
x²	9	4	1	0	1	4	9
4	4	4	4	4	4	4	4
y	13	8	5	0	5	8	13

2 $y = x^2 - 7$

x	-3	-2	-1	0	1	2	3
x²	9	4	1	0	1	4	9
-7	-7	-7	-7	-7	-7	-7	-7
y	2	-3	-6	-7	-6	-3	2

3 $y = x^2 + 3x$

x	-4	-3	-2	-1	0	1	2
x²	16	9	4	1	0	1	4
3x	-12	-9	-6	-3	0	3	6
y	4	0	-2	-2	0	4	10

4 $y = x^2 + 5x$

x	-4	-3	-2	-1	0	1	2
x²	16	9	4	1	0	1	4
5x	-20	-15	-10	-5	0	5	10
y	-4	-6	-6	-4	0	6	14

5 $y = x^2 - 2x$

x	-3	-2	-1	0	1	2	3
x²	9	4	1	0	1	4	9
-2x	6	4	2	0	-2	-4	-6
y	15	8	3	0	-1	0	3

6 $y = x^2 - 4x$

x	-2	-1	0	1	2	3	4
x²	4	1	0	1	4	9	16
-4x	8	4	0	-4	-8	-12	-16
y	15	5	0	-3	-4	-3	0

7 $y = x^2 + 2x + 3$

x	-3	-2	-1	0	1	2	3
x²	9	4	1	0	1	4	9
-2x	-6	-4	-2	0	2	4	6
3	3	3	3	3	3	3	3
y	6	3	2	3	6	11	18

8 $y = x^2 + 3x - 2$

x	-4	-3	-2	-1	0	1	2
x²	16	9	4	1	0	1	4
3x	-12	-9	-6	-3	0	3	6
-2	-2	-2	-2	-2	-2	-2	-2
y	2	-2	-4	-4	-2	2	8

9 $y = x^2 + 4x - 5$

x	-3	-2	-1	0	1	2	3
x²	9	4	1	0	1	4	9
+4x	-12	-8	-4	0	4	8	12
-5	-5	-5	-5	-5	-5	-5	-5
y	-8	-9	-8	-5	0	7	16

10 $y = x^2 - 2x + 6$

x	-3	-2	-1	0	1	2	3
x²	9	4	1	0	1	4	9
-2x	6	4	2	0	-2	-4	-6
6	6	6	6	6	6	6	6
y	21	14	9	6	5	6	9

11

x	-3	-2	-1	0	1	2	3
y	-3	-4	-3	0	5	12	21

12

x	-4	-3	-2	-1	0	1	2
y	40	27	16	7	0	-5	-8

13

x	-3	-2	-1	0	1	2	3
y	17	12	9	8	9	12	17

14

x	-4	-3	-2	-1	0	1	2
y	5	1	-1	-1	1	5	11

15

x	-3	-2	-1	0	1	2	3
y	27	17	9	3	-1	-3	-3

16

x	-2	-1	0	1	2	3	4
y	5	-1	-5	-7	-7	-5	-1

17

x	-5	-4	-3	-2	-1	0	1
y	22	13	6	1	-2	-3	-2

18

x	-3	-2	-1	0	1	2	3
y	19	9	3	1	3	9	19

page 306 **Exercise 26** 2

1

x	-3	-2	-1	0	1	2	3
y	11	6	3	2	3	6	11

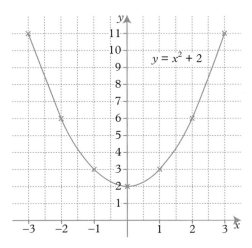

y-intercept at 2 $(0, 2)$, no real roots, minimum

2

x	-3	-2	-1	0	1	2	3
y	14	9	6	5	6	9	14

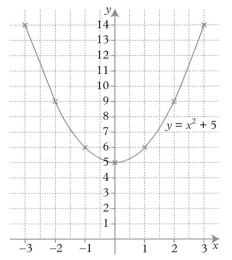

5, $(0, 5)$, no real roots, minimum

3

x	-3	-2	-1	0	1	2	3
y	5	0	-3	-4	-3	0	5

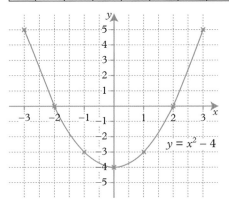

-4, $(0, -4)$, $x = -2$, $x = 2$, minimum

4

x	-3	-2	-1	0	1	2	3
y	1	-4	-7	-8	-7	-4	1

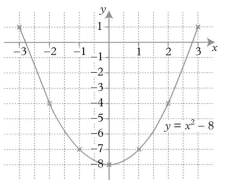

-8, $(0, -8)$, $x = 2 \cdot 8$, $x = -2 \cdot 8$, minimum

5

x	-4	-3	-2	-1	0	1	2
y	8	3	0	-1	0	3	8

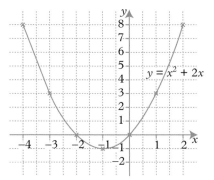

$y = x^2 + 2x$

0, (−1, 1), $x = 2$, $x = 0$, minimum

6

x	-5	-4	-3	-2	-1	0	1
y	5	0	-3	-4	-3	0	5

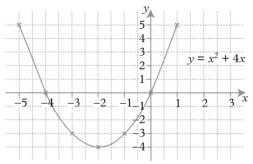

$y = x^2 + 4x$

0, (−2, −4), $x = 4$, $x = 0$, minimum

7

x	-2	-1	0	1	2	3	4
y	-5	-4	-1	4	11	20	31

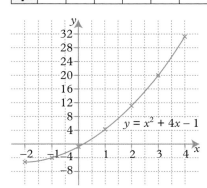

$y = x^2 + 4x - 1$

−1, (−2, −5), $x = 4·24$, $x = −0·24$, minimum

8

x	-4	-3	-2	-1	0	1	2
y	3	-2	-5	-6	-5	-2	3

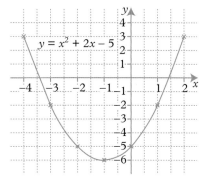

$y = x^2 + 2x - 5$

−5, (−1, −6), $x = 3·45$, $x = −1·45$, minimum

9

x	-4	-3	-2	-1	0	1	2
y	5	1	-1	-1	1	5	11

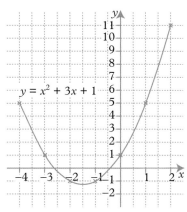

$y = x^2 + 3x + 1$

1, (−1·5, −1·25), $x = 0·38$, $x = 2·62$, minimum

page 307 Exercise 27 2

1 **a** $x = −1·6$ or $3·6$
   b $x = −0·4$ or $2·4$
   c $x = −1$ or 3

2 **a** $x = −0·4$ or $2·4$ **b** $x = 0$ or 2

3 **a** $x = 0$ or 3 **b** $x = −0·8$ or $3·8$

4 **a** $x = −1·6$ or $3·6$ **b** $x = −0·4$ or $2·4$

5 **a** $x = 0·5$ or $6·5$
   b The curve $y = x^2 − 7x$ does not meet the line $y = −14$.

6 a b

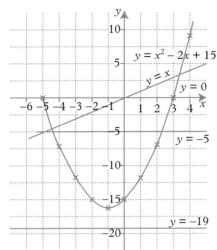

$y = x^2 - 2x + 15$

$y = x$

$y = 0$

$y = -5$

$y = -19$

c i −4·4, 2·4   **ii** −5, 3

    iii No solutions   **iv** −4·4, 3·4

7 a b

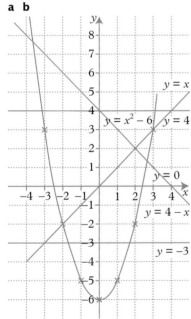

$y = x$

$y = x^2 - 6$ $y = 4$

$y = 0$

$y = 4 - x$

$y = -3$

c i $x = \pm3\cdot2$ **ii** $x = \pm2\cdot4$

    iii $x = \pm1\cdot7$ **iv** $x = -2$ or 3

    v $x = -3\cdot7$ or $2\cdot7$

8 $x^2 + 5x - 24 = 0$ **9** $x^2 + 4x - 32 = 0$

   $x = 3, -8$ $x = -8, 4$

10 $x^2 + 2x - 24$

   $x = -6, 4$

page 309 **Exercise 28** ②

1 $y = 2\ x = 5$ **2** $y = 3\ x = 4$ **3** $y = 5\ x = 3$

4 $y = 4\ x = 2$ **5** $y = 1\ x = 7$ **6** $y = 10\ x = 6$

7 $y = 7\ x = 5$ **8** $y = 2\ x = 2$ **9** $y = 2\ x = 10$

10 $x = 9\ y = 4$

page 310 **Exercise 29** ①

1 4   **2** 12   **3** −7   **4** −5

5 −5x   **6** 4y   **7** 3x   **8** 0

9 −8y   **10** 3x   **11** 0   **12** 13x

13 0   **14** 7   **15** 7x   **16** −10

page 310 **Exercise 30** ②

1 $x = 2, y = 4$ **2** $x = 1, y = 4$ **3** $x = 2, y = 5$

4 $x = 3, y = 7$ **5** $x = 5, y = 2$ **6** $a = 3, b = 1$

7 $x = -2, y = 3$ **8** $x = 4, y = 1$ **9** $x = \frac{5}{7}, y = 4\frac{3}{7}$

10 $x = 1, y = 2$ **11** $x = 2, y = 3$ **12** $x = 4, y = -1$

13 $x = 1, y = 2$ **14** $a = 4, b = 3$ **15** $x = 4\ y = 3$

16 $x = 5\ y = -2$

page 311 **Exercise 31** ②

1 $x = 2, y = 1$ **2** $x = 4, y = 2$ **3** $x = 3, y = 1$

4 $x = -\frac{1}{3}, y = -2\frac{1}{3}$ **5** $x = 3, y = 2$ **6** $x = 5, y = -2$

7 $x = 2, y = 1$ **8** $x = 5, y = 3$ **9** $x = 3, y = -1$

10 $a = 2, b = -3$ **11** $a = 5, b = \frac{1}{4}$ **12** $a = 1, b = 3$

13 $m = \frac{1}{2}, n = 4$ **14** $w = 2, x = 3$ **15** $x = 6, y = 3$

16 $x = \frac{1}{2}, z = -3$ **17** $m = 2, n = 1$ **18** $c = \frac{39}{23}, d = -\frac{58}{23}$

page 312 **Exercise 32** ②

1 $9\frac{1}{2}$ and $5\frac{1}{2}$ **2** 6 and 3   **3** 5 and 2

4 8 and 5   **5** $? = 10\frac{1}{2}, \star = 7\frac{1}{2}$ **6** 54°, 63°, 63°

7 $c = £3, p = £4$

8 TV costs £200, DVD player costs £450

9 $b = 3\frac{1}{2}$ oz, $w = 2$ oz   **10** £240

11 $m = 4, c = -3$

12 15 two pence coins; 25 five pence coins

13 14 ten pence coins; 7 fifty pence coins

14 20

page 314 **Test yourself**

1 $y(3 + y)$

2 a $x \le 4$ **b**

                −2 −1 0 1 2 3 4 5

3 a i $3p$ **ii** $8x + 3y - 4$

   b $B = 9$ **c** $K = 2$

4 a

        −2 −1 0 1 2 3 4   **b** $x \le 2$

5 a −10, 2   **b** graph of $y = 4x - 2$

   c $x = \frac{3}{2}$

6 a 26   **b** odd

7 a $2a + 4c$ **b** $S = \frac{9}{8}$

   c $x(x - 5)$ **d** $x = 2\cdot5$

8 Graph of $y = 3 - 2x$

9 a i 30, 30, 32, 34, 36, 40

    ii Graph of values

  b i y values at: 18, 21, 24, 27, 30, 33, 36, 39, 42

    ii 120 envelopes

10a $(4, 0)$ **b** 2   **c** $y = 2x + 1$

11a $2x + 4y$ **b** $12t - 4$ **c** $y = 32$ **d** $p = 6$

12a -1 **b** Graph of $y = x^2 + x - 3$

  c $x = -2 \cdot 3, 1 \cdot 3$

13a $2, -1, -1, 2$ **b** Graph of $y = x^2 - 2$

  c $x = -1 \cdot 4, 1 \cdot 4$

7   Geometry 2

page 319 **Exercise 1** ❶

1

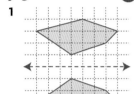

2

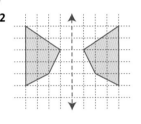

3

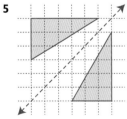

4

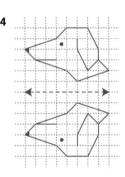

5

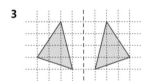

6

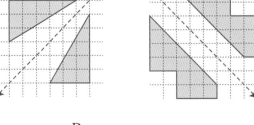

7

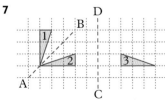

page 319 **Exercise 2** ❶

1

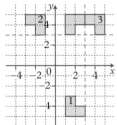

2 a, b, c

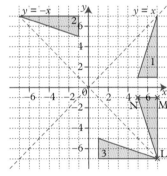

  d i $(7, -7)$ **ii** $(-5, 5)$ **iii** $(5, 7)$

3 a, b, c

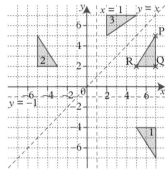

  d i $(7, 7)$ **ii** $(-7, 7)$ **iii** $(7, -7)$

4 a–f

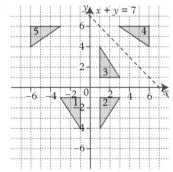

  g $(-3, 6), (-6, 6), (-6, 4)$

5 a–f

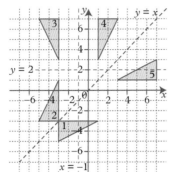

g (3, 1), (7, 1), (7, 3)

6 a $y = 0$ **b** $x = 1$ **c** $y = 1$
   d $y = -x$

page 321 **Exercise 3** ❶

1   **2**

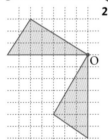

3   **4**

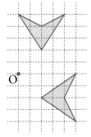

5   **6**

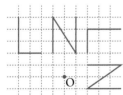
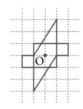

7 Shape 1: Centre C, 90°, clockwise
     Shape 2: Centre B, 180°, clockwise
     Shape 3: Centre A, 90°, anticlockwise
     Shape 4: Centre E, 90°, clockwise
     Shape 5: Centre F, 180°, clockwise

page 322 **Exercise 4** ❶

1

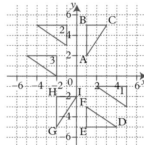

2 a–d

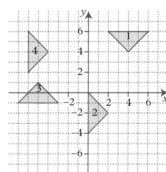

e (−1, −4), (3, −4), (2, −7)

3 a–d

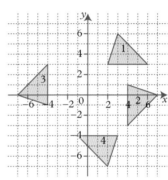

e (−6, 2), (−4, 4), (−6, 6)

page 323 **Exercise 5** ❶

1

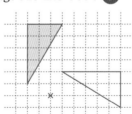

2

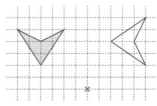

3

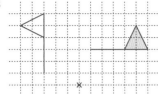

4 a (0, 0) **b** (1, 2) **c** (0, 0) **d** (−1, 1)

5 a

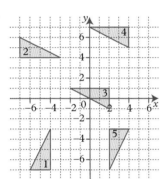

b i (3, 3)
 ii (0, 0)
 iii (2, 0)
 iv (−1, 1)

6 a

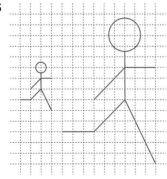

b i (0, 0)
 ii (−5, 0)
 iii (−7, 4)
 iv (−1, −5)

page 325 **Exercise 6** **1** **2**

1 a Yes **b** No **c** Yes **d** No
2 $x = 75\,\text{mm}$ **3 a** $y = 66\,\text{mm}$ **b** $z = 18\,\text{mm}$
4

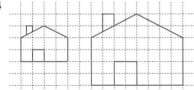

5

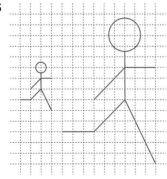

6 a unchanged **b** not congruent
7 OA′ = 2 × OA, OB′ = 2 × OB
8

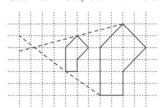

9 a **b** 1·5

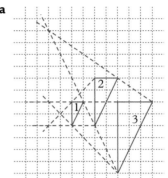

page 328 **Exercise 7** **1** **2**

1 **2**

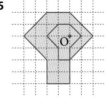

3 **4**

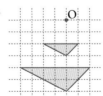

5 **6**

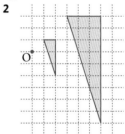

7 a-d

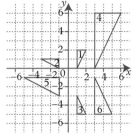

e (3, 0), (−5, −1), (3, −1)

f 27°, They are the same.

8 a-d

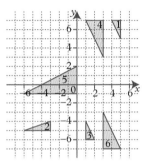

e (3, 3), (−6, −1), (3, −3)

9 a-d

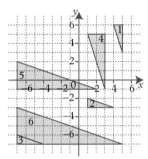

e (3, −1), (2, −1), (5, −7)

page 330 **Exercise 8** ②

1   **2**

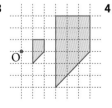 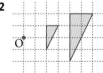

3   **4**

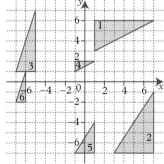

page 332 **Exercise 9** ①

1 a $\begin{pmatrix} 4 \\ 6 \end{pmatrix}$ **b** $\begin{pmatrix} 6 \\ 4 \end{pmatrix}$ **c** $\begin{pmatrix} 0 \\ 3 \end{pmatrix}$ **d** $\begin{pmatrix} 6 \\ 0 \end{pmatrix}$ **e** $\begin{pmatrix} 5 \\ -2 \end{pmatrix}$

   f $\begin{pmatrix} 1 \\ 2 \end{pmatrix}$ **g** $\begin{pmatrix} -2 \\ 5 \end{pmatrix}$ **h** $\begin{pmatrix} 2 \\ -2 \end{pmatrix}$ **i** $\begin{pmatrix} -4 \\ -3 \end{pmatrix}$ **j** $\begin{pmatrix} 2 \\ -6 \end{pmatrix}$

   k $\begin{pmatrix} 1 \\ -8 \end{pmatrix}$ **l** $\begin{pmatrix} -6 \\ -1 \end{pmatrix}$ **m** $\begin{pmatrix} 0 \\ -4 \end{pmatrix}$ **n** $\begin{pmatrix} 6 \\ 1 \end{pmatrix}$

page 334 **Exercise 10** ②

1 d   **2** 2c   **3** 3c   **4** 3d

5 5d   **6** 3c   **7** −2d   **8** −2c

9 −3c   **10** −c   **11** c + d   **12** c + 2d

13 2c + d   **14** 3c + d   **15** 2c + 2d

16 $\overrightarrow{QI}$ **17** $\overrightarrow{QU}$ **18** $\overrightarrow{QH}$ **19** $\overrightarrow{QB}$

20 $\overrightarrow{QF}$ **21** $\overrightarrow{QJ}$

22 a 2a + b   **b** 2a + 2b   **c** −a − b

    d 4a + 2b   **e** 2a − 2b   **f** 2a + b

23 a $\overrightarrow{CO}$ **b** $\overrightarrow{TN}$ **c** $\overrightarrow{FT}$ **d** $\overrightarrow{KC}$

24 a −a   **b** a + b   **c** 2a − b   **d** b − a

25 a a + b   **b** a − 2b   **c** b − a   **d** −a − b

26 a −a − b   **b** 3a − b   **c** 2a − b   **d** b − 2a

27 a a − 2b   **b** a − b   **c** 2a d   **3**b − 2a

28 a or **d**   **29** A and D, B and E, C and F

page 335 **Exercise 11** ① ②

1 a   Radius = 6 cm, diameter = 12 cm

  b   Radius = 9 cm, diameter = 18 cm

  c   Radius = 5 cm, diameter = 10 cm

  d   Radius = 2 m, diameter = 4 m

2 Radius = 1·4 cm, diameter = 2·8 cm

3 Radius = 1·1 cm, diameter = 2·2 cm

4 45°

5 a   chord   **b**   tangent   **c**   ninety degrees

  d   segment   **e**   arc   **f**   sector

page 337 **Exercise 12** ①

1 34·6 cm   **2** 25·1 cm   **3** 37·7 cm   **4** 15·7 cm

5 31·4 cm   **6** 40·8 cm   **7** 6·28 cm   **8** 13·8 cm

9 9π cm   **10** 34π m   **11** 14·2π m   **12** 46π m

13 1·9 cm   **14** 2·7π m   **15** 400 m   **16** 90·4 cm

17 823 m   **18** 212

19 A = 10π cm **B** = 13π cm   **20** 642·74 cm

page 339 **Exercise 13** ①

1 95·0 cm²   **2** 78·5 cm²   **3** 28·3 m²

4 38·5 m²   **5** 113 cm²   **6** 201 cm²

7 19·6 m²   **8** 380 cm²   **9** 29·5 cm²

10 125 m²   **11** 21·46 cm²   **12** 4580 g

13 30 if not staggered, 508 cm²   **14** 40·8 m²

15 13·73 cm²   **16** 42·92 cm²

page 341 **Exercise 14** **1** **2**

1 23·1 cm **2** 38·6 cm **3** 20·6 m **4** 8·23 cm
5 28·6 cm **6** 39·4 m **7** 17·9 cm **8** 28·1 m
9 24·8 cm **10** 46·3 cm **11** 28·8 cm

page 343 **Exercise 15** **2**

Answers are given to an appropriate degree of accuracy.

1 a 2·09 cm **b** 7·85 cm **c** 4·36 cm
2 15·36 cm
3 a 8·7 cm² **b** 40·6 cm² **c** 9·4 cm²
4 a 6·28 cm **b** 61·1 cm² **c** 11·64 cm
  d 150·5 cm² **e** 3945 cm²
5 19 cm
6 a 7·1 cm² **b** 9·5 cm²
7 17·8 cm **8** 74·2 cm³
9 a 12 cm **b** 30°
10 a 30° **b** 10·5 cm

page 345 **Exercise 16** **2**

1 a 12·57 cm² **b** 4·71 cm² **c** 49·09 cm²
2 a 4·71 cm **b** 10·47 cm **c** 11·0 cm
3 a 6·25 cm **b** 6·37 cm **c** 13·09 cm²
4 $PQ = \dfrac{40}{360} \times 2 \times \pi \times 9 = 2\pi$ **5** $x = 57·3°$

page 347 **Exercise 17** **1**

1 48 cm³, 88 cm² **2** 30 cm³, 62 cm²
3 60 cm³, 104 cm² **4** 120 cm³, 164 cm²
5 100 cm³, 174 cm² **6** 27 cm³, 54 cm²
7 27 cm³ **8** 4 cm³ **9** 8 cm³
10 9 cm³ **11** 6 cm³ **12** 12 cm³
13 9 or 10 cm³ **14** 1 hour 40 minutes
15 27 cm³
16 a 3 cm **b** 4 cm **c** 2 cm
17 333 **18** 1 000 000
19 No, the garage would be solid brick inside.

page 350 **Exercise 18** **1** **2**

1 a 150 cm³ **b** 60 cm³ **c** 110 cm³
  d 94·5 cm³ **e** 57 cm³ **f** 32 cm³
2 2400 cm³ **3 a** 200 m² **c** 2400 m³
4 a 62·8 cm³ **b** 113·1 cm³ **c** 502·7 cm³
5 a 502·7 cm³ **b** 760·3 cm³
6 10 litres **7** 7·1 kg (2 sf)
8 a 141·4 cm³ **b** 25·1 cm³ **9** 769·7 cm³
10 $x = 7\dfrac{7}{15}$ g/cm³ (or 7·46°) **11** 10 litres

page 352 **Exercise 19** **2**

1 40 cm³ **2** 33·5 cm³ **3** 66·0 cm³ **4** 89·8 cm³
5 339 cm³ **6** 144 cm³ **7** 4·71 kg **8** 262 cm³
9 359 cm³ **10** 235 cm³ **11** 20 m³ **12** 37·7 mm³
13 415 cm³ **14** 1·8 mins

page 353 **Exercise 20** **2**

1 a 36π cm² **b** 40π cm² **c** 60π cm² **d** 1·4π cm²
2 a 40π **b** 28π **c** 32π **d** 24π
3 £3 870 **4** £2205 **5** 675 cm²
6 a 1520·5 cm² **b** 153·9 cm²
7 a 125·7 cm² **b** 175·9 cm²

page 355 **Exercise 21** **1**

Agnes 035°, Belinda 070°, Carlo 155°,
Derek 220°, Ernie 290°, Louise 340°,
Ann 040°, Bill 065°, Chris 130°,
Duncan 160°, Eric 230°, Fred 330°

page 356 **Exercise 22** **1**

1 a 147° **b** 122° **c** 090°
2 a 285° **b** 225° **c** 154°
3 a 061° **b** 327°
4 a 302° **b** 344° **c** 046°

page 357 **Exercise 23** **1**

1 10 km **2** 5 km **3** 11·5 km
4 14·1 km **5 a** 12·5 km **b** 032°

page 357 **Exercise 24** **1**

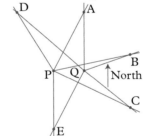

page 358 **Exercise 25** **1**

1 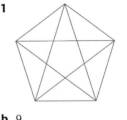 **2**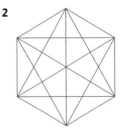

b 9
3 a Equilateral triangle **b** square
4 a, b

page 359 **Exercise 26** ❶

1 $a = 80°$, $b = 140°$, $c = 170°$, $d = 100°$,
$e = 100°$, $f = 110°$, $g = 120°$

2 a $36°$

b Angle ECB = $108° - 36° = 72°$
Angle ABC = $108°$
So angle ABC + angle ECB = $180°$
Therefore EC is parallel to AB.

3 a $1080°$ **b** $1440°$

4 a $3240°$ **b** $162°$

5 a $1800°$ **b** $30°$

page 361 **Exercise 27** ❶

1 $360°$ **2** $72°$

3 $a = 80°$, $b = 70°$, $c = 65°$, $d = 86°$, $e = 59°$

4 a $36°$ **b** $144°$

5 a $40°$ **b** $20°$ **c** $6°$

6 $p = 101°$, $q = 79°$, $x = 70°$, $m = 70°$, $n = 130°$

7 a $30°$ **b** $150°$

8 24 **9** 20 **10** 9

page 363 **Exercise 28** ❷

Students' own constructions

page 364 **Exercise 29** ❷

1 Circle, radius 3 cm centred on X

2 Perpendicular bisector of PQ

3 Angle bisector of angle BAC

4 a   **b**

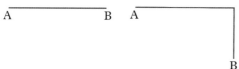

c

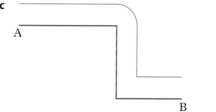

5

6 Intersection of circle radius 40 km centred on L and the circle radius 20 km centred on C.

7 b i Perpendicular bisector of LN

     ii Angle bisector of angle LNM

     iii Circle radius 4 cm centred on M

8 Students' own construction

9 Line perpendicular to AB through P

10

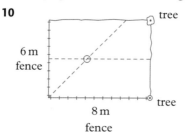

page 366 **Exercise 30** ❷

1 a 7·21 cm   **b** 5 cm   **c** 5·39 cm   **d** 12·21 cm

    e 13 cm   **f** 5·20 cm   **g** 6·26 cm   **h** 12·73 cm

2 11·31 cm

3 a 8·32 cm   **b** 12·32 cm   **4** 6·32 m   **5** 114·13 m

page 367 **Exercise 31** ❷

1 a 5·74 cm   **b** 4·47 cm   **c** 5·66 cm   **d** 8·49 cm

    e 4 cm   **f** 7·5 cm   **g** 5 cm   **h** 4·65 cm

2 3·46 m   **3** 7·07 cm

page 368 **Exercise 32** ❷

1 10 cm   **2** 8·94 cm   **3** 5·66 cm   **4** 5·66 cm

5 4·24 cm   **6** 9·90 cm   **7** 4·58 cm   **8** 5·20 cm

9 9·85 cm   **10** 5·39 units **11** 9·80 cm   **12** 4·27 m

13 40·31 km   **14** 5·70 cm

page 369 **Exercise 33** ❷

1 a 5 units   **b** 40 square units   **2** 9·49 cm

3 32·56 cm   **4 a** 5 units   **b** 4·12 units

5 6·9 m   **6** 6·63 units   **7** 5·57 units

8 8·7 units   **9** 5·66 units   **10** 6·63 units

11 4·69 units

page 371 **Exercise 34** ❷

2 $\sin x = \dfrac{O}{H} = \dfrac{5}{13}$ **3** $\sin y = \dfrac{O}{H} = \dfrac{6}{10}$

4 $\sin z = \dfrac{O}{H} = \dfrac{7}{25}$ **5** $\cos w = \dfrac{A}{H} = \dfrac{4}{5}$

6 $\cos x = \dfrac{A}{H} = \dfrac{12}{13}$ **7** $\cos y = \dfrac{A}{H} = \dfrac{8}{10}$

8 $\tan w = \dfrac{O}{A} = \dfrac{3}{4}$ **9** $\tan x = \dfrac{O}{A} = \dfrac{5}{12}$

10 $\tan y = \dfrac{O}{A} = \dfrac{6}{8}$ **11** $\cos z = \dfrac{A}{H} = \dfrac{24}{25}$

12 $\tan z = \dfrac{O}{A} = \dfrac{7}{24}$

page 372 Exercise 35

1 3·01 cm	**2** 5·35 cm	**3** 3·13 cm
4 7·00 cm	**5** 73·1 cm	**6** 15·4 cm
7 5·31 cm	**8** 7·99 cm	**9** 11·6 cm
10 11·4 cm	**11** 961 cm	**12** 19·7 cm
13 46·0 cm	**14** 34·9 cm	**15** 9·39 cm
16 8·23 cm	**17** 35·6 cm	**18** 80·2 cm

page 373 Exercise 36 2

1 18·4 cm	**2** 9·15 cm	**3** 10·7 cm
4 17·1 cm	**5** 13·7 cm	**6** 126 cm

7

	30°	45°	60°	120°	135°	150°
sin	$\frac{1}{2}$	$\frac{1}{\sqrt{2}}$	$\frac{\sqrt{3}}{2}$	$\frac{\sqrt{3}}{2}$	$\frac{1}{\sqrt{2}}$	$\frac{1}{2}$
cos	$\frac{\sqrt{3}}{2}$	$\frac{1}{\sqrt{2}}$	$\frac{1}{2}$	$\frac{1}{2}$	$\frac{-1}{\sqrt{2}}$	$\frac{\sqrt{3}}{2}$
tan	$\frac{1}{\sqrt{3}}$	1	$\sqrt{3}$	$\sqrt{3}$	−1	$\frac{1}{\sqrt{3}}$

page 374 Exercise 37 2

1 38·7°	**2** 48·6°	**3** 31·0°	**4** 54·5°
5 38·7°	**6** 17·5°	**7** 38·9°	**8** 59·0°
9 41·3°	**10** 62·7°	**11** 54·3°	**12** 66·0°
13 48·2°	**14** 12·4°	**15** 72·9°	**16** 56·9°

page 376 Exercise 38 2

1 68·0°	**2** 3·65 m	**3** 14 m	**4** 20·6°
5 56·7°	**6** 15 m	**7** 7·66 cm	**8** 66 km
9 189 km	**10** 36·4°	**11** 10·3 cm	**12** 180 km

page 378 Test yourself

1 a 8   **b** Net of cube   **c** 54 cm^2
2 a 207·57 cm^3 **b** 1000 cuboids
3 a 24 m^3 **b** 7 tins
4 1240 cm^3
5 a 150°   **b** 18 km
  c (7·5 cm) Student's own construction
6 a 050°, 12 km
  b

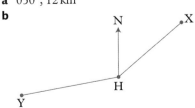

7 320π cm^3
8 a i trapezium   **ii** 6 cm^2

b

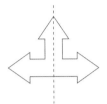

c

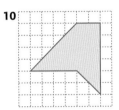

9 a

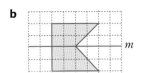

b i order 4   **ii**

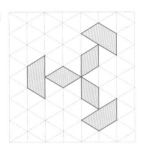

10

11 30·9 cm^2
12 a 79·8 cm^2 **b** 788·2 cm^2
13 73·5 cm   **14** Student's own construction
15 a 32 cm
  b i 360° ÷ 6 = 60° **ii** 188·5 cm
  c £2·40
16 Student's own construction
17 a Pentagon **b** 5 **c** 5 **d** 5x cm
18 a Angles on a straight line add to 180°.
  b $x = 20°$
19 $19·4^2 \neq 11·5^2 + 14·7^2$ **20** 3·52 m

8   Statistics 2

page 388 Exercise 1 1

1 Not satisfactory. She will not get the opinions of students that no longer have school dinners.
2 Not satisfactory. People travelling abroad will have different opinions to those that have never been abroad.
3 Not satisfactory. Not all the people may be drivers.
4 She will not get the views of people at work or school.

page 390 **Exercise 2** ①

1 No choices, such as beach, winter, activity.
2 Too vague: Do you like the new head teacher? yes/no
3 Choices should include less than 2 hours and more than 6 hours. They also shouldn't overlap. No option for 4–5 hours.
4 There should be a choice for between £1 and £2·50.
5 Leading question. Which is the most important subject at school? Maths, English, Science, other.
6 Not clear what 'often' means. How often do you or your parents hire DVDs from a shop? Never, 1–2 times a month, 3–4 times a month, 5 or more times a month.
7 Leading question. Do you get too much homework? yes/no
8 Name is not needed. There should be a choice of age ranges to choose from. The choices for question 1 are not specific enough. Question 2 should have some options including Other. Question 3 is leading. Question 4 is too vague. A better question would be 'How often do you watch nature programs?' with some numeric choices.
9 What type of TV program do you most watch? Comedy, sport, news, soaps, films, factual, other.
10 Many possible answers.

page 392 **Exercise 3** ①

1 a 5   b 19   c 23

   d 55   e $\frac{6}{23}$

2

Score	Tally	Frequency
1	卌 I	6
2	IIII	4
3	IIII	4
4	III	3
5	卌 III	8
6	卌	5

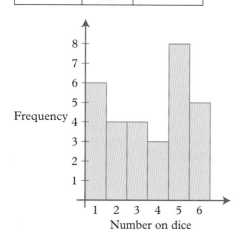

3 a

Number of letters	Tally	Frequency
1	卌 I	6
2	卌 II	7
3	卌 III	8
4	卌 卌 II	12
5	卌 I	6
6	卌 III	8
7	IIII	4
8	II	2
9	II	2
10	I	1
	Total	56

   b 4

4 a

Score	Tally	Freq.	Score	Tally	Freq.
2	II	2	8	卌 I	6
3	IIII	4	9		0
4	卌	5	10	卌	5
5	卌 II	7	11	卌 I	6
6	卌 IIII	9	12	III	3
7	卌 卌 III	13		TOTAL	60

   b

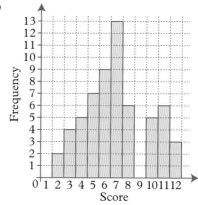

5 a £3,400
   b The shop's profits increase drastically in the run-up to Christmas. However, in January, sales drop-off as no more presents are being bought. Sales begin to improve as the year progresses, but a loss is still being made as late as April.

page 395 **Exercise 4** ①

1 a 5   b 24 c   35
2 a D   b A c   A
   d D   e B

3

Weight (gm)	Tally	Frequency
60 ≤ w < 70	||||	4
70 ≤ w < 80	ⅢⅢ |	6
80 ≤ w < 90	ⅢⅢ	5
90 ≤ w < 100	ⅢⅢ ⅢⅢ	10
100 ≤ w < 110	ⅢⅢ ⅢⅢ ||||	14
110 ≤ w < 120	ⅢⅢ ⅢⅢ |	11
		50

Field A was treated with the new fertilizer.

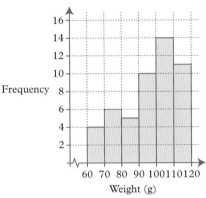

4 Should go to Isola, as it has 16 days with more than 40 cm of snow compared to only 3 days in Les Jets.

5 a Multiple answers possible.
b Multiple answers possible.
c Check conclusion against answer **(b)**.

page 399 Exercise 5 ① ②

1 Many possible answers.
2 a Strong positive correlation **b** No correlation

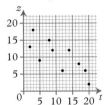

c Weak negative correlation

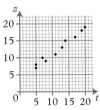

3 a No correlation
b Strong negative correlation
c No correlation
d Weak positive correlation

4 a 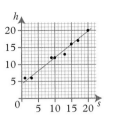 **b** 9

5 a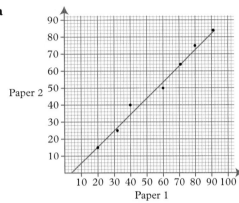

b 44
6 a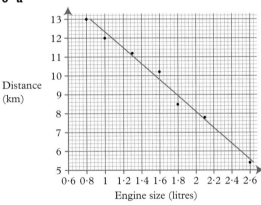

b 6·8 km
7 a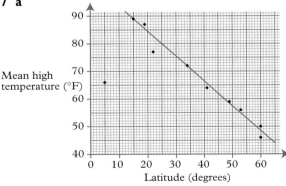

b Bogota, high altitude **c** 75°F
8 a Positive correlation **b** Positive correlation
c No correlation **d** Negative correlation
e Positive correlation

page 402 **Test yourself**

1 a Correct graph plotted **b** 41 tonnes (line of best fit)
  c Yes, positive correlation.
2 a 171 cm, 90 kg **b** Positive   **c** 65 kg
  d No, as extends beyond the data.
3 a Duplicates responses for 4 and 8.
    No response for more than 12.
    Most people may only buy one or two.
  b Student's own answer
  c Bias, as not a sample of total population.
4 a Method of transport may vary over the week.
  b Duplicated numbers, no option for more than 50.
5 a Too few options
  b Limited sample, prone to bias.
6 a 21, 37, 77, 53   **b** 8:7
7 a

30-49	5
50-69	7
70-89	4
90-109	6
110-129	6

  b 12 days

8 7, 7, 7, 10, 33, 53, 22

9   Number 3

page 407 **Exercise 1** ① ②

1 a $0 \cdot 25$ **b** $0 \cdot \dot{4}$ **c** $0 \cdot 375$
  d $0 \cdot 41\dot{6}$ **e** $0 \cdot 1\dot{6}$ **f** $0 \cdot \dot{2}8571\dot{4}$
2 a $\frac{1}{5}$ **b** $\frac{9}{20}$ **c** $\frac{9}{25}$
  d $\frac{1}{8}$ **e** $1\frac{1}{20}$ **f** $\frac{7}{1000}$
3 a 25%   **b** 10%   **c** 72%
  d 7·5%   **e** 2%   **f** $3\dot{3}\cdot\dot{3}\%$
4 a 45 000   **b** 396 000   **c** $\frac{1}{30}$

5

Fraction	Decimal	Percentage
$\frac{1}{4}$	0.25	25%
$\frac{1}{5}$	0.2	20%
$\frac{4}{5}$	0.8	80%
$\frac{1}{100}$	0.01	1%
$\frac{3}{10}$	0.3	30%
$\frac{1}{3}$	0.3	33.3%

6 a $\frac{1}{5}$ **b** $\frac{4}{10}$ **c** $\frac{11}{33}$ **d** $\frac{7}{12}$

7 A   **8** $\frac{1}{3}$ off is cheaper

9 $\frac{1}{6}$ of £5000   **10** 25%   **11** 20%

page 409 **Exercise 2** ① ②

1 a 110 miles   **b** 340 m   **c** 315 km
  d 60 cm
2 a 14 mph   **b** 45·5 km/h   **c** 25 m/s
  d 10·5 cm/min
3 120 miles   **4** 500 m   **5** 300 m
6 360 m   **7** 2000 km   **8** 12 km
9 61 km/h   **10** 7·5 m/s   **11** 83 km/h
12 0·5 cm/s   **13** 3 hours   **14** 6 hours
15 250 km
16 a 18 hours   **b** 7 hours   **c** 5 seconds
  d 30 minutes
17 88 mph   **18** $4\frac{1}{2}$ km

page 410 **Exercise 3** ②

1 a $8 \, \text{g/cm}^3$ **b** $2 \cdot 5 \, \text{g/cm}^3$ **c** $5 \cdot 6 \, \text{g/cm}^3$
2 8 kg
3 a $2400 \, \text{cm}^3$ **b** 21·6 kg
4 a $4 \cdot 52 \, \text{cm}^2$ **b** $0 \cdot 9 \, \text{cm}^3$ **c** 7·24 g
5 80 g; $2 \, \text{g/cm}^3$; $10 \, \text{cm}^3$; $3 \, \text{g/cm}^3$; 2250 g
6 $6 \, \text{g/cm}^3$ **7** 375 g
8 4264 g   **9** £11 760
10 £17·05
11 a 2993 g > 2257 g   **b** 3584 g > 3467·5 g
12 $469 \, 800 \, \text{cm}^3$

page 412 **Exercise 4** ②

1 £90   **2** £15 000   **3** £700
4 £329 000   **5** £30 000
6 a 924 people/square mile   **b** 43 400

page 413 **Exercise 5** ①

1 400 cm   **2** 2400 cm   **3** 63 cm   **4** 0·25 cm
5 7 cm   **6** 20 mm   **7** 1200 m   **8** 8·15 m
9 0·65 km   **10** 2·5 cm   **11** 5000 g   **12** 4200 g
13 6400 g   **14** 3000 g   **15** 800 g   **16** 0·4 kg
17 2000 kg   **18** 0·25 kg   **19** 500 kg   **20** 620 kg
21 7000 g   **22** 1·5 kg   **23** 8000 ml   **24** 2000 ml
25 1 litre   **26** 4500 ml   **27** 320 cm   **28** 5·5 cm
29 1400 g   **30** 110 mm   **31** $2 \, \text{m}^2$ **32** $125 \, 000 \, \text{cm}$
33 $500 \, 000 \, \text{cm}^3$ **34** $0 \cdot 001 \, \text{m}^3$
35 a kilometres **b** millilitres **c** grams
  d millimetres **e** tonnes   **f** square metres

page 413 **Exercise 6 ① ②**

1 24	**2** 16	**3** 30	**4** 4480	**5** 48
6 17 600	**7** 36	**8** 70	**9** 4	**10** 880
11 3	**12** 2	**13** 3520	**14** 80	**15** 140
16 12	**17** 48	**18** 22 400	**19** 5280	**20** 2

page 414 **Exercise 7 ① ②**

1 25 cm	**2** 50 litres	**3** 6 pounds
4 8 pounds	**5** 5 miles	**6** 1 kg
7 20 miles	**8** 10 cm	**9** 4 kg
10 5 litres	**11** 25 miles	**12** 20 gallons
13 6 pounds	**14** 8 km	**15** 80 gallons
16 2 ounces	**17** 300 g	**18** 1 gallon
19 40 pounds	**20** 10 inches	**21** 30 cm
22 5000 miles	**23** 50 mph	

24 Super market B

25 Bake in a 30 cm dish, 500 g flour, 120 g sugar, 8 large apples (about 10 cm in diameter.)

26 10 g	**27** 7 m	**28** 500 ml
29 100 miles	**30** 3 mm	

31 a 2·75 kg   **b** 15 litres   **c** 3 pounds

32 $\frac{1}{8}$

page 417 **Exercise 8 ②**

1 195·5 cm	**2** 36·5 kg	**3** 95·55 m
4 3·25 kg	**5** 28·65 s	

6 a 1·5 °C, 2·5 °C   **b** 2·25 g, 2·35 g
   c 63·5 m, 64·5 m   **d** 13·55 s, 13·65 s

7 B   **8** C

9 a No   **b** 1 cm

10 Yes, the largest the card could be is 11·25 cm, the smallest the envelope could be is 11·5 cm

11 16·5 kg, 17·5 kg	**12** 255·5 km, 256·5 km
13 2·35 m, 2·45 m	**14** 0·335 g, 0·345 g
15 2·035 m/s, 2·045 m/s	**16** 7·5 cm, 8·5 cm
17 81·35 °C, 81·45 °C	**18** 0·25 kg, 0·35 kg
19 4·795 cm, 4·805 cm	**20** 0·065 cm, 0·075 cm

21 0·65 tonne, 0·75 tonne

22 614·5 seconds, 615·5 seconds

23 7·125 m, 7·135 m   **24** 51 500, 52 500

page 419 **Exercise 9 ①**

1 4p per can	**2** 11p per ruler
3 9p per card	**4** 10p per can
5 4p per paper	**6** 18p per box
7 6p per kg	**8** 13p per kg
9 £1·75 per T-shirt	**10** 36p per dozen
11 2·5p per orange	**12** £7·25
13 15p per 100 ml	**14** 5p per kg
15 16p per m	**16** 89p per kg
17 14p per m	**18** 2p per apple
19 £3·60	**20** 15p per tin

page 419 **Exercise 10 ① ②**

1 a
```
    5 7 3 2
  +   2 6 9 6
  ‾‾‾‾‾‾‾‾‾‾
    8 4 2 8
```
    b
```
      8 3 5
  −   2 6 2
  ‾‾‾‾‾‾‾‾‾
      5 7 3
```

  c 245 ÷ 7 = 35

2 6p   **3** 15 litres

4 a 36   **b** £11 111

5 a £1600   **b** 8%

6 70 mm   **7** 24

8 One

9 a 0·3   **b** 0·003   **c** 0·115   **d** 70 000

10 1·29

page 421 **Exercise 11 ① ②**

1 20	**2** 200 g	**3** Six
4 400	**5** 3854	

6 a 323 g   **b** 23   **c** 67p   **d** 29

7 20 cm^2 **8** 22 222   **9** 234 − 65

10 a £131·50   **b** £45·20

page 422 **Exercise 12 ① ②**

1 a
```
    4 3 2
  +   2 3 1
  ‾‾‾‾‾‾‾‾
  1 3 9 8
```
    b
```
    4 3 6
    5 2 1
  +   1 1 5
  ‾‾‾‾‾‾‾‾
  1 0 7 2
```

2 1221

3 a $x = 1$ **b** $x = 15$

4 £18

5

	Passed	Failed	Total
Boys	350	245	595
Girls	416	180	596
Total	766	425	1191

6 a 18, 19   **b** 24, 25, 26, 27
  c 11, 37   **d** 3, 16

7 8   **8** 120

9 125°   **10** 57%

page 423 **Test yourself**

1

Fraction	Decimal	Percentage
$\frac{1}{2}$	0·5	50%
$\frac{3}{4}$	0·75	75%
$\frac{1}{4}$	0·25	25%
$\frac{9}{100}$	0·09	9%

2 a g or kg   **b** cm or mm   **c** tonne
3 a km   **b** kg   **c** cm^3 or litres **d** m
4 a 850   **b** 900   **c** 854·3
5 a centimetres   **b** grams
   c millilitres **d** kilometres **e** kilograms
6 a 7am Thursday morning **b** 36 km/h
7 a i   kilometres   **ii** litres
   b i   50 mm   **ii** 4 kg
8 19·5 kg; 20·49 kg; 745 ml; 754 ml
9 16 km
10 a 30 cm/s   **b** 90 m
11 a 1·5   **b** 0·17   **c** 87   **d** £1·26
12 a i 24°C   **ii** 24·2 cm   **b** 5 jugs
13 a 114·5 cm, 115·5 cm
   b The space could be 1·15 m wide, but the wardrobe could be 1·115 m wide.
14 a 800 g under   **b** 22·7 kg > 20 kg. Over limit.

10   Probability

page 428 Exercise 1 ❶

1 a {3, 6, 9, 12, ...}   {3, 6, 9}
   b M   **c** $9 \notin M$
2 a Boys' names with the initial J
   b Multiples of 5
   c Square numbers less than 82
   d The initials of the colours of the rainbow
3 a 6, 8, 10, 12, 14   **b** a, b, c, d
   c 1, 8, 27, 64   **d** A, C, E, H, I, M, S, T
4 true **a, b, f**   false **c, d, e**
5 a

   c

6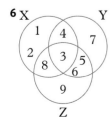

1 16   **2 a** 12   **b** 3
3 a HH, HT, TH, TT
4 b DD, DC, DR, CD, CC, CR, RD, RC, RR
   c 5   **d** $\dfrac{5}{9}$
5 a 11   **b** 6   **c** 5   **d** 25
   e $\dfrac{10}{36} = \dfrac{5}{18}$ **f** $\dfrac{1}{36}$
6 a BBB, BBG, BGB, BGG, GBB, GBG, GGB, GGG
   b Frequencies are 1, 3, 3, 1.
   c A grid has only two axes. A three-child family would need three axes which are difficult to draw on paper in 2-D.
7 a 1   **b** 3   **c** $\dfrac{3}{8}$ **d** $\dfrac{1}{8}$
8 1, 5   1, 10   1, 20   2, 5   2, 10   2, 20
9 a AY AZ BY BZ CY CZ DY DZ; 8 ways
   b KP KQ KR LP LQ LR MP MQ MR; 9 ways
10 a 20   **b** 30   **c** 420   **d** 525

1–12 Many possible answers.

1 $\dfrac{1}{6}$ **a, b** Many possible answers.
2 a 25%   **b** Higher number of trials
3 Many possible answers.
4 $\dfrac{6}{36} = \dfrac{1}{6}$
5 a 67   **b** 200

1 a $\dfrac{1}{3}$ **b** $\dfrac{1}{3}$
2 a $\dfrac{1}{4}$ **b** $\dfrac{1}{4}$ **c** $\dfrac{1}{4}$
3 a $\dfrac{2}{3}$ **b** $\dfrac{1}{3}$
4 a $\dfrac{1}{6}$ **b** $\dfrac{1}{6}$ **c** 0
5 a $\dfrac{1}{8}$ **b** $\dfrac{1}{8}$ **c** $\dfrac{1}{2}$
6 a $\dfrac{3}{4}$ **b** $\dfrac{1}{2}$ **c** $\dfrac{1}{4}$
   d $\dfrac{1}{3}$ **e** $\dfrac{1}{8}$ **f** $\dfrac{1}{2}$
   g $\dfrac{11}{12}$ **h** $\dfrac{3}{5}$
7 a True   **b** False   **c** False

8 a $\frac{1}{13}$ **b** $\frac{1}{52}$ **c** $\frac{1}{4}$

9 a $\frac{1}{9}$ **b** $\frac{1}{3}$ **c** $\frac{4}{9}$ **d** $\frac{2}{9}$

10 a $\frac{5}{11}$ **b** $\frac{2}{11}$ **c** $\frac{4}{11}$

11 a $\frac{4}{17}$ **b** $\frac{3}{17}$ **c** $\frac{11}{17}$

12 a $\frac{4}{17}$ **b** $\frac{8}{17}$ **c** $\frac{5}{17}$

13 a $\frac{2}{9}$ **b** $\frac{2}{9}$ **c** $\frac{1}{9}$

   d 0   **e** $\frac{5}{9}$

14 a $\frac{1}{13}$ **b** $\frac{2}{13}$ **c** $\frac{1}{52}$ **d** $\frac{5}{13}$

page 439 Exercise 6 ①

1 a $\frac{1}{7}$ **b** $\frac{4}{7}$ **c** $\frac{6}{7}$

2 a $\frac{1}{6}$ **b** $\frac{1}{6}$ **c** $\frac{1}{6}$

3 a $\frac{1}{5}$ **b** $\frac{1}{5}$ **c** $\frac{2}{5}$

4 a $\frac{1}{10}$ **b** $\frac{3}{10}$ **c** $\frac{3}{10}$

5 a $\frac{3}{13}$ **b** $\frac{5}{13}$ **c** $\frac{8}{13}$

6 a i $\frac{5}{13}$ **ii** $\frac{6}{13}$

   b i $\frac{5}{12}$ **ii** $\frac{1}{12}$

7 a $\frac{1}{12}$ **b** $\frac{1}{40}$ **c** $\frac{1}{4}$

8 $\frac{9}{20}$ **9** Megan

10 3 white, 6 black   **11** $\frac{1}{7}$

12 a 1   **b** He may have just not picked the blue ball.

13 a i $\frac{1}{4}$ **ii** $\frac{1}{4}$ **iii** $\frac{1}{4}$

   b $\frac{1}{4}$ **c** $\frac{6}{27}$

14 15

page 442 Exercise 7 ①

1 a Student's own diagram   **b** $\frac{1}{8}$

2 16 different outcomes, $\frac{1}{16}$

3 a 36   **b** 4 **c** $\frac{1}{9}$

4 a (1, 1), (1, 2), (1, 3), (2, 1), (2, 2), (2, 3), (3, 1), (3, 2), (3, 3), (4, 1), (4, 2), (4, 3)

   b 3   **c** $\frac{1}{4}$

	Red Spinner			
	1	**2**	**3**	**4**
White Spinner 1	1,1 2	1,2 3	1,3 4	1,4 5
2	2,1 3	2,2 4	2,3 5	2,4 6
3	3,1 4	3,2 5	3,3 6	3,4 7

5 $(w, x), (w, y), (w, z), (x, y), (x, z), (y, z)$

6 a   Dice   **b** $\frac{1}{2}$

	1	**2**	**3**	**4**	**5**	**6**
1	1,1	1,2	1,3	1,4	1,5	1,6
Spinner 2	2,1	2,2	2,3	2,4	2,5	2,6
3	3,1	3,2	3,3	3,4	3,5	3,6
4	4,1	4,2	4,3	4,4	4,5	4,6
5	5,1	5,2	5,3	5,4	5,5	5,6

7 a (A, C, F), (B, C, F), (A, D, F), (B, D, F), (A, E, F), (B, E, F), (A, C, G), (B, C, G), (A, D, G), (B, D, G), (A, E, G), (B, E, G)

   b $\frac{1}{12}$

8 a 15   **b** 676

9 a 24 different ways   **b** 12

10 Yes   **11** No

page 445 Exercise 8 ①

1 $\frac{4}{5}$

2 a $\frac{1}{13}$ **b** $\frac{12}{13}$ **c** $\frac{3}{13}$ **d** $\frac{10}{13}$

3 $\frac{35}{36}$ **4** 0·76   **5** 0·494

6 a $\frac{1}{4}$ **b** $\frac{3}{4}$ **c** $\frac{1}{4}$ **d** $\frac{3}{4}$

   e 0   **f** 1

7 a $\frac{1}{17}$ **b** $\frac{16}{17}$ **c** $\frac{4}{17}$

8 a 0·3   **b** 0·9

9 a i 0·24   **ii** 0·89   **b** 575

page 448 Exercise 9 ②

1 a $p(A) = \frac{1}{13}$, $p(B) = \frac{1}{6}$ **b** $\frac{1}{78}$

2 a $\frac{1}{2}$ **b** $\frac{1}{2}$ **c** $\frac{1}{4}$

3 $\frac{1}{10}$

4 a $\frac{1}{78}$ **b** $\frac{1}{104}$ **c** $\frac{1}{24}$

5 a $\frac{1}{16}$ **b** $\frac{1}{169}$ **c** $\frac{9}{169}$

6 a $\frac{1}{16}$ **b** $\frac{25}{144}$

7 a $\frac{1}{121}$ **b** $\frac{9}{121}$ **8** $\frac{8}{1125}$

9 a $\frac{1}{288}$ **b** $\frac{1}{72}$

10 a $\frac{1}{9}$ **b** $\frac{4}{27}$ **11** $\frac{1}{24}$

12 $\frac{1}{128}$ **13** $\frac{1}{144}$

14 a $\left(\frac{1}{6}\right)^{20} = 2\cdot7 \times 10^{-16}$ **b** $\left(\frac{5}{6}\right)^{n}$ **c** $1 - \left(\frac{5}{6}\right)^{n}$

page 452 **Exercise 10** ②

1 a $\frac{1}{8}$ **b** $\frac{3}{8}$ **2 a** $\frac{1}{4}$ **b** $\frac{1}{4}$

3 a $\frac{49}{100}$ **b** $\frac{9}{100}$ **4 a** $\frac{9}{64}$ **b** $\frac{15}{64}$

5 a $\frac{1}{4}$ **b** $\frac{3}{50}$ **c** $\frac{2}{50}$

6 a $\frac{7}{15}$ **b** $\frac{1}{15}$

7 a $\frac{1}{12}$ **b** $\frac{1}{6}$ **c** $\frac{1}{3}$ **d** $\frac{2}{9}$

8 a $\frac{1}{216}$ **b** $\frac{125}{216}$ **c** $\frac{25}{72}$ **d** $\frac{91}{216}$

9 a $\frac{1}{64}$ **b** $\frac{5}{32}$ **c** $\frac{27}{64}$

10 a $\frac{1}{6}$ **b** $\frac{1}{30}$ **c** $\frac{1}{30}$ **d** $\frac{29}{30}$

11 a $\frac{9}{16}$ **b** $\frac{1}{16}$ **12 a** 6 **b** $\frac{1}{3}$

13 a $\frac{4}{9}$ **b** $\frac{1}{24}$ **14 a** $\frac{3}{20}$ **b** $\frac{9}{20}$

15 a $\frac{6}{6840}$ **b** $\frac{60}{6840}$ **c** $\frac{120}{116280}$

page 455 **Exercise 11** ②

1 a $\frac{90}{999000}$ **b** $\frac{979\,110}{999\,000}$ **c** $\frac{19\,800}{999\,000}$

2 a $\frac{3}{20}$ **b** $\frac{7}{20}$ **c** $\frac{1}{2}$

3 a 5 **b** $\frac{1}{64}$

4 a $\frac{1}{220}$ **b** $\frac{1}{22}$ **c** $\frac{3}{11}$ **d** 5

5 a $\frac{3}{5}$ **b** $\frac{1}{3}$ **c** $\frac{2}{15}$

d $\frac{2}{21}$ **e** $\frac{1}{7}$

6 a 0·00781 **b** 0·511

7 a $\frac{21}{506}$ **b** $\frac{455}{2024}$ **c** $\frac{945}{2024}$

8 a $\frac{x}{x+y}$ **b** $\frac{x(x-1)}{(x+y)(x+y-1)}$

c $\frac{2xy}{(x+y)(x+y-1)}$ **d** $\frac{y(y-1)}{(x+y)(x+y-1)}$

9 a $\frac{x}{z}$ **b** $\frac{x(x-1)}{z(z-1)}$ **c** $\frac{2x(z-x)}{z(z-1)}$

10 a $\frac{1}{125}$ **b** $\frac{1}{125}$ **c** $\frac{1}{10\,000}$ **d** $\frac{3}{500}$

11 a $\frac{1}{49}$ **b** $\frac{1}{7}$ **12 a** $\frac{1}{52}$ **b** 1

page 458 **Exercise 12** ②

1 a $\frac{23}{50}$ **b** $\frac{11}{50}$ **c** $\frac{13}{25}$ **d** $\frac{13}{24}$

2 a $\frac{2}{15}$ **b** $\frac{4}{45}$ **c** $\frac{1}{10}$ **d** $\frac{2}{3}$

3 a $\frac{1}{24}$ **b** $\frac{2}{3}$ **c** $\frac{1}{10}$ **d** $\frac{6}{7}$

4 a $\frac{1}{5}$ **b** $\frac{13}{20}$ **c** $\frac{9}{13}$ **d** $\frac{2}{5}$

5 a $\frac{11}{25}$ **b** $\frac{8}{25}$ **c** $\frac{8}{11}$ **d** $\frac{1}{7}$

6 a $\frac{18}{25}$ **b** $\frac{3}{10}$ **c** $\frac{5}{12}$ **d** $\frac{6}{7}$

page 460 **Test yourself**

1 a fifty-fifty **b** likely **c** unlikely **d** certain

2 a A Impossible, B Unlikely, C Even, D Likely, E Certain

b Likely

3 a i $\frac{1}{6}$ **ii** 0 **b** 20

4 a $\frac{1}{520}$ **b** $\frac{7}{13}$

5 a 12 Outcomes

b i $\frac{1}{4}$ **ii** $\frac{5}{12} > \frac{1}{4}$

6 0·2, 0·3, 0·3

7 a $\frac{3}{8} = 0\cdot375$ **b i** $\frac{7}{20} = 0\cdot35$

ii Each experiment will usually have different results.

c 250 times

8 a 6 outcomes **b** $\frac{2}{3}$

9 a 0·25

b Rain and sunny are not mutually exclusive out comes.

10 a 12, 14, 20, 22; 11, 13, 19, 21…

b $\dfrac{3}{16}$ **c** 15

11 a 0·18, 0·32   **b** 0·3

12 a 0·08   **b** 0·57   **c** 875

13 a 0·35   **b** 42

11 Problems, proofs and puzzles

page 465 **11.1.1 Opposite corners**

The difference in an $n \times n$ square is $(3n - 3)^2$.

page 466 **11.1.2 Hiring a car**

Hav-a-car is cheapest up to 950 miles. Snowdon is cheapest between 950 and 1300 miles. Gibson is cheapest over 1300 miles.

page 466 **11.1.3 Half-time score**

12 different scores were possible at half-time. The general rule is $(a + 1)(b + 1)$ for a final score $a - b$.

page 466 **11.1.4 Squares inside squares**

Area is always $a^2 + b^2$.

page 468 **11.1.5 Maximum box**

The corner squares should be $\dfrac{1}{6}$th the size of the square sheet.

page 468 **11.1.6 Diagonals**

889 squares

page 469 **11.1.7 Painting cubes**

27 cubes: 8 have 3 red faces, 12 have 2 red faces, 6 have 1 red face, 1 has 0 red faces.
n^3 cubes: 8 have 3 red faces, $12(n - 2)$ have 2 red faces, $6(n - 2)^2$ have 1 red face, $(n - 2)^3$ have 0 red faces.

page 469 **Exercise 1**

1 False, $n = 2$ **2** False, $x = \dfrac{1}{2}$ **3** False, $n = 0$

4 False, $n = -3$ **5** False, 1, 2, 2   **6** False, $x = \dfrac{1}{2}$

7 False, a square

8 Not proven

page 469 **11.2.1 Crossnumbers**

A

[1]5	2	[2]5		[3]3	5		[4]2
2		[5]2	4	1	7		1
[6]1	7	2		0		[7]6	4
8		[8]1	[9]7	0		8	
	[10]1	0	1		[11]5	4	9
[12]8	7		[13]5	[14]5	5		
7		[15]4		[16]3	0	[17]4	[18]6
[19]7	2	9		5		[20]6	6

B

[1]5	8	[2]4		[3]3	5		[4]3
8		[5]3	4	2	7		6
[6]3	9	9		8		[7]3	4
6		[8]1	[9]8	3		1	
	[10]7	6	4		[11]9	5	0
[12]8	7		[13]5	[14]6	7		
5		[15]9		[16]6	8	[17]7	[18]5
[19]8	2	4		7		[20]3	6

C

[1]1	1	[2]5		[3]3	9		[4]1
8		[5]1	4	4	0		2
[6]6	5	0		7		[7]6	5
0		[8]5	[9]7	0		4	
	[10]2	1	9		[11]6	6	6
[12]1	4		[13]3	[14]4	1		
2		[15]4		[16]9	0	[17]5	[18]0
[19]7	0	5		0		[20]8	1

D

[1]1	7	[2]5		[3]2	4		[4]0
2		[5]9	9	6	0		0
[6]9	9	9		9		[7]3	7
5		[8]9	[9]0	5		8	
	[10]1	6	0		[11]8	0	7
[12]3	6		[13]5	[14]7	2		
0		[15]2		[16]2	3	[17]4	[18]5
[19]5	0	1		9		[20]2	2

12 Revision

page 473 **12.1 Revision exercises**
page 473 **Exercise 1**

1 £25·60, £6·70, 4, £55·30

2 a 30, 37   **b** 12, 10   **c** 7, 10
   d 8, 4   **e** 26, 33

3 £6·50   **4** £172

5 a 1810 seconds   **b** 72·4 seconds

6 0·8 cm

7 a £13   **b** £148   **c** £170

8 a 5·9   **b** 6   **c** 7

9 a −11   **b** 23   **c** −10
   d −20   **e** 6   **f** −14

10 a 3   **b** 5   **c** −6   **d** −7

11 a $x = 9$ **b** $x = 11$ **c** $x = 3$ **d** $x = 7$

12 Net **c** or **a**

13 a

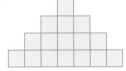

                        b 1, 4, 9, 16

   c Square numbers   **d** 49

page 474 **Exercise 2**

1 a $x = 7$ **b** $x = \dfrac{1}{4}$ **c** $x = \dfrac{4}{5}$

2 a 7·21 cm   **b** 9·22 cm   **c** 7·33 cm

3 a $\dfrac{3}{8}$ **b** $\dfrac{5}{8}$

4 a $\dfrac{2}{11}$ **b** $\dfrac{5}{11}$ **c** $\dfrac{9}{11}$

5 a 2·088   **b** 3·043

6 a Reflection in line $y = 0$ (x-axis)
   b Reflection in line $x = -1$
   c Reflection in line $y = x$
   d Rotation 90° clockwise about (0, 0)
   e Reflection in line $y = -1$
   f Rotation 180° about (0, −1)

7

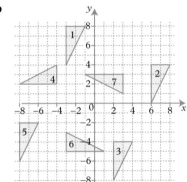

a Enlargement with centre (1, −4), scale factor 1·5

b Rotation 90° clockwise about centre (0, −4)

c Reflection in line $y = -x$

d Translation $\begin{pmatrix} 1 & 1 \\ 1 & 0 \end{pmatrix}$

e Enlargement with centre (−3, 8), scale factor $\dfrac{1}{2}$

f Rotation 90° anticlockwise about centre $\left(\dfrac{1}{2}, 6\dfrac{1}{2}\right)$

g Enlargement with centre (−2, 5), scale factor 3

page 475 **Exercise 3**

1 a £10·40   **b** 29 minutes

2 a 2 cm   **b** 8 m

3 i 9   **ii** 50
   iii $(7 \times 11) - 6 = 72 - 1$

4 a, b

c Δ 2: (6, 0), (6, 4), (8, 4)
   Δ 3: (2, −8), (2, −4), (4, −4)
   Δ 4: (−4, −2), (−8, 2), (−4, 4)
   Δ 6: (−3, −3), (−3, −5), (1, −5)
   Δ 7: (3, 1), (3, 3), (−1, 3)

5 17·7 cm

6 a 197·9 cm³   **b** 1·4 cm³   **c** 145

page 476 **Exercise 4**

1 a £66 000   **b** 198
   c £330   **d** £65 340

2 a 4   **b** 19

3 a 99p   **b** 760 miles

4 A: Cross-channel swimmer
   B: Car ferry from Calais
   C: Hovercraft from Dover
   D: Train from Dover
   E: Marker buoy outside harbour
   F: Car ferry from Dover

5 a 560 kg   **b** 57 kg

6 a A: 28 274   B: £79·15   C: £85·42
  b November

page 477 **Exercise 5**

1 a $\frac{1}{8}$ **b** $\frac{3}{4}$ **c** 2

  d $\frac{1}{2}$ **e** $\frac{1}{64}$

2 a 0·05 m/s   **b** 1·6 seconds   **c** 172·8 km
3 a 14   **b** 18   **c** 28
4 A: $y = 6$ B: $y = \frac{1}{2}x - 3$
  C: $y = 10 - x$ D: $y = 3x$
5 a $s = rt + 3t$ **b** $r = \frac{S - 3t}{t}$
6 12 cm²   **7** 9·9 cm

page 478 **Exercise 6**

1 20·7 litres
2 a 8   **b** 140   **c** 29   **d** 42
3 25
4 a

  b 6, 10, 14, 18, 22, 26
  c i 42   **ii** 62
  d $n = 4x + 2$
5 a $t + t + t = 3t$ **b** $a^2 \times a^2 = a^4$
  c Correct
6 a 5·45   **b** 5   **c** 5
7 $a = 45°$, $b = 67·5°$
8 a $z = x - 5y$ **b** $k = \frac{11 - 3m}{m}$ **c** $z = \frac{C}{T}$

page 479 **Exercise 7**

1 a i Consett   **ii** Durham   **iii** Consett
  b i 55 km   **ii** 40 km
  c i 80 km/h   **ii** 55 km/h   **iii** 70 km/h
     iv 80 km/h
  d $1\frac{3}{4}$ hours
2 One ton is 17 kg heavier than one tonne.
3 3 hours

4 a 0·198   **b** 0·0160
  c 64·9   **d** 0·0585
5 a 6   **b** 7   **c** 2
6 a $(600 \times 50) \div 50 = 600$
  b $(10 + 1000) \times 9 = 9090$ **c** $\sqrt{\frac{90}{10}} = 3$
  d $3 \times \sqrt{(5^2 + 20^2)} = 3 \times \sqrt{425} \approx 60$
7 a 40°   **b** 100°
8 $\frac{x}{x + 5}$
9 A: $4y = 3x - 16$ B: $2y = x - 8$ C: $2y + x = 8$
  D: $4y + 3x = 16$

page 480 **Exercise 8**

1 1 lb jar is better value.
2 a $\frac{5}{3}$ **b** 20 cm
3 a $6x + 15 < 200$ **b** 29
4 28 cm²
5 SB = 100 km
6 a 0·5601   **b** 3·215   **c** 0·6954
  d 0·4743
7 a 84   **b** 19·2
8 54

page 482 **12·2 Multiple choice tests**
page 482 **Test 1**

1 C	2 A	3 A	4 C
5 C	6 A	7 B	8 D
9 B	10 B	11 C	12 A
13 D	14 C	15 C	16 B
17 A	18 C	19 B	20 D
21 A	22 B	23 C	24 B
25 B			

page 484 **Test 2**

1 B	2 C	3 B	4 A
5 D	6 C	7 A	8 D
9 B	10 C	11 A	12 D
13 A	14 C	15 C	16 C
17 B	18 B	19 B	20 B
21 C	22 D	23 B	24 A
25 C			

Index